MATHEMATICS IN ACTION

Prealgebra Problem Solving

MATHEMATICS IN ACTION

Prealgebra Problem Solving

SECOND EDITION

The Consortium for Foundation Mathematics

Ralph Bertelle	*Columbia-Greene Community College*
Judith Bloch	*University of Rochester*
Roy Cameron	*SUNY Cobleskill*
Carolyn Curley	*Erie Community College—South Campus*
Ernie Danforth	*Corning Community College*
Brian Gray	*Howard Community College*
Arlene Kleinstein	*SUNY Farmingdale*
Kathleen Milligan	*Monroe Community College*
Patricia Pacitti	*SUNY Oswego*
Rick Patrick	*Adirondack Community College*
Renan Sezer	*LaGuardia Community College*
Patricia Shuart	*Polk Community College—Winter Haven, Florida*
Sylvia Svitak	*Queensborough Community College*
Assad J. Thompson	*LaGuardia Community College*

Boston San Francisco New York
London Toronto Sydney Tokyo Singapore Madrid
Mexico City Munich Paris Cape Town Hong Kong Montreal

Publisher Greg Tobin
Editor in Chief Maureen O'Connor
Project Editor Katie Nopper
Assistant Editor Caroline Case
Senior Managing Editor Karen Wernholm
Senior Production Supervisor Kathleen A. Manley
Senior Designer/Cover Designer Dennis Schaefer
Media Producer Michelle Small
Software Development Mary Durnwald, TestGen; Malcolm Litowitz, MathXL
Marketing Manager Jay Jenkins
Marketing Assistant Alexandra Waibel
Senior Author Support/Technology Specialist Joe Vetere
Senior Prepress Supervisor Caroline Fell
Manufacturing Manager Evelyn Beaton
Text Design Leslie Haimes
Production Coordination and Composition Pre-PressPMG
Illustrations Pre-PressPMG and Jim McLaughlin
Cover photo © Gary Gerovac / Masterfile

Library of Congress Cataloging-in-Publication Data

Mathematics in action: prealgebra problem solving / Consortium for Foundation Mathematics. — 2nd ed.

p. cm.

Includes index.

ISBN 0-321-44612-7 (pbk.)

1. Mathematics. I. Consortium for Foundation Mathematics.

QA39.3.M384 2007
510–dc22 2006052464

3 4 5 6 7 8 9 10—VHG—11 10 09 08

Contents

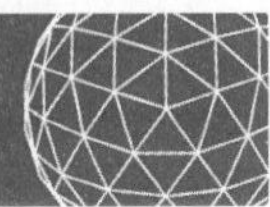

Preface xv

CHAPTER 1 WHOLE NUMBERS 1

ACTIVITY 1.1 Education Pays 1

Objectives:
1. Read and write whole numbers.
2. Compare whole numbers using inequality symbols.
3. Round whole numbers to specified place values.
4. Use rounding for estimation.
5. Classify whole numbers as even or odd, prime, or composite.
6. Solve problems involving whole numbers.

ACTIVITY 1.2 Bald Eagle Population Increasing Again 10

Objectives:
1. Read tables.
2. Read bar graphs.
3. Interpret bar graphs.
4. Construct graphs.

ACTIVITY 1.3 Bald Eagles Revisited 19

Objectives:
1. Add whole numbers by hand and mentally.
2. Subtract whole numbers by hand and mentally.
3. Estimate sums and differences using rounding.
4. Recognize the associative property and the commutative property for addition.
5. Translate a written statement into an arithmetic expression.

ACTIVITY 1.4 Summer Camp 31

Objectives:
1. Multiply whole numbers and check calculations using a calculator.
2. Multiply whole numbers using the distributive property.
3. Estimate the product of whole numbers by rounding.
4. Recognize the associative and commutative properties for multiplication.

ACTIVITY 1.5 College Supplies 39

Objectives:
1. Divide whole numbers by grouping.
2. Divide whole numbers by hand and by calculator.
3. Estimate the quotient of whole numbers by rounding.
4. Recognize the noncommutative property for division.

Activity 1.6 **Reach for the Stars** 49

Objectives:

1. Use exponential notation.
2. Factor whole numbers.
3. Determine the prime factorization of a whole number.
4. Recognize square numbers and roots of square numbers.
5. Recognize cubed numbers.
6. Apply the multiplication rule for numbers in exponential form with the same base.

Activity 1.7 **You and Your Calculator** 59

Objectives:

1. Use order of operations to evaluate arithmetic expressions.
2. Use order of operations to evaluate formulas involving whole numbers.

What Have I Learned? 67

How Can I Practice? 71

Chapter 1 Summary 77

Chapter 1 Gateway Review 83

CHAPTER 2 Variables and Problem Solving 91

Activity 2.1 **How Much Do I Need to Buy?** 91

Objectives:

1. Recognize and understand the concept of a variable in context and symbolically.
2. Translate a written statement (verbal rule) into a statement involving variables (symbolic rule).
3. Evaluate variable expressions.
4. Apply formulas (area, perimeter, and others) to solve contextual problems.

Activity 2.2 **How High Will It Go?** 101

Objectives:

1. Recognize input/output relationship between variables in a formula or equation (two variables only).
2. Evaluate variable expressions in formulas and equations.
3. Generate a table of input and corresponding output values from a given equation, formula, or situation.
4. Read, interpret, and plot points in rectangular coordinates that are obtained from evaluating a formula or equation.

Activity 2.3 **Are You Balanced?** 110

Objectives:

1. Translate contextual situations and verbal statements into equations.
2. Apply the fundamental principle of equality to solve equations of the forms $x + a = b$, $a + x = b$ and $x - a = b$.

Activity 2.4 **How Far Will You Go? How Long Will It Take?** 117
Objectives:
1. Apply the fundamental principle of equality to solve equations of the form $ax = b, a \neq 0$.
2. Translate contextual situations and verbal statements into equations.
3. Use the relationship rate · time = amount in various contexts.

Activity 2.5 **Web Devices for Sale** 124
Objectives:
1. Identify like terms.
2. Combine like terms using the distributive property.
3. Solve equations of the form $ax + bx = c$.

Activity 2.6 **Make Me an Offer** 130
Objectives:
1. Use the basic steps for problem solving.
2. Translate verbal statements into algebraic equations.
3. Use the basic principles of algebra to solve real-world problems.

What Have I Learned? 138
How Can I Practice? 139
Chapter 2 Summary 143
Chapter 2 Gateway Review 147

CHAPTER 3 **Problem Solving with Rational Numbers: Addition and Subtraction of Integers, Fractions, and Decimals** 151

Cluster 1 **Adding and Subtracting Integers**

Activity 3.1 **On the Negative Side** 151
Objectives:
1. Identify integers.
2. Represent quantities in real-world situations using integers.
3. Compare integers.
4. Calculate absolute value of integers.

Activity 3.2 **Maintaining Your Balance** 159
Objectives:
1. Add and subtract integers.
2. Identify properties of addition and subtraction of integers.

Activity 3.3 **What's the Bottom Line?** 168
Objectives:
1. Write formulas from verbal statements.
2. Evaluate expressions in formulas.
3. Solve equations of the form $x + b = c$ and $b - x = c$.
4. Solve formulas for a given variable.

ACTIVITY 3.4 **Riding in the Wind** 175

Objectives: 1. Translate verbal rules into equations.

2. Determine an equation from a table of values.

3. Use a rectangular coordinate system to represent an equation graphically.

What Have I Learned? 183

How Can I Practice? 185

Cluster 2 **Adding and Subtracting Fractions**

ACTIVITY 3.5 **Are You Hungry?** 191

Objectives: 1. Identify the numerator and denominator of a fraction.

2. Determine the greatest common factor (GCF).

3. Determine equivalent fractions.

4. Reduce fractions to equivalent fractions in lowest terms.

5. Convert mixed numbers to improper fractions and improper fractions to mixed numbers.

6. Determine the least common denominator (LCD) of two or more fractions.

7. Compare fractions.

ACTIVITY 3.6 **Food for Thought** 203

Objective: 1. Add and subtract fractions and mixed numbers with the same denominators.

ACTIVITY 3.7 **Math Is a Trip** 210

Objectives: 1. Determine the least common denominator (LCD) for two or more fractions.

2. Add and subtract fractions with different denominators.

3. Solve equations in the form $x + b = c$ and $x - b = c$ that involve fractions.

What Have I Learned? 217

How Can I Practice? 219

Cluster 3 **Adding and Subtracting Decimals**

ACTIVITY 3.8 **What Are You Made Of?** 223

Objectives: 1. Identify place values of numbers written in decimal form.

2. Convert a decimal to a fraction or a mixed number.

3. Compare decimals.

4. Read and write decimals.

5. Round decimals.

Activity 3.9 **Think Metric** 231
Objectives: 1. Know the metric prefixes and their decimal values.
2. Convert measurements between metric quantities.

Activity 3.10 **Dive into Decimals** 235
Objectives: 1. Add and subtract decimals.
2. Compare and interpret decimal numbers.

Activity 3.11 **Boiling, Freezing, and Financial Aid** 240
Objectives: 1. Add and subtract positive and negative decimal numbers.
2. Interpret and compare decimal numbers.
3. Solve equations of the type $x + b = c$ and $x - b = c$ involving decimal numbers.

What Have I Learned? 246
How Can I Practice? 247
Chapter 3 Summary 251
Chapter 3 Gateway Review 255

CHAPTER 4 **Multiplication and Division of Rational Numbers** **263**

Cluster 1 **Multiplying and Dividing Integers**

Activity 4.1 **Are You Physically Fit?** 263
Objectives: 1. Multiply and divide integers.
2. Perform calculations involving a sequence of operations.
3. Apply exponents to integers.
4. Identify properties of calculations involving multiplication and division with zero.

Activity 4.2 **Integers and Tiger Woods** 275
Objectives: 1. Use order of operations with expressions involving integers.
2. Apply the distributive property.
3. Evaluate algebraic expressions and formulas using integers.
4. Solve equations of the form $ax = b$, where $a \neq 0$, involving integers.

What Have I Learned? 283
How Can I Practice? 285

Cluster 2 **Multiplying and Dividing Fractions**

Activity 4.3 **Get Your Homestead Land** 288
Objectives: 1. Multiply and divide fractions.
2. Recognize the sign of a fraction.
3. Determine the reciprocal of a fraction.
4. Solve equations of form $ax = b, a \neq 0$, that involve fractions.

Activity 4.4 **Tiling the Bathroom** 299
Objectives: 1. Multiply and divide mixed numbers.
2. Evaluate expressions with mixed numbers.
3. Calculate the square root of a mixed number.

What Have I Learned? 309
How Can I Practice? 310

Cluster 3 **Multiplying and Dividing Decimals**

Activity 4.5 **Quality Points and GPA: Tracking Academic Standing** 315
Objectives: 1. Multiply and divide decimals.
2. Estimate products and quotients involving decimals.

Activity 4.6 **Tracking Temperature** 325
Objectives: 1. Use the order of operations to evaluate expressions that include decimals.
2. Use the distributive property in calculations involving decimals.
3. Evaluate formulas that include decimals.
4. Solve equations of the form $ax = b$ that include decimals.

What Have I Learned? 333
How Can I Practice? 335
Chapter 4 Summary 339
Chapter 4 Gateway Review 343

CHAPTER 5 PROBLEM SOLVING WITH RATIOS, PROPORTIONS, AND PERCENTS 349

Activity 5.1 **Everything Is Relative** 349
Objectives: 1. Understand the distinction between actual and relative measure.
2. Write a ratio in its verbal, fraction, decimal and percent formats.

Activity 5.2 **The Devastation of AIDS in Africa** 360
Objectives: 1. Use proportional reasoning to apply a known ratio to a given piece of information.
2. Write an equation using the relationship "ratio · total = part" and then solve the resulting equation.

Activity 5.3 **Who Really Did Better?** 365
Objectives: 1. Define actual and relative change.
2. Distinguish between actual and relative change.

Activity 5.4 **Don't Forget the Sales Tax** 370
Objectives: 1. Define and determine growth factors.
2. Use growth factors in problems involving percent increases.

Activity 5.5 **It's All on Sale!** 377
Objectives: 1. Define and determine decay factors.
2. Use decay factors in problems involving percent decreases.

Activity 5.6 **Take an Additional 20% Off** 384
Objective: 1. Apply consecutive growth and/or decay factors to problems involving two or more percent changes.

Activity 5.7 **Fuel Economy** 391
Objectives: 1. Apply rates directly to solve problems.
2. Use unit analysis (dimensional analysis) to solve problems that involve consecutive rates.

Activity 5.8 **Four out of Five Dentists Prefer the Brooklyn Dodgers?** 399
Objectives: 1. Recognize that equivalent fractions lead to proportions.
2. Use proportions to solve problems involving ratios and rates.

What Have I Learned? 405
How Can I Practice? 407
Chapter 5 Summary 411
Chapter 5 Gateway Review 413

CHAPTER 6 **PROBLEM SOLVING WITH GEOMETRY** 417

Cluster 1 **The Geometry of Two-Dimensional Plane Figures**

Activity 6.1 **Walking around Bases, Gardens, Trusses, and Other Figures** 417
Objectives: 1. Recognize perimeter as a geometric property of plane figures.
2. Write formulas for, and calculate perimeters of, squares, rectangles, triangles, parallelograms, trapezoids, and polygons.
3. Use unit analysis to solve problems involving perimeter.

Activity 6.2 **Circles Are Everywhere** 429
Objectives: 1. Measure lengths of diameters and circumferences of circles.
2. Develop and use formulas for calculating circumferences of circles.

Activity 6.3 **Lance Armstrong and You** 433
Objectives: 1. Calculate perimeters of many-sided plane figures using formulas and combinations of formulas.
2. Use unit analysis to solve problems involving perimeters.

Activity 6.4 **Baseball Diamonds, Gardens, and Other Figures Revisited** 437
Objectives: 1. Write formulas for areas of squares, rectangles, parallelograms, triangles, trapezoids, and polygons.
2. Calculate areas of polygons using appropriate formulas.

Activity 6.5 **How Big Is That Circle Really?** 446
Objectives: 1. Develop a formula for the area of a circle.
2. Use the formula to determine areas of circles.

Activity 6.6 **A New Pool and Other Home Improvements** 452
Objectives: 1. Solve problems in context using geometric formulas.
2. Distinguish between problems that require area formulas and those that require perimeter formulas.

Laboratory Activity 6.7 **How Big Is That Angle?** 457
Objectives: 1. Measure sizes of angles with a protractor.
2. Classify triangles as equiangular, equilateral, isosceles, or scalene.

Laboratory Activity 6.8 **How About Pythagoras?** 464
Objectives: 1. Develop and use the Pythagorean Theorem for right triangles.
2. Calculate the square root of numbers other than perfect squares.
3. Apply the Pythagorean Theorem in context.

Activity 6.9 **Moving Up with Math** 471
Objectives: 1. Recognize the geometric properties of similar triangles.
2. Use similar triangles in indirect measurement.

What Have I Learned? 475

How Can I Practice? 477

Cluster 2 **The Geometry of Three-Dimensional Space Figures**

Activity 6.10 Painting Your Way through Summer 482
Objectives:
1. Recognize geometric properties of three-dimensional figures.
2. Write formulas for and calculate surface areas of boxes (rectangular prisms), cans (right circular cylinders), and balls (spheres).

Activity 6.11 Truth in Labeling 486
Objectives:
1. Write formulas for and calculate volumes of boxes and cans.
2. Recognize geometric properties of three-dimensional figures.

Activity 6.12 Analyzing an Ice Cream Cone 491
Objectives:
1. Write formulas for and calculate volumes of balls (spheres) and cones.

Laboratory Activity 6.13 Summertime 497
Objectives:
1. Use geometry formulas to solve problems.
2. Use scale drawings in problem-solving.

What Have I Learned? 500
How Can I Practice? 503
Chapter 6 Summary 507
Chapter 6 Gateway Review 513

CHAPTER 7 More Problem Solving with Algebra and Mathematical Models 519

Activity 7.1 Leasing a Copier 519
Objectives:
1. Describe a mathematical situation as a set of verbal statements.
2. Translate verbal rules into symbolic equations.
3. Solve problems involving equations of the form $y = ax + b$.
4. Solve equations of the form $y = ax + b$ for the input x.
5. Evaluate expressions $ax + b$ in equations of the form $y = ax + b$ to obtain an output y.

Activity 7.2 Windchill 529
Objectives:
1. Evaluate expressions to determine the output for a formula.
2. Solve formulas for a specific variable.

Activity 7.3 **Comparing Energy Costs** 536
Objectives: 1. Write symbolic equations from information organized in a table.
2. Produce tables and graphs to compare outputs from two different mathematical models.
3. Solve equations of the form $ax + b = cx + d$.

Activity 7.4 **Volume of a Storage Tank** 544
Objectives: 1. Use property of exponents to multiply powers having the same base.
2. Use property of exponents to raise a power to a power.
3. Use the distributive property and properties of exponents to write an expression as an equivalent expression in expanded form.

Activity 7.5 **Math Magic** 550
Objectives: 1. Recognize an algebraic expression as a code of instructions to obtain an output.
2. Simplify algebraic expressions.

Activity 7.6 **Mathematical Modeling** 557
Objectives: 1. Develop an equation to model and solve problems.
2. Solve problems using formulas as models.
3. Recognize patterns and trends between two variables using tables as models.
4. Recognize patterns and trends between two variables using graphs as models.

What Have I Learned? 568
How Can I Practice? 570
Chapter 7 Summary 577
Chapter 7 Gateway Review 579

APPENDIXES

Learning Math Opens Doors: Twelve Keys to Success A-1
Selected Answers A-15
Glossary A-33
Index I-1

Preface

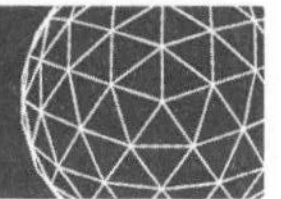

Our Vision

Mathematics in Action: Prealgebra Problem Solving, Second Edition, is intended to help college mathematics students gain mathematical literacy in the real world and simultaneously help them build a solid foundation for future study in mathematics and other disciplines.

Our authoring team used the AMATYC *Crossroads* standards to develop a three-book series to serve a large and diverse population of college students who, for whatever reason, have not yet succeeded in learning mathematics. It became apparent to us that teaching the same content in the same manner to students who have not previously comprehended it is not effective, and this realization motivated us to develop a new approach.

Mathematics in Action is based on the principle that students learn mathematics best by doing mathematics within a meaningful context. In keeping with this premise, students solve problems in a series of realistic situations from which the crucial need for mathematics arises. *Mathematics in Action* guides students toward developing a sense of independence and taking responsibility for their own learning. Students are encouraged to construct, reflect on, apply, and describe their own mathematical models, which they use to solve meaningful problems. We see this as the key to bridging the gap between abstraction and application and as the basis for transfer learning. Appropriate technology is integrated throughout the books, allowing students to interpret real-life data verbally, numerically, symbolically, and graphically.

We expect that by using the *Mathematics in Action* series, all students will be able to achieve the following goals:

- Develop mathematical intuition and a relevant base of mathematical knowledge.
- Gain experiences that connect classroom learning with real-world applications.
- Prepare effectively for further college work in mathematics and related disciplines.
- Learn to work in groups as well as independently.
- Increase knowledge of mathematics through explorations with appropriate technology.
- Develop a positive attitude about learning and using mathematics.
- Build techniques of reasoning for effective problem solving.
- Learn to apply and display knowledge through alternative means of assessment, such as mathematical portfolios and journal writing.

Our vision for you is to join the growing number of students using our approaches who have discovered that mathematics is an essential and learnable survival skill for the twenty-first century.

Pedagogical Features

The pedagogical core of *Mathematics in Action* is a series of guided-discovery activities in which students work in groups to discover mathematical principles embedded in realistic situations. The key principles of each activity are highlighted and summarized at the activity's conclusion. Each activity is followed by exercises that reinforce the concepts and skills revealed in the activity.

The activities are clustered within each chapter. Each cluster's activities all relate to a particular subset of topics addressed in the chapter. The clusters in Chapter 6 contain lab activities in addition to regular activities. The lab activities require more than just paper, pencil, and calculator—they often require measurements and data collection and are ideal for in-class group work. For specific suggestions on how to use the two types of activities, we strongly encourage instructors to refer to the *Printed Test Bank/Instructor's Resource Guide* that accompanies this text.

Each cluster concludes with two sections: What Have I Learned? and How Can I Practice? The What Have I Learned? exercises are designed to help students pull together the key concepts of the cluster. The How Can I Practice? exercises are designed primarily to provide additional work with the mathematical skills of the cluster. Taken as a whole, these exercises give students the tools they need to bridge the gaps between abstraction, skills, and application.

Additionally, each chapter ends with a Summary that briefly describes key concepts and skills discussed in the chapter, plus examples illustrating these concepts and skills. The concepts and skills are also referenced to the activity in which they appear, making the format easier to follow for those students who are unfamiliar with our approach. Each chapter also ends with a Gateway Review, providing students with an opportunity to check their understanding of the chapter's concepts and skills, as well as prepare them for a chapter assessment.

Changes from the First Edition

This newly revised edition of *Prealgebra Problem Solving* has been substantially rewritten to reflect updated data and to provide more contextual applications. Two new activities have been added. In Chapter 3, a new activity introduces the metric system, and in Chapter 5 a new activity addresses separately the use of proportion equations to solve problems involving ratios. Chapter 7 was significantly restructured and revised to showcase formulas, symbolic rules, graphs and tables as mathematical models. Many new exercises were added throughout the text to better reference and reinforce previously introduced concepts.

Supplements

Instructor Supplements

Annotated Instructor's Edition

ISBN-10 0-321-44638-0

ISBN-13 978-0-321-44638-1

This special version of the student text provides answers to all exercises directly beneath each problem.

Printed Test Bank/Instructor's Resource Guide

ISBN-10 0-321-44869-3

ISBN-13 978-0-321-44869-9

This valuable teaching resource includes the following materials:

- Sample syllabi suggesting ways to structure the course around core and supplemental activities.

- Teaching notes for each chapter.
- Extra practice worksheets for topics with which students typically have difficulty.
- Sample chapter tests and final exams for in-class and take-home use by individual students and groups.
- Information about incorporating technology in the classroom, including sample graphing calculator assignments.

TestGen®

ISBN-10 0-321-45029-9

ISBN-13 978-0-321-45029-6

TestGen enables instructors to build, edit, print, and administer tests using a computerized bank of questions developed to cover all the objectives of the text. TestGen is algorithmically based, allowing instructors to create multiple but equivalent versions of the same question or test with the click of a button. Instructors can also modify test bank questions or add new questions. Tests can be printed or administered online. The software is available on a dual-platform Windows/Macintosh CD-ROM.

Instructor Training Video on CD

ISBN-10 0-321-50134-9

ISBN-13 978-0-321-50134-9

This innovative video discusses effective ways to implement the teaching pedagogy of the *Mathematics in Action* series, focusing on how to make collaborative learning, discovery learning, and alternative means of assessment work in the classroom.

Student Supplements

MathXL® Tutorials on CD

ISBN-10 0-321-45567-3

ISBN-13 978-0-321-45567-3

This interactive tutorial CD-ROM provides algorithmically generated practice exercises that are correlated at the objective level to the exercises in the textbook. Every practice exercise is accompanied by an example and a guided solution designed to involve students in the solution process. The software provides helpful feedfack for incorrect answers and can generate printed summarises of students' progress.

InterAct Math Tutorial Website www.interactmath.com

Get practice and tutorial help online! This interactive tutorial Web site provides algorithmically generated practice exercises that correlate directly to the exercises in the textbook. Students can retry an exercise as many times as they like with new values each time for unlimited practice and mastery. Every exercise is accompanied by an interactive guided solution that provides helpful feedfack for incorrect answers, and students can also view a worked-out sample problem that steps them through an exercise similar to the one they're working on.

Addison-Wesley Math Tutor Center www.aw-bc.com/tutorcenter

The Addison-Wesley Math Tutor Center is staffed by qualified mathematics instructors who tutor students on examples and exercises from the textbook. Tutoring is provided via toll-free telephone, toll-free fax, e-mail, and the Internet.

White Board technology allows students and tutors to view and listen to live instruction in real-time over the Internet.

Supplements for Instructors and Students

MathXL®

MathXL® is a powerful online homework, tutorial, and assessment system that accompanies Addition-Wesley textbooks in mathematics or statistics. With MathXL, instructors can create, edit, and assign online homework and tests using algorithmically generated exercises correlated at the objective level to the textbook. They can also create and assign their own online exercises and import TestGen tests for added flexibility. All student work is tracked in MathXL's online gradebook. Students can take chapter tests in MathXL and receive personalized study plans based on their test results. The study plan diagnoses weaknesses and links students directly to tutorial exercises for the objectives they need to study and retest. Students can also access supplemental animations and textbook pages directly from selected exercises. MathXL is available to qualified adopters. For more information, visit our Web site at www.mathxl.com, or contact your Addison-Wesley sales representative.

MyMathLab®

My MathLab® is a series of text-specific, easily customizable online courses for Addison-Wesley textbooks in mathematics and statistics. Powered by CourseCompass™ (Pearson Education's online teaching and learning environment) and MathXL® our online homework, tutorial, and assessment system), MyMathLab gives you the tools you need to deliver all or a portion of your course online, whether your students are in a lab setting or working from home. MyMathLab provides a rich and flexible set of course materials, featuring free-response exercises that are algorithmically generated for unlimited practice and mastery. Students can also use online tools, such as animations and a multimedia textbook, to independently improve their understanding and performance. Instructors can use MyMathLab's homework and test managers to select and assign online exercises correlated directly to the textbook, and they can also create and assign their own online exercises and import TestGen tests for added flexibility. MyMathLab's online gradebook—designed specifically for mathematics and statistics—automatically tracks students' homework and test results and gives the instructor control over how to calculate final grades. Instructors can also add offline (paper-and-pencil) grades to the gradebook. MyMathLab is available to qualified adoptors. For more information, visit our Web site at www.mymathlab.com or contact your Addison-Wesley sales representative.

Acknowledgments

The consortium would like to acknowledge and thank the following people for their invaluable assistance in reviewing and testing material for this text in the past and current editions:

Michele Bach, *Kansas City Community College*

Kathleen Bavelas, *Manchester Community College*

Vera Brennan, *Ulster County Community College*

Jennifer Dollar, *Grand Rapids Community College*

Marion Glasby, *Anne Arundel Community College*

Bob Hervey, *Hillsborough Community College*

Ashok Kumar, *Valdosta State University*

Rob Lewis, *Linn Benton Community College*

Jim Matovina, *Community College of Southern Nevada*

Janice McCue, *College of Southen Maryland*

Bobbi Righi, *Seattle Central Community College*

Jody Rooney, *Jackson Community College*

Janice Roy, *Montcalm Community College*

Amy Salvati, *Adirondack Community College*

Carolyn Spillman, *Georgia Perimeter College*

Janet E. Teeguarden, *Ivy Technical Community College*

Sharon Testone, *Onondaga Community College*

Cheryl Wilcox, *Diablo Valley College*

Jill C. Zimmerman, *Manchester Community College*

Cathleen Zucco-Teveloff, *Trinity College*

We would also like to thank our accuracy checkers, Cheryl Cantwell and Scott Fallstrom.

Finally, a special thank you to our families for their unwavering support and sacrifice, which enabled us to make this text a reality.

The Consortium for Foundation Mathematics

To the Student

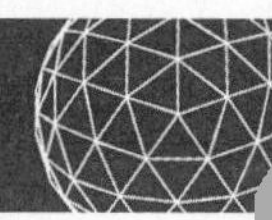

The book in your hands is most likely very different from any mathematics book you have seen before. In this book, you will take an active role in developing the important ideas of arithmetic and beginning algebra. You will be expected to add your own words to the text. This will be part of your daily work, both in and out of class and for homework. It is the belief of the authors that students learn mathematics best when they are actively involved in solving problems that are meaningful to them.

The text is primarily a collection of situations drawn from real life. Each situation leads to one or more problems. By answering a series of questions and solving each part of the problem, you will be led to use one or more ideas of introductory college mathematics. Sometimes, these will be basic skills that build on your knowledge of arithmetic. Other times, they will be new concepts that are more general and far reaching. The important point is that you won't be asked to master a skill until you see a real need for that skill as part of solving a realistic application.

Another important aspect of this text and the course you are taking is the benefit gained by collaborating with your classmates. Much of your work in class will result from being a member of a team. Working in small groups, you will help each other work through a problem situation. While you may feel uncomfortable working this way at first, there are several reasons we believe it is appropriate in this course. First, it is part of the learn-by-doing philosophy. You will be talking about mathematics, needing to express your thoughts in words—this is a key to learning. Secondly, you will be developing skills that will be very valuable when you leave the classroom. Currently, many jobs and careers require the ability to collaborate within a team environment. Your instructor will provide you with more specific information about this collaboration.

One more fundamental part of this course is that you will have access to appropriate technology at all times. Technology is a part of our modern world, and learning to use technology goes hand in hand with learning mathematics. Your work in this course will help prepare you for whatever you pursue in your working life.

This course will help you develop both the mathematical and general skills necessary in today's workplace, such as organization, problem solving, communication, and collaborative skills. By keeping up with your work and following the suggested organization of the text, you will gain a valuable resource that will serve you well in the future. With hard work and dedication you will be ready for the next step.

The Consortium for Foundation Mathematics

CHAPTER 1

WHOLE NUMBERS

Do you remember when you first started learning about numbers? From those early days, you went on to learn more about numbers—what they are, how they are related to one another, and how you operate with them.

Whole numbers are the basis for your further study of arithmetic and introductory algebra used throughout this book. In Chapter 1, we will see whole numbers in real-life use, clarify what you already know about them, and learn more about them.

ACTIVITY 1.1

Education Pays

OBJECTIVES

1. Read and write whole numbers.
2. Compare whole numbers using inequality symbols.
3. Round whole numbers to specified place values.
4. Use rounding for estimation.
5. Classify whole numbers as even or odd, prime, or composite.
6. Solve problems involving whole numbers.

The U.S. Bureau of the Census tracks information yearly about educational levels and income levels of the population in the United States (http://www.census.gov). In 2004, the bureau reported that more than one in four adults holds a bachelor's degree. The bureau also presented data on average 2004 earnings and education level for all workers, aged 18 and older. Some of the data is given in the table below.

1. a. What is the average income of those workers who had some college? An associate's degree? A high school graduate? A bachelor's degree?

   b. Which group of workers earned the most income in 2004? Which group earned the least income?

   c. What does the table indicate about the value of an education in the United States?

Learning to Earn

AVERAGE 2004 EARNINGS BY EDUCATIONAL LEVEL: WORKERS 18 YEARS OF AGE AND OLDER

EDUCATIONAL LEVEL	All workers combined	Not high school graduates	High school graduates	Some college	Associate's degree	Bachelor's degree	Master's degree	Doctorate degree	Professional degree
AVERAGE INCOME LEVEL	$37,897	$19,041	$28,631	$30,173	$36,021	$51,568	$67,073	$93,033	$114,878

d. How does the census information relate to your decision to attend college?

Whole Numbers

The earnings listed in the preceding table are represented by **whole numbers.** The set of whole numbers consists of zero and all the counting numbers, 1, 2, 3, 4, and so on. Whole numbers are used to describe "how many" (for example, the dollar values in Problem 1). Each **whole number** is represented by a **numeral,** which is a sequence of symbols called *digits.* The relative placement of the **digits** (0, 1, 2, 3, 4, 5, 6, 7, 8, and 9) in our standard base-10 system determines the value of the number that the numeral represents.

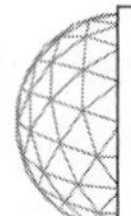

EXAMPLE 1 ***What does the numeral 3547 represent? What is the place value of 3 in the number 3547?***

SOLUTION

3547 is the numeral representing 3 thousands, 5 hundreds, 4 tens, and 7 ones. It is read and written in words as "three thousand five hundred forty-seven." In this number, the digit 3 has a **place value** of one thousand (1000). This means that the digit 3 represents 3000 of the units in the number 3547.

2. What are the place values of the other digits in the number 3547?

3. a. Write the number 30,928 in words.

b. What are the place values of the digits 8 and 9 in the number 30,928?

For ease in reading a number in the base-10 system, digits are grouped in threes with each grouping of three separated by a comma. The triples are named as shown in the following table. Beginning with the second triple from the right and moving to the left, the triples are named thousands, millions, billions, etc. For example, the number 548,902,473,150 is written in the following table.

GROUP	Billions			Millions			Thousands			Ones		
TRIPLES	Hundreds	Tens	Ones	Hundreds	Tens	Ones	Hundreds	Tens	Ones	Hundreds	Tens	Ones
EXAMPLE	5	4	8	9	0	2	4	7	3	1	5	0

The number in the preceding table is read "five hundred forty-eight billion, nine hundred two million, four hundred seventy-three thousand, one hundred fifty."

4. Write the earnings from Problem 1a in words.

5. The digit 0 occurs twice in the number in the preceding table. What is the place value of each occurrence?

Comparing Whole Numbers

The place value system of writing numbers makes it easy to compare numbers. For example, it is easy to see that an income of $30,173 is less than an income of $36,021 by comparing the values 0 and 6 in the thousands place. You can represent the relationship by writing $30{,}173 < 36{,}021$. Alternatively, you can say that $36,021 is greater than $30,173 and write $36{,}021 > 30{,}173$. Such statements involving the symbols $<$ (less than) and $>$ (greater than) are called **inequalities**.

6. In the 2004 census, Oregon's population was counted as 3,594,586 and Oklahoma's was 3,523,553. Which state had the greater population?

PROCEDURE

Comparing Two Whole Numbers

1. Use the symbol $>$ to write that one number is *greater than* another. For example, $8 > 3$ is read from left to right as "eight is greater than three."
2. Use the symbol $<$ to write that one number is *less than* another. For example, $3 < 8$ is read from left to right as "three is less than eight."
3. Use the symbol $=$ to write that two numbers *are equal.* For example, $2 + 1 = 3$ is read from left to right as "two plus one is equal to three."
4. Compare two numbers by reading each of them from left to right to find the first position where they differ. For example, 7,180,597 and 7,180,642 first differ in the hundreds place. Since $6 > 5$, write $7{,}180{,}642 > 7{,}180{,}597$. Alternatively, $7{,}180{,}597 < 7{,}180{,}642$ is also correct because $5 < 6$.

Rounding Whole Numbers

The U.S. Bureau of the Census provides United States population counts on the Internet on a regular basis. For example, the bureau estimated the population at 297,963,037 persons on January 23, 2006. Not every digit in this number is meaningful because the population is constantly changing. Therefore, a reasonable approximation is usually sufficient. For example, it would often be good enough to say the population is about 298,000,000, or two hundred ninety-eight million.

The process of determining an approximation to a number is known as **rounding.**

PROCEDURE

Rounding a Whole Number to a Specified Place Value

1. Underline the digit with the place value to which the number will be rounded, such as "to the nearest million" or "to the nearest thousand."
2. If the digit directly to its right is *less than 5*, keep the digit underlined in step 1 and replace all the digits to its right with zeros.
3. If the digit directly to its right is *5 or greater*, increase the digit underlined in step 1 by 1 unit and replace all the digits to its right with zeros.

EXAMPLE 2 ***Round* 37,146 *to the nearest ten thousand.***

SOLUTION

The digit in the ten thousands place is 3. The digit to its right is 7. Therefore, increase 3 to 4 and insert zeros in place of all the digits to the right. The rounded value is 40,000.

7. On another date in 2004, the U.S. Bureau of the Census gave the U.S. population as 285,691,501 persons.

   a. Approximate this count by rounding to the nearest million.

   b. Approximate the count by rounding to the nearest ten thousand.

Classifying Whole Numbers: Even or Odd, Prime or Composite

At times, it can be useful to classify whole numbers that share certain common features. One way to classify whole numbers is as even or odd. **Even numbers** are those that are evenly divisible by 2. That is, an even number is any number that when divided by 2 has a remainder equal to 0. A whole number that is not even is **odd.** Odd numbers are those that when divided by 2 leave a remainder of 1.

8. When an odd number is divided by 2, what is its remainder? Give an example.

9. Is 0 an even number? Explain.

10. By examining the digits of a whole number, how can you determine if the number is even or odd?

Whole numbers can also be classified as prime or composite. Any whole number greater than 1 that is divisible *only* by itself and 1 is called **prime.** For example, 5 is divisible only by itself and 1, so 5 is prime. A whole number greater than 1 that is not prime is called **composite.** For example, 6 is divisible by itself and 1, but also by 2 and 3. Therefore, 6 is a composite number. Also, 1, 2, 3, and 6 are called the **factors** of 6.

11. Is 21 prime or composite? Explain.

12. Is 2 prime or composite? Explain.

13. a. List all the prime numbers between 1 and 30.

    b. How many even numbers are included in your list?

    c. How many even prime numbers do you think there are? Explain.

Problem Solving with Whole Numbers

The United States Bureau of Labor Statistics provides data for various occupations and the education levels that they usually require. The bureau lists 19 occupations that typically require an associate's degree. The occupations are listed in the table on page 6.

The table also provides **median** yearly salaries for 2001. A median value in a data set is a value that divides the set into an upper half and a lower half of values. One-half of all salaries for a given occupation are below the median salary and the other half are above the median salary. For instance, one-half of males working in health technology earned less than $33,950 and the other half earned more than $33,950 in 2001.

On the Job...

OCCUPATION	2001 MALE MEDIAN ANNUAL SALARY ($)	2001 FEMALE MEDIAN ANNUAL SALARY ($)
Managerial and professional specialty occupation	52,260	35,160
Management-related occupations	48,550	34,100
Technical, sales, and administrative support	31,950	20,040
Health technology and technicians	33,950	23,850
Technologies and technicians except health	45,110	29,740
Sales occupations	32,730	13,040
Administrative support, including clerical work	26,740	21,140
Protective service occupations	34,340	17,560
Professional specialty occupations	51,400	34,410
Farming, forestry, and fishing occupations	13,710	7,060
Precision, production, craft, and repair occupations	31,040	21,700
Operators, fabricators, laborers	23,590	16,330
Armed forces	31,920	33,890

14. a. The salaries in the preceding table are rounded to what place value?

b. Is the median income of males in precision and craft occupations more than that of males in the armed forces? Explain how you obtained your answer.

c. Which two female occupations are closest in median annual salaries?

d. Which occupation has the lowest median salary for women? For men? The highest median salary for women? For men?

e. If you rounded the median salary of male health technicians to the nearest thousand dollars, how would you report the salary? Would you be overestimating or underestimating the salary?

f. If you rounded the median salary of female administrative support to the nearest thousand dollars, how would you report the salary? Would you be overestimating or underestimating the salary?

SUMMARY
ACTIVITY 1.1

1. A **whole number** is represented by a **numeral** consisting of a sequence of **digits.** The relative placement of the **digits** determines the value of the number that the numeral represents.
2. Each digit in a numeral has a **place value** determined by its relative placement in the numeral. Numbers are compared for size by comparing their corresponding place values. The symbols $<$ (read "less than") and $>$ (read "greater than") are used to compare the size of the numbers.
3. **Rounding** a given number to a specified place value is used to approximate its value. Rules for rounding are provided on page 4.
4. Whole numbers are classified as even or odd. **Even numbers** are those that are divisible by 2 leaving no remainder. Any number that is not even is an **odd number.**
5. Whole numbers are also classified as **prime** or **composite.** A number is **prime** if it is greater than 1 and divisible only by itself and 1. A whole number greater than 1 that is not prime is called **composite.** The **factors** of a number are all the numbers that divide evenly into the given number.

EXERCISES
ACTIVITY 1.1

1. The sticker price of a new Lexus is thirty-four thousand two hundred eighty-five dollars. Write this number as a numeral.

2. The total population of the United States on December 21, 2005, was estimated at 297,922,224. Write this population count in words.

3. China has the largest population on Earth, with a population of approximately one billion, three hundred six million, three hundred thirteen thousand eight hundred twelve. Write this population estimate as a numeral.

4. The average earned income for a person with a master's degree is $67,073. Round this value to the nearest thousand. To the nearest hundred.

5. One estimate of the world population toward the end of 2005 was 6,486,411,859 people. Round this value to the nearest million. The nearest billion.

6. You use a check to purchase this semester's textbooks. The total is $243.78. How will you write the amount in words on your check?

7. Explain why 90,210 is less than 91,021.

8. Determine whether each of the following numbers is even or odd. In each case, give a reason for your answer.

   a. 22,225   b. 13,578   c. 1500

   a.

   b.

   c.

9. Determine whether each of the following numbers is prime or composite. In each case, give a reason for your answer.

   a. 35   b. 31   c. 51

Exercise numbers appearing in color are answered in the Selected Answers appendix.

10. Some of the occupations that earn the highest income in the twenty-first century require a bachelor's degree. They are listed in the following table with their median annual salaries for 2001.

OCCUPATION	MALE MEDIAN ANNUAL SALARY ($)	FEMALE MEDIAN ANNUAL SALARY ($)
Managerial and professional specialty occupation	52,255	35,160
Management-related occupations	48,553	34,104
Professional specialty occupations	51,399	34,411

a. Round each median salary for males to the nearest thousand.

b. Round each median salary for females to the nearest thousand.

c. Is the median salary for females in management-related occupations as much as that of females in professional specialty occupations? Justify your answer.

11. You are researching information on buying a new sports utility vehicle (SUV). A particular SUV that you are considering has a manufacturer's suggested retail price (MSRP) of $31,310. The invoice price to the dealer for the SUV is $28,707. Do parts a and b to estimate how much bargaining room you have between the MSRP and the dealer's invoice price.

a. Round the MSRP and the invoice price each to the nearest thousand.

b. Use the rounded values from part a to estimate the difference between the MSRP and invoice price.

ACTIVITY 1.2

Bald Eagle Population Increasing Again

OBJECTIVES

1. Read tables.
2. Read bar graphs.
3. Interpret bar graphs.
4. Construct graphs.

The bald eagle has been the national symbol of the United States since 1782, when its image with outspread wings was placed on the country's Great Seal. Bald eagles were in danger of becoming extinct about thirty-five years ago, but efforts to protect them are working. In 1999, President Clinton announced a proposal to remove this majestic bird from the list of threatened species, but as of 2005, the bald eagle remains on the list.

Bar Graphs

The following bar graph displays the numbers of nesting bald eagle pairs in the lower 48 states for the years from 1963 to 1999. The horizontal direction represents the years from 1963 to 1999. The vertical direction represents the number of nesting pairs.

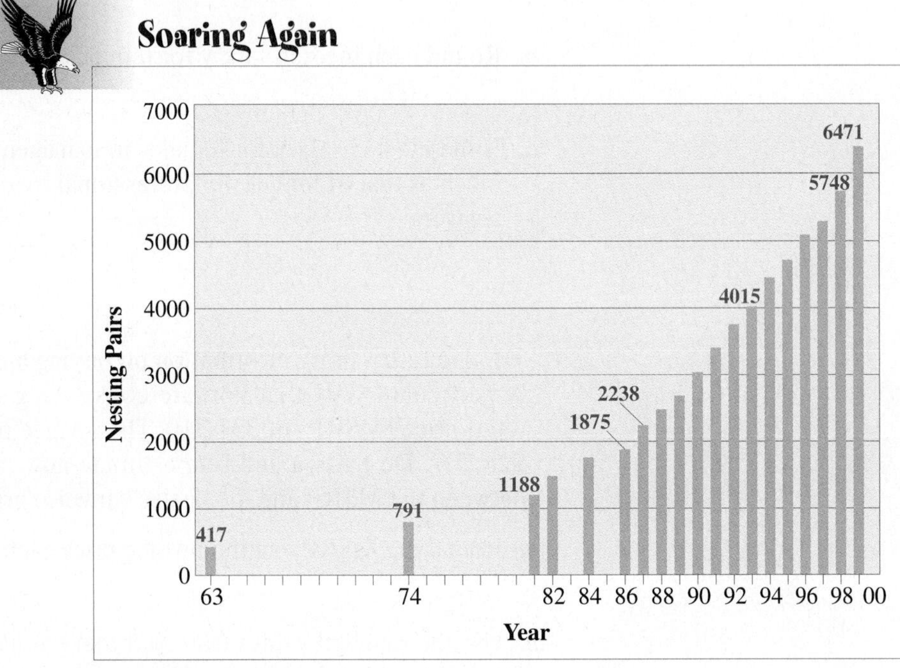

Source: U.S. Fish and Wildlife Service

1. a. What was the number of nesting pairs in 1963? In 1986? In 1998? In 1999?

   b. Write the numbers from part a in words.

   c. Explain how you located these numbers on the bar graph.

2. a. Estimate the number of nesting pairs in 1982. In 1988. In 1996.

b. Explain how you estimated the numbers from the bar graph.

c. To what place value did you estimate the number of nesting pairs in each case?

d. Compare your estimates with the estimates of some of your classmates. Briefly describe the comparisons.

3. a. Estimate the number of nesting pairs in 1983 and 1985.

b. Explain how you determined your estimate from the graph.

c. Estimate to the nearest thousand the number of nesting pairs in 1977.

d. Compare the growth in the number of nesting pairs from 1965 to 1986 and from 1986 to 1999.

Graphing and Coordinate Systems

Note that you can represent the 791 nesting pairs in 1974 symbolically by (1974, 791). Two paired numbers listed in parentheses and separated by a comma are called an **ordered pair.** The first number in an ordered pair is always found or given along the horizontal direction on a graph and is called an **input** value. The second number is found or given in the vertical direction and is known as an **output** value.

4. a. Write the corresponding ordered pairs for the years 1987 and 1993, where the first number represents the year and the second number represents the number of nesting pairs.

b. What is the input value in the ordered pair (1989, 2680)? What does the value represent in this situation?

c. What is the output value in the ordered pair (1992, 3749)? What does the value represent in this situation?

The following is an example of a basic graphing grid used to display paired data values.

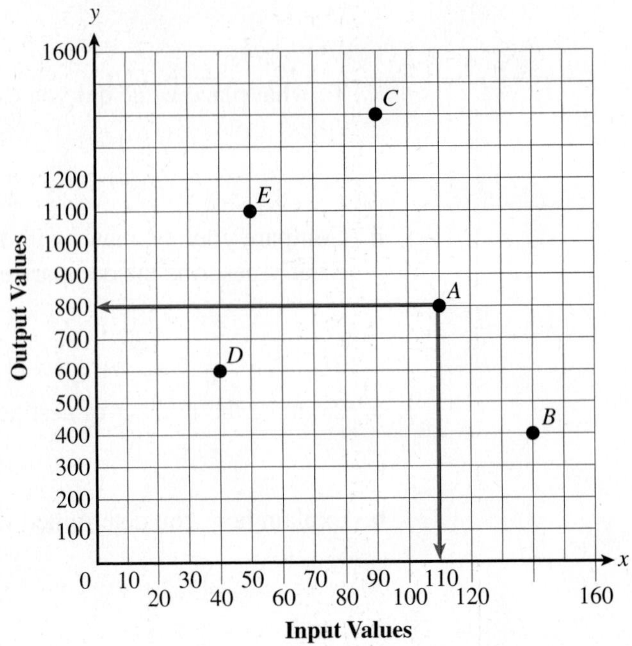

The horizontal line where the input values are referenced is called the **horizontal axis.** The vertical line where the output values are referenced is called the **vertical axis.** The scale (the number of units per block) in each direction should be appropriate for the given data. Each block in the grid above represents 10 units in the horizontal direction and 100 units in the vertical direction. It is important when making a graph to label the units in each direction.

5. Label the remaining three units on the horizontal axis on the graphing grid. Do the same for the vertical axis.

When units are given on the axes, you can determine the input and output values of a given point. For example, you can determine the input value of point *A* by following the vertical line straight down from point *A* to where the line crosses the horizontal axis. Read the input value 110 at the intersection. Similarly, read the output value 800 by following the horizontal line straight across from point *A* to the vertical axis.

6. Determine the input and output values of the points *B* and *C* on the graphing grid. Write each answer as ordered pairs on the grid next to its point.

The input and output values of an ordered pair are also referred to as the **coordinates** of the point that represents the pair on the graph. The letter x is frequently used to denote the input and y is used to denote the output. In such a case, the input value is called the x-coordinate and the horizontal axis is referred to as the x-axis. The output value is called the y-coordinate and the vertical axis is referred to as the y-axis.

7. a. What are the coordinates of the point C on the grid?

   b. What is the x-coordinate of the point D?

   c. What is the y-coordinate of the point E?

You can **plot**, or place points on the grid after you determine the scale and label the units. For example, the point with coordinates (20, 300) is located as follows:

i. Start at the lower right-hand corner point labeled 0 (called the origin) and count 20 units to the right.

ii. Then count 300 units up and mark the spot with a dot.

8. a. Plot the point with coordinates (70, 900) on the graphing grid. Write the ordered pair next to the point.

   b. The x-coordinate of a point is 150 and the y-coordinate is 100. Plot and label the point.

   c. Plot and label the points (75, 350) and (122, 975).

   d. Plot and label the points (0, 0), (0, 500), and (80,0).

You may have noticed that a **bar chart** is a common variation of the basic grid that is used when inputs are **categories.** Categories on a bar chart are represented by intervals of equal length on the horizontal axis. The rectangular bars drawn from the horizontal axis have equal widths and vary in height according to their outputs.

9. Graph the data in the table on page 14 as a bar chart on the accompanying grid. Notice that the inputs are educational levels (categories).

   a. Write the name of the output along the vertical axis and the name of the input along the horizontal axis.

   b. List the input along the horizontal axis. The first two inputs are placed for you.

   c. List the units along the vertical axis. The first three units are given.

d. Draw the bars corresponding to the given input/output pairs.

AVERAGE 2004 EARNINGS BY EDUCATIONAL LEVEL: WORKERS 18 YEARS OF AGE AND OLDER									
EDUCATIONAL LEVEL	Not high school graduate	High school graduate	Some college	Associate's degree	Bachelor's degree	Master's degree	Doctor-ate degree	Professional degree	All workers combined
AVERAGE INCOME LEVEL	$19,041	$28,631	$30,173	$36,021	$51,568	$67,073	$93,033	$114,878	$37,897

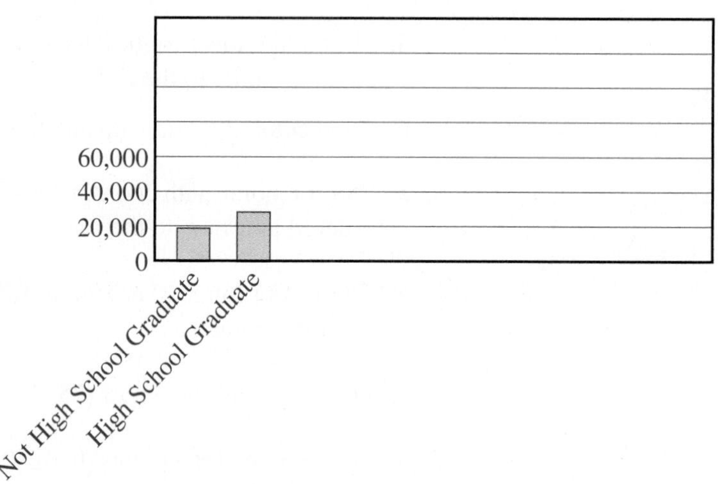

SUMMARY ACTIVITY 1.2

1. A **graphing grid** displays paired data (input value, output value).
2. The horizontal direction is marked in units along a line called the **horizontal axis.** The vertical direction is marked in units along a line called the **vertical axis.**
3. Paired values are represented by points on the grid. The input value is read by following a vertical line down from the point to the horizontal axis. The output value is read by following a horizontal line from the point across to the vertical axis.
4. To graph a paired value (input, output), start at the point 0 and move the given number of input units to the right and then move the given number of output units up. Mark the point.

EXERCISES
ACTIVITY 1.2

In Exercises 1–3, use the following grid to answer the questions.

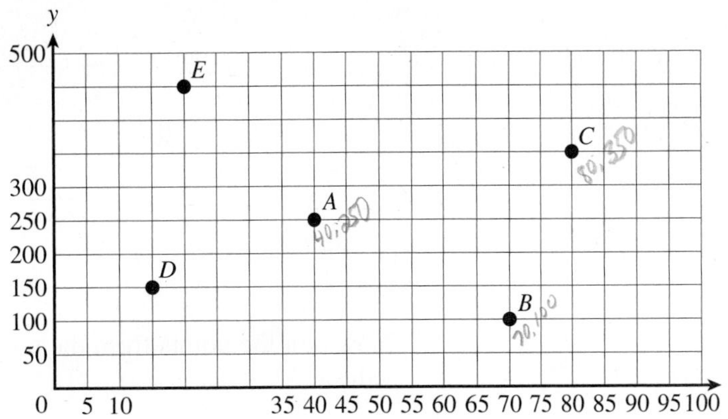

1. a. What is the size of the input unit on the grid shown above? ______
   b. Write in the missing input units on the grid.
   c. What is the size of the output unit? ______
   d. Write in the missing output units on the grid.

2. a. What are the coordinates of the points *A*, *B*, and *C* on the given grid? Write your answers as ordered pairs next to the points on the grid.
   b. What is the x-coordinate of the point *D*?
   c. What is the y-coordinate of the point *E*?

3. a. Plot the point with coordinates (60, 100) on the given grid. Write the ordered pair next to the point.
   b. Plot the point with coordinates (10, 75) on the given grid. Write the ordered pair next to the point.
   c. The x-coordinate of a point is 95 and the y-coordinate is 100. Plot and label the point.
   d. The x-coordinate of a point is 55 and the y-coordinate is 225. Plot and label the point.
   e. Plot and label the points (0, 0), (0, 200), (40, 0), and (0, 325).

4. Graph the ordered pairs given in the table. Use the following grid as given; do not extend the graph in either direction.

INPUT	2	4	7	8	10	12	14	15
OUTPUT	5	10	17	20	25	30	35	38

   a. What size unit would be reasonable for the input?
   b. What size unit would be reasonable for the output?

Exercise numbers appearing in color are answered in the Selected Answers appendix.

c. List the units along the horizontal axis; along the vertical axis.

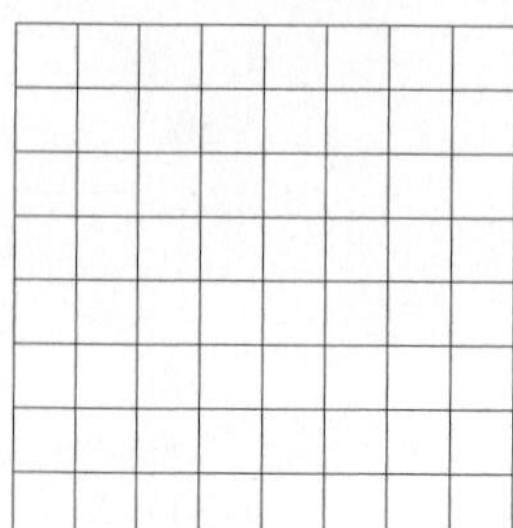

d. Now plot the points from the table. Label each point with its input/output value.

5. Graph the data in the following table as a bar chart on the accompanying grid.

a. Write the name of the output along the vertical axis and the name of the input along the horizontal axis.

b. List the input categories along the horizontal axis.

c. List the units along the vertical axis.

d. Draw the bars corresponding to the given input/output pairs.

OCCUPATION	2004 MEDIAN ANNUAL SALARY ($)
Computer engineer	81,100
Systems analyst	67,500
Database administrator	61,900
Physician assistant	69,200

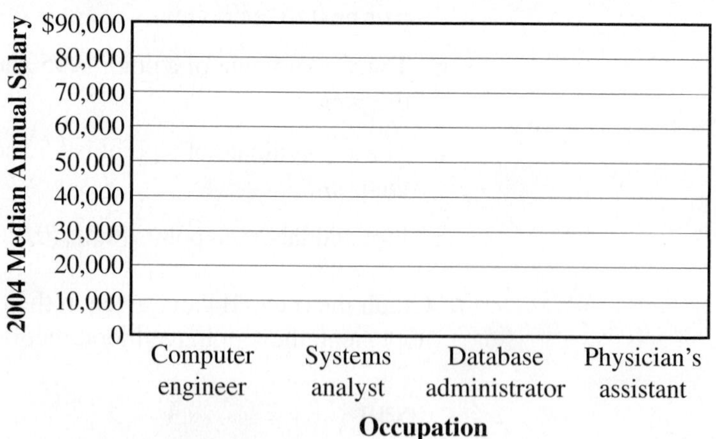

6. In the twentieth century, the dumping of hazardous waste throughout the United States polluted waterways, soil, and air. Statistical evidence was one factor that led to federal laws aiming to protect human health in the environment. For example, the following chart presents some historical evidence on the number of abandoned hazardous waste sites found in various regions in the United States in the 1990s. Use the information in the chart to answer questions a–g that follow it.

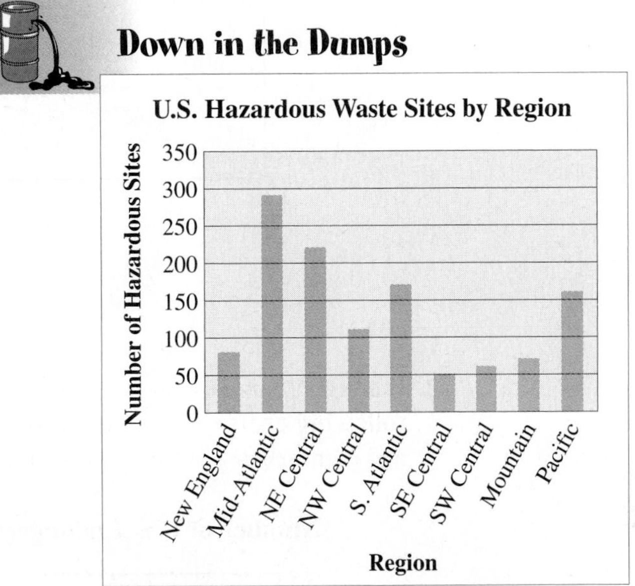

a. How many regions are represented in the chart?

b. Which region has the greatest number of hazardous waste sites? Estimate the number.

c. Which region has the least number of hazardous waste sites? Estimate the number.

d. Which region would you say is in the middle in terms of waste sites?

e. What feature of this chart aids in estimating the number of waste sites for a given region?

f. Based on this data, in what region(s) would you advise someone to live (or not live)?

g. From the chart, estimate the total number of waste sites found in the United States by the 1990s.

7. Bar graphs can be oriented either vertically or horizontally. Use the following chart to estimate how many more bald eagle pairs were observed in Louisiana compared to Texas in 2000.

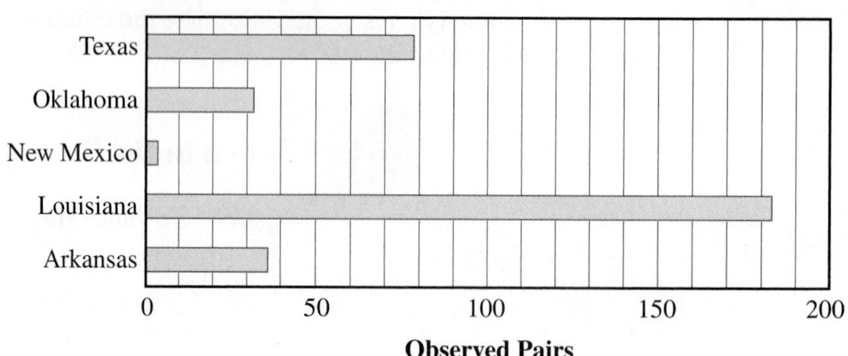

8. Bar graphs can also show, for comparison, data collected at different times. The following chart shows data for four categories of animals: fish, reptiles, birds and mammals.

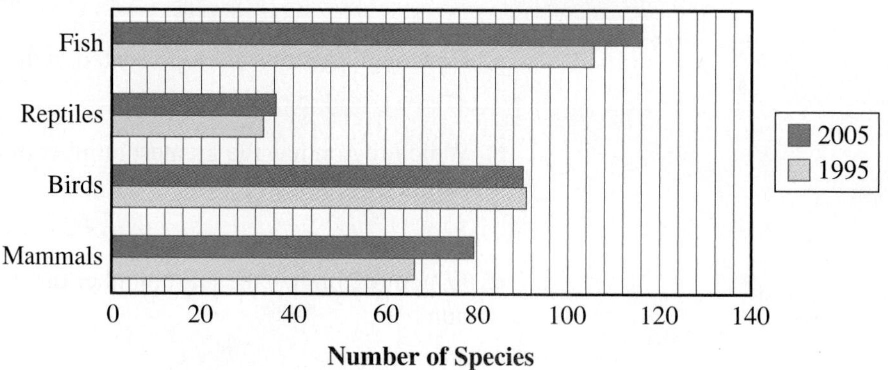

a. Which category of animal actually saw a drop in the number of endangered and threatened species between 1995 and 2005? Explain your answer.

b. Which category of animal showed the largest increase, and by approximately how many species?

ACTIVITY 1.3

Bald Eagles Revisited

OBJECTIVES

1. Add whole numbers by hand and mentally.
2. Subtract whole numbers by hand and mentally.
3. Estimate sums and differences using rounding.
4. Recognize the associative property and the commutative property for addition.
5. Translate a written statement into an arithmetic expression.

In the 1700s, there were an estimated 25,000 to 75,000 nesting bald eagle pairs in what are now the contiguous 48 states. By the 1960s, there were less than 450 nesting pairs due to the destruction of forests for towns and farms, shooting, and DDT and other pesticides. In 1972, the federal government banned the use of DDT. In 1973, the bald eagle was formally listed as an endangered species. By the 1980s, the bald eagle population was clearly increasing. The following map of the contiguous 48 states displays the number of nesting pairs of bald eagles for each state. There are two numbers for each state. The first is for 1982, and the second is for 2000.

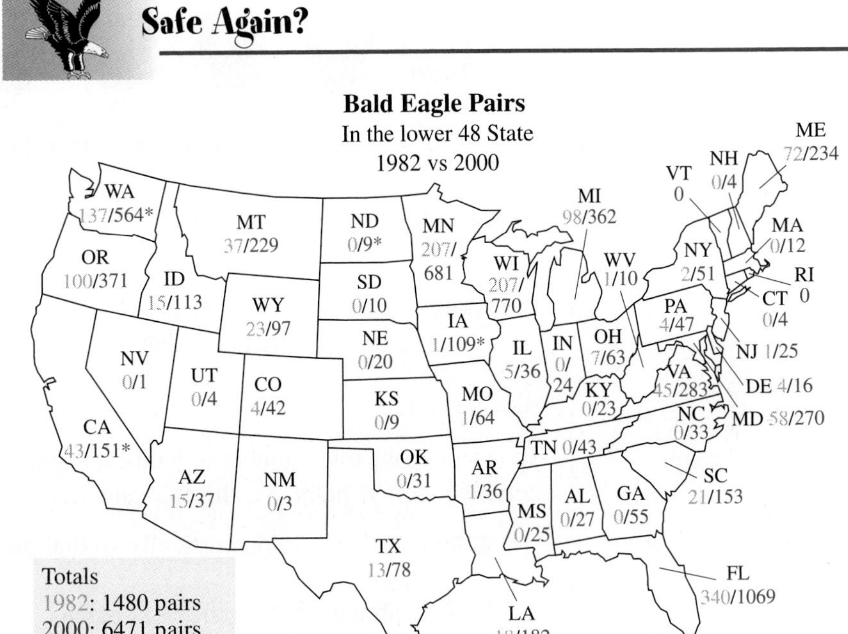

Addition of Whole Numbers

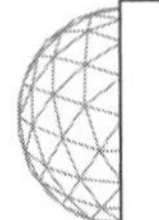

EXAMPLE 1 ***What was the total number of nesting bald eagle pairs in New York and Pennsylvania in 1982?***

SOLUTION

Calculate the total number of bald eagle nesting pairs in 1982 in New York and Pennsylvania by combining the two sets. Note that each eagle on the next page represents a bald eagle nesting pair.

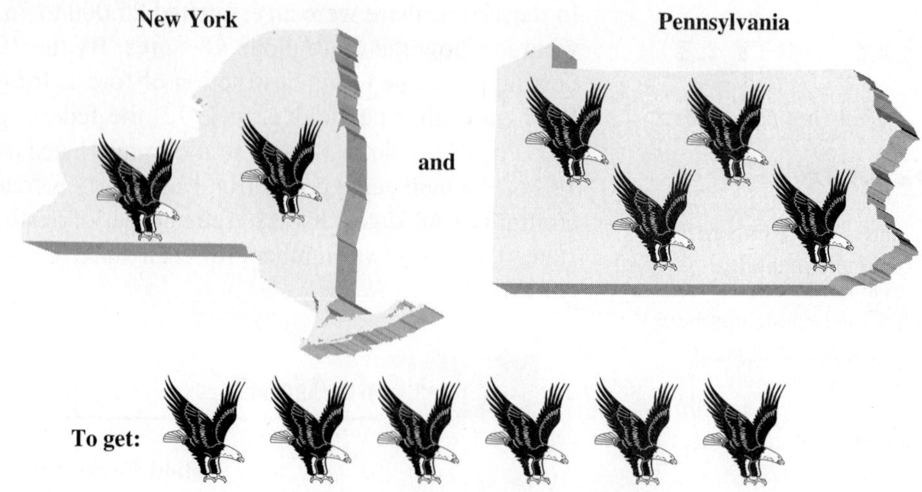

In whole-number notation, $2 + 4 = 6$.

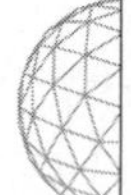

EXAMPLE 2 ***Calculate the total number of bald eagle pairs in California and Colorado in 2000.***

SOLUTION

To calculate the total number of bald eagle pairs in California and Colorado in 2000, combine the 151 pairs in California and 42 pairs in Colorado using addition.

- **a.** Set up the addition vertically so that the place values are aligned vertically.
- **b.** Add all the digits in the ones place.
- **c.** Add all the digits in the tens place.

Tens: Add the 5 + 4 in tens place to get 9.

Hundreds: Bring down the 1.

Ones: Add the 1 + 2 in the ones place to get 3.

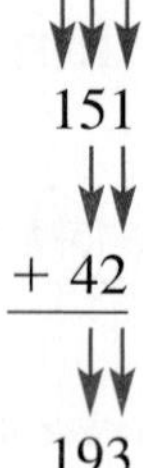

$$\begin{array}{r} 151 \\ +\ 42 \\ \hline 193 \end{array}$$

The numbers that are being added, 151 and 42, are called **addends.** Their total, 193, is called the **sum.**

1. **a.** Calculate the total number of bald eagle pairs in New York and Pennsylvania in 2000.

   b. Calculate the total number of bald eagle pairs in Idaho and Minnesota in 2000.

c. In part b, does it make a difference whether you set up the addition $113 + 681$ or $681 + 113$?

The order in which you add two numbers does not matter. This property of addition is called the **commutative property.** For example,

$$17 + 32 = 32 + 17$$

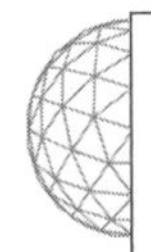

EXAMPLE 3 ***Calculate the total number of bald eagle pairs in South Carolina and Texas in 2000 by setting up the addition vertically.***

SOLUTION

Set up the addition vertically. There are two methods for determining the sum.

Method 1:

Sums of Digits in Place Value

$$\begin{array}{r} 153 \\ +\ 78 \\ \hline 11 \\ 120 \\ 100 \\ \hline 231 \end{array}$$

11 — adding 3 ones + 8 ones

120 — adding 5 tens + 7 tens

Method 2:

Regroup to Next Higher Place Value

$$\begin{array}{r} {\scriptstyle 1\,1} \\ 153 \\ +\ \ 78 \\ \hline 231 \end{array}$$

Regroup 1 to hundreds place.

Regroup 1 to tens place.

$3 + 8 = 11$, write 1 in ones place.

2. a. Calculate the total number of bald eagle pairs in New York and Ohio in 2000 using method 1.

b. Calculate the total number of bald eagle pairs in New York and Ohio in 1998 using method 2.

c. Calculate the total number of bald eagle pairs in Texas, New Mexico, and Arizona in 2000 using method 1.

d. Calculate the total number of bald eagle pairs in Texas, New Mexico, and Arizona in 2000 using method 2. Compare this result to the one obtained in part c.

3. a. Mentally calculate the total number of bald eagle pairs in California, New Mexico, and Arizona in 2000 by determining the sum of two numbers, then adding the sum to the third number.

b. Does it make a difference which two numbers you add together first?

When calculating the sum of three whole numbers, it makes no difference whether the first two numbers or the last two numbers are added together first. This property of addition is called the **associative property.** For example,

$$(143 + 4) + 36 = 143 + (4 + 36).$$

4. a. Mentally calculate the total number of bald eagle pairs in Washington, Oregon, and California in 1982.

b. Mentally calculate the total number of bald eagle pairs in Washington, Oregon, and California in 2000.

c. Explain the process you used to add these numbers.

5. a. Choose five states and calculate the total number of nesting bald eagle pairs in 1982.

b. Calculate the total number of nesting bald eagle pairs in 2000 for the five states you chose in part a.

Estimating Sums of Whole Numbers

Estimation is useful for adding several numbers quickly and for checking that a given sum is reasonable. One way to estimate is to round each number (addend) to the same place value.

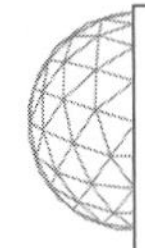

EXAMPLE 4 ***Estimate the total number of nesting pairs in Florida, South Carolina, and Virginia in 2000. Then calculate the exact sum.***

SOLUTION

Estimated Sum:		*Exact Sum:*
1100 ←	Florida →	1069
200 ←	South Carolina →	153
+ 300 ←	Virginia →	+ 283
1600		1505

In Example 4, the estimate is higher than the exact sum. An estimate may be higher, lower, or occasionally equal to the exact sum. Notice that the exact sum 1505 is reasonable for the given data because it is close in value to the estimated sum 1600.

6. a. Estimate the number of nesting pairs in Louisiana, Maine, and Arizona in 1982.

b. Calculate the exact number of pairs in 1982.

c. Estimate the number of nesting pairs in Louisiana, Maine, and Arizona in 2000.

d. Calculate the exact number of pairs in 2000.

e. Compare the exact results to your estimates. State whether the estimates are higher, lower, or the same as the exact results. Do your estimates indicate that your exact results are reasonable for the data given?

EXAMPLE 5 Subtraction of Whole Numbers

The number of bald eagle nesting pairs in 1982 for New York and the combined total for New York and Pennsylvania are given in the following graphic. Calculate the number of nesting pairs in Pennsylvania.

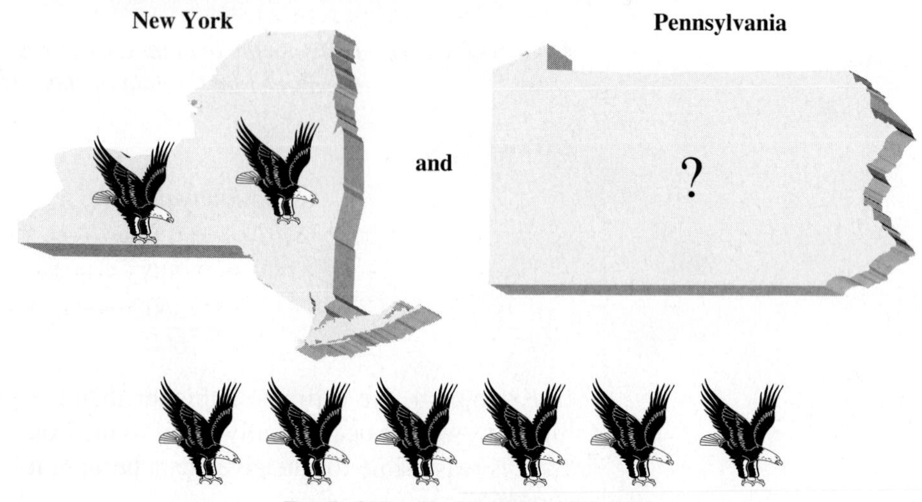

The picture suggests that the problem is to find the **missing addend.** In symbols, the calculation can be written in terms of addition as $\mathbf{2 + ? = 6}$. By thinking of a number that added to 2 gives 6, you see that the answer is 4 pairs of bald eagles.

Alternatively, by thinking of taking away **(subtracting)** 2 from 6, the answer is the same, 4 pairs of bald eagles. In symbols, the calculation is written as $\mathbf{6 - 2 = ?}$

Subtraction is finding the difference between two numbers. The operation of subtraction involves taking away. In Example 5, take 2 bald eagle nesting pairs away from 6 bald eagle nesting pairs to get the difference, 4 nesting pairs.

7. a. How many more bald eagle nesting pairs were there in 2000 than in 1982 in Oregon?

   b. Set up the calculation using addition.

   c. Set up the calculation using subtraction.

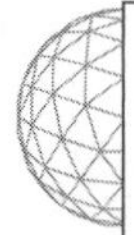

EXAMPLE 6 ***How many more nesting pairs were in Georgia than were in Alabama in 2000?***

SOLUTION

$$\begin{array}{r} {}^{4}\,{}^{15} \\ \not{5}\,\not{5} \\ -2\,7 \\ \hline 2\,8 \end{array}$$

Convert 1 ten to 10 ones.
add ones to 5 to obtain 15 ones.
Subtract 7 from 15, and subtract 2 from 4.

Check using addition:

$$\begin{array}{r} 28 \\ +\ 27 \\ \hline 55 \end{array}$$

Formally, the number that is subtracted is called the **subtrahend.** The number subtracted from is called the **minuend.** The result is called the **difference.** To check subtraction, add the difference and the subtrahend. The result should be the minuend.

		Check using addition:
55	minuend	28
− 27	subtrahend	+ 27
28	difference	55

Here, addition is used to check subtraction. Addition is called the **inverse** operation for subtraction.

8. a. Determine the difference between the number of nesting pairs in Louisiana in 2000 and 1982.

b. What number is being subtracted? Why?

9. a. Calculate the increase in the population of nesting pairs in Michigan from 1982 to 2000.

b. Determine the difference between the number of nesting pairs in Wisconsin and in Minnesota in 2000.

c. Determine the difference between the number of nesting pairs in Wisconsin and Minnesota in 1982.

10. a. Estimate the difference between the number of nesting pairs in Florida and in Washington in 2000.

b. Determine the exact difference.

c. Was your estimate higher or lower than the exact difference?

11. a. To what place value should you round to estimate the difference between the number of nesting pairs in Louisiana and in California in 2000?

b. The highest place value is the hundreds. Would it make sense to round to the hundreds place? Why or why not?

Add or Subtract?

When you set up a problem, it is sometimes difficult to decide if you need to do addition or subtraction. It is helpful if you can recognize some key phrases so that you will write a correct arithmetic expression. An **arithmetic expression** consists of numbers, operation signs $(+, -, \cdot, \div)$, and sometimes parentheses.

The following tables contain some typical key phrases, examples, and corresponding arithmetic expressions for addition and subtraction.

ADDITION		
KEY PHRASE	EXAMPLE	ARITHMETIC EXPRESSION
sum of	sum of 3 and 5	3 + 5
increased by	7 increased by 4	7 + 4
plus	12 plus 10	12 + 10
more than	5 more than 6	6 + 5
total of	total of 13 and 8	13 + 8
added to	45 added to 50	50 + 45

SUBTRACTION		
KEY PHRASE	EXAMPLE	ARITHMETIC EXPRESSION
difference of	difference of 12 and 7	12 − 7
decreased by	95 decreased by 10	95 − 10
minus	57 minus 26	57 − 26
less than	5 less than 23	23 − 5
subtracted from	12 subtracted from 37	37 − 12
subtract	8 subtract 5	8 − 5

12. Translate each of the following into an arithmetic expression.

a. 34 plus 42

b. difference of 33 and 22

c. 100 minus 25

d. total of 25 and 19

e. 17 more than 102

f. 14 subtracted from 28

g. 81 increased by 16

h. 50 less than 230

i. 250 decreased by 120

j. sum of 18 and 21

k. 101 added to 850

SUMMARY
ACTIVITY 1.3

1. Numbers that are added together are called **addends.** Their total is the **sum.**
2. **To add numbers:** Align them vertically according to place value. Add the digits in the ones place. If their sum is a two-digit number, write down the ones digit and carry the tens digit to the next column as a number to be added. Repeat with the next higher place value.
3. The order in which you add two numbers does not matter. This property of addition is called the **commutative property.**
4. When calculating the sum of three whole numbers, it makes no difference whether the first two numbers or the last two numbers are added together first. This property of addition is called the **associative property.**
5. **Estimation** is useful for adding or subtracting numbers quickly and to check the reasonableness of an exact calculation. One way to estimate is to round each number to its highest place value. In most cases, it may be better to round to a place value lower than the highest one.
6. **Subtraction** is used to find the difference between two numbers. The operation of subtraction involves *taking away.*

   The number that is subtracted is called the **subtrahend.** The number being subtracted from is called the **minuend.** The result is called the **difference.** To check subtraction, add the difference and the subtrahend. The result should be the minuend.

   To subtract: Align the subtrahend under the minuend according to place value. Subtract digits having the same place value. Regroup from a higher place value, if necessary.

EXERCISES
ACTIVITY 1.3

1. Determine the sum using method 1 (sum of the digits by place values) as shown in Example 3.

a. $\begin{array}{r} 256 \\ +\ 35 \\ \hline \end{array}$

b. $\begin{array}{r} 617 \\ +149 \\ \hline \end{array}$

c. $\begin{array}{r} 51 \\ 382 \\ +\ 77 \\ \hline \end{array}$

2. Determine the sum using method 2 (regroup to next higher place value) as shown in Example 3.

a. $\begin{array}{r} 159 \\ +\ 27 \\ \hline \end{array}$

b. $\begin{array}{r} 924 \\ +138 \\ \hline \end{array}$

c. $\begin{array}{r} 51 \\ 382 \\ +\ 77 \\ \hline \end{array}$

3. Determine the sum of 67 and 75.

a. 67 + 75

b. 75 + 67

c. Are the sums in parts a and b the same?

d. What property of addition is demonstrated?

4. Determine the sum. Do the addition in the parentheses first.

a. 34 + (15 + 71)

b. (34 + 15) + 71

c. Are the sums in parts a and b the same?

d. What property of addition is demonstrated?

5. a. Estimate the sum: 171 + 90 + 226

b. Determine the actual sum.

c. Was your estimate higher, lower, or the same as the actual sum?

6. a. Estimate the sum: 326 + 474

b. Determine the actual sum.

c. Was your estimate higher, lower, or the same as the actual sum?

7. Evaluate.

a. 123 − 91 **b.** 543 − 125 **c.** 78 − 49

d. 1002 − 250 **e.** 2001 − 1962 f. 696 − 384

8. a. Subtract the year in which you were born from this year.

b. Is the difference you obtain your age?

c. Have you had your birthday yet this year? Does this affect your answer in part b?

9. This week, you took home $96 from your part-time job. You owe your mother $39.

a. Estimate the amount of money that you will have after you pay your mother.

b. Determine the actual amount of money you will have after you pay your mother.

10. In 2000, there were 113 bald eagle nesting pairs in Idaho, 770 in Wisconsin, and 564 in Washington.

a. Estimate the total number of nesting pairs in all three states.

b. Determine the actual total.

c. Is your estimate higher or lower than the actual total?

11. In 2000, there were 371 nesting pairs in Oregon and 362 nesting pairs in Michigan.

a. Estimate the difference between the nesting pairs in Oregon and Michigan.

b. If you round both numbers to the hundreds place, what is your estimate?

c. If you round both numbers to the tens place, what is your estimate?

d. Which is the better "estimate"? Explain.

12. Translate each of the following into an arithmetic expression.

a. 13 plus 23

b. 108 minus 15

c. difference of 70 and 58

d. total of 45 and 79

e. 7 more than 12

f. 13 subtracted from 28

g. 85 increased by 8

h. 52 less than 300

i. 25 decreased by 12

ACTIVITY 1.4

Summer Camp

OBJECTIVES

1. Multiply whole numbers and check calculations using a calculator.
2. Multiply whole numbers using the distributive property.
3. Estimate the product of whole numbers by rounding.
4. Recognize the associative and commutative properties for multiplication.

You accept a job working in the kitchen at a small, private summer camp in New England. One hundred children attend the 8-week program under the supervision of 24 staff members. The job pays well, and it includes room and board with every other weekend off. One of your responsibilities is to pick up supplies twice a week at a local wholesale food club. Some of the items listed on this week's order form appear in the following receipt from the food club.

QUANTITY	ITEM	UNIT PRICE ($)
8	1 CASE (24 BOTTLES) 10-OZ BOTTLES OF JUICE	8.99
20	36 1.55-OZ MILK CHOCOLATE BARS	11.19
12	36 1-OZ SERVINGS OF CREAM CHEESE	6.39
18	32 1.25-OZ GRANOLA BARS	6.99
6	1 BOX OF 15 CARTONS OF 1 DOZ. LARGE EGGS	8.99
10	4-LB PACKAGE HOT DOGS (40 COUNT)	6.69
12	10-LB PACKAGE HAMBURGER (40 COUNT)	12.99
16	24-PACK HOT DOG ROLLS	2.39
18	24-PACK HAMBURGER ROLLS	2.39
30	2-LOAF PACK 20-OZ BREAD (20 SLICES)	2.39
12	42-COUNT VARIETY PACKAGE OF CHIPS	7.99
8	200-COUNT PACKAGE 9" PLATES	6.99
1	1500-COUNT PACKAGE DISPENSER NAPKINS	10.39
4	500-COUNT PACKAGE SPOONS	5.69
2	500-COUNT PACKAGE KNIVES	5.69
3	500-COUNT PACKAGE FORKS	5.69

1. The total number of bottles of juice purchased can be represented by the following sum.

$$24 + 24 + 24 + 24 + 24 + 24 + 24 + 24 = \_\_\_\_\_\_$$

a. Calculate this sum directly.

b. Calculate this sum using your calculator.

c. Is there a more efficient (shorter) way to do this calculation?

Multiplication of whole numbers is repeated addition. The sum $24 + 24 + 24 + 24 + 24 + 24 + 24 + 24$ can be rewritten as the **product** $8 \cdot 24$ or 8×24. The operation sign "×" is the multiplication symbol generally used in arithmetic courses, but the symbol " · " is more common in algebra. The whole numbers 8 and 24 are called **factors** of the product 192. To do the multiplication by hand, set up the calculation vertically as follows.

$$\begin{array}{r l} 24 & \\ \times\ 8 & \\ \hline 32 & \text{Multiply 8 times 4.} \\ 160 & \text{Multiply 8 times 20.} \\ \hline 192 & \text{Add 32 and160.} \end{array}$$

Multiplication works when set up vertically because 24 can be written as 20 + 4 and the factor 8 multiplies both the 20 and the 4. Set up horizontally, the calculation is written as follows:

$$8 \cdot 24 = 8 \cdot (20 + 4) = 8 \cdot 20 + 8 \cdot 4 = 160 + 32 = 192$$

Rewriting $8 \cdot (20 + 4)$ as $8 \cdot 20 + 8 \cdot 4$ is an example of the **distributive property of multiplication over addition.**

2. a. Calculate the total number of cartons of eggs you are to purchase this week.

   b. Calculate the number of 9-inch plates you will purchase.

The multiplication for the total number of hamburger rolls can be set up vertically or horizontally.

Vertically:

$$\begin{array}{r l} 24 & \\ \times\ 18 & \\ \hline 32 & \text{Multiply 8 times 4.} \\ 160 & \text{Multiply 8 times 20.} \\ 240 & \text{Multiply 10 times 24.} \\ \hline 432 & \end{array}$$

Horizontally:

$$18 \cdot 24 = \underbrace{18 \cdot (20 + 4) = 18 \cdot 20 + 18 \cdot 4}_{\text{Distributive property}} = 360 + 72 = 432$$

3. a. How many ounces of juice are there in one case of juice from the wholesale club?

   b. Use your answer in part a to calculate the total number of ounces of juice needed this week.

   c. How many bottles of juice are needed?

d. Use your answer in part c to calculate the total number of ounces of juice needed this week.

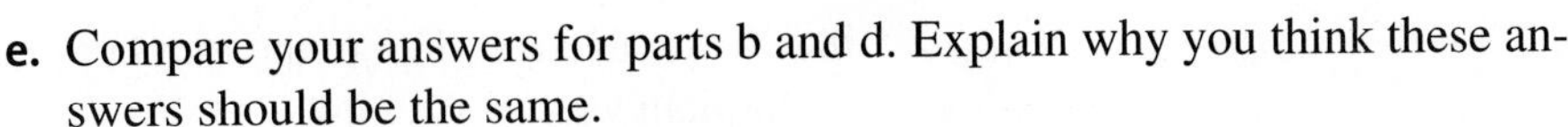

e. Compare your answers for parts b and d. Explain why you think these answers should be the same.

The property illustrated in Problems 3b and 3d is the **associative property of multiplication.** For example,

$$5 \cdot (4 \cdot 7) = (5 \cdot 4) \cdot 7$$

4. Describe the associative property of multiplication in your own words.

5. Calculate the total number of eggs on the list in two ways using the associative property.

6. The multiplication to determine the total number of hamburger rolls can be set up two ways.

$$\begin{array}{r} 24 \\ \times 18 \\ \hline \end{array} \quad \text{or} \quad \begin{array}{r} 18 \\ \times 24 \\ \hline \end{array}$$

a. Determine the product for each of the two multiplication problems.

b. Explain why the answers are the same.

The mathematical property illustrated in Problem 6a is the **commutative property of multiplication.** For example,

$$8 \cdot 4 = 4 \cdot 8$$

7. Describe the commutative property of multiplication in your own words.

8. a. Determine the total number of hamburgers and the total number of hamburger rolls you need to purchase.

b. Do you need to change the number of packages of hamburger rolls? Explain why or why not.

9. One 1500-count package of dispenser napkins is purchased.

a. Using multiplication, calculate the total number of napkins purchased.

b. If you multiply any whole number by 1, what is the result?

c. If you multiply any whole number by zero, what is the result?

Estimation

You may not have easy access to a calculator at summer camp. So you may need to multiply or check multiplication mentally or by hand. To calculate the total number of individual 1-ounce servings of cream cheese, you need to multiply 12 times 36. The product can be estimated by rounding.

- Round 12 down to 10.
- Round 36 up to 40.
- Multiply 10 times 40.
- The estimated product is ______

10. a. Multiply 12 and 36 to determine the actual number of servings of cream cheese.

b. What is the difference between the actual number of servings and the estimated number of servings?

11. a. Estimate the total number of granola bars in the order.

b. Determine the actual number of granola bars.

c. What is the difference between the actual number and the estimated number of granola bars?

PROCEDURE

Estimating Products

- Round each factor to a large enough place value so that you can do the multiplication mentally.
- There is no one correct answer when estimating, only a reasonable answer.

12. Estimate the total number of chocolate bars.

13. Estimate the total number of individual snack packs of chips.

SUMMARY
ACTIVITY 1.4

1. **Multiplication properties of whole numbers**

   Any whole number times 1 remains the same.

   Any whole number times 0 is 0.

2. **Distributive property**

   A whole number placed in front of a set of parentheses containing a sum or difference of two numbers multiplies each of the inside numbers.

   $$3 \cdot (7 + 2) = 3 \cdot 7 + 3 \cdot 2 \quad \text{or} \quad 5 \cdot (10 - 4) = 5 \cdot 10 - 5 \cdot 4$$

3. **Associative property**

   When multiplying three whole numbers, it makes no difference which two numbers are multiplied first.

   $$(2 \cdot 5) \cdot 7 = 2 \cdot (5 \cdot 7)$$

4. **Commutative property**

   Changing the order of two whole numbers when multiplying them produces the same product.

   $$3 \cdot 6 = 6 \cdot 3$$

5. **Estimating products**

   Round each factor to a large enough place value so that you can do the multiplication mentally.

   There is no one correct answer when estimating, only reasonable answers.

EXERCISES
ACTIVITY 1.4

1. Multiply vertically. Verify your answer using a calculator.

a. $\begin{array}{r} 34 \\ \times\ 4 \\ \hline \end{array}$

b. $\begin{array}{r} 529 \\ \times\ 8 \\ \hline \end{array}$

c. $\begin{array}{r} 67 \\ \times\ 5 \\ \hline \end{array}$

d. $\begin{array}{r} 807 \\ \times\ 9 \\ \hline \end{array}$

e. $\begin{array}{r} 125 \\ \times\ 8 \\ \hline \end{array}$

f. $\begin{array}{r} 2001 \\ \times\ 25 \\ \hline \end{array}$

g. $\begin{array}{r} 75 \\ \times 52 \\ \hline \end{array}$

h. $\begin{array}{r} 1967 \\ \times 105 \\ \hline \end{array}$

2. a. Multiply 8 and 47 by rewriting 47 as $40 + 7$ and use the distributive property to obtain the result.

b. Multiply 8 and 47 vertically.

3. a. Multiply 12 and 36 by rewriting 36 as $30 + 6$ and use the distributive property to obtain the result.

b. Multiply 12 and 36 vertically.

4. Three 500-count packages of forks are purchased.

a. Use addition to determine the total number of forks.

b. Use multiplication to determine the total number of forks.

5. Ten 40-count packages of hot dogs are purchased.

a. Determine the total number of hot dogs by calculating $10 \cdot 40$.

b. Calculate: $40 \cdot 10$.

c. What property of multiplication is demonstrated by the fact that the answers to parts a and b should be the same?

6. a. Evaluate: $72 \cdot 23$

b. Evaluate: $23 \cdot 72$

c. Are the answers to parts a and b the same?

d. What property do the results of this exercise demonstrate?

7. Evaluate by finding the product in parentheses first.

a. $7 \cdot (13 \cdot 20)$

b. $(7 \cdot 13) \cdot 20$

c. Are the answers to parts a and b the same?

d. What property do the results in parts a and b demonstrate?

8. You purchase twelve 42-count packages of variety chips for the summer camp.

a. Estimate the total number of individual packages of chips purchased.

b. Determine the actual number of individual packages you purchased.

c. Is the estimated total higher or lower than the actual total?

9. You purchase sixteen 24-pack hot dog rolls this week.

a. Estimate the total number of hot dog rolls.

b. Determine the actual number of hot dog rolls.

c. Is the estimated total higher or lower than the actual total?

ACTIVITY 1.5

College Supplies

OBJECTIVES

1. Divide whole numbers by "grouping."
2. Divide whole numbers "by hand" and by calculator.
3. Estimate the quotient of whole numbers by rounding.
4. Recognize the noncommutative property for division.

School Supplies

It is the beginning of a new semester and time to purchase supplies. You and five fellow students decide to shop at a discount office supply store. The six of you purchase the following items.

QUANTITY	ITEM	UNIT PRICE ($)
3	8-PACK NUMBER 2 PENCILS	2.89
3	5-PACK MECHANICAL PENCILS	3.95
3	4-PACK REFILLABLE MECHANICAL PENCILS	3.58
2	10-PACK ASSORTED GEL RETRACTABLE PENS	9.98
3	2-PACK BALLPOINT PENS	6.18
2	4-PK LIQUID PAPER	6.98
3	10-PACK ASSORTED HIGHLIGHTERS	9.98
3	5-PACK PERMANENT MARKERS	3.49
6	400-COUNT PACKAGE 8.5" X 11" COLLEGE-RULED PAPER	3.09
20	POCKET FOLDERS	0.18
12	THREE-SUBJECT SPIRAL NOTEBOOK	4.72
4	5-COUNT PACK REPORT COVERS	5.18
3	500-COUNT PACK 3" X 5" INDEX CARDS	2.78

There are three 8-count packages of number-two pencils or a total of 24 pencils, illustrated below.

These 24 pencils need to be divided among the six of you. To set this up as a division calculation, write either $24 \div 6$, $24/6$, or $6\overline{)24}$.

DEFINITION

The number *being divided* is called the **dividend.** The number that divides the dividend is called the **divisor.** Here, the number 6 is the divisor, and the number 24 is the dividend. The **quotient** is the result of the division. In this case, 4 is the quotient.

$$\begin{array}{r} 4 \leftarrow \text{quotient} \\ \text{divisor} \rightarrow 6\overline{)24} \leftarrow \text{dividend} \end{array}$$

If 24 pencils are distributed evenly among the six of you, you will each get 4 pencils.

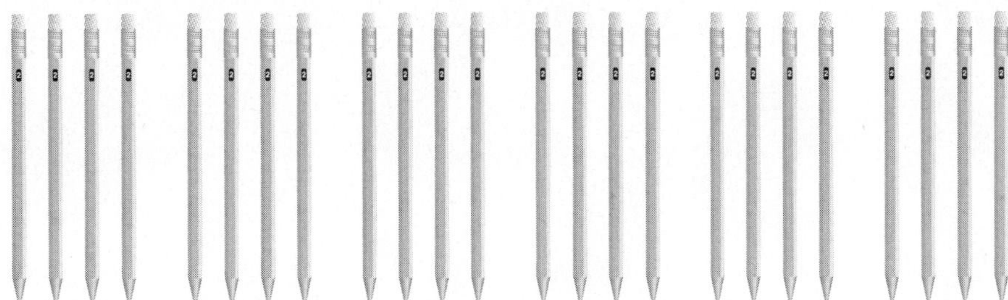

You used division to distribute the 24 pencils equally among yourselves. The result was 4 pencils per student. In general, you use division to separate items into a specified number of equal groupings.

The two ways the division process can be stated:

$$\text{dividend} \div \text{divisor} = \text{quotient, with a possible remainder,}$$

or, in long division format,

$$\text{divisor}\overline{)\text{dividend}}^{\text{quotient with a possible remainder}}$$

Multiplication and division are **inverse operations.** This means that the division $24 \div 6 = 4$ can be written as the multiplication $4 \cdot 6 = 24$, and vice versa.

1. a. Your group buys three 5-packs of mechanical pencils. Distribute these 15 pencils evenly among the six of you. How many pencils will each of you receive? Are there any pencils left over?

b. Identify the divisor and the dividend. What is the quotient? What is the remainder, if any?

2. a. How many gel pens are there in the two packages of 10-count assorted gel retractable pens?

b. Distribute the gel pens evenly among the six of you. Set up as a division problem. What is the quotient? What is the remainder, if any?

To check division, multiply the quotient and the divisor, then add the remainder. The result should be the dividend.

$$(\text{quotient} \times \text{divisor}) + \text{remainder} = \text{dividend}$$

3. Check the division you did in Problem 2.

Division can be considered as repeated subtraction of the divisor from the dividend, with a remainder left over.

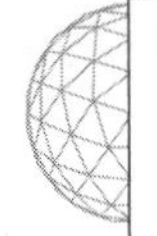

EXAMPLE 1 ***You buy 20 pocket folders. Distribute them evenly among the six of you.***

SOLUTION

a. Use long division.

$$\begin{array}{r} 3 \\ 6\overline{)20} \\ \underline{18} \\ 2 \end{array}$$

b. Use repeated subtraction.

20	
− 6	6 can be subtracted from 20 three times.
14	
− 6	
8	3 is the quotient.
−6	
2	2 is the remainder.

In general, when you divide, you are repeatedly subtracting multiples of the divisor from the dividend until no whole multiples remain.

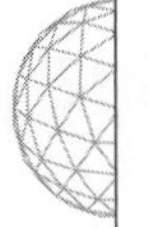

EXAMPLE 2 ***Your group purchased three 500-count packs of 3" × 5" index cards. Distribute these 1500 index cards evenly among the six of you. How many index cards do you each get?***

SOLUTION

Use a long division problem process to determine that each of you receives 250 index cards with none left over.

250	
6)1500	6 does not divide into 1. 6 does divide into 15 two times.
−12	Write down the product $2 \cdot 6 = 12$. Subtract 12 from 15.
30	Bring down 0, the next digit. 6 divides into 30 five times.
−30	Write down the product $5 \cdot 6 = 30$. Subtract 30 from 30.
00	Bring down the last 0. 6 divides into 0 zero times.
	The remainder is 0.

4. a. Your group purchased six 400-count packages of 8.5" × 11" college-ruled paper. If these packages are distributed equally among the six of you, how many packages will you each receive?

b. Suppose 52 packages are distributed equally among 52 students. How many packages will each student receive?

c. How many times can you subtract 52 from 52?

d. What result do you get when you divide any nonzero number by itself?

5. There was a stapler in the shopping cart, but you returned it to the shelf because all of you already had one.

a. How many staplers will you receive from this shopping expedition?

b. Use your calculator to divide zero by six. What is the result? Try dividing zero by another nonzero whole number. What is the result?

c. Use your calculator to divide 6 by 0. What happens?

If you try to divide by zero, a basic or scientific calculator will display the letter E to signify an error. Graphing calculators usually display the word ERROR with a message. The reason is that if you change $6 \div 0 = ?$ to a multiplication calculation, it becomes $0 \cdot ? = 6$. No whole number works because zero times any whole number is zero, not 6. This means that $6 \div 0$ has no answer. Another way to say this is that $6 \div 0$ is **undefined.**

d. Explain why 9 divided by 0 is undefined.

DEFINITION

Division Properties Involving 0 and 1	**Example:**
• Any whole number divided by itself is 1.	$34 \div 34 = 1$
• Any whole number divided by 1 is itself.	$6 \div 1 = 6$
• 0 divided by any nonzero whole number is 0.	$0 \div 10 = 0$
• Division by 0 is undefined.	$7 \div 0$ is undefined.

6. Three 10-count packs of assorted highlighters contain a total of 30 highlighters.

a. By doing the division $30 \div 6$, determine the number of highlighters each student will receive.

b. Do you get the same result by doing the division, $6 \div 30$?

c. Does the commutative property hold for division? That is, does $6 \div 30 = 30 \div 6$?

Rent and Utilities

This year, you decide to rent an apartment near campus with three other students. The rent is $1150 per month, basic telephone service is $41 per month, and the average monthly utility bill is $173.

To estimate your share of the monthly telephone service charge, you would round $41 to the tens place and get $40. Dividing $40 by 4 (students), you estimate your share to be $10 per month.

To estimate a quotient, first estimate the divisor and the dividend using numbers that allow for easier division mentally or by hand. For example, estimate $384 \div 6$ by rounding 384 to 400 and replacing 6 by 5. The estimate is $400 \div 5 = 80$, which is 16 more than the exact result, 64.

7. **a.** Estimate your share of the rent by rounding $1150 to the thousands place and then dividing by 4.

b. Is your estimate lower or higher than the actual amount that you owe?

8. **a.** Estimate your share of the utility bill by rounding $173.

b. Is your estimate lower or higher than the actual amount that you owe?

9. Estimate your share of the monthly telephone service charge by rounding.

10. a. Add the monthly charges for rent, telephone, and utilities. Calculate your actual share of this sum. Verify your calculation using your calculator.

b. Compare your actual share of the monthly expenses with the sum of the individual estimates for the monthly rent, telephone, and utility costs.

11. a. Estimate: $19{,}500 \div 78$

b. Determine the exact answer.

c. Is your estimate lower, higher, or the same?

12. a. Estimate: $5880 \div 120$

b. Determine the exact answer.

c. Is your estimate lower, higher, or the same?

13. a. Estimate: $30{,}380 \div 490$

b. Determine the exact answer.

c. Is your estimate lower, higher, or the same?

SUMMARY
ACTIVITY 1.5

Division Properties of Whole Numbers

1. In general, use division to separate items into a specified number of equal groupings.
2. The numbers involved in the division process have specific names: dividend ÷ divisor = quotient with a possible remainder, or, in long division format,

$$\text{divisor}\overline{)\text{dividend}}^{\text{quotient with a possible remainder}}.$$

3. To check division, multiply the quotient by the divisor, then add the remainder. The result should be the dividend.

$$(\text{quotient} \times \text{divisor}) + \text{remainder} = \text{dividend}$$

4. Multiplication and division are **inverse operations.**
5. Division is *not* commutative. Example: $10 \div 5 \neq 5 \div 10$

Division Properties Involving 0 and 1

6. Any whole number divided by itself is 1.
7. Any whole number divided by 1 remains the same.
8. 0 divided by any nonzero whole number is 0.
9. Division by 0 is undefined.

Estimating a Quotient

10. To estimate a quotient, first estimate the divisor and the dividend by numbers that provide a division easily done mentally or by hand.

EXERCISES
ACTIVITY 1.5

1. Calculate the following. As part of your answer, identify the quotient and remainder (if any).

a. $56 \div 7$

b. $112 \div 4$

c. $95 \div 3$

d. $222 \div 11$

e. $506 \div 13$

f. $587 \div 23$

g. $0 \div 15$

h. $15 \div 0$

2. Six students purchase seven 5-packs of report covers to distribute evenly among themselves.

a. Determine the total number of report covers to be distributed.

b. How many report covers will each student receive?

c. Set the calculation up as a long division and divide.

d. Identify the divisor and the dividend. What is the quotient? What is the remainder, if any?

e. Are there any report covers left over?

f. Do this problem again using "repeated subtraction."

Exercise numbers appearing in color are answered in the Selected Answers appendix.

3. Four students will share equally three 500-count packages of 3" × 5" index cards.

   a. Determine the total number of index cards to be distributed.

   b. How many index cards will each student receive?

   c. Set up the calculation as a long division and divide.

   d. Identify the divisor and the dividend. What is the quotient? What is the remainder, if any?

   e. Are there any index cards left over?

4. a. Divide: $24 \div 8$.

   b. Do you get the same result by doing the division $8 \div 24$?

   c. Does the commutative property hold for division? That is, does $24 \div 8 = 8 \div 24$?

5. Your college campus has many more students who drive to campus than it has parking spaces. Even if you arrive early, it is difficult to find a parking space on any Monday, Wednesday, or Friday. The college is planning for future growth and in assessing the current parking problem estimates that 825 additional parking spaces will be needed. There are several parcels of land that will each accommodate a 180-car parking lot.

   a. Estimate the number of parking lots that are needed.

b. Determine the actual number of parking lots that are needed by first dividing 825 by 180.

c. Was your estimate too high or too low?

6. a. Estimate: $3850 \div 52$

b. Find the exact answer.

c. Is your estimate lower, higher, or the same?

7. a. Estimate: $28{,}800 \div 314$

b. Find the exact answer.

c. Is your estimate lower, higher, or the same?

ACTIVITY 1.6

Reach for the Stars

OBJECTIVES

1. Use exponential notation.
2. Factor whole numbers.
3. Determine the prime factorization of a whole number.
4. Recognize square numbers and roots of square numbers.
5. Recognize cubed numbers.
6. Apply the multiplication rule for numbers in exponential form with the same base.

Astronomical Distances

On a clear night, thousands of stars are visible. Some stars appear larger and brighter than others. Their distances from Earth also vary greatly. For instance, the distance to the star Altair is about 100,000,000,000,000 miles. These distances are so large that they are unmanageable as written whole numbers. It is easier to write very large numbers using exponents. The distance from Earth to Altair is approximately one hundred trillion miles and can be written in exponential form as 10^{14} miles. The exponent, 14, means to use the base, 10, as a factor 14 times.

$$\overset{\text{exponent}}{}\underbrace{10}_{\text{base}}{}^{14} = \underbrace{10 \cdot 10 \cdot 10 \cdot 10 \cdot 10 \cdot 10 \cdot 10 \cdot 10 \cdot 10 \cdot 10 \cdot 10 \cdot 10 \cdot 10 \cdot 10}_{\textbf{10 is used as a factor 14 times.}}$$

$$= 100{,}000{,}000{,}000{,}000$$

A whole-number **exponent** indicates the number of times the **base** is used as a factor. An exponent is also called a **power.** Note that 10 can be written as 10^1. When there is no exponent, it is understood to be 1. A number such as 10^{14} is in **exponential form** and is read as "ten to the fourteenth power."

1. The distance from Earth to the Great Whirlpool Galaxy is 10^{20} miles. Write this distance as a whole number.

2. a. Refer to the chart in Problem 3 and write the distance from Earth to Barnard's Galaxy as a whole number.

   b. Now write the distance to Barnard's Galaxy in base 10 using an exponent.

   c. Is it easier to write this distance as a whole number or as a number in base 10 using an exponent?

   d. Discuss the relationship between the number of zeros that make up this whole number and the exponent when you write this number as a power of 10.

3. Fill in the missing whole numbers and powers of 10.

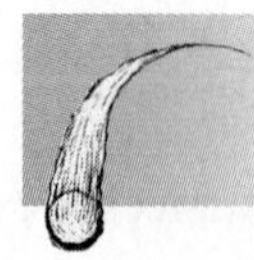

Way Out There...

CELESTIAL BODY	DISTANCE FROM EARTH IN MILES	WRITTEN USING AN EXPONENT
Barnard's Galaxy, first known dwarf galaxy, discovered in 1882	10,000,000,000,000,000,000	10^{19}
Brightest quasar, 3C 273		10^{22}
Comet Hale-Bopp on April 6, 2091		10^{10}
Double star Shuart 1	1,000,000,000,000,000	10^{15}
First magnitude star, Altair	100,000,000,000,000	
First near-Earth asteroid, Eros, at its closest	10,000,000	
Great Whirlpool Galaxy	100,000,000,000,000,000,000	
Ionosphere		10^{2}
Mars on Nov. 2, 2001	100,000,000	
Million-star globular cluster Omega Centauri	100,000,000,000,000,000	
Nothing known about things at this distance		10^{12}
Russian *Molnyia* (Lightning) communications satellites at highest altitude		10^{4}
Saturn on Oct. 17, 2015	1,000,000,000	
Space shuttle when you lose sight of it	1,000	
Stratosphere	10	
Typical near-Earth asteroid when it flies by	1,000,000	

EXAMPLE 1 ***A space shuttle is no longer visible to the naked eye when it is $1000 = 10^3$ miles away from an observer. Since 10 can be written as the product $2 \cdot 5$, you can rewrite 10^3 as $(2 \cdot 5)^3$. By the associative and commutative properties of multiplication $(2 \cdot 5)^3$ is equal to $2^3 5^3$. Therefore, 1000 can be written as***

$$1000 = 10^3 = (2 \cdot 5)^3 = (2 \cdot 5)(2 \cdot 5)(2 \cdot 5) = (2 \cdot 2 \cdot 2)(5 \cdot 5 \cdot 5) = 2^3 5^3$$

4. Use the associative and commutative properties of multiplication to show how or explain why $(3 \cdot 5)^2$ is equal to $3^2 \cdot 5^2$.

When a number is written as a product of its factors, it is called a **factorization** of the number. When all the factors are prime numbers, the product is called the **prime factorization** of the number. (Recall that a prime number is a whole number greater than 1 whose only whole-number factors are itself and 1.)

Fundamental property of whole numbers: For any whole number, there is only one prime factorization.

Notice that $2^3 5^3$ is a factorization of 1000 written in exponential form. It is also a prime factorization of 1000 because the factors, 2 and 5, which appear as base numbers, are both prime numbers.

5. a. $10 \cdot 10 \cdot 10 = 10^3$ is a factorization of 1000. Is 10^3 a prime factorization of 1000? Explain.

b. In Example 1, 10^3 was rewritten as $(2 \cdot 5)^3$. Is $(2 \cdot 5)^3$ a prime factorization of 1000?

c. If you ignore the order of the factors, how many prime factorizations of 1000 are there?

6. a. List the prime numbers that are less than 50.

b. What is the smallest prime number?

7. a. Determine the prime factorization of 90.

b. Describe a way to find the prime factorization of any whole number.

c. Use your method in part b to find the prime factorization of 100.

8. The distance to the double star Shuart 1 is 1,000,000,000,000,000, or 10^{15}, miles. Determine the prime factorization of this number. Write your result in both types of exponential form similar to those in Problem 5.

9. The National Collegiate Athletic Association (NCAA) Men's Basketball Tournament is a major TV event each year. Sixty-four colleges are invited to participate in the tournament based on their seasonal records and performance in their conference playoffs. The teams are then paired and half of the teams are eliminated after each round of play. After round one, there are 32 teams, then 16, 8, 4, 2, and, finally, 1.

a. Determine the prime factorization of the numbers 64, 32, 16, 8, 4, and 2 and write each number using an exponent. The first entry, 64, is done for you.

NUMBER OF TEAMS	PRIME FACTORIZATION	WRITTEN USING AN EXPONENT
64	$2 \cdot 2 \cdot 2 \cdot 2 \cdot 2 \cdot 2$	2^6
32		
16		
8		
4		
2		
Winner		

b. Describe the pattern for the exponents in the last column of this chart.

c. If you continue the pattern in the last column and write 1 in terms of base 2, what exponent would you attach to base 2?

Any nonzero whole number raised to the zero power is equal to 1.

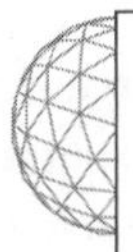

EXAMPLE 2 **The Square of a Number** ***In geometry, a square is a rectangle in which all the sides have equal length. The area of a square is a product of two factors, each equal to the length of a side. In exponential form, you say that the length is squared.***

A square with sides 1 unit in length has an area equal to 1 square unit.	(1 × 1 square)	$1^2 = 1$
A square with sides 2 units in length has an area equal to 4 square units.	(2 × 2 square)	$2^2 = 4$
A square with sides 3 units in length has an area equal to 9 square units.	(3 × 3 square)	$3^2 = 9$

10. a. Determine the area of a square whose sides are 5 units in length.

b. Determine the area of a square whose sides are 11 units in length.

11. Explain how to determine the square of any number.

A whole number is **square** (sometimes called a *perfect square*) if it is the product of a whole number times itself. For example, 36 and 100 are both square numbers because $36 = 6^2$ and $100 = 10^2$.

12. a. Draw a square that has an area of 49 square units on the grid.

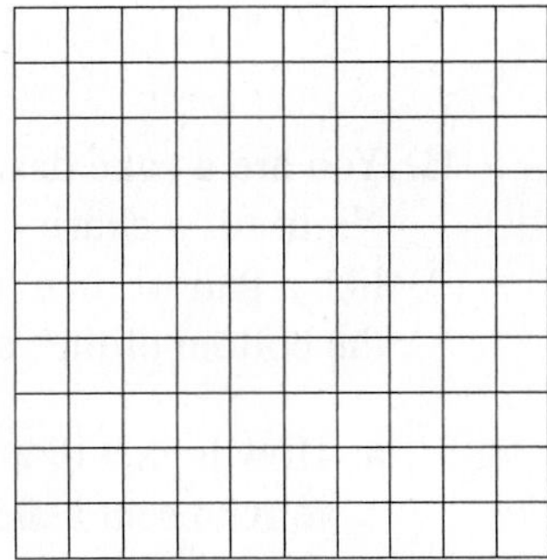

b. What is the length of each side?

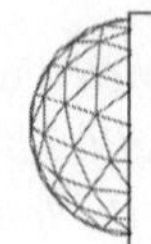

EXAMPLE 3 **The Cube of a Number** ***In geometry, a cube is a box in which all the edges have equal length. The volume or space inside the cube is the product of three factors, each equal to the length of an edge. We say the volume is the length of the edge cubed. The following pictures illustrate this idea.***

A cube with edges 1 unit in length has a volume equal to 1 cubic unit.	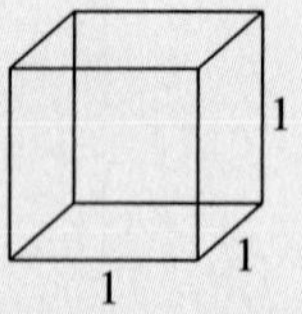	$1^3 = 1$
A cube with edges 2 units in length has a volume equal to 8 cubic units.	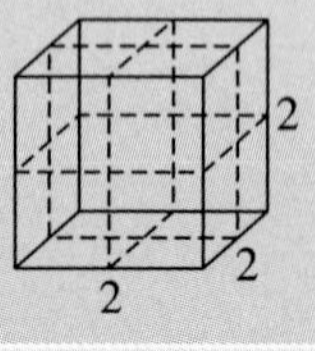	$2^3 = 8$
A cube with edges 3 units in length has a volume equal to 27 cubic units.	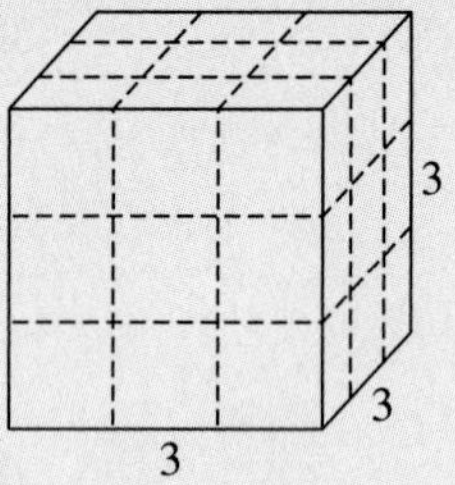	$3^3 = 27$

13. a. Determine the volume of a cube whose edges are 4 units in length.

b. Determine the volume of a cube whose edges are 8 units in length.

14. Explain how to determine the cube of any number.

15. You are a cake designer and have a client who is organizing a Monte Carlo Night for a charity benefit. The client wants several pairs of cakes that look like a pair of dice (cubes) . You have 6-inch square pans. The square area of the bottom of the cake in each pan will be 36 square inches.

a. How high will a cake have to be to represent a die (cube)?

b. What will be the volume of the cube cake in cubic inches?

Perhaps you showed the calculation in Problem 15.b as $6^2 \cdot 6^1 = 6 \cdot 6 \cdot 6 = 216$ cubic inches. Note that $6^2 \cdot 6^1$ is equivalent to 6^3 in value and that the sum of the exponents is $2 + 1 = 3$.

When multiplying two numbers written in exponential form that have the same base, add the exponents. This sum becomes the new exponent attached to the original base. For example,

$$5^4 \cdot 5^3 = 5^7,$$

since the product is the result of multiplying seven factors of 5.

$$(5 \cdot 5 \cdot 5 \cdot 5)(5 \cdot 5 \cdot 5) = 5^7$$

16. The Great Whirlpool Galaxy is one million times farther away from Earth than the first-magnitude star Altair. Use exponents to express this relationship (refer to chart in Problem 3).

17. Rewrite the following numbers using a single exponent. Check with your calculator.

a. $10^5 \cdot 10^4$

b. $4^7 \cdot 4^5$

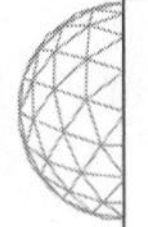

EXAMPLE 4 **Ceramic Tile** ***Ceramic floor tiles can be square and measure 1 foot by 1 foot in size. If you want to tile a square space that is 5 feet by 5 feet as shown in the figure, you would need 25 ceramic tiles. The 5-foot sides are called the* dimensions *of the 25 square foot area.***

Numerically, you express the relationship between the dimensions and the area as $5 \cdot 5 = 5^2 = 25$. The factor, 5, appears twice in this factorization of 25, and is called the* square root *of 25. Using symbols, $\sqrt{25} = 5$, which is read as "the square root of 25 is 5."

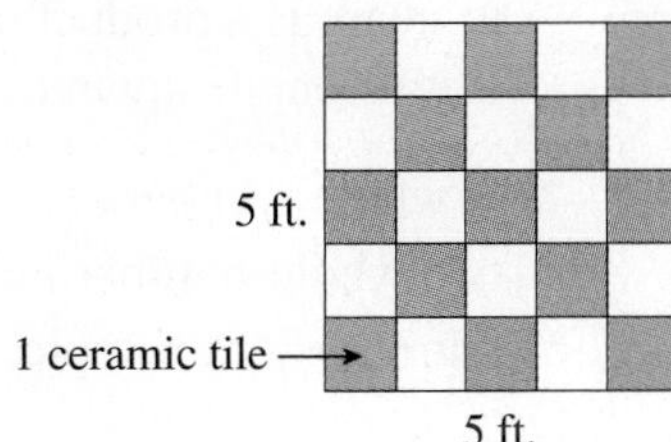

18. a. Determine the dimensions of the square area that you could tile with 81 ceramic tiles.

b. Determine the dimensions of the square area that you could tile with 144 ceramic tiles.

19. a. Determine the square roots of 64 and 225.

b. If your calculator has a square root key (0), check your answers in part a.

20. a. Can you tile a square area with 100 1-foot-square tiles? Explain.

b. Can you tile a square area with 24 of the 1-foot-square tiles? Explain.

SUMMARY
ACTIVITY 1.6

1. A whole-number **exponent** indicates the number of times to use the **base** as a factor. A number written as 10^{14} is in **exponential form.** The expression 10^{14} is called a power of 10.

2. Writing a number as a product of its factors is called **factorization.** When all the factors are prime numbers, the product is called the **prime factorization** of the number. A **prime number** is a whole number greater than 1 whose only whole number factors are itself and 1.

3. **Fundamental property of whole numbers:** For any whole number, there is only one prime factorization.

4. Any nonzero whole number raised to the **zero power** equals 1.

5. When multiplying numbers written in exponential form that have the same base, add the exponents. This sum becomes the new exponent attached to the original base. For example,

$$9^4 \cdot 9^3 = 9^7.$$

6. A **square** is a rectangle in which all the sides have equal length. The area of a square is a product of two factors, each equal to the length of a side, that is, the length squared.

7. A whole number is a perfect **square** if it can be rewritten as the product of two whole-number factors that are equal, that is, as the square of a whole number. For example, 36 is a perfect square because $6^2 = 36$.

8. The **square root** of a whole number is one of the two equal factors whose product is the whole number. For example, 6 is the square root of 36 because $6^2 = 36$. Using symbols, $\sqrt{36} = 6$.

EXERCISES
ACTIVITY 1.6

1. The distance from Earth to Alnilam, the center star in Orion's Belt, is 10^{16} miles. Write 10^{16} as a whole number.

2. The distance from Earth to the Large Magellanic Cloud, a satellite galaxy of the Milky Way, is 10,000,000,000,000,000,000 miles. Write this distance in exponential form.

3. Determine two prime numbers between 50 and 60.

4. How many prime numbers are there between 60 and 70? List them.

5. List all possible factorizations of the following numbers.

   a. 6

   b. 15

   c. 35

   d. 22

   e. How many different factorizations do each of these numbers have?

6. a. List all possible factorizations of 30 and 105.

   b. How many different factorizations do each of these numbers have?

7. a. List all possible factorizations of 4, 9, and 25.

   b. What do the factorizations of these three numbers share in common?

Exercise numbers appearing in color are answered in the Selected Answers appendix.

8. Determine the prime factorizations of each number.

a. 12

b. 75

c. 42

d. 96

9. Computers come with 1024, 512, 256, 128 or 64 MB of RAM (random access memory). Write 1024, 512, 256, 128, and 64 as powers of 2.

10. Write each exponential form as a whole number.

a. 3^0

b. 9^2

c. 5^4

d. 2^5

e. 12^2

11. Write each expression using a single exponent. Check your answers with a calculator.

a. $5^3 \cdot 5^8$

b. $9^2 \cdot 9^5$

c. $7^4 \cdot 7^7$

d. $7^5 \cdot 7^0$

12. Determine the square root of each number.

a. 64

b. 81

c. 121

d. 169

e. 225

f. 400

13. Determine if the given area is the area of a square that has a whole-number length.

a. 144 square feet

b. 160 square feet

c. 664 square feet

d. 256 square feet

ACTIVITY 1.7

You and Your Calculator

OBJECTIVES

1. Use order of operations to evaluate arithmetic expressions.
2. Use order of operations to evaluate formulas involving whole numbers.

A calculator is a powerful tool for problem solving. Calculators come in many sizes and shapes and with varying capabilities. Some calculators perform only basic operations such as addition, subtraction, multiplication, division, and square roots. Others also handle operations with exponents, perform operations with fractions, and do trigonometry and statistics. There are also calculators that graph equations and generate tables of values; some even manipulate algebraic symbols.

Unlike people, however, calculators do not think for themselves and can only perform tasks in the way that you instruct them (or program them). Therefore, if you understand the properties of numbers, you will understand how a calculator operates with numbers. In particular, you will learn the order in which your calculator performs the operations you request.

If you do not have a calculator for this course, perform the calculations using paper and pencil. There are many skills in this activity that are important to your understanding of whole numbers.

1. **a.** Use your calculator to determine the sum $126 + 785$.

   b. Now, input $785 + 126$ into your calculator and evaluate. How does this sum compare to the sum in Problem 1?

   c. If you use numbers other than 126 and 785, does reversing the order of the numbers change the result? Explain by giving examples.

   d. What property is demonstrated in this problem?

2. Is the commutative property true for the operation of subtraction? Multiplication? Division? Explain by giving examples for each operation.

Mental Arithmetic

It is sometimes necessary to do mental arithmetic (that is, *without* your calculator or paper and pencil). For example, to evaluate $3 \cdot 29$ without the aid of your calculator, think about the multiplication as follows: 29 can be written as $20 + 9$. Therefore, $3 \cdot 29$ can be written as $3 \cdot (20 + 9)$, which can be evaluated as $3 \cdot 20 + 3 \cdot 9$. The product $3 \cdot 29$ can now be thought of as $60 + 27$, or 87. To summarize,

$$3 \cdot 29 = 3 \cdot (20 + 9) = 3 \cdot 20 + 3 \cdot 9 = 60 + 27 = 87.$$

3. What property did the above calculation demonstrate?

4. Another way to express 29 is $25 + 4$ or $30 - 1$.

a. Express 29 as $25 + 4$ and use the distributive property to multiply $3 \cdot 29$.

b. Express 29 as $30 - 1$ and use the distributive property to multiply $3 \cdot 29$.

5. Evaluate mentally the following multiplication problems using the distributive property. Verify your answer using your calculator.

a. $6 \cdot 72$

b. $3 \cdot 109$

6. a. Evaluate $10 + 7 \cdot 3$ mentally and record the result. Verify your answer using your calculator.

b. What operations are involved in the calculation in part a?

c. In what order did you and your calculator perform the operations to get the answer?

d. Evaluate $(10 + 7) \cdot 3$ and record your result. Verify using your calculator.

e. Why is the result in part d different from the result in part a?

Order of Operations

Operations on numbers are performed in a universally accepted order. Scientific and graphing calculators are programmed to perform operations in this order. Part of the order of operations priority convention is as follows.

1. Perform multiplication and division before addition and subtraction.
2. If both multiplication and division are present, perform the operations in order from left to right.
3. If both addition and subtraction are present, perform the operations in order, *from left to right.*

EXAMPLE 1 *Evaluate* $12 - 2 \cdot 4 + 5$ *without a calculator.*

SOLUTION

$12 - 2 \cdot 4 + 5$	Do multiplication before addition and subtraction.
$= 12 - 8 + 5$	Subtract 8 from 12 since you encounter it first as you read from left to right.
$= 4 + 5$	Add.
$= 9$	

7. Perform the following calculations *without* a calculator. Then use your calculator to verify your result.

   a. $24 \div 4 + 8$

   b. $24 \div 4 - 2 \cdot 3$

   c. $6 + 24 - 4 \cdot 3 - 2$

   d. $6 + 2 \cdot 9 - 16 \div 4$

Notice the importance of the "from left to right" rule for both multiplication/division and addition/subtraction. For example, $12 \div 4 \cdot 3 = 3 \cdot 3 = 9$ by performing the operations left to right. If multiplication is performed before division, the result is 1 (try it and see). This shows the need for a decision on which of these calculations is correct. All the arithmetic experience that people had over hundreds of years led to the decision to do multiplication/division from left to right. That decision became part of the order of operations agreement.

Check that $12 \div 4 \cdot 3 = 9$ by entering $12 \div 4 \cdot 3$ all at once on your calculator.

8. Perform the following calculations without a calculator. State which operations you must perform first, and why. Then use your calculator to verify your result.

   a. $15 \div 5 \cdot 3$

   b. $7 \cdot 8 \div 4$

   c. $15 - 6 \div 2 + 4$

   d. $20 + 3 - 5 + 8$

Some expressions involve parentheses. For example, the expression $15 \div (1 + 2)$ means 15 divided by the sum of 1 and 2. The calculation is $15 \div (1 + 2) = 15 \div 3 = 5$. This observation leads to a fourth convention for order of operations priorities.

4. Parentheses are grouping symbols that are used to override the standard order of operations. Operations contained in parentheses are performed first.

9. a. Evaluate $24 \div (2 + 6)$ without your calculator.

b. Use your calculator to evaluate $24 \div (2 + 6)$. Did you obtain 3 as a result? If not, then perhaps you entered the expression $24 \div 2 + 6$ and your answer is 18.

c. Explain why the result of $24 \div 2 + 6$ is 18.

EXAMPLE 2 *Evaluate* $2 \cdot (3 + 4 \cdot 5)$ *without a calculator.*

SOLUTION

$2 \cdot (3 + 4 \cdot 5)$	Evaluate the arithmetic expression in parentheses first using order of operations.
$= 2 \cdot (3 + 20)$	First do the multiplication inside the parentheses, then the addition.
$= 2 \cdot 23$	
$= 46$	Multiply the result by the 2 that was outside of the parentheses.

10. Evaluate the following mentally and verify on your calculator.

a. $6/(3 + 3)$

b. $(2 + 8)/(4 - 2)$

c. $24 \div (4 + 8)$

d. $24 \div (4 - 2) \cdot 3$

e. $(6 + 24) \div (4 \cdot 3 - 2)$

f. $(6 + 2) \cdot 9 - 16 \div 4$

g. $5 + 2 \cdot (4 \div 2 + 3)$

h. $10 - (12 - 3 \cdot 2) \div 3$

Exponentiation

Recall that $5 \cdot 5$ can be written as 5^2 (read "5 squared"). Besides multiplying 5 times 5, there are two additional ways to square a number on your calculator. Try Problem 11 if you have a calculator. Otherwise, go to Example 3.

11. a. One way to evaluate 5^2 is to use the [0] key. Input 5 and then press the [0] key. Do this now and record your answer.

b. Another way to evaluate 5^2 is to use the exponent key. Depending on your calculator, the exponent key may resemble [x^y], [y^x], or [^]. To calculate 5^2, input 5, press the exponent key, then enter [2] and press [ENTER]. Do this now and record your answer.

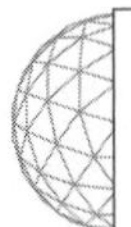

EXAMPLE 3 Order of Operations Involving Exponential Expressions

a. Evaluate 5^3.

SOLUTION

5^3 can be written as $5 \cdot 5 \cdot 5 = 125$. Verify your answer using a calculator.

b. Evaluate the expression $20 - 2 \cdot 3^2$.

SOLUTION

To evaluate the expression $20 - 2 \cdot 3^2$, follow the steps:

$20 - 2 \cdot 3^2$	Evaluate all exponents as you read the arithmetic expression from left to right.
$= 20 - 2 \cdot 9$	Do all multiplication and division as you read the expression from left to right.
$= 20 - 18$	Do all addition and subtraction as you read the expression from left to right.
$= 2$	

An exponential expression such as 5^3 is called a **power** of 5, as you will recall from Activity 1.6. The **base** is 5 and the **exponent** is 3. When a power is contained in an expression, it is evaluated *before* any multiplication or division, but only after operations in parentheses.

12. If you have a calculator, enter the expression $20 - 2 \cdot 3^2$ into your calculator and verify the result in Example 3 above.

13. Evaluate the following numerical expressions by hand. Verify using a calculator.

a. $6 + 3 \cdot 4^3$

b. $2 \cdot 3^4 - 5^3$

c. $2^2 \cdot 3^2 \div 3 - 2$

d. $3^2 \cdot 2 + 3 \cdot 2^3$

14. Evaluate each of the following arithmetic expressions mentally or by hand. Perform the operations in the appropriate order and then use your calculator to check your results.

a. $18 - 2 \cdot (8 - 2 \cdot 3) + 3^2$

b. $3^4 + 5 \cdot 4^2$

c. $128/(16 - 2^3)$

d. $(17 - 3 \cdot 4)/5$

e. $5 \cdot 2^3 - 6 \cdot 2 + 5$

f. $5^2 \cdot 5^3$

g. $2^3 \cdot 3^2$

h. $4^2 + 4^3$

i. $500 \div 25 \cdot 2 - 3 \cdot 2$

j. $(3^2 - 6)^2$

SUMMARY ACTIVITY 1.7

1. The **commutative property** states that the order in which you add or multiply two whole numbers gives the same result. The commutative property does *not* hold for subtraction or division.
2. $3(10 - 2) = 3 \cdot 10 - 3 \cdot 2$ is an example of the **distributive property.**
3. An exponential expression such as 5^3 is called a **power** of the base number 5. The **base** is 5 and the **exponent** is 3. The exponent indicates how many times the base is written as a factor. When a power is contained in an arithmetic expression, it is evaluated before any multiplication or division.
4. **Order of operations** for arithmetic expressions containing parentheses, addition, subtraction, multiplication, division, and exponentiation:
   a. Operations contained within parentheses are performed *first* before any operations outside the parentheses. All operations are performed in the following order.
   b. Evaluate all exponents as you read the expression from left to right.
   c. Do all multiplication and division as you read the expression from left to right.
   d. Do all addition and subtraction as you read the expression from left to right.

EXERCISES ACTIVITY 1.7

1. Evaluate each expression. Check your answers with a calculator.

a. $7(20 + 5)$

b. $7 \cdot 20 + 5$

c. $7 \cdot 20 + 7 \cdot 5$

d. $20 + 7 \cdot 5$

e. Which two of the preceding arithmetic expressions have the same answer?

f. State the property that produces the same answer for that pair of expressions.

Exercise numbers appearing in color are answered in the Selected Answers appendix.

2. Evaluate each expression. Check your answers with a calculator.

a. $20(100 - 2)$

b. $20 \cdot 100 - 2$

c. $20 \cdot 100 - 20 \cdot 2$

d. $100 - 20 \cdot 2$

e. Which two of the preceding arithmetic expressions have the same answer?

f. State the property that produces the same answer for that pair of expressions.

3. Evaluate each expression using order of operations.

a. $17 \cdot (52 - 2)$

b. $(90 - 7) \cdot 5$

4. Evaluate each expression using the distributive property.

a. $17 \cdot (52 - 2)$

b. $(90 - 7) \cdot 5$

5. Perform the following calculations *without a calculator*. After solving, use your calculator to check your answer.

a. $45 \div 3 + 12$

b. $54 \div 9 - 2 \cdot 3$

c. $12 + 30 \div 2 \cdot 3 - 4$

d. $26 + 2 \cdot 7 - 12 \div 4$

6. a. Explain why the result of $72 \div 8 + 4$ is 13.

b. Explain why the result of $72 \div (8 + 4)$ is 6.

7. Evaluate the following expressions. Check your answers with a calculator.

a. $48/(4 + 4)$

b. $(8 + 12)/(6 - 2)$

c. $120 \div (6 + 4)$

d. $64 \div (6 - 2) \cdot 2$

e. $(16 + 84) \div (4 \cdot 3 - 2)$

f. $(6 + 2) \cdot 20 - 12 \div 3$

g. $39 + 3 \cdot (8 \div 2 + 3)$

h. $100 - (81 - 27 \cdot 3) \div 3$

8. Evaluate the following expressions. Check your answers with a calculator.

a. $15 + 2 \cdot 5^3$

b. $5 \cdot 2^4 - 3^3$

c. $5^2 \cdot 2^3 \div 10 - 6$

d. $5^2 \cdot 2 - 5 \cdot 2^3$

9. Evaluate each of the following arithmetic expressions by performing the operations in the appropriate order. Check your answers with a calculator.

a. $37 - 2 \cdot (18 - 2 \cdot 5) + 1^2$

b. $3^5 + 2 \cdot 10^2$

c. $243/(36 - 3^3)$

d. $(75 - 2 \cdot 15)/9$

e. $7 \cdot 2^3 - 9 \cdot 2 + 5$

f. $2^5 \cdot 5^2$

g. $2^3 \cdot 2^5$

h. $5^3 + 2^6$

i. $1350 \div 75 \cdot 5 - 15 \cdot 2$

j. $(3^2 - 4)^2$

What Have I Learned?

Write your explanations in full sentences.

1. Would you prefer to win $1,050,000 or $1,005,000? Use the idea of place value to explain how you determined your answer.

2. Suppose you want to get a good deal on leasing a car for 3 years (36 months) and you do some checking. The following is some preliminary information that you found in car ads in your local newspaper. Note that these costs do not include other fees such as tax. Those fees are ignored in this problem.

CAR MODEL	DOWN PAYMENT*	MONTHLY FEE
Honda Pilot	$1,995	$277
Toyota Tundra	$2,999	$229
Mercury Mountaineer	$1,594	$299
Ford Explorer	$3,268	$249
Dodge Durango	$2,254	$259

*The down payment includes the first monthly fee.

a. At first, you round off the down payments to the nearest thousand and the monthly fee to the nearest 100 so you can mentally estimate the total payments for each car. Does your estimate allow you to say which car is the most expensive to lease and which is the least expensive to lease? Explain.

CAR MODEL	DOWN PAYMENT*	DOWN PAYMENT ESTIMATE TO THE NEAREST THOUSAND	MONTHLY FEE	MONTHLY FEE ESTIMATE TO THE NEAREST HUNDRED	ESTIMATED TOTAL COST
Honda Pilot	$1,995		$277		
Toyota Tundra	$2,999		$229		
Mercury Mountaineer	$1,594		$299		
Ford Explorer	$3,268		$249		
Dodge Durango	$2,254		$259		

b. What would be better choices for rounding off the down payment costs and the monthly fee to estimate the leasing costs for each car? Use your choices to get new estimates.

CAR MODEL	DOWN PAYMENT*	DOWN PAYMENT ESTIMATE TO THE NEAREST HUNDRED	MONTHLY FEE	MONTHLY FEE ESTIMATE TO THE NEAREST TEN	ESTIMATED TOTAL COST
Honda Pilot	$1,995		$277		
Toyota Tundra	$2,999		$229		
Mercury Mountaineer	$1,594		$299		
Ford Explorer	$3,268		$249		
Dodge Durango	$2,254		$259		

c. Use your calculator to determine the actual total cost for each car from the actual down payment and the monthly fee. Do the actual costs show the same cars as most expensive and least expensive that you named in part b?

CAR MODEL	DOWN PAYMENT*	MONTHLY FEE	TOTAL COST
Honda Pilot	$1,995	$277	
Toyota Tundra	$2,999	$229	
Mercury Mountaineer	$1,594	$299	
Ford Explorer	$3,268	$249	
Dodge Durango	$2,254	$259	

d. From this exercise, what conclusions can you make about the usefulness of rounding off numbers in calculations that you need to make so you can compare costs.

3. a. Explain a procedure you could use to determine if 37 is a composite or prime number.

b. Check to see if your procedure works for a number greater than 100, say 101. Explain why it works or does not work.

c. What is the largest prime number that *you know*? Do you think it is the largest prime number there is? Give a reason for your answer.

d. How many prime numbers do you think there are in all? Give a reason for your answer. Compare your answer with those of your classmates.

4. a. For which two operations does the commutative property hold? Give an example in each case.

b. For which two operations does the commutative property fail to hold? Give an example in each case.

5. a. Is it true that $(69 + 21) + 17 = 69 + (21 + 17)$? Justify your answer by calculating each side of the statement. In each case, add the numbers in the parentheses first.

b. What arithmetic property did you demonstrate in part a?

c. Does the same property hold true for $(69 - 21) - 17 = 69 - (21 - 17)$? Justify your answer.

6. Try this experiment: Ask a friend or classmate to calculate 9×999 by the usual vertical method. Then ask the person to calculate $9(1000 - 1)$ by using the distributive property.

a. What is the correct answer in each case?

b. Which calculation do you each think is "easier"? Why?

c. Show how you would calculate 9×9990 by using the distributive property.

7. Division of whole numbers can be considered as repeated subtraction.

a. Use this idea to divide 12 by 4. Show your calculation and the result.

b. Does this idea work when you try to divide 12 by 0? Explain.

How Can I Practice?

1. You bought a laptop computer and wrote a check for two thousand one hundred six dollars. The price tag read $2016.

   a. Write the amount of the check in numeral form.

   b. Did you pay the correct amount, too much, or too little? Explain.

2. Currently, the disease diabetes affects an estimated 21,000,000 Americans, and about 1,500,000 new cases are diagnosed each year. What are the place values to which each estimate apparently is rounded?

3. In 2001, the National Institutes of Health (NIH) spent about 690 million dollars for diabetes research. However, the cost of diabetes to the nation is more staggering. In 2002, it was estimated that 132 billion dollars was spent on health care and other costs related to diabetes. How many times more is spent in health care costs than in research? Write your answer to the nearest whole number.

4. In 2004, the median household income level in the United States was $44,389. Round this amount to the nearest hundred dollars.

5. On a web site, the distance from Earth to the Sun was given as 92,955,807 miles. Round this distance to

   a. the nearest thousand miles.

   b. the nearest million miles.

6. Calculate each of the following by hand. Check your answer with a calculator.

   a. $\begin{array}{r} 523 \\ +\ 108 \\ \hline \end{array}$

   b. $\begin{array}{r} 1052 \\ +\ 957 \\ \hline \end{array}$

   c. $\begin{array}{r} 3051 \\ 1282 \\ +\ 327 \\ \hline \end{array}$

Exercise numbers appearing in color are answered in the Selected Answers appendix.

7. Evaluate each of the following by hand. Check your answer with a calculator.

   a. $283 - 95$ b. $233 - 145$

   c. $67 - 39$ d. $1003 - 349$

8. Translate each of the following into an arithmetic expression and calculate by hand. Check your answer with a calculator.

   a. 67 plus 25 b. the difference between 24 and 18

   c. 98 minus 15 d. total of 104 and 729

   e. 25 more than 495 f. 33 subtracted from 67

   g. 145 increased by 28 h. 34 less than 156

   i. 95 decreased by 25

9. Multiply by hand. Check your answer with a calculator.

   a. $\begin{array}{r} 25 \\ \times\ 9 \\ \hline \end{array}$ b. $\begin{array}{r} 347 \\ \times\ 6 \\ \hline \end{array}$ c. $\begin{array}{r} 167 \\ \times\ 17 \\ \hline \end{array}$ d. $\begin{array}{r} 227 \\ \times\ 109 \\ \hline \end{array}$

10. a. Multiply 3 times 45 in a vertical format.

    b. Use the distributive property to calculate $3(40 + 5)$.

    c. Use the distributive property to calculate $3 \cdot 49$.

11. Calculate. As part of your answer, identify the quotient and remainder (if any).

    a. $126 \div 4$ b. $312 \div 4$

    c. $195 \div 13$ d. $224 \div 12$

12. a. Estimate $3212 \div 414$.

    b. Find the exact answer.

    c. Is your estimate lower, higher, or the same?

13. Determine and list all the prime numbers between 30 and 50.

14. Determine and list all factors of the following numbers.

    a. 12

    b. 21

    c. 71

    d. 18

15. Determine the prime factorizations of the following numbers.

    a. 24

    b. 63

16. Determine the prime factorizations of the following numbers. Write your answers in exponent form.

    a. 27

    b. 125

    c. What do the two prime factorizations have in common?

17. Write as whole numbers.

    a. 8^0

    b. 12^2

    c. 3^4

    d. 2^6

18. Write each numerical expression using a single exponent.

    a. $7^3 \cdot 7^9$

    b. $11^5 \cdot 11^7$

    c. $13^2 \cdot 13^0$

    d. $9 \cdot 9^2 \cdot 9^3$

19. Determine the square root without using a calculator. Approximate if necessary. Check your answer with a calculator.

a. 4
b. 24
c. 225
d. 36
e. 90

20. Evaluate each of the following arithmetic expressions by performing the operations in the appropriate order. Use your calculator to check your results.

a. $7 - 3 \cdot (8 - 2 \cdot 3) + 2^2$
b. $3 \cdot 2^5 + 2 \cdot 5^2$
c. $144/(24 - 2^3)$
d. $(36 - 2 \cdot 9)/6$
e. $9 \cdot 5 - 5 \cdot 2^3 + 5$
f. $1^5 \cdot 5^1$
g. $2^3 \cdot 2^0$
h. $7^2 + 7^2$
i. $(3^2 - 4 \cdot 0)^2$

21. Determine the arithmetic property expressed by each numerical statement.

a. $25(30) = 30(25)$
b. $15(9) = 15(10 - 1) = 15 \cdot 10 - 15 \cdot 1$
c. $(11 + 21) + 39 = 11 + (21 + 39)$

22. Determine if each of the following numerical statements is true or false. In each case, justify your answer.

a. $7(20 + 2) = 7(22)$
b. $4(11)(10) = 10(11)4$

c. $25 - 10 - 4 = 25 - (10 - 4)$

d. $1/0 = 1$

23. The New York City Department of Environment Protection keeps records of noise complaints. New York County (Manhattan) records the greatest number of complaints of all the five counties that make up New York City. The following table lists the complaints for 2000–2001 by the Community Planning Boards in Manhattan.

a. Round each number of complaints to the nearest tens and list in column 3 in the table.

Keep It Down!

COMMUNITY PLANNING BOARD (NAMED BY NUMBER) [INPUT]	NUMBER OF NOISE COMPLAINTS [OUTPUT]	NUMBER OF NOISE COMPLAINTS TO NEAREST TENS
1	263	
2	713	
3	684	
4	602	
5	639	
6	517	
7	791	
8	761	
9	103	
10	67	
11	84	
12	150	

Source: New York City Department of Environment Protection

b. Plot the points that represent the number of complaints for each planning board.

i. Use the rounded numbers you listed in column 3 of the table.

ii. List the name of the input along the horizontal axis and the name of the output along the vertical axis.

iii. List the units along the appropriate axes.

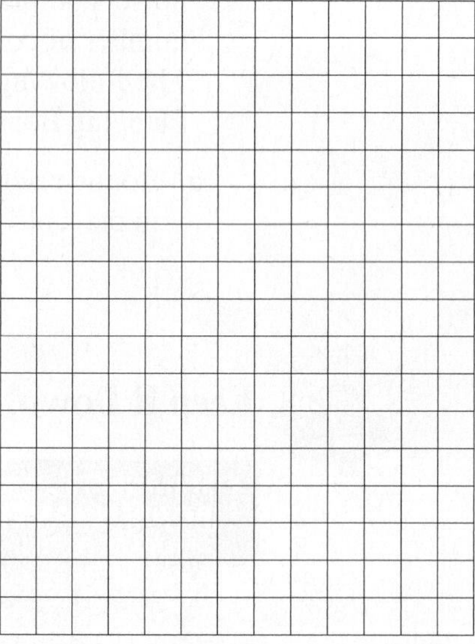

CHAPTER 1 SUMMARY

The bracketed numbers following each concept indicate the activity in which the concept is discussed.

CONCEPT / SKILL	DESCRIPTION	EXAMPLE
Place value [1.1]	MILLIONS: Hundreds, Tens, Ones; THOUSANDS: Hundreds, Tens, Ones; ONES: Hundreds, Tens, Ones	
Read and write whole numbers [1.1]	Start at the left and read each digit and its place value.	2345 is read "two thousand three hundred forty-five."
Round whole numbers to specified place value [1.1]	If the digit immediately to the right of the specified place value is *5 or more*, add 1 to the digit in the specified place and change all the digits to its right to zero. If the digit immediately to the right of the specified place value is *less than 5*, do not change the digit in the specified place but change all the digits to its right to zero.	157 rounded to the tens place yields 160. 152 rounded to the tens place yields 150.
Prime numbers [1.1], [1.6]	A **prime number** is a whole number greater than 1 whose only whole-number factors are itself and 1.	2, 3, 7, 11, 13, 17, 19, 23,. . .
Read bar graphs [1.2]	Each bar along the horizontal axis is associated with a number along the vertical axis.	Bar graph: bar at 1 has height 1; bar at 2 has height 2
Interpret bar graphs [1.2]	Find and read input values along the horizontal axis and the corresponding output values along the vertical axis in context.	
Addend, sum [1.3]	The numbers being added are called **addends.** Their total is called the **sum.**	$\underbrace{12}_{\text{addend}} + \underbrace{11}_{\text{addend}} = \underbrace{23}_{\text{sum}}$
Addition property of zero [1.3]	The sum of any whole number and zero is the same whole number.	$5 + 0 = 5$
Commutative property of addition [1.3]	Changing the **order** of the addends yields the same sum.	$9 + 8 = 8 + 9 = 17$
Associative property of addition [1.3]	Given three addends, it makes no difference whether the first two numbers or the last two numbers are added first.	$5 + (6 + 7)$ $= (5 + 6) + 7 = 18$

CONCEPT / SKILL	DESCRIPTION	EXAMPLE
Add whole numbers by hand and mentally [1.3]	***To add:*** Align numbers vertically according to place value. Add the digits in the ones place. If their sum is a two-digit number, write down the ones digit and regroup the tens digit to the next column as a number to be added. Repeat with next higher place value.	1 129 Regroup 1 to tens place. + 17 146 9 + 7 = 16; write 6 in ones place.
Subtrahend, minuend, difference [1.3]	The number being subtracted is called the **subtrahend.** The number it is subtracted from is called the **minuend.** The result is called the **difference.**	32 (minuend) − 18 (subtrahend) = 14 (difference)
Subtract whole numbers by hand and mentally [1.3]	***To subtract:*** Align the subtrahend under the minuend according to place value. Subtract digits having the same place value. Regroup from a higher place value, if necessary.	5 12 62 Regroup from tens. 2 changes to 12. −15 Subtract 5 from 12. 47
Estimate sums and differences using rounding [1.3]	One way to estimate is to round each number (addend) to its highest place value. In some cases, it may be better to round to a place value lower than the highest one.	*Exact Sum:* 780 + 219 + 164 = 1163 *Estimate:* 800 + 200 + 200 = 1200
Missing addend approach to subtraction [1.3]	Minuend − subtrahend = difference can be written as subtrahend + difference = minuend.	10 − 6 = ? can be written 6 + ? = 10.
Key phrases for addition [1.3]	• Sum of • Increased by • More than • Plus • Total of • Added to	15 *increased by* 10: 15 + 10 3 *more than* 7: 7 + 3 Total of 10 and 25: 10 + 25
Key phrases for subtraction [1.3]	• Difference between • Minus • Decreased by • Less than • Subtracted from	50 *decreased by* 16: 50 − 16 6 *less than* 92: 92 − 6 7 *subtracted from* 15: 15 − 7
Factor, product [1.4]	**Factors** are numbers being multiplied. The result is called the **product.**	12 (factor) · 8 (factor) = 96 (product)
Multiplication as repeated addition [1.4]	4 times 30 can be thought of as 30 + 30 + 30 + 30.	9 · 5 = 9 + 9 + 9 + 9 + 9 = 45 5 · 9 = 5 + 5 + 5 + 5 + 5 + 5 + 5 + 5 + 5 = 45

CONCEPT / SKILL	DESCRIPTION	EXAMPLE
Multiply whole numbers [1.4]	***To multiply:*** Align numbers vertically by place value. Starting with the rightmost digit of the bottom factor, multiply each place value of the top factor. Repeat using the next rightmost digit of the bottom factor. Add the resulting products, to get the final product.	$\begin{array}{r} 37 \\ \times\ 4 \\ \hline 28 \\ 120 \\ \hline 148 \end{array}$ 28: product of 7 times 4; 120: product of 30 times 4; 148: sum of 28 and 120
Multiplication by 1 [1.4]	Any whole number multiplied by 1 remains the same.	$134 \cdot 1 = 134$
Multiplication by 0 [1.4]	Any whole number multiplied by 0 is 0.	$45 \cdot 0 = 0$
Distributive property of multiplication over addition [1.4]	A whole number placed immediately to the left or right of a set of parentheses containing the sum or difference of two numbers multiplies each of the inside numbers.	$8 \cdot 24 = 8 \cdot (20 + 4) = 8 \cdot 20 + 8 \cdot 4 = 160 + 32 = 192$ or $32 \cdot 5 = (30 + 2) \cdot 5 = 30 \cdot 5 + 2 \cdot 5 = 150 + 10 = 160$
Commutative property of multiplication [1.4]	Changing the **order** of the factors yields the same product.	$5 \cdot 12 = 12 \cdot 5 = 60$
Associative property of multiplication [1.4]	When multiplying three factors, it makes no difference whether the first two numbers or the last two numbers are multiplied first. The same product results.	$(5 \cdot 7) \cdot 12 = 5 \cdot (7 \cdot 12) = 420$
Estimate products [1.4]	Round each factor to a large enough place value so that you can do the multiplication mentally. There is no one correct answer when estimating.	284 ⟶ 300; 525 ⟶ × 500; estimate: 150,000
Divisor, dividend [1.5]	The number being divided is called the **dividend.** The number that divides is called the **divisor.**	$3\overline{)45}$ (3: divisor; 45: dividend)
Quotient [1.5]	A whole number representing the number of times the divisor can be subtracted from the dividend is called a **quotient.**	$\begin{array}{r} 15 \\ 3\overline{)45} \end{array}$ (15: quotient; 3: divisor; 45: dividend)
Remainder [1.5]	The **remainder** is the whole number left over after the divisor has been subtracted from the dividend as many times as possible.	$\begin{array}{r} 15 \text{ remainder } 2 \\ 3\overline{)47} \\ 45 \\ \hline 2 \end{array}$ (15: quotient; 3: divisor; 47: dividend)

CONCEPT / SKILL	DESCRIPTION	EXAMPLE
Divide whole numbers [1.5]	$\begin{array}{r} 17 \\ 6\overline{)104} \\ \underline{6} \\ 44 \\ \underline{42} \\ 2 \end{array}$	**6 does not divide into 1. 6 does divide into 10 one time.** **Write down the product $1 \cdot 6 = 6$. Subtract 6 from 10.** **Bring down the next digit, 4. 6 divides into 44 seven times.** **Write down the product $7 \cdot 6 = 42$. Subtract 42 from 44.** **remainder**
Check division [1.5]	To check division, multiply the quotient times the divisor, then add the remainder. The result should be the dividend. (quotient $\times$ divisor) + remainder = dividend	$\begin{array}{r} 3 \\ 7\overline{)25} \\ \underline{21} \\ 4 \end{array}$ Check: $3 \cdot 7 + 4 = 25$
Inverse operations [1.5]	Multiplication and division are inverse operations	$3 \cdot 4 = 12$ $12 \div 4 = 3$
Division properties involving 0 [1.5]	0 divided by any nonzero whole number is 0. Division by 0 is undefined.	$0 \div 42 = 0$ $42 \div 0 =$ *undefined!*
Division properties involving 1 [1.5]	Any nonzero whole number divided by itself is equal to 1. Any whole number divided by 1 remains the same.	$71 \div 71 = 1$ $23 \div 1 = 23$
Division is not commutative [1.5]	The dividend and the divisor can *not* be interchanged without changing the quotient (unless the dividend and the divisor are the same whole number).	$9\overline{)81} \neq 81\overline{)9}$
Estimate a quotient [1.5]	To estimate a quotient, first estimate the divisor and the dividend by numbers that provide a division easily done mentally or by hand.	$18\overline{)680} \longrightarrow 20\overline{)700}$ with estimate 35
Exponent, base [1.6]	A whole-number **exponent** indicates the number of times to use the **base** as a factor.	$10^4 = 10 \cdot 10 \cdot 10 \cdot 10$ $= 10{,}000$
A whole number raised to the zero power [1.6]	Any nonzero whole number raised to the zero power equals 1.	$2^0 = 1,\ 50^0 = 1,\ 200^0 = 1$
Exponential form of a whole number [1.6]	A number written as 10^4 is in **exponential form.** 10 is the base and 4 is the exponent. The exponent indicates how many times the base appears as a factor.	**10 is used as a factor 4 times.** $10^4 = \overbrace{10 \cdot 10 \cdot 10 \cdot 10}$ $= 10{,}000$

CONCEPT / SKILL	DESCRIPTION	EXAMPLE
Multiplication rule for numbers in exponential form [1.6]	When multiplying numbers written in exponential form with the *same base*, add the exponents. This sum becomes the new exponent attached to the original base.	$7^2 7^6 = 7^8$ $3^2 3^5 3^7 = 3^{14}$
Factor whole numbers [1.6]	Writing a number as a product of its factors is called **factorization.**	$12 = 1 \cdot 12 = 12 \cdot 1$ $12 = 2 \cdot 6 = 6 \cdot 2$ $12 = 4 \cdot 3 = 3 \cdot 4$
Prime factorization form of a whole number [1.6]	A number written as a product of its prime factors is called the **prime factorization** of the number.	$60 = 2 \cdot 2 \cdot 3 \cdot 5$
Finding the prime factorization of a whole number [1.6]	Determine the prime factors of the number and write the number as a product of the prime factors.	$60 = 2 \cdot 2 \cdot 3 \cdot 5$
Recognizing square whole numbers [1.6]	A whole number is a **perfect square** if it can be rewritten as the square of a whole number.	$36 = 6^2$ and $100 = 10^2$
Recognizing roots of square numbers [1.6]	13 is the **square root** of 169 because 13 times 13 is 169.	$\sqrt{25} = 5$ because $5^2 = 25$
Order of operations [1.7]	**Order of operations** for arithmetic expressions containing parentheses, addition, subtraction, multiplication, division, and exponentiation: **a.** Operations contained within parentheses are performed *first*—before any operations *outside* the parentheses. All operations are performed in the following order. **b.** Evaluate all exponents as you read the expression *from left to right.* **c.** Do all multiplication and division as you read the expression *from left to right.* **d.** Do all addition and subtraction as you read the expression from *left to right.*	$2 \cdot (3 + 4^2 \cdot 5) - 9$ $= 2 \cdot (3 + 16 \cdot 5) - 9$ $= 2 \cdot (3 + 80) - 9$ $= 2 \cdot (83) - 9$ $= 166 - 9$ $= 157$

CHAPTER 1 GATEWAY REVIEW

1. When you move into a new apartment, you and your roommates stock up on groceries. The bill comes to $143. How will you write this amount in words on your check?

2. The chance of winning in New York State Lotto is 1 out of 22,528,737. Write 22,528,737 in words.

3. Put the following whole numbers in order from smallest to largest: 108,901; 180,901; 108,091; 108,910; and 109,801.

4. What are the place values of the digits 0 and 9 in the number 40,693?

5. Determine whether each of the following numbers is even or odd. In each case, give a reason for your answer.

   a. 333 **b.** 1378 **c.** 121

6. Determine whether each of the following numbers is prime or composite. In each case, give a reason for your answer.

   a. 145 **b.** 61

   c. 2 **d.** 121

7. **a.** Round 1,252,757 to the thousands place.

   b. Round 899 to the tens place.

8. The following table provides the six most popular names for girls and for boys in California in the year 2004. The names are listed in alphabetical order.

GIRLS' NAMES	NUMBER IN YEAR 2004	BOYS' NAMES	NUMBER IN YEAR 2004
Alyssa	1795	Andrew	3425
Ashley	2872	Anthony	3739
Emily	3358	Daniel	4095
Isabella	2394	Jacob	3290
Natalie	1921	Jose	3312
Samantha	1795	Joshua	3224

Answers to all Gateway exercises are included in the Selected Answers appendix.

a. What was the most popular girl's name? Most popular boy's name?

b. Which boy's name was the sixth most popular?

c. Which girl's name was the third most popular?

9. The following bar graph provides the annual rainfall in San Antonio, Texas, from 1934 to 2000.

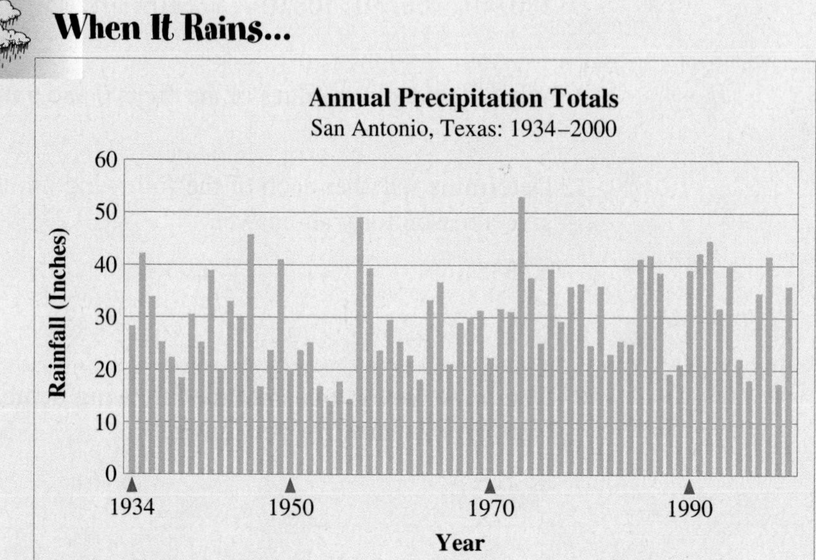

Source: U.S. Weather Bureau

a. In which year(s) was there the greatest rainfall? Estimate the number of inches.

b. In which year(s) was there the least rainfall? Estimate the number of inches.

c. How many years did San Antonio have 20 to 30 inches of rain per year? Is this more or less than the number of years that San Antonio had 30 to 40 inches of rain per year? Explain.

10. Use the following grid as given to plot the points listed in the accompanying table. Do not extend the grid in either direction.

a. What size unit would be reasonable for the input?

b. List the units along the horizontal axis.

c. What size unit would be reasonable for the output?

d. List the units along the vertical axis.

e. Plot the points from the following table on the grid that you labeled. Label each point with its input/output values.

Input	1	4	7	8	10	16	17	20
Output	5	10	20	15	25	30	5	40

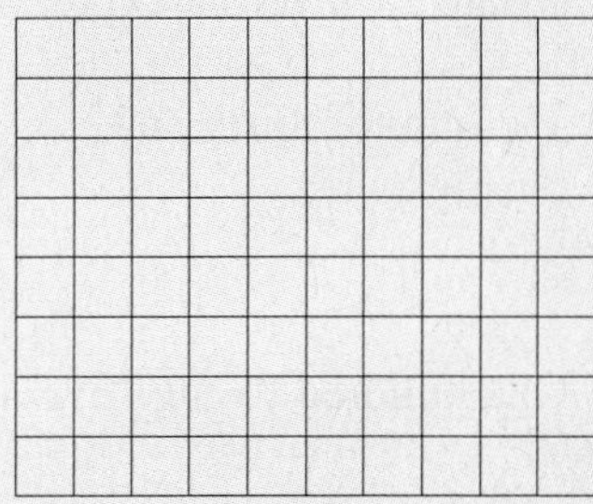

11. Perform the indicated operations. Check your answers with a calculator.

a. $\begin{array}{r} 7982 \\ +\ 969 \\ \hline \end{array}$

b. $\begin{array}{r} 235 \\ 45 \\ 59 \\ 210 \\ +347 \\ \hline \end{array}$

c. $\begin{array}{r} 231 \\ -\ 54 \\ \hline \end{array}$

d. $\begin{array}{r} 5344 \\ -4682 \\ \hline \end{array}$

e. $654 - 179$

12. Rewrite the statement $113 - 41 = 72$ as a statement in terms of addition.

13. a. What property of addition does the statement $(8 + 5) + 5 = 8 + (5 + 5)$ demonstrate?

b. What property of addition does the statement $4 + (7 + 6) = 4 + (6 + 7)$ demonstrate?

c. Why is the expression $11 + 89 + 0$ equal to the expression $11 + 89$?

14. Are the numerical expressions $5 - 2$ and $2 - 5$ equal? What does this mean in terms of a commutative property for subtraction?

15. **a.** Estimate the sum: $510 + 86 + 120 + 350$.

b. Determine the actual sum.

c. Was your estimate higher or lower than the actual sum? Explain.

16. Translate each verbal expression into an arithmetic expression.

a. The sum of thirty and twenty-two

b. The difference between sixty-seven and fifteen

c. One hundred twenty-five decreased by forty-four

d. 175 less than 250

e. The product of twenty-five and thirty-six

f. The quotient of fifty-five and eleven

g. The square of thirteen

h. Two to the fifth power

i. The square root of forty-nine

j. The product of twenty-seven and the sum of fifty and seventeen

17. **a.** Write $5 \cdot 61$ as an addition calculation and find the sum.

b. How does the sum you obtained in part a compare to $5 \cdot 61$ when calculated as a product?

c. Write 61 as $60 + 1$ and use the distributive property to evaluate the product of 5 and 61.

18. **a.** Determine the product of $4 \cdot 29$ by rewriting 29 as $30 - 1$ and then using the distributive property.

b. Is it faster to mentally multiply $4 \cdot 29$ or $4 \cdot 30 - 4 \cdot 1$?

c. Multiply $6 \cdot 98$ by using the distributive property.

19. **a.** What property of multiplication does the statement $8 \cdot (6 \cdot 12) = (8 \cdot 6) \cdot 12$ demonstrate?

b. What property of multiplication does the statement $4 \cdot (7 \cdot 6) = 4 \cdot (6 \cdot 7)$ demonstrate?

c. Why is the expression $1 \cdot (4 + 5)$ equivalent to the expression $(4 + 5)$?

20. **a.** Determine the quotient of $72 \div 12$ by repeatedly subtracting 12 from 72.

b. Determine the quotient and remainder of $86 \div 16$ by repeatedly subtracting 16 from 86.

21. Calculate $12 \div 6$ and then $6 \div 12$. Use the results to determine if the commutative property holds for division.

22. In each case, perform the indicated operation. Check your answer with a calculator.

a. $\begin{array}{r} 2096 \\ \times \quad 87 \\ \hline \end{array}$

b. $15\overline{)231}$

c. $(41 \cdot 3) \cdot 7$

d. $41 \cdot (3 \cdot 7)$

e. $27 \div 27$

f. $0 \div 27$

g. $27 \div 0$

h. $24\overline{)6572}$

23. **a.** Estimate $329 \cdot 75$.

b. Determine the exact answer.

c. Is your estimate lower, higher, or the same?

24. **a.** Estimate $1850 \div 42$.

b. Determine the exact answer.

c. Is your estimate lower, higher, or the same?

25. In converting 63 feet into yards, suppose you divided 63 by 3 and obtained 20 yards. Check to see if you calculated correctly.

26. Explain the difference between the expressions $1 \div 0$ and $0 \div 1$.

27. **a.** If you divide any whole number by 1, what is the result? Give an example.

b. If you divide the number 5634 by itself, what is the result? Does this property apply to any whole number you choose? Explain.

28. List all whole-number factors of each of the following numbers.

a. 350

b. 81

c. 36

29. Determine the prime factorization of each of the following numbers.

a. 350

b. 81

c. 36

30. Write each number as a whole number in standard form.

a. 6^0

b. 7^2

c. 2^5

d. 25^0

31. **a.** Write 81 as a power of 9.

b. Write 81 as a power of 3.

c. Write 625 as a power of 25.

d. Write 625 as a power of 5.

32. Write each number using a single exponent.

a. $3^3 \cdot 3^7$

b. $5^{33} \cdot 5^{12}$

c. $21^0 \cdot 21^{19}$

33. Determine which of the following numbers are perfect-square numbers. In each case, explain your answer.

a. 25

b. 125

c. 121

d. 100

e. 200

34. Determine the square root of each of the following numbers.

a. 16

b. 36

c. 289

35. Which of the following numbers are perfect squares?

a. 40

b. 400

c. 10

d. 10,000

e. 1000

36. Evaluate each of the following arithmetic expressions by performing the operations in the appropriate order. Verify using your calculator.

a. $48 - 3 \cdot (18 - 2 \cdot 7) + 3^2$

b. $2^4 + 4 \cdot 2^2$

c. $243/(35 - 2^3)$

d. $(160 - 2 \cdot 25)/10$

e. $7 \cdot 2^3 - 9 \cdot 2 + 5$

f. $2^3 \cdot 3^2$

g. $6^2 + 2^6$

h. $(3^2 - 2^3)^2$

37. Do the expressions $6 + 10 \div 2$ and $(6 + 10) \div 2$ have the same result when calculated? Explain.

CHAPTER 2

VARIABLES AND PROBLEM SOLVING

In Chapter 1, you saw how the properties of whole numbers and arithmetic are used to solve problems.

In Chapter 2, you will be introduced to basic algebra, where the properties of arithmetic are extended to solve a wider range of problems.

ACTIVITY 2.1

How Much Do I Need to Buy?

OBJECTIVES

1. Recognize and understand the concept of a variable in context and symbolically.
2. Translate a written statement (verbal rule) into a statement involving variables (symbolic rule).
3. Evaluate variable expressions.
4. Apply formulas (area, perimeter, and others) to solve contextual problems.

To earn money for college, you and a friend run a home-improvement business during the summer. Most of your jobs involve interior work, such as painting walls and ceilings, and installing carpeting, floor tiles, and floor molding. One of the first tasks in any job is to calculate the quantity of material to buy.

Your newest clients need floor molding installed in their living room. You measure the floor of the rectangular room. It is 24 feet long and 15 feet wide. The grid below represents a floor plan of the room.

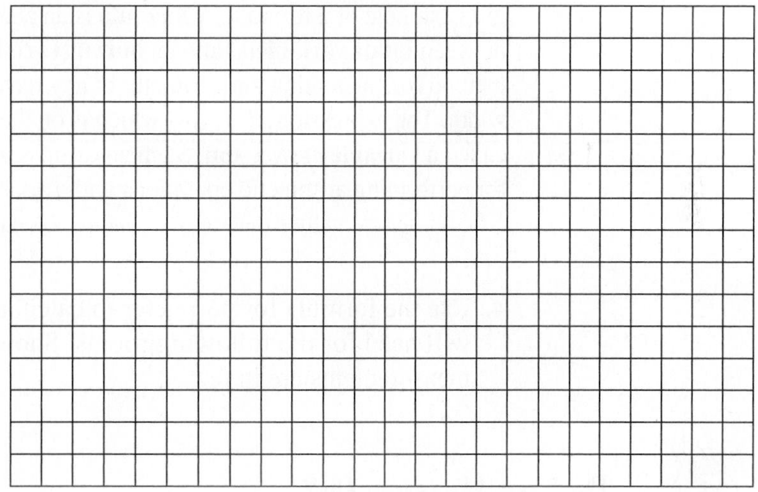

1. How much floor molding will you need for the living room? (You may ignore any door openings for now.) Describe or show the calculation you used.

DEFINITION

The distance around the edge of a rectangle is the **perimeter** of the rectangle. Since perimeter is a distance, it is measured in feet, meters, miles, etc.

2. a. For any rectangle, describe in your own words how to determine the perimeter if you know its length and width.

b. A description of a rule in words is called a **verbal rule**. A verbal rule can be translated into an equation that uses letters to represent the quantities. For example, you can represent the **perimeter** of the rectangle with the letter ***P***, the **length** with the letter ***l***, and the **width** with the letter ***w***. Write an equation that translates your verbal rule in part a into an equation using the letters ***P***, ***l***, and ***w***. The equation is called a **symbolic rule**.

Since quantities like width, length, and perimeter may **vary**, or change in value, from one situation to another, they are called **variables**. A variable is usually represented by a letter (or some other symbol) as a simple shorthand notation. In Problem 2, ***P***, ***l***, and ***w*** are shorthand notations for the variables perimeter, length, and width, respectively.

3. The clients have a rectangular dining room, where they also want floor molding installed. The dining room measures 16 feet long by 12 feet wide. List the variables and state their values in this situation.

The equation $P = 2l + 2w$ connecting the variables perimeter, length, and width is an example of a formula. A **formula** is an equation that represents the relationship between **input variable(s)** and an **output variable**. In this case, the formula is a method for calculating the perimeter of any rectangle, once you know the length and width. The expression $2l + 2w$, which is on the right-hand side of the equals sign, is called a **variable expression**. Such an expression is a symbolic code of instructions for performing arithmetic operations with variables and numbers. The numbers, specific numerical values that do not change, are called **constants**.

4. Use the formula for perimeter to calculate the amount of floor molding you will need for the following rooms. Show the calculations in each case. (All measurements are in feet.)

Do You Measure Up?

ROOM	LENGTH, l (feet)	WIDTH, w (feet)	CALCULATIONS: $2l + 2w = P$	PERIMETER, P (feet)
Living	24	15	$2(24) + 2(15) = 48 + 30 = 78$	78
Dining	16			
Family	30			
Bedroom 1	18			
Bedroom 2	14			
Bedroom 3	13			

In Problem 4, you determined the perimeter, P, by replacing l and w with the numerical values in the formula $P = 2l + 2w$. You then performed the arithmetic operations to determine the perimeter. This process is called **evaluating a variable expression** for the given input values. Note that replacing the variables (l and w in this case) with numerical values is also referred to as substituting given values for the variables.

5. You find a deal on floor molding at the Home Depot. The molding costs $2.00 for a 10-foot-long piece. In all the rooms you have measured, there are 10 doorways, each 3 feet wide. How much will the molding cost for all six rooms?

6. Your client decides to replace the molding in the den with a more expensive oak molding. The floor of the den is a square, measuring 12 feet by 12 feet.

   a. Write a verbal rule to describe how to determine the perimeter of a square if you know the length of one of the sides.

   b. Let s represent the length of one side of a square. Translate the verbal rule in part a into a symbolic rule (a formula) for the perimeter of a square.

   c. Use the formula to determine the perimeter of the square den.

Area of a Rectangle

7. For the living room in Problem 1, your client wants you to buy enough carpeting to cover the entire floor (wall to wall).

   a. How many square feet (squares that measure 1 foot by 1 foot) will it take to cover the living room floor? You may want to refer to the floor plan on page 91.

   b. Describe or show the calculation you used.

DEFINITION

The **area** of a geometric figure is the measure of the size of the region bounded by the sides of the figure. Area is measured in square units, such as square feet, square meters, square miles, etc.

c. Write a verbal rule to describe how to determine the area of a rectangle if you know the length and width of the rectangle.

d. Represent the area by the letter A. Translate the verbal rule in part c into a formula for the area of a rectangle. Use l to represent length and w to represent width.

The expression lw means l times w, also written as $l \cdot w$ or $(l)(w)$. Generally, you do not use the multiplication symbol × in a variable expression, since it might be confused with the letter x.

e. Use the formula in part d to verify your result in part a.

8. Use the formula for the area of a rectangle to calculate the number of square feet of carpeting you will need for the following rooms. Show the calculations in each case.

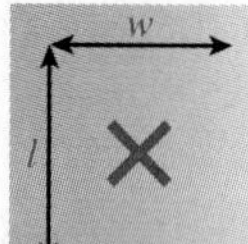

Wide Open Spaces

ROOM	LENGTH, l (feet)	WIDTH, w (feet)	CALCULATIONS: $lw = A$	AREA, A (square feet)
Living	24	15	$24 \cdot 15 = 360$	360
Dining	16	12		
Family	30	16		
Bedroom 1	18	15		
Bedroom 2	14	14		
Bedroom 3	13	13		

9. Carpeting is usually measured by the square foot but is sold by the square yard. Therefore, you have to convert your measurements from square feet to square yards.

a. How many feet are in 1 yard?

b. How many square feet are in 1 square yard? (*Hint*: Draw a 3 feet by 3 feet square, and count the number of 1-foot by 1-foot squares.)

c. Calculate the square yards of carpeting you will need for the living room.

10. Use s to represent the length of one side of a square and write the formula for the area, A, of a square. Use your formula to determine the area of the den in Problem 6.

SUMMARY
ACTIVITY 2.1

1. A **variable** is a quantity that changes (or varies) in value from one situation to another. A single letter usually represents a variable.

2. A **variable expression** is a symbolic code giving instructions for performing arithmetic operations with variables and specific numerical values called constants.

3. A **formula** is an equation (symbolic rule) consisting of an output variable, usually on the left-hand side of the equals sign, and an algebraic expression of input variables, usually on the right-hand side of the equals sign.

4. Evaluating an algebraic expression in a formula for given input values determines the corresponding output value.

5. Multiplication is always the operation to perform when you see a number next to a variable or two variables next to each other.

6. Some formulas from geometry:

a. Perimeter of a Rectangle

$P = 2l + 2w$

b. Area of a Rectangle

$A = lw$

c. Perimeter of a Square

$P = 4s$

d. Area of a Square

$A = s^2$

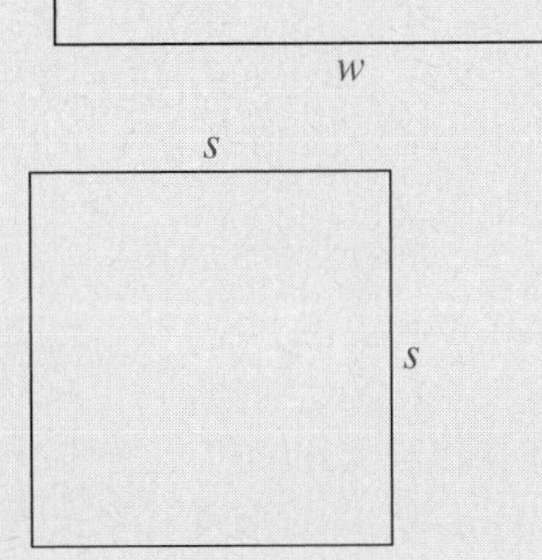

EXERCISES
ACTIVITY 2.1

1. Use the formulas for perimeter and area to answer the following questions.

   a. The side of a square is 5 centimeters (cm). What is its perimeter?

   b. A rectangle is 23 inches long and 35 inches wide. What is its area?

   c. The adjacent sides of a rectangle are 6 and 8 centimeters, respectively. What is the perimeter of the rectangle?

   d. A square measures 15 inches on each side. What is the area of this square?

2. The area of a triangle can be found using the formula $A = b \cdot h \div 2$. In this formula, b represents the base of the triangle and h represents the height.

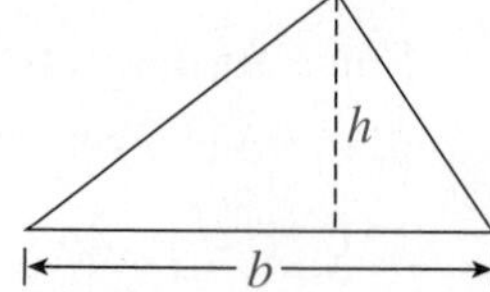

Use this formula to determine the areas of the following triangles.

   a. The base of a triangle is 10 inches and its height is 5 inches.

   b.

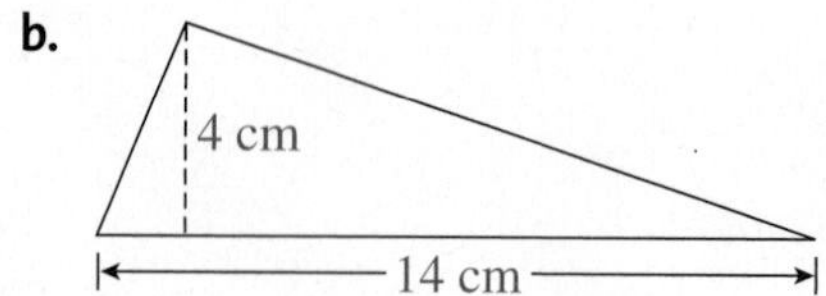

3. Let a, b, and c be the lengths of the sides of a triangle, as shown.

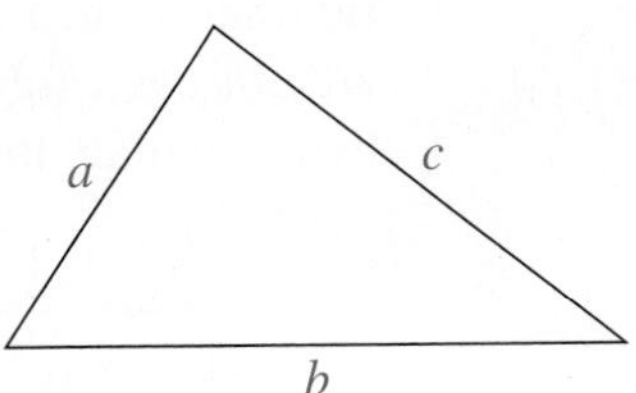

   a. Write a formula for the perimeter of a triangle.

   b. Use the formula you wrote in part a to determine the perimeter of the following triangles.

      i. The sides of the triangle are 4 feet, 7 feet, and 10 feet.

      ii. $a = 24$ cm, $b = 31$ cm, and $c = 47$ cm

      iii.

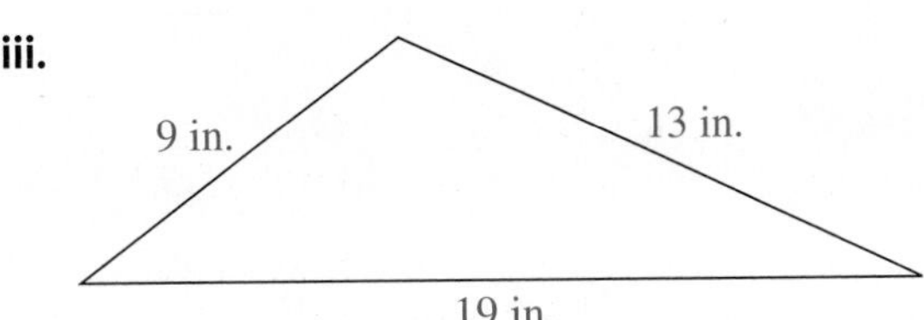

4. You are coating your rectangular driveway with blacktop sealer. The driveway is 60 feet long and 10 feet wide. One can of sealer costs $15 and covers 200 square feet.

   a. Determine the area of your driveway.

   b. Determine how many cans of blacktop sealer you need.

   c. Determine the cost of sealing the driveway.

5. Identify each geometric figure as a square, rectangle, or triangle. Use a metric ruler to measure the lengths that you need to determine the perimeter and area of each figure. Measure each length to the nearest whole centimeter, then calculate the perimeter and area.

a.

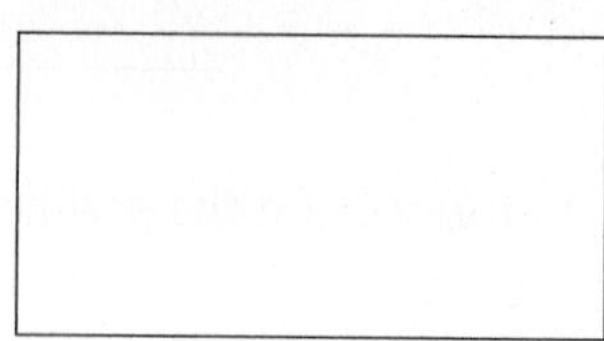

b.

c.

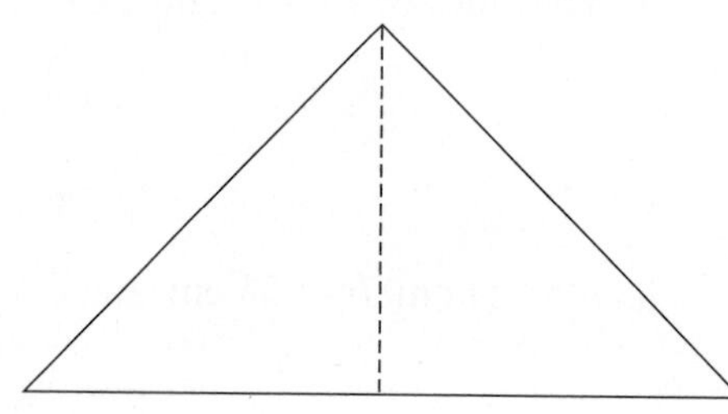

d.

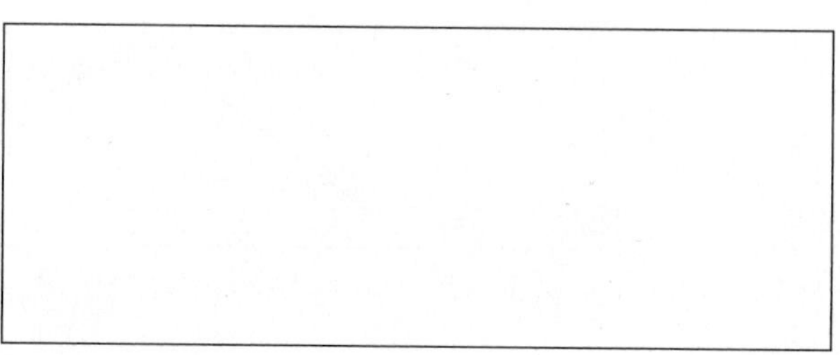

6. You built your house on a rectangular piece of land with dimensions of 80 feet by 100 feet.

a. What is the area of the lot?

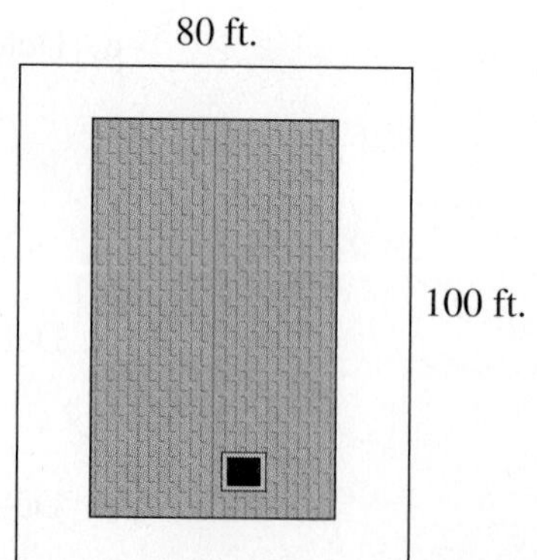

b. The rectangular house measures 50 feet by 79 feet. You wish to lay sod on the remainder of your property. How many square feet of sod are needed?

c. Sod is sold by the square yard. How many square yards do you have to buy?

Formulas are useful in many business applications. For Exercises 7–9,

a. *choose appropriate letters to represent each variable quantity and write what each letter represents*

b. *use the letters to translate each verbal rule into a symbolic formula*

c. *use the formula to determine the result*

7. Profit is equal to the total revenue from selling an item minus the cost of producing the item. Determine the profit if the total revenue is \$400,000 and the cost is \$156,800.

8. Net pay is the difference between a worker's gross income and his or her deductions. A person's gross income for the year was \$65,000 and the total deductions were \$12,860. What was the net pay for the year?

9. Annual depreciation equals the difference between the original cost of an item and its remaining value, divided by its estimated life in years. Determine the annual depreciation of a new car that costs \$25,000, has an estimated life of 10 years, and has a remaining value of \$2000.

Formulas are also found in the sciences. In Exercises 10–11, use the given formulas to determine the results.

10. In general, temperature is measured using either the Celsius (C) or Fahrenheit (F) scales. To convert from degrees Celsius to degrees Fahrenheit, use the following formula.

$$F = (9C \div 5) + 32$$

a. Use the formula to convert 20° Celsius into degrees Fahrenheit.

b. 100° Celsius is equal to how many degrees Fahrenheit?

11. To convert from degrees Fahrenheit to degrees Celsius, use the following formula.

$$C = 5(F - 32) \div 9$$

a. Convert 86° Fahrenheit into degrees Celsius.

b. What is a temperature of 41° Fahrenheit equal to in degrees Celsius?

12. The distance an object travels is determined by how fast it goes and for how long it moves. If an object's speed remains constant, the formula is $d = rt$, where d is the distance, r is the speed, and t is the time.

a. If you drive at 50 miles per hour for 5 hours, how far will you have traveled?

b. A satellite is orbiting Earth at the rate of 25 kilometers per minute. How far will it travel in 2 hours?

Orbits at 25 km per minute

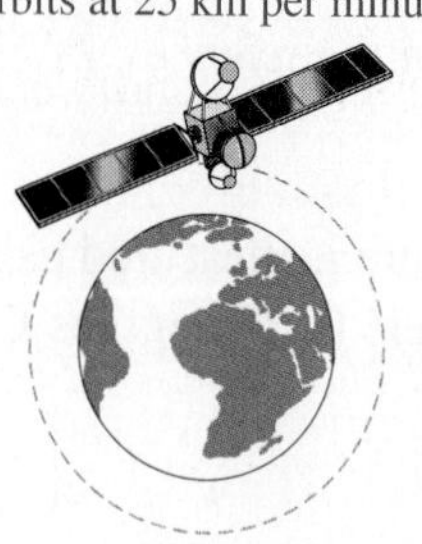

ACTIVITY 2.2

How High Will It Go?

OBJECTIVES

1. Recognize input/output relationship between variables in a formula or equation (two variables only).
2. Evaluate variable expressions in formulas and equations.
3. Generate a table of input and corresponding output values from a given equation, formula, or situation.
4. Read, interpret, and plot points in rectangular coordinates that are obtained from evaluating a formula or equation.

You gave your niece a model rocket for her birthday. The instruction booklet states that the rocket will have a speed of 160 feet per second at launch. The instructions also provide a formula that determines the height of the rocket after launching. The variables are time, t, measured in seconds after launch, and height, h, measured in feet above the ground. Using these variables, the formula is

$$h = 160t - 16t^2.$$

Note that the formula contains only one input variable, namely time, t.

Use this formula to answer Problems 1–8 in this activity.

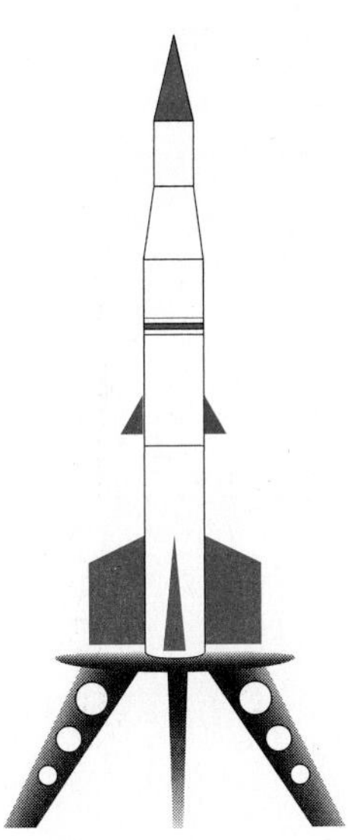

EXAMPLE 1 ***Determine the rocket's height above the ground at 5 seconds after launch, based on the formula given above.***

SOLUTION

The rocket's height above the ground at 5 seconds after launch is determined by replacing the variable t in the formula with the number 5 and doing the arithmetic. Recall that a number next to the variable in a formula means to multiply. Also, remember to follow the order of operations when doing the arithmetic.

Substituting 5 for t, the formula becomes

$$h = 160t - 16t^2$$
$$h = 160 \cdot 5 - 16 \cdot 5^2$$
$$h = 160 \cdot 5 - 16 \cdot 25$$
$$h = 800 - 400$$
$$h = 400 \text{ ft.}$$

The rocket's height above the ground at 5 seconds after launch is 400 feet.

1. How high will the rocket be after 1 second?

2. How high will the rocket be after 2 seconds? After 3 seconds?

Input/Output Table

In Problems 1 and 2, you started with values for the input variable time, *t*, and calculated corresponding values for the output variable height, *h*. A table is useful to show and record a set of such calculations.

Each row in a **table of values** displays a value for the **input variable** and the corresponding value for the **output variable.** Keep in mind that in a formula, the variable involved in the variable expression is the input and the variable by itself on the other side of the equal sign is the output.

3. Record your results from Problems 1 and 2 in the following table. Then calculate the remaining heights to complete the table of values.

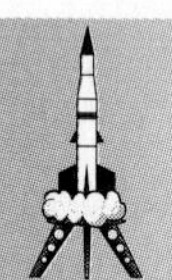

3, 2, 1 Blastoff

INPUT, t (time in seconds)	CALCULATIONS $h = 160 \cdot t - 16 \cdot t^2$	OUTPUT, h (height in feet)
0	$h = 160(0) - 16(0)^2$	0
1		
2		
3		
4		
5		

Visual Display of Input/Output Pairs

As discussed in Activity 1.2, Bald Eagle Population Increasing Again, a useful way to display the information in your table is visually with a graph. Each pair of input/output numbers is represented by a plotted point on a rectangular grid. The input value is referenced on a horizontal number line, the horizontal axis. The output value is referenced on a vertical number line, the vertical axis. For example, in the following grid, one of the (input, output) pairs of numbers from Problem 3, (1, 144), is indicated by a small labeled dot.

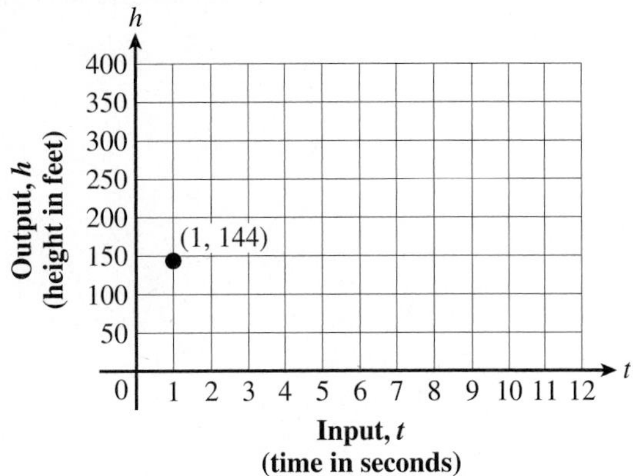

When plotted as a point, a pair of input/output numbers is called the **coordinates** of the point and can be written next to the point in parentheses, separated by a comma. Notice that the numbers on each axis, the **scales,** are different. The choice of scale for an axis is determined by the range of the coordinate values the axis represents.

4. **a.** Use the table of values from Problem 3 to plot the points that represent each input/output pair of values. The grid has been scaled for you. Estimate the heights on the vertical axis as best you can.

b. Notice that the scale on each axis is uniform. This means, for example, that on the time axis the distance between adjacent gridlines always represents 1 second. What distance is represented between adjacent gridlines on the height axis?

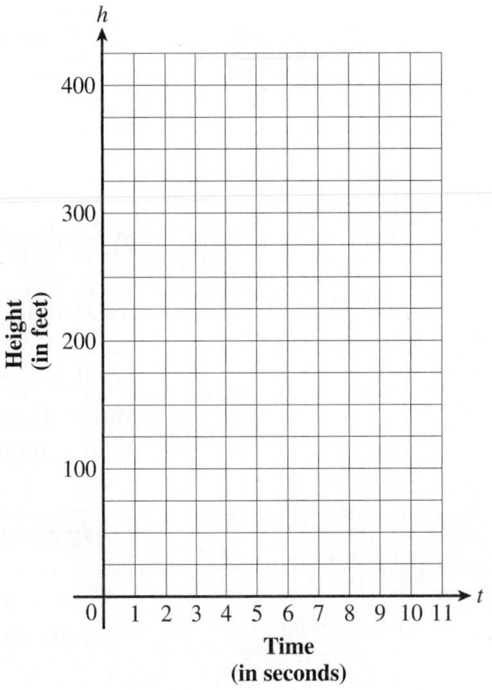

5. Extend the table in Problem 3 for the times given below and plot the additional points on the graph in Problem 4.

Getting Grounded

INPUT, t (time in seconds)	CALCULATIONS $h = 160t - 16t^2$	OUTPUT, h (height in feet)
6		
7		
8		
9		
10		

6. a. Using the output values in the table, how high does the rocket get?

   b. How long does it take to reach this height?

7. What happens at 10 seconds?

8. Draw a curved line to connect the points on the graph in Problem 4. How might this be useful?

A Second Rocket Launch

To visually see the points determined by the rocket height formula, you created a table of input/output values, selected some of the ordered pairs of numbers and plotted them on a grid (see Problems 4 and 5). However, a formula usually determines infinitely many points. By connecting the selected points you plotted on the grid with a smooth curve, you can estimate many other points given by the formula.

DEFINITION

The collection of all the points determined by a formula results in a curve called the **graph** of the formula or equation.

The following graph shows the height of a different rocket after launch, given by the formula $h = 192t - 16t^2$. Use the graph below to answer Problems 9–14.

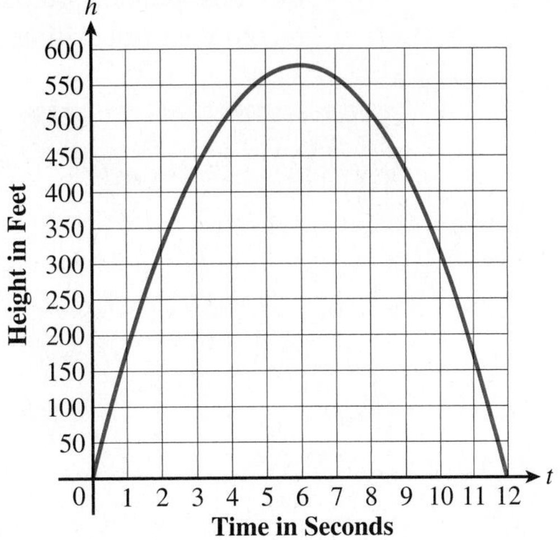

9. What is the input variable and the output variable?

10. Using the graph, estimate the height of the rocket 3 seconds after launch.

11. At what time(s) is the rocket approximately 200 feet above the ground?

12. Approximately how high above the ground does the rocket get, and how many seconds after launch is this highest point reached?

13. How many seconds after launch will the rocket hit the ground?

14. Give the coordinates of the points you used to answer Problems 10–13.

15. a. Identify the input variable and output variable in the formula for the second rocket launch, $h = 192t - 16t^2$.

b. Check the accuracy of your estimates of the input/output pairs you listed in Problem 14 by substituting the input values into the formula and calculating the corresponding output values. Use the following table to show and record your calculations.

INPUT, t (time in seconds)	CALCULATIONS $h = 192t - 16t^2$	OUTPUT, h (height in feet)	PROBLEM 14 ESTIMATE
3			
1			
11			
6			
12			

c. How close were your estimates compared to the calculated output values?

SUMMARY
ACTIVITY 2.2

1. A **formula** can be used to represent the relationship between a single **input variable** and an **output variable.** For example, h is the output variable and t is the input variable in the formula $h = 160t - 16t^2$.

2. A vertical **table of values** is a listing of **input values** with their corresponding **output values** in the same row. Alternatively, a horizontal table of values is a listing of input values with their corresponding output values in the same column.

3. An input value and its corresponding output value can be written as an **ordered pair** of numbers within a set of parentheses, separated by a comma. By tradition, the input value is always written first and the output value is second. For example, (6, 576) represents an input of 6 with a corresponding output of 576.

4. An ordered pair of input/output values from a formula can be represented as a point on a rectangular coordinate grid. If x represents the input, then on the grid, the input value refers to the horizontal or x-axis and is called the ***x*-coordinate** of the point. Similarly, if the output is represented by y, then the output value refers to the vertical or y-axis and is called the ***y*-coordinate** of the point. The point can be labeled on the grid using the ordered-pair notation.

5. When selected points from a formula or equation are plotted on a grid and then smoothly connected by a curve, the result is called a **graph** of the formula or equation.

EXERCISES
ACTIVITY 2.2

1. Use the formulas for the perimeter and area of a square from Activity 2.1 to complete the input/output tables for squares of varying sizes.

s, LENGTH OF SIDE	*P*, PERIMETER OF SQUARE
1	
2	
3	
4	
5	
6	
7	

l, LENGTH OF SIDE	*A*, AREA OF SQUARE
1	
2	
3	
4	
5	
6	
7	

2. Consider the formula $y = 2x + 1$.

   a. Complete the following table of values for this formula.

   b. Which variable is the input variable?

x	y
0	
2	
4	
6	
8	
10	

   c. Plot the points that represent the input/output pairs in your table on the following coordinate system. Label each point with its coordinates.

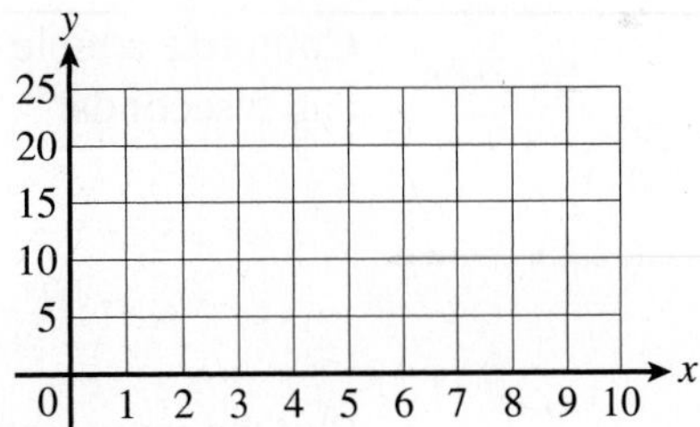

3. Consider the formula $y = x^2 + 3x + 2$.

   a. Complete the following table of values for this formula.

   b. Which variable is the output variable?

x	y
0	
1	
2	
3	
4	
5	

c. Plot the points that represent the input/output pairs of your table on the following coordinate system.

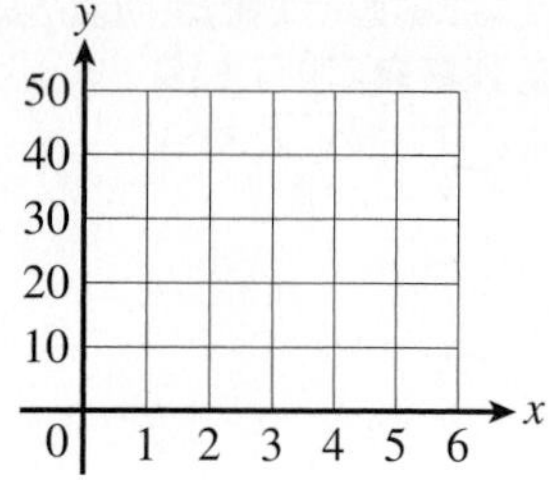

4. You work at a local restaurant. You are paid \$8.00 per hour up to the first 40 hours per week, and \$12.00 per hour for any hours over 40 (overtime hours).

a. Let h represent the hours you work in one week and p your total pay for one week. Complete the input/output table.

INPUT, *h*	20	25	30	35	40	45	50	55	60
OUTPUT, *p*									

b. Plot the input/output pairs from the table in part a on the following coordinate system.

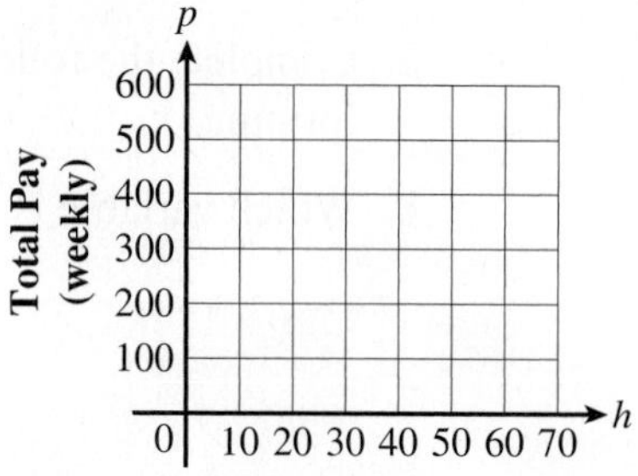

5. The formula $h = 80t - 16t^2$ gives the height above the ground for a ball that is thrown straight up in the air with an initial velocity of 80 feet per second at time t.

a. Complete a table of values for the following input values: $t = 0, 1, 2, 3, 4,$ and 5 seconds.

t	0	1	2	3	4	5
h						

b. Plot the input/output pairs from the table in part a on the following coordinate system, and connect them in a smooth curve.

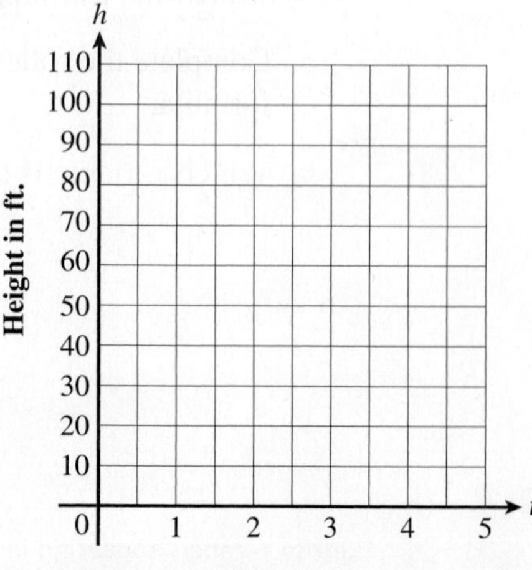

c. Use the graph to estimate how high above the ground the ball will get.

d. From the graph, when will the ball be 100 feet above the ground?

e. When will the ball hit the ground? Give a reason for your answer.

6. The formula $d = 50t$ determines the distance, in miles, that a car travels in t hours at a steady speed of 50 miles per hour.

a. Complete a table of values for the input values: $t = 0, 1, 2, 3, 4,$ and 5 hours.

t	0	1	2	3	4	5
d						

b. Plot the input/output pairs from the table in part a on the following grid. Choose uniform scales on each axis that will allow you to clearly plot all your points.

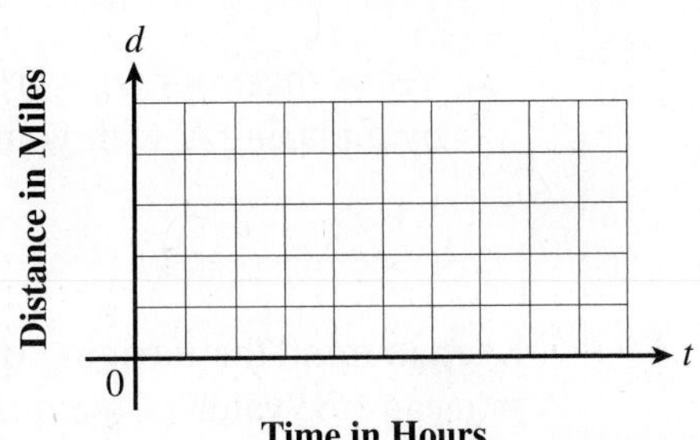

c. How long does it take to travel 200 miles?

7. Use the following formulas to fill in the table indicating which letter represents the input variable and which letter represents the output variable.

	INPUT VARIABLE	OUTPUT VARIABLE
a. $T = 40x$		
b. $F = 2x^2 + 4x - 5$		
c. $C = 5(F - 32)/9$		
d. $A = 13s + 7$		

ACTIVITY 2.3

Are You Balanced?

OBJECTIVES

1. Translate contextual situations and verbal statements into equations.
2. Apply the fundamental principle of equality to solve equations of the forms $x + a = b$, $a + x = b$ and $x - a = b$.

Keeping track of your growing CD collection can be a big job. You would like to know how close you are to owning 200 CDs. Counting them shows you have 117 CDs.

1. How many CDs do you need to reach your goal of 200 CDs? (State the arithmetic operation you performed.)

Problem 1 was easy to solve. It is a simple introduction to one of the most powerful tools in mathematics, solving equations that represent real-world problems.

You can model the CDs problem by the verbal rule:

The number of CDs you have plus the number of CDs you need is equal to the number of CDs in your goal.

Each quantity in the verbal rule can be replaced by either a known number that is given in the problem or by a letter that represents an unknown quantity. If you let N represent the number of CDs you need, the verbal rule becomes the equation

$$117 + N = 200.$$

The value of N that makes this equation a true statement is called the **solution** of the equation.

2. Verify that your answer to Problem 1 is the solution to the equation. (Do this by replacing N with your answer for N in Problem 1 and doing the addition.)

Keep in mind the meaning of the equals symbol, =. In any equation, the symbol = means the value of the quantity on the left side must be *exactly the same* as the value of the quantity on the right side, although the two sides may look quite different. When the values of both sides of an equation are exactly the same, you say the equation is true.

Solving Equations of the Form $x + a = b$ and $a + x = b$

In the equation $117 + N = 200$, the letter N represents an unknown quantity. Solving the equation means to determine the value of the unknown quantity, N. The strategy is to do one (or more) operations on both sides of the equation to isolate the letter N on one side of the equation. The goal is to be able to read the solution as "N is equal to (the value you determined)".

The **fundamental principle of equality** is a powerful guide to solving equations. It states that in performing the same operation on both sides of the equals sign in a true equation, the resulting equation is still true.

For example, if you add 12 to both sides of $13 = 13$, you get $13 + 12 = 13 + 12$, which is $25 = 25$, still true.

This fundamental principle can be viewed visually. Think of a true equation as a scale in balanced. A balanced scale will be horizontal as long as both sides are equal in weight (see below).

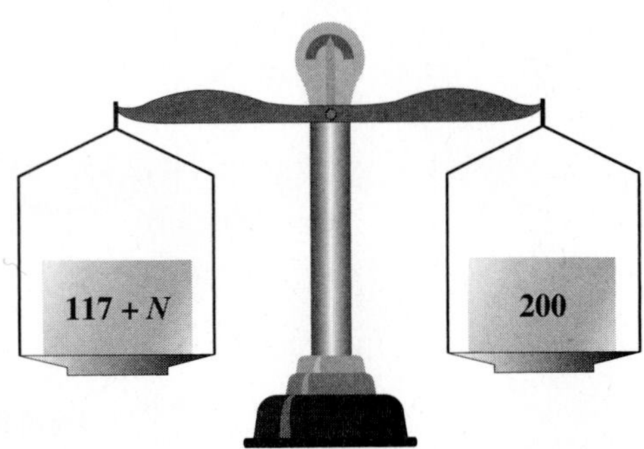

The fundamental principle states that if an equal amount is added to or subtracted from both sides, the scale will remain in balance.

EXAMPLE 1 ***Apply the fundamental principle of equality to solve the equation*** **$117 + N = 200$** ***for N.***

SOLUTION

When solving equations, it is good practice to record your work very carefully. To isolate the variable N, you need to undo the addition of 117. This can be accomplished by subtracting 117 from both sides of the equation.

$$\begin{array}{rcr} 117 + N & = & 200 \\ -117 & & -117 \\ \hline 0 + N & = & 83 \\ N & = & 83 \end{array}$$

To isolate N, subtract 117 from both sides of the equation.

Note that $117 - 117 = 0$.

Check: $117 + 83 = 200$

$200 = 200$

Check by substituting your solution back into the *original* equation.

Using the fundamental principle of equality to solve an equation is referred to as an algebraic approach to solving equations.

In solving $117 + N = 200$, you can imagine the following series of scales.

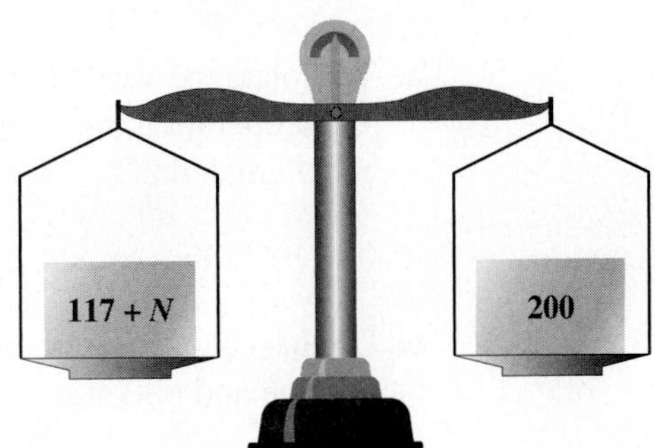

Subtracting 117 from both the left and right sides keeps the scale in balance. The end result gives the solution.

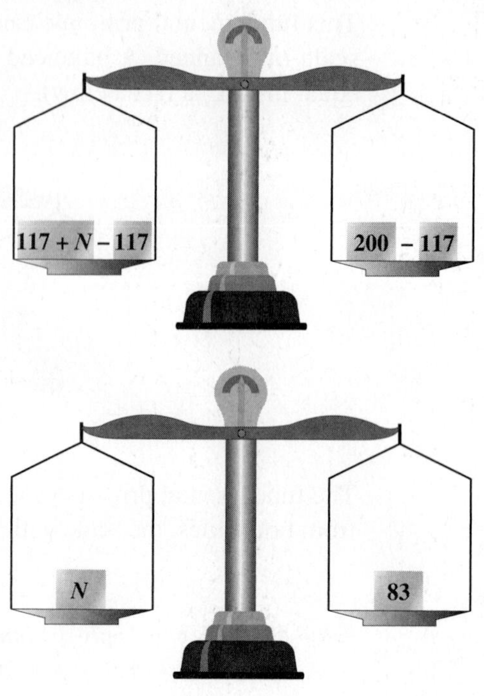

3. Solve the equation $N + 129 = 350$.

a. To isolate the variable N on one side of the equation, what operation must you perform on both sides of $N + 129 = 350$?

b. Use the result from part a to solve the equation. Check your solution.

Solving Equations of the Form $x - a = b$

4. Solve the equation $x - 35 = 47$ for x.

a. To isolate the variable x, you need to undo the subtraction of 35 from x. What operation must you perform on each side of the equation to accomplish this?

b. Do this calculation on both sides of the equation to keep the equation in balance and obtain a solution for x.

c. Check by substituting your solution back into the *original* equation.

5. Suppose you have a \$15 coupon that you can use on any purchases over \$100 at your favorite music store.

a. Write a verbal rule to determine the amount you will pay on a purchase of CDs worth over \$100.

b. Let x represent the cost (amount is over \$100) of the CDs you are buying. Let p represent the amount you pay (ignore sales tax). Translate the verbal rule in part a into an equation.

c. If you purchase \$167 worth of CDs, determine how much you pay using your coupon.

d. Suppose you pay \$205 at the register. Write an equation that can be used to determine the value of the CDs (before the discount coupon is applied).

e. Solve the equation in part d using an algebraic approach.

In general, you can use an algebraic approach to solve equations of the form: $x + a = b$, or $x - a = b$, where x is the unknown quantity and a and b are numbers. In each case, the strategy is to isolate the variable on one side of the equation by undoing the given operation. To solve $x + a = b$, subtract a from both sides. To solve $x - a = b$, add a to both sides. Operations that undo each other, like addition and subtraction, are called *inverse* operations.

6. Use an algebraic approach to solve each of the following equations for x.

a. $100 = x + 27$

b. $16 = x - 7$

c. $x + 4 = 53$

d. $x - 13 = 45$

SUMMARY
ACTIVITY 2.3

1. An **equation** is a statement that says that two expressions are equal. In the equation $10 + x = 25$, $10 + x$ is an expression that is stated to be equal to the expression 25.
2. The **fundamental principle of equality** states that performing the same operation on both sides of a true equation will result in an equation that is also true. The two equations are said to be equivalent.
3. The **solution** to an **equation** is a number that when substituted for the unknown quantity in the equation will result in a true statement when all the arithmetic has been performed.
4. An **equation** is **solved** when the unknown quantity is determined. When the equation $10 + x = 25$ is solved, the solution is $x = 15$.
5. **Expressions** are **evaluated.** For example, the expression $12 + x$ is evaluated when x is replaced with any numerical value and the operation is then performed. If $x = 30$, then $12 + 30 = 42$.
6. To solve equations of the form $x + a = b$ or $a + x = b$ (where a and b are numbers and x is the unknown quantity), "undo" the addition by subtracting a from both sides of the equation.
7. To solve equations of the form $x - a = b$ (where a and b are numbers and x is the unknown quantity), "undo" the subtraction by adding a to both sides of the equation.
8. Because addition and subtraction "undo" each other, they are called *inverse operations* of each other.

EXERCISES
ACTIVITY 2.3

1. Have you ever seen the popular TV game show *The Price is Right*? The idea is to guess the price of an item. You win the item by coming closest to guessing the correct price of the item without going over. If your opponent goes first, a good strategy is to overbid her regularly by a small amount, say \$20. Then your opponent can win only if the price falls in that \$20 region between her bid and yours.

   a. Write a verbal rule to determine your bid if your opponent bids first.

   b. Which bid is the input variable and which is the output variable in the verbal rule? Choose letters to represent your bid and your opponent's bid.

   c. Translate your verbal rule into an equation, using the letters you chose in part b to represent the input and output variables.

d. Use your equation to determine your bid if your opponent bid \$575 on an item.

e. If you bid \$1245 on a laptop computer, use your equation to determine your opponent's bid if he went first.

2. You are looking for a new car and the dealer claims the price of the model you like has been reduced by \$1150. The window sticker shows the new price as \$12,985.

a. Represent the original price by a letter and then write an equation for the new price.

b. Solve your equation to determine the original price. (Remember to check your result as a last step.)

3. Use an algebraic approach to solve each of the following equations.

a. $x + 43 = 191$	**b.** $315 + s = 592$	**c.** $w - 37 = 102$
d. $428 = y + 82$	**e.** $t + 100 = 100$	**f.** $31 = y - 14$
g. $643 + x = 932$	**h.** $z - 56 = 244$	**i.** $W + 2495 = 8340$
j. $y - 259 = 120$	**k.** $4200 + x = 6150$	**l.** $251 = t - 250$
m. $y + 788 = 788$	**n.** $32{,}900 = z + 29{,}240$	**o.** $x - 648 = 0$

4. Your goal is to save enough money to purchase a new portable DVD player costing \$390. In the past 5 weeks, you have saved \$40, \$25, \$50, \$35, and \$20.

a. Write an equation for the cost of the DVD player, using a letter for the dollar amount you still need to save.

b. Solve your equation to determine the amount you still need to save.

5. Monthly salaries for the managers in your company have been increased by \$1500.

a. Write a verbal statement to determine a manager's new salary after the \$1500 increase?

b. Translate your verbal statement in part a into an equation. State what each letter in your equation represents.

c. Which variable in your equation in part b is the input variable? Which is the output variable?

d. Complete this input/output table.

OLD SALARY x	NEW SALARY y
2200	
2400	
2600	
2800	
3000	

e. Use the results from the table to solve $x + 1500 = 4100$.

f. Plot the input/output pairs from your table on the coordinate system below.

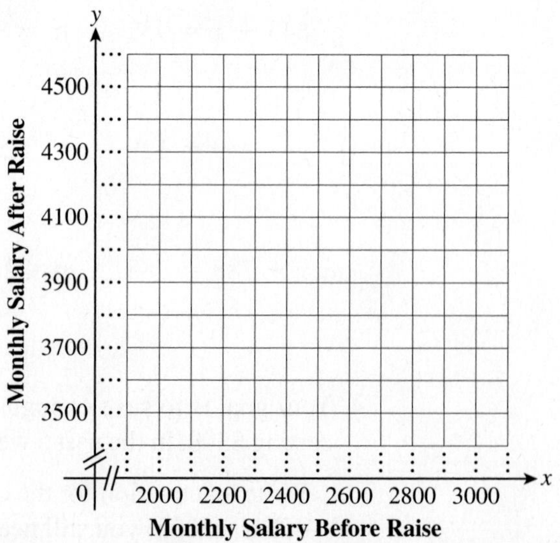

g. What pattern do the points make if you were to connect them?

ACTIVITY 2.4

How Far Will You Go? How Long Will It Take?

OBJECTIVES

1. Apply the fundamental principle of equality to solve equations in the form $ax = b, a \neq 0$.
2. Translate contextual situations and verbal statements into equations.
3. Use the relationship rate • time = amount in various contexts.

Imagine that you are about to take a trip and plan to drive at a constant speed of 60 miles per hour (mph).

1. How far will you go in 2 hours? In 5 hours?

The answers to Problem 1 are based on an important basic relationship:

rate • time = distance

In this situation, the speed of 60 mph is the constant rate or how fast you are driving, and 2 hours is the time or how long you drove. Therefore, the distance you traveled in 2 hours is 60 mph · 2 hours = 120 miles. In 5 hours you will travel 60 mph · 5 hours = 300 miles.

2. a. Let t represent the time traveled (in hours) and d the distance (in miles). Write a formula to determine the distance traveled for various times at an average speed of 60 mph.

b. Use the formula from part a to complete an input/output table. Note that t is the input variable and d is the output variable.

t, TIME IN HOURS	*d*, DISTANCE IN MILES
2	
3	
5	
6	
8	

3. Plot the points that represent the input/output pairs of the table in Problem 2b. Connect the points. What pattern do the points make?

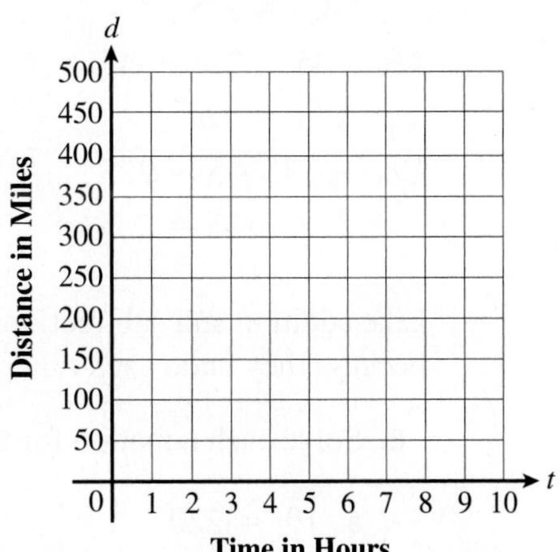

4. Now suppose you are traveling a constant 45 mph.

a. Write a formula to calculate the distance traveled in this case.

b. How far will you go in 3 hours? 4 hours? 6 hours? 10 hours? Summarize your results in the following table.

t, TIME IN HOURS	*d*, DISTANCE IN MILES
3	
4	
6	
10	

c. Plot the points from this table on the same coordinate system as used for Problem 3. Connect the points.

d. How does the pattern of points from part c compare to that for the points in Problem 3, driving at 60 mph?

Now, instead of asking the question "How far will you go?", you might ask the question "How long will it take?" For example, suppose you are driving down the highway at 45 mph, and you ask "How long will it take to go 225 miles?" To answer this question, you can again use the formula: $d = 45t$.

5. Use the formula $d = 45t$ to write an equation to answer the question "How long will it take to travel 225 miles if you drive at an average speed of 45 mph?"

To solve the equation in Problem 5 for the variable t, apply the fundamental principle you used in Activity 2.3. That is, if you perform the same operation to both sides of a true equation, the result will still be a true equation.

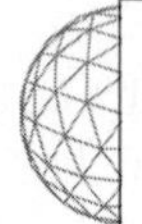

EXAMPLE 1 ***Solve the equation that you obtained in Problem 5, $45t = 225$, for t.***

SOLUTION

$$45t = 225$$

$$45t \div 45 = 225 \div 45$$

To isolate variable t, divide both sides of the equation by 45, since division will undo the multiplication. Note that $45 \div 45 = 1$.

$$t = 5 \text{ hr.}$$

Check: $45 \cdot 5 = 225$

$$225 = 225$$

The solution checks.

Like addition and subtraction, multiplication and division are inverse operations because they undo each other.

6. Solve each equation for the given variable. Check your answer in each case.

a. $10t = 220$ **b.** $4x = 12$ **c.** $3q = 27$

There are other situations that are modeled by the relationship

$$\text{rate} \cdot \text{time} = \text{amount}.$$

7. Suppose you are pumping gasoline into your car's gas tank. Assume that the gas pump delivers 5 gallons per minute.

   a. How much gasoline will you pump in 2 minutes?

   b. Why did you multiply to get the correct answer?

   c. How much gasoline will be pumped in 3 minutes? In 5 minutes?

   d. Let g represent the number of gallons of gasoline pumped in t minutes and write a formula that determines g when you pump at a rate of 5 gallons per minute.

As in Problem 4, you can turn things around to ask "How long will it take?" rather than "How much is pumped?"

8. a. Assume that you continue to pump the gas at 5 gallons per minute. Write an equation to determine how long it will take to pump 20 gallons into your SUV's gas tank.

   b. Solve the equation you wrote in part a.

9. a. Write a formula for the number of gallons of water, w, pumped into a bathtub in t minutes, when the water is flowing in at 3 gallons per minute.

   b. How much water is in the bathtub after 9 minutes?

c. How long does it take to fill the bathtub with 48 gallons? First write an equation and then solve it.

10. a. Write a formula for the number of rabbits, R, added to a population after t months if the rabbit population grows at a rate of 150 rabbits per month.

b. How many rabbits are added over 4 months?

c. How long will it take to add 900 rabbits? First write an equation and then solve it.

11. Many conversion formulas are of the form $ax = b$. For example, in measuring length, to convert from feet to inches, you use

12 inches per foot multiplied by the number of feet = the number of inches.

a. Use i for the distance in inches and f for the number of feet to write the formula for converting feet into inches.

b. Convert 15 feet to inches by substituting 15 into the formula and evaluating the resulting numerical expression to determine the number of inches.

c. Convert 156 inches to feet by substituting into the formula and solving the resulting equation to determine the number of feet.

SUMMARY
ACTIVITY 2.4

1. To solve an equation, apply the fundamental principle of equality, which states: Performing the same operation on *both sides* of a true equation will result in a true equation.
2. To solve equations of the form $ax = b$ (where a is a non-zero number and x is the unknown quantity), "undo" the multiplication by dividing both sides of the equation by a, thus isolating x.
3. Because division "undoes" multiplication and multiplication "undoes" division, the two operations are inverse operations of each other.

EXERCISES
ACTIVITY 2.4

1. Use the fundamental principle of equality to solve each of the following equations.

   a. $6x = 54$ b. $12t = 156$ c. $240 = 3y$ d. $15w = 660$

   e. $169 = 13z$ f. $5x = 285$ g. $374w = 748$ h. $2y = 5874$

   i. $382t = 0$ j. $381 = 3x$ k. $8y = 336$ l. $875 = 25x$

2. a. Write a formula for the area of grass, A, you can mow in t seconds if you can mow 12 square feet per second.

   b. How many square feet can be mowed in 30 seconds?

   c. How long will it take to mow 600 square feet? Write an equation and then solve.

Exercise numbers appearing in color are answered in the Selected Answers appendix.

3. Recall the remodeling job in Activity 2.1. One gallon of paint will cover 400 square feet. (This is the rate of coverage, 400 square feet per gallon.) Use g for the number of gallons of paint and A for the area to be covered to write an equation for each of the following problems. Determine the unknown in each equation.

   a. How much area can be covered with 5 gallons of paint?

   b. How many gallons would be needed to cover 2400 square feet?

   c. All the walls in your house are rectangular and 8 feet high. The table shows the size of each room you wish to paint. How many gallons of paint will you need? (Ignore all window and door openings.)

Painting by Numbers

ROOM	LENGTH	WIDTH	HEIGHT	WALL AREA
Living	24	15	8	624
Dining	16	12	8	
Family	30	16	8	
Bedroom 1	18	15	8	
Bedroom 2	14	14	8	
Bedroom 3	13	13	8	
Total				

4. For each of the following conversions, write an equation of the form $ax = b$. Use letters of your own choosing. Determine the unknown in each equation to answer each question.

   a. There are 5280 feet in one mile. How many feet are in 4 miles?

   b. There are 16 ounces in 1 pound. How many pounds are in 832 ounces?

c. $25t - 15t = 4200$

d. $326s + 124s = 3600$

e. $6n + 2n - 7n = 15$

SUMMARY
ACTIVITY 2.5

1. **Like terms** are terms with exactly the same variable factor. The numeric factor of the term is called the **coefficient** of the variable factor.

   For example, $3x$ and $8x$ are like terms. 3 and 8 are coefficients of x. $3x$ and $3y$ are *not* like terms.

2. **Combining like terms** into a single term means applying the distributive property to add or subtract the coefficients of the like terms.

   For example, $3x + 8x = (3 + 8)x = 11x$ and $9x - 8x = (9 - 8)x = 1x = x$.

3. To solve an equation like $3x + 8x = 22$, first combine like terms to get $11x = 22$. Then apply the fundamental principle of equality by dividing each side of the equation by the coefficient of x. In this example,

$$11x \div 11 = 22 \div 11$$

$$x = 2.$$

EXERCISES
ACTIVITY 2.5

1. a. How many terms are in the expression $5t + 3s + 2t$?

   b. What are the coefficients?

   c. What are the like terms?

2. Rewrite each expression by combining like terms.

   a. $14x + 7y - 8x + 5y$

   b. $2s + 11t + 9s - 3t + 14$

Exercise numbers appearing in color are answered in the Selected Answers appendix.

c. $9x + 3y - 5x + 2y$

d. $5b + 12t + t + 9b - 6$

e. $27p + 39n + 12p - 2n + 5n$

f. $252w + 130z + 175y - 181w + 65z - 88y$

3. Solve the following equations. Check your answers in the original equation.

a. $3x + 15x = 720$

b. $14y - 7y = 49$

c. $320 = 21s - 13s$

d. $0 = 34w + 83w$

e. $5x + 17x - 4x = 270$

f. $32 = 4t - 3t + 7t$

g. $y + 3y - 2y = 178$

h. $13x + 24x - 31x = 144$

i. $2368 = 41t + 58t - 95t$

j. $3w + 2w + 5w - 4w + w - 3w + w = 1635$

k. $36 = 14x + 8 - 13x$

l. $65y - 51 - 64y = 13$

4. In your part-time job, your hours vary each week. Last month you worked 22 hours the first week, 25 hours the second week, 14 hours during week 3, and 19 hours in week 4. Your gross pay for those 4 weeks was $960.

a. Let p represent your hourly pay rate. Write the equation for your gross pay over these 4 weeks.

b. Solve this equation to determine your hourly pay rate.

5. Over the first 3 months of the year you sold 12, 15, and 21 SuperPix digital cameras. The total in sales was $18,960.

a. Let p represent the price of one SuperPix digital camera. Write the equation for your total in sales over these 3 months.

b. Solve the equation in part a to determine the retail price of the SuperPix digital camera.

ACTIVITY 2.6

Make Me an Offer

OBJECTIVES

1. Use the basic steps for problem solving.
2. Translate verbal statements into algebraic equations.
3. Use the basic principles of algebra to solve real-world problems.

Many students believe "I can do math. I just can't do word problems." In reality, you learn mathematics so you can solve practical problems in everyday life and in science, business, technology, medicine, and most other fields.

On a personal level, you solve problems every day while balancing your checkbook, remembering an anniversary, or figuring out how to get more exercise. Most everyday problems don't require algebra, or even arithmetic, to be solved. But the basic steps and methods for solving any problem can be discussed and understood. In the process you will, with practice, become a better problem solver.

Solve the following problem any way you can, making notes of your thinking as you work. In the space provided, record your work on the left side. On the right side, jot down in a few sentences what you are thinking as you work on the problem.

"Night Train" Henson's Contract

1. "Night Train" Henson is negotiating a new contract with his team. He wants \$800,000 for the year with an additional \$6000 for every game he starts. His team offered \$10,000 for every game he starts, but only \$700,000 for a base salary. How many games would he have to start to make more with the team's offer?

SOLUTION TO THE PROBLEM	YOUR THINKING

2. Talk about and compare your methods for solving the problem above with classmates or your group. As a group, write down the steps you all went through from the very beginning to the end of your problem solving.

Experience has shown that there is a set of basic steps that serve as a guide to solving problems.

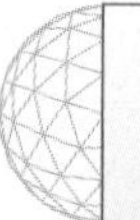

PROCEDURE

Basic Steps for Problem Solving Solving any problem generally requires the following four steps.

1. Understand the problem.
2. Develop a strategy for solving the problem.
3. Execute your strategy to solve the problem.
4. Check your solution for correctness.

In Problems 3–6 that follow, compare your group's strategies for solving "Night Train" Henson's problem with each of the four basic steps for problem solving.

Let's consider each of the steps 1 through 4.

Step 1. Understand the problem. This step may seem obvious, but many times this is the most important step, and is the one that commonly leads to errors. Read the problem carefully, as many times as necessary. Draw some diagrams to help you visualize the situation. Get an explanation from available resources if you are unsure of any details.

3. List some strategies you could use to make sure you understand the "Night Train" Henson problem.

Step 2. Develop a strategy for solving the problem. Once you understand the problem, it may not be at all clear what strategy is required. Practice and experience is the best guide. Algebra is often useful for solving a math problem but is not always required.

4. How many different strategies were tried in your group for solving the "Night Train" problem? Describe one or two of them.

Step 3. Execute your strategy to solve the problem. Once you have decided on a strategy, carrying it out is sometimes the easiest step in the process. If you really understand the problem, and are clear on your strategy, the solution should happen almost automatically.

5. When you decided on a strategy, how confident did you feel about your solution? Give a reason for your level of confidence.

Step 4. Check your solution for correctness. This last step is *critically important*. Your solution must be reasonable, it must answer the question, and most importantly, it must be correct.

6. Are you absolutely confident your solution is correct? If so, how do you know?

Solution: "Night Train" Henson's Contract

As a beginning problem solver, you may find it useful to develop a consistent strategy. Examine the following way you can solve "Night Train's" problem by applying some algebra tools as you follow the four basic steps of problem solving.

Step 1. Understand the problem. If you understand the problem, you will realize that "Night Train" needs to make \$100,000 more with the team's offer (800,000 − 700,000) to make up for the difference in base salary. So he must start enough games to make up for this difference. Otherwise, he should not want the team's offer.

Step 2. Develop a strategy for solving the problem. With the observation above, you could let the variable x represent the number of games "Night Train" has to start to make the same amount of money with either offer. Then we note that he would make \$4000 more (10,000−6000) for each game with the team's offer. So, \$4000 for each game started times the number of games started must equal \$100,000.

Step 3. Execute your strategy for solving the problem. You can translate the statement above into an equation, then apply the fundamental principle of equality to obtain the number of games started. Let x represent the number of games started.

\$4000 times the number of games started equals \$100,000.

$$\begin{aligned} 4000x &= 100{,}000 \\ 4000x \div 4000 &= 100{,}000 \div 4000 \\ x &= 25 \end{aligned}$$

The solution to your equation says that "Night Train" must start in 25 games to make the same money with each deal.

Your answer would be: "Night Train" needs to play in more than 25 games to make more with the team's offer.

Step 4. Check your solution for correctness. This answer seems reasonable, but to check it you should refer back to the original statement of the problem. If "Night Train" plays in exactly 25 games his salary for the two offers can be calculated.

"Night Train's" demand: $\$800{,}000 + \$6000 \cdot 25 = \$950{,}000$

Team's offer: $\$700{,}000 + \$10{,}000 \cdot 25 = \$950{,}000$

If he starts in 26 games, the team's offer is \$960,000, but "Night Train's" demand would give him only \$956,000. These calculations confirm the solution.

There are other ways to solve this problem. But the above solution gives you an example of how algebra can be used. Use algebra in a similar way to solve the following problems. Be sure to follow the four steps of problem solving.

Additional Problems to Solve

7. You need to open a checking account and decide to shop around for a bank. Acme Bank has an account with a \$10 monthly charge, plus 25 cents per check. Farmer's Bank will charge you \$12 per month, with a 20-cent charge per check. How many checks would you need to write each month to make Farmer's Bank the better deal?

   Step 1. Understand the problem.

Step 2. Develop a strategy for solving the problem.

Step 3. Execute your strategy to solve the problem.

Step 4. Check your solution for correctness.

8. The perimeter of a rectangular pasture is 2400 feet. If the width is 800 feet, how long is the pasture?

Step 1. Understand the problem.

Step 2. Develop a strategy for solving the problem.

Step 3. Execute your strategy to solve the problem.

Step 4. Check your solution for correctness.

Translating Statements into Algebraic Equations

The key to setting up and solving many word problems is recognizing the correct arithmetic operations. If there is a known formula, like the perimeter of a rectangle, then the arithmetic is already determined. Sometimes there is *only* the language to guide you.

In Problems 9–12, translate each statement into an equation. In each statement, let the unknown number be represented by the letter x.

9. The sum of 35 and a number is 140.

10. 144 is the product of 36 and a number.

11. What number times 7 equals 56?

12. The difference between the regular price and sale price of $245 is $35.

13. In Problems 9, 10, and 12 above, the verb "is" translates into what part of the equation?

14. Solve each of the equations you wrote in Problems 9–12.

    a.

    b.

    c.

    d.

SUMMARY
ACTIVITY 2.6

The Four Steps of Problem Solving

Step 1. Understand the problem.

a. Read the problem completely and carefully.

b. Draw a sketch of the problem, if possible.

Step 2. Develop a strategy for solving the problem.

a. Identify and list everything you know about the problem, including relevant formulas. Add labels to the diagram if you have one.

b. Identify and list what you want to know.

Step 3. Execute your strategy to solve the problem.

a. Write an equation that includes the known quantities and the unknown quantity.

b. Solve the equation.

Step 4. Check your solution for correctness.

a. Is your answer reasonable?

b. Is your answer correct? (Does it answer the original question? Does it agree with all the given information?)

EXERCISES
ACTIVITY 2.6

In Exercises 1–10, translate each statement into an equation, then solve the equation for the unknown number.

1. An unknown number plus 425 is equal to 981.

2. The product of some number and 40 is 2600.

3. A number minus 541 is 198.

4. The sum of an unknown number and 378 is 2841.

Exercise numbers appearing in color are answered in the Selected Answers appendix.

5. The quotient of some number and 9 is 63.

6. The prime factors of 237 are 79 and an unknown number.

7. 13 plus an unknown number is equal to 51.

8. The difference between a number and 41 is 105.

9. 642 is the product of 6 and an unknown number.

10. The sum of an unknown number and itself is 784.

In Exercises 11–17, solve the following problems by applying the four steps of problem solving. Use the strategy of solving an algebra equation for each problem.

11. In preparing for a family trip, your assignment is to make the travel arrangements. You can rent a car for $75 per day with unlimited mileage. If you have budgeted $600 for car rental, how many days can you drive?

12. You need to drive 440 miles to get to your best friend's wedding. How fast must you drive to get there in 8 hours?

13. Your goal is to save $1200 to pay for next year's books and fees. How much must you save each month if you have 5 months to accomplish your goal?

14. You have enough wallpaper to cover 240 square feet. If your walls are 8 feet high, how wide a wall can you paper?

15. In your part-time job selling kitchen knives, you have two different sets available. The better set sells for $35, the cheaper set for $20. Last week you sold more of the cheaper set, in fact twice as many as the better set. Your receipts for the week totaled $525. How many of the better sets did you sell?

16. A rectangle that has an area of 357 square inches is 17 inches wide. How long is the rectangle?

17. A rectangular field is 5 times longer than it is wide. If the perimeter is 540 feet, what are the dimensions (length and width) of the field?

w

$\ell = 5w$

What Have I Learned?

Write your responses in complete sentences.

1. What is the difference between an expression and an equation?

2. Describe each step required to evaluate the expression $5t^2 - 3t$ for $t = 2$.

3. In your own words, describe the fundamental principle of equality.

4. Describe the step required to solve the equation $6x = 96$.

5. Describe the step required to solve the equation $w - 38 = 114$.

6. Generate an input/output table for $y = x^2 - 5$, using input values $x = 3, 4, 5$, and 6.

x	$y = x^2 - 5$	y
3		
4		
5		
6		

7. Which of the following are like terms: $4y$, x, $7x$, $4w$, 3, $3y$? Why?

8. Give an example of how the distributive property allows you to combine like terms.

9. What are the four steps of problem solving?

10. What would you do to check your answer after solving a problem?

How Can I Practice?

1. Determine the area and perimeter of a rectangle that is 13 centimeters (cm) wide and 25 centimeters long.

2. What is the perimeter of a triangle, in inches, that has sides that are 6 inches, 9 inches, and 1 foot long?

3. What is the area of the triangle shown?

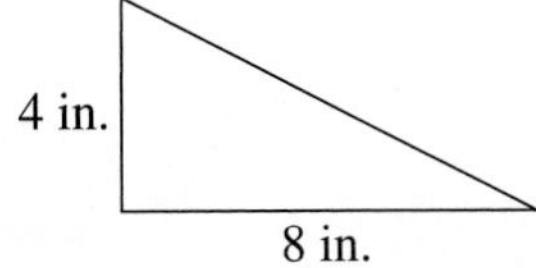

4. You need to buy grass seed for a new lawn. The lawn will cover a rectangular plot that is 60 feet by 90 feet. Each ounce of grass seed will cover 120 square feet.

   a. How many ounces of grass seed will you need to buy?

   b. If the grass seed you want comes in 1-pound bags, how many bags will you need to buy?

5. Translate each statement into a formula. State what each letter in your fomula represents.

   a. The total weekly earnings equal $12 per hour times the number of hours worked.

Exercise numbers appearing in color are answered in the Selected Answers appendix.

b. The total time traveled equals the distance traveled divided by the average speed.

c. The distance between two cities, measured in feet, is equal to the distance measured in miles times 5280 feet per mile.

6. Use your formulas in Exercise 5 to answer the following questions. State your final answer in a complete sentence.

a. If you make \$12 per hour, what will your earnings be in a week when you work 35 hours?

b. If you earned \$336 last week (still assuming \$12 per hour), how many hours did you work?

c. If you travel 480 miles at an average speed of 40 miles per hour, for how many hours were you traveling?

d. The distance between Dallas and Houston is 244 miles. How far is the distance in feet?

7. What will the temperature be in degrees Fahrenheit when it is 40° Celsius outside? Recall the formula $F = (9C \div 5) + 32$.

8. In the following formulas, which letter represents the input variable and which letter represents the output variable?

a. $w = 13x + 5$

b. $6(9 - 2x) = S$

9. a. Complete the following table for the formula $y = 2x^2 + x - 2$.

x	y
1	
2	
3	
4	
5	
6	

b. Which variable is the output variable?

c. Plot the points that represent the input/output pairs in the table.

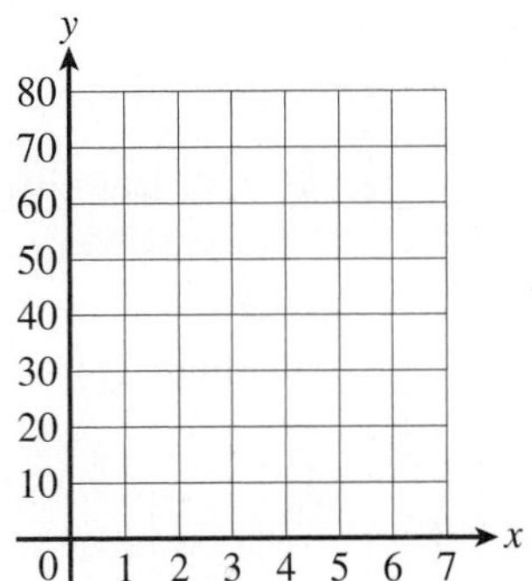

10. Solve the following equations for the unknown value.

a. $42x = 462$

b. $x + 33 = 214$

c. $518 = 14s$

d. $783 = 5y + 4y$

e. $6t - 14 - 5t = 90$

f. $413 = w - 46$

g. $39x - 24x = 765$

h. $340 = 2x + 3x$

11. a. Determine a formula for the height of a tree, H, that is growing at the rate of 3 feet per year, where x is the number of years since planting.

b. How tall is the tree after 10 years?

c. How long will it take the tree to grow to 75 feet?

12. At your town's community college, tuition is \$105 per credit for less than 12 credits. Your aunt decided to register for classes on a part-time basis and paid a tuition bill of \$945. Describe how you will determine from this information how many credits your aunt is taking.

CHAPTER 2 SUMMARY

The bracketed numbers following each concept indicate the activity in which the concept is discussed.

<table>
<tr><th>CONCEPT / SKILL</th><th>DESCRIPTION</th><th>EXAMPLE</th></tr>
<tr><td>Variable [2.1]</td><td>A quantity that varies from one situation to another, usually represented by a letter.</td><td>P, l, and w can be shorthand notation for the variables perimeter, length, and width of a rectangle.</td></tr>
<tr><td>Formula [2.1]</td><td>An equation or symbolic rule that is a recipe for performing a specific calculation.</td><td>The formula for the perimeter of a rectangle, $P = 2w + 2l$, is a recipe for calculating the perimeter of any rectangle, once you know the length and width.</td></tr>
<tr><td>Variable expression [2.1, 2.2]</td><td>A symbolic code of instructions for performing arithmetic operations with variables and numbers.</td><td>The right side of the above formula, $2w + 2l$, is an algebraic expression.</td></tr>
<tr><td>Evaluating an expression [2.1, 2.2]</td><td>Replacing the variable in the expression with a numerical value and then performing all the arithmetic.</td><td>If $w = 10$ and $l = 15$, then $2w + 2l = 2(10) + 2(15)$ $= 20 + 30 = 50$.</td></tr>
<tr><td>Input variable [2.2]</td><td>The variable(s) in a formula whose value(s) is given first.</td><td>In the formula $d = 25t$, t is typically the input variable.</td></tr>
<tr><td>Output variable [2.2]</td><td>The variable in a formula whose value depends on the value of the input variable(s).</td><td>In the formula $d = 25t$, d is typically the output variable.</td></tr>
<tr><td>Table of values [2.2]</td><td>A listing of input variable values with their corresponding output variable values in the same row.</td><td><table><tr><th>t (input)</th><th>d (output)</th></tr><tr><td>2</td><td>50</td></tr><tr><td>4</td><td>100</td></tr></table></td></tr>
<tr><td>Rectangular coordinate system [2.2]</td><td>A grid scaled by two coordinate axes (number lines) drawn at right angles to each other. The horizontal axis corresponds to the input variable and the vertical axis corresponds to the output variable.</td><td>y
5
4
3
2 (3, 2)
1
0 1 2 3 4 5 x</td></tr>
</table>

CONCEPT / SKILL	DESCRIPTION	EXAMPLE
Coordinates of a point [2.2]	Identify the location of an input/output ordered pair in a rectangular coordinate system. Written as (input, output).	See (3, 2) on the graph at the bottom of previous page.
Solution of an equation [2.3]	A value for the unknown quantity that makes the equation a true statement.	$x = 5$ is a solution to $3x = 15$ because $3 \cdot 5 = 15$ is true.
Fundamental principle of equality [2.3]	In general, performing the same operation on both sides of the equals sign in a true equation results in a true equation.	Adding 7 to both sides of $3 = 3$, you get $3 + 7 = 3 + 7$, which is $10 = 10$, still a true equation.
Solving equations of the form $a + x = b$ or $x + a = b$ [2.3]	To solve equations of the form $x + a = b$ or $a + x = b$, (where a and b are numbers and x is the unknown quantity), "undo" the addition by subtracting a from both sides of the equation.	If $x + 14 = 29$ is true, then subtracting 14 from both sides of the equation will result in a true equation. $x + 14 = 29$ $-14 \quad -14$ $x = 15$
Solving equations of the form $x - a = b$ [2.3]	To solve equations of the form $x - a = b$ (where a and b are numbers and x is the unknown quantity), "undo" the subtraction by adding a to both sides of the equation.	If $x - 23 = 31$ is true, then adding 23 to both sides of the equation will result in a true equation. $x - 23 = 31$ $+23 \quad +23$ $x = 16$
Solving equations of the form $ax = b$ [2.4]	To solve equations of the form $ax = b$ (where a and b are numbers, $a \neq 0$, and x is the unknown quantity), undo the multiplication by dividing both sides of the equation by a, the factor that multiplies the variable.	If $9x = 144$ is true, then dividing both sides of the equation by 9 will result in a true equation. $9x = 144$ $9x \div 9 = 144 \div 9$ $x = 16$
Terms of an expression [2.5]	Parts of an expression that are separated by addition or subtraction signs are called **terms.**	The expression $45p + 75p$ has two terms, $45p$ and $75p$, which are separated by an addition sign.
Like terms [2.5]	Terms in an expression with exactly the same variable factor.	$3x$ and $8x$ are like terms.
Coefficient [2.5]	The factor that multiplies the variable portion of a term.	3 is the coefficient of x in the term $3x$.

CONCEPT / SKILL	DESCRIPTION	EXAMPLE
Combining like terms [2.5]	Use the distributive property to add or subtract the coefficients of like terms to obtain a single term.	$3x + 8x = (3 + 8)x = 11x$ and $9x - 8x = (9 - 8)x = 1x = x$
The four steps of problem solving [2.6]	**Step 1.** Understand the problem.	Read the problem completely and carefully, draw a diagram, ask questions.
	Step 2. Develop a strategy for solving the problem.	Identify and list every quantity, known and unknown. Write an equation that relates the known quantities and the unknown quantity.
	Step 3. Execute your strategy to solve the problem.	Solve the equation.
	Step 4. Check your solution for correctness.	Is your answer reasonable? Is your answer correct?

CHAPTER 2 GATEWAY REVIEW

1. Use the appropriate formulas for perimeter and area to answer the following questions.

a. The side of a square is 24 inches. What is its perimeter?

b. A rectangle is 41 centimeters long and 28 centimeters wide. What is its area?

c. The adjacent sides of a rectangle are 3 feet, and 11 feet respectively. What is the perimeter of the rectangle?

d. A square measures 33 centimeters on each side. What is the area of this square?

2. Recall the formula for calculating the temperature in degrees Fahrenheit if you know the temperature in degrees Celsius: $F = 9C \div 5 + 32$.

a. Which variable is the input variable?

b. Complete the following table of values.

TEMPERATURE (in degrees Celsius)	TEMPERATURE (in degrees Fahrenheit)
0	
10	
20	
30	
40	
50	
60	
70	
80	
90	
100	

Answers to all Gateway exercises are included in the Selected Answers appendix.

c. Plot the points that represent the input/output pairs of your table.

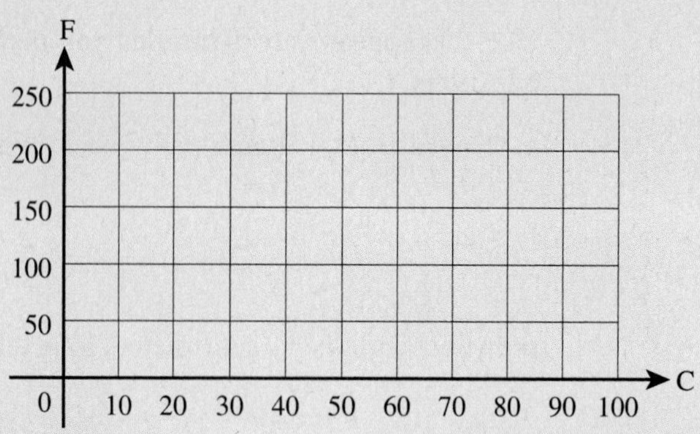

3. The formula $h = 200t - 16t^2$ gives the height of a ball that is tossed in the air with an initial velocity of 200 feet per second.

a. Complete a table of values for the following input values: $t = 0, 1, 2, 3, 4, 5, 6, 7, 8, 9$, and 10 seconds.

t											
h											

b. Plot the points that represent the input/output pairs of your table and then connect the points with a smooth curve.

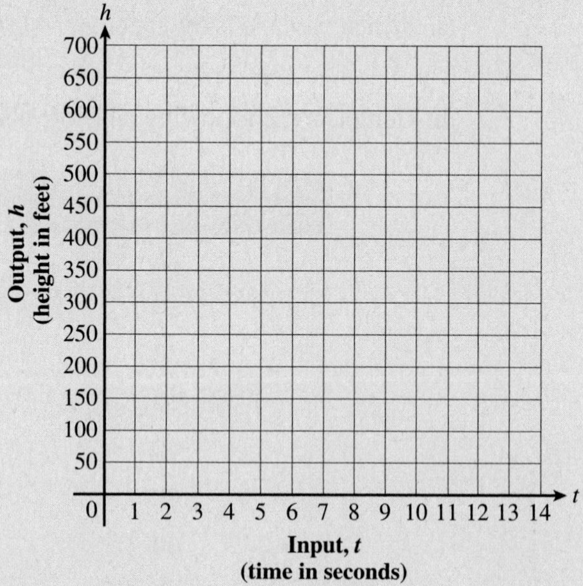

c. When will the ball be 400 feet above the ground?

d. Approximately how high do you think the ball will get?

e. Approximately when will the ball hit the ground?

4. Solve each of the following equations for the unknown value.

a. $35 + x = 121$

b. $25y = 1375$

c. $603 = 591 + t$

d. $14x + 7x - 17x = 216$

e. $58 + x = 63$

f. $504 = 3w$

g. $91 = 3y + 12y - 8y$

h. $3w - 2w + 11 = 58$

5. You have a goal to buy a used car for \$2500. If you have saved \$240, \$420, and \$370 in the previous 3 months, how much more must you save to reach your goal? Write and solve an equation for the cost of the used car.

6. Your study group can review 20 pages of text in 1 hour. Use the relationship rate · time = amount to write and solve an equation to answer the following question. How long will it take your group to review 260 pages of text?

7. You ordered 3 pizzas this week, 6 pizzas last week, and 5 pizzas the week before that. Assuming all these pizzas were the same price, how much did each pizza cost if your total bill for all the pizzas came to $126?

8. Translate each statement into an equation, then solve. Let x represent the unknown number.

a. The sum of 351 and some number is 718.

b. 624 is the product of 48 and an unknown number.

c. The product of 27 and what number is equal to 405?

d. The difference between some number and 38 is 205.

e. The sum of an unknown number and twice itself is 273.

9. Solve the following problem by applying the four steps of problem solving. Use the strategy of solving an algebra equation.

Your average reading speed is 160 words per minute. There are approximately 800 words on each page of the textbook you need to read. Approximately how long will it take you to read 50 pages of your textbook?

CHAPTER 3

PROBLEM SOLVING WITH RATIONAL NUMBERS: ADDITION AND SUBTRACTION OF INTEGERS, FRACTIONS, AND DECIMALS

Chapters 1 and 2 deal with counting numbers that represent quantities greater than zero, and for this reason counting numbers are also called *positive numbers*. However, quantities like debits in business or electrical charges in physics and chemistry require numbers that represent negative quantities. Such numbers are called negative numbers. Positive and negative counting numbers and zero, taken together, form a number system called the *integers*.

In some situations, a quantity is divided into parts; or in other situations, two quantities are compared. These situations require numbers called fractions. Another name for a fraction is *rational number*. In this chapter, you will solve problems involving integers and fractions. In Chapter 5, you will learn why fractions are also called ratios or rational numbers.

CLUSTER 1 Adding and Subtracting Integers

ACTIVITY 3.1

On the Negative Side

OBJECTIVES

1. Identify integers.
2. Represent quantities in real-world situations using integers.
3. Compare integers.
4. Calculate absolute values of integers.

The Hindus introduced negative numbers to represent debts, the first known use was by Brahmagupta in about 628 A.D. However, the history of mathematics shows that it took a long time for negative numbers to be accepted by everyone, including mathematicians, and that didn't happen until the 1600s.

1. Show with an example how you can represent a temperature less than 0 degrees Fahrenheit.

2. Can a number be used to represent a checking account balance less than $0? Explain.

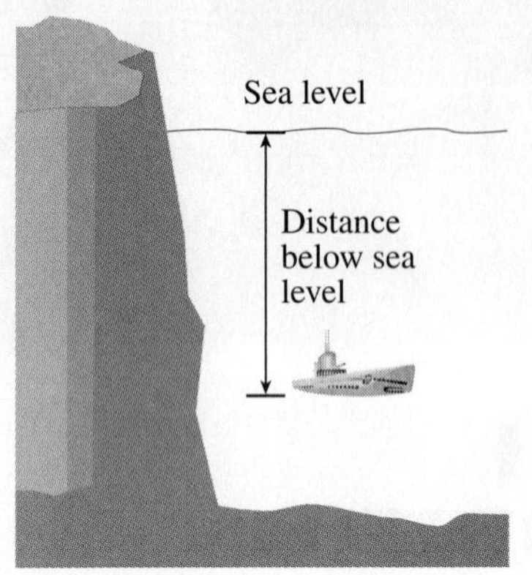

3. Sometimes you might want to know how far above or below sea level an object is located (for example, an airplane or a school of fish).

   a. What number would you use to represent sea level? Explain.

   b. How would you represent the depth of an object that is below sea level? Describe an example.

A number less than zero is a negative number. Negative numbers are indicated by a dash, −, to the left of the number such as −20, −100, and −6.25. In a similar way, a positive number, for example positive three, can be written as +3. However, most of the time, a positive three is simply written as 3. This is true for positive numbers in general.

4. If you gain 10 pounds you can represent your change in weight by the number 10. What number represents a loss of 10 pounds?

Integers and the Number Line

The collection of all of the counting numbers, zero, and the negatives of the counting numbers is called the set of **integers**: {. . .−4, −3, −2, −1, 0, 1, 2, 3, 4, . . .}. The positive counting numbers are called *positive integers* and the negative counting numbers are called *negative integers*. Note that the terms *counting numbers* and *positive whole numbers* mean the same collection of numbers.

A good technique for visualizing integers is to use a number line. A number line is a line with evenly spaced tick marks. A tick mark is a small line segment perpendicular to the number line. Each tick mark on the following number line represents an integer. Notice that the negative integers are to the left of zero and the positive integers are to the right of zero. The numbers increase as you read from left to right.

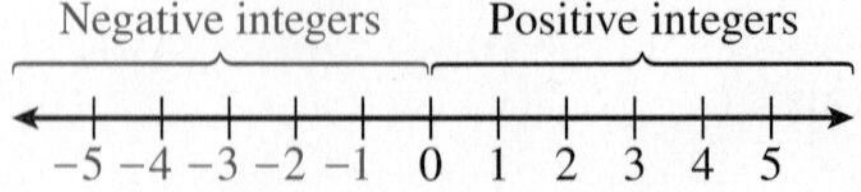

5. a. Give an example of a positive integer, a negative integer, and an integer that is neither negative nor positive, and place these integers on a number line.

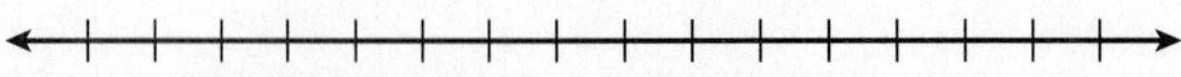

b. Describe a situation in which integers might be used. What would zero represent in the situation you describe?

6. You start the month with \$225 in your checking account with no hope of increasing the balance during the month. You place the balance on the following number line.

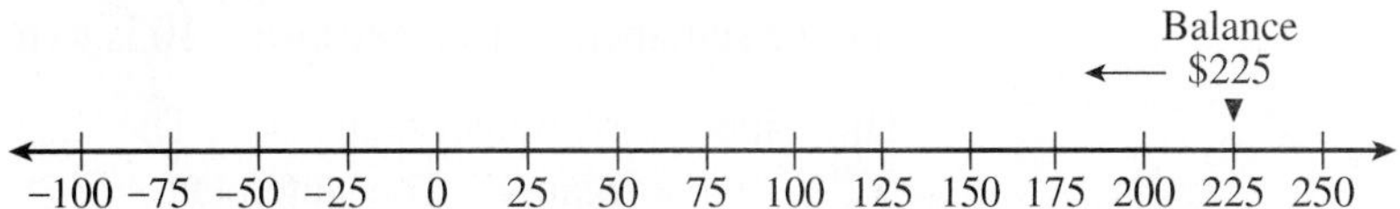

a. During the first week, you write checks totaling \$125. What is the new balance in your checking account after week 1? Place that balance on the number line above.

b. During week 2, your checks added up to \$85. Compute the amount left in your checking account after 2 weeks and place that amount on the number line.

c. During week 3, you wrote a check for \$15. What is the new balance in your checking account after 3 weeks? Place that amount on the number line.

d. During the last week in the month your car is towed and you must write a check for \$50 to get it back. What is the new balance in your checking account after 4 weeks? Place that amount on the number line.

Comparing Integers

Sometimes it is important to decide if one integer is greater than or less than another integer. The number line is one way to compare integers. Recall that on a number line as you move from *left to right* integers *increase* in value.

7. a. Which is warmer: -10°F or -16°F? Use a number line to help explain your answer.

b. Which checking account balance would you prefer, $-\$60$ or $-\$100$? Explain.

c. Which is closer to sea level, a depth of -20 feet or -80 feet? Explain.

d. On a number line the larger number is to the _______ (right/left) of the smaller number.

Symbols for Comparing Integers

The symbol for "less than" is $<$. The statement $3 < 5$ is read as "3 is less than 5" or "3 is smaller than 5". The statement $-10 < -2$ is read as "-10 is less than -2", and the statement is true because -10 is to the left of -2 on the number line.

The symbol for "greater than" is $>$. The statement $6 > 4$ is read "6 is greater than 4" or "6 is larger than 4". The statement $-7 > -12$ is read "-7 is greater than -12", and the statement is true because -7 is to the right of -12 on the number line.

The symbols $<$ and $>$ are called **inequality symbols.**

8. Identify which of the following statements are true and which are false.

a. $-3 > -5$

b. $20 < -100$

c. $0 > -40$

d. $-30 < -50$

Absolute Value on the Number Line

The absolute value of a number is a useful idea in working with positive and negative numbers. You may have noticed that on the number line an integer has two parts: its distance from zero and its direction (to the right or left of 0). For example, the 7 in -7 represents the number of units that -7 is from 0 and the dash ("$-$") represents the direction of -7 to the left of 0. Note that $+7$ or 7 is also 7 units from 0, but to the right.

9. a. What is the increase in temperature (in °F) if the temperature moves from 0°F to 8°F? How many degrees does the temperature decrease if the temperature moves from 0°F to -10°F?

b. Which is further away from a \$0 balance in a checking account, \$150 or $-$\$50? Explain.

c. How far from sea level (0 feet altitude) is an altitude of 200 feet? How far from sea level (0 feet altitude) is an altitude of -20 feet?

On the number line, the **absolute value** of a number is represented by the distance the number is from zero. Since distance is always considered positive or zero, the absolute value of a number is *always* positive or zero.

For example, the absolute value of -5 is 5 since -5 is 5 units from zero on the number line. The absolute value of 8 is 8 since 8 is 8 units from zero on the number line.

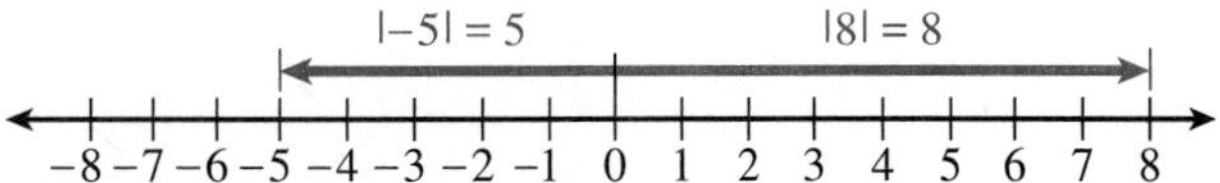

Note that this distance from zero is always positive or zero whether the number is to the left or the right of zero on the number line.

10. Determine the absolute value of each of the following integers.

a. 47

b. -30

c. -64

d. 56

To represent the absolute value of a number in symbols, enclose the number in vertical lines. For example, $|7|$ represents the absolute value of 7, which is 7; $|-32|$ represents the absolute value of -32, which is 32.

11. Determine the following.

a. $|-12|$

b. $|127|$

c. $|0|$

12. a. What number has the same absolute value as -26?

b. What number has the same absolute value as 45?

c. Is there a number whose absolute value is zero?

d. Can the absolute value of a number be negative?

Two numbers that are the same distance from 0 are called **opposites**. For example, in Problem 12a and b, -26 and 26 are opposites and 45 and -45 are opposites. 0 is its own opposite.

On the number line, opposites are two numbers that are the same distance from zero but are on opposite sides of zero. The numbers 12 and -12 are opposites because each number is 12 units from zero, but they are on opposite sides of zero.

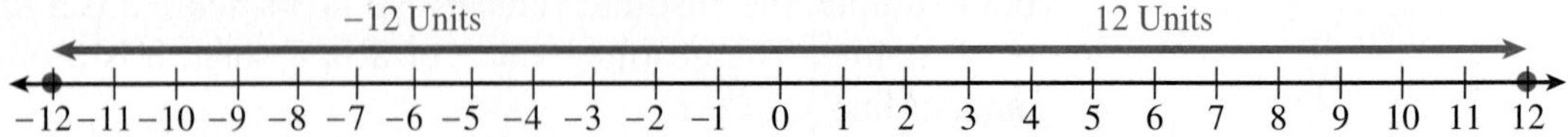

13. a. What number is the opposite of 22?

b. What number is the opposite of -15 ?

c. What number is the opposite of $|-7|$?

SUMMARY
ACTIVITY 3.1

1. **Negative numbers** are numbers that are less than zero.
2. The set of **integers** includes all the whole numbers (the counting numbers and zero) and their opposites (the negatives of the counting numbers).
3. If $a < b$, then a is to the left of b on a number line. If $a > b$, then a is to the right of b on a number line.
4. On the number line, the **absolute value** of a number a, written $|a|$, is the distance of a from zero. The absolute value of a number is always positive or zero.
5. Two numbers that are the same distance from zero on the number line but are on opposite sides of zero are called **opposites**. 0 is its own opposite.

EXERCISES
ACTIVITY 3.1

1. The ground floor of an apartment building is numbered zero.

a. Describe the floor numbered -2.

b. If you are on the floor numbered -3, would you take the elevator up or down to get to the floor labeled -1?

Exercise numbers appearing in color are answered in the Selected Answers appendix.

2. Express the quantities in each of the following as an integer.

   a. The Dow-Jones stock index lost 120 points today.

   b. The scuba diver is 145 feet below the surface of the water.

   c. A deposit of $50 into your checking account.

   d. The Oakland Raiders lost 15 yards on a penalty.

   e. A withdrawal of $75 from your checking account.

3. Write $<$ or $>$ between each of the following to make the statement true.

   a. $7 \quad 5$ b. $-5 \quad 4$ c. $-10 \quad -15$

   d. $0 \quad -2$ e. $-4 \quad -3$ f. $-50 \quad 0$

4. Determine the value of each of the following.

   a. $|4|$ b. $|-13|$ c. $|32|$

   d. $|-7|$ e. $|0|$

5. a. What number is the opposite of -6?

   b. What number is the opposite of 8?

6. a. Which is warmer: -3°F or 0°F?

   b. Which is farther below sea level: -20 feet or -100 feet?

7. Data on exports and imports by countries around the world are often shown by graphs for comparison purposes. The following graph is one example. The number line in the graph represents net exports in billions of dollars for six countries. Net exports are obtained by subtracting total imports from total exports; a negative net export means the country imported more goods than it exported.

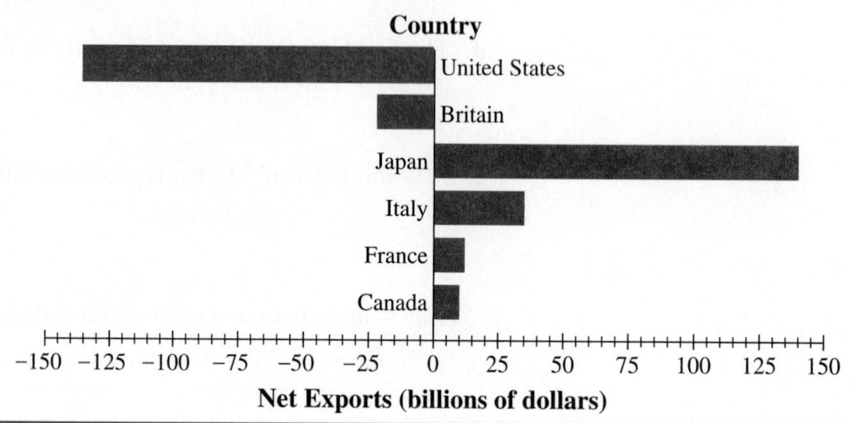

a. Estimate the net amount of exports for Japan. Is your answer a positive or negative integer? Explain what this number tells you about the imports and exports of Japan.

b. Estimate the net amount of exports for the United States. Is your answer a positive or negative integer? Explain what this number tells you about the imports and exports of the United States.

c. Determine the absolute value of the net amount of exports for Britain.

d. What is the opposite of the net amount of exports for Canada?

ACTIVITY 3.2

Maintaining Your Balance

OBJECTIVES

1. Add and subtract integers.
2. Identify properties of addition and subtraction of integers.

This activity guides you to learn the addition and subtraction of integers using concise and efficient rules. Through practice, you will make these rules your own.

Adding Integers

Suppose you use your checking account to manage all your expenses. You keep very accurate records. There are times when you must write a check for an amount that is greater than your balance. When you record this check, your new balance is represented by a negative number.

CHECK NUMBER	DATE	DESCRIPTION	SUBTRACTIONS (−)		ADDITIONS (+)		BALANCE	
							64	
	3/2/07	Paycheck			120			
381	3/4/07	Car repairs	167					
382	3/15/07	Clothes	47					

1. a. You deposit a paycheck for \$120 from your part-time job. If the balance in your account before the deposit was \$64, what is the new balance? Write the new balance in the check record above.

b. Determine the balance after making your car repair payment and record it in the check record above.

2. a. After writing a check in the amount of \$47 for clothing, you realize that your account is overdrawn by \$30. Explain why.

b. What integer would represent the balance at this point? Record that integer in the check record above.

c. Because you have a negative balance in your account, you decide to borrow \$20 from a friend for expenses until you get an additional paycheck. What integer represents what you owe your friend?

d. What is the total of your checking account balance and the money you owe your friend?

In Problems 1 and 2, the integers being added had the same sign.

Problem 1: $64 + 120 = 184$

Problem 2: $-30 + (-20) = -50$

Note that when you add two negative numbers such as -30 and -20, the sign of the answer is negative. The 50 in the answer is obtained by neglecting the sign of -30 and -20 and simply adding 30 to 20. This is the same as adding the absolute values of -30 and -20.

This leads to the following rule.

Rule 1: To add two numbers with the same sign, add the absolute values of the numbers. The sign of the sum is the same as the sign of the numbers being added.

3. Combine the following integers.

a. $-18 + (-26)$ **b.** $17 + 108$ **c.** $-48 + (-9)$

4. a. You finally get paid. You cash the check for \$120 and give \$10 to your friend. Now, what do you owe your friend?

b. You deposit the remaining \$110 into your checking account. What is the new balance? Recall that you were overdrawn by \$30.

In Problem 4a and 4b, the integers being added have different signs.

Problem 4a: $-20 + 10 = -10$

Problem 4b: $-30 + 110 = 80$

Note that when you add two numbers with different signs, such as -30 and 110, the sign of the result will be the same as the sign of the number with the greater absolute value. Since $|110| = 110$, $|-30| = 30$, and $110 > 30$, the sign of the result will be positive. To calculate the answer, ignore the signs of the numbers and subtract the smaller absolute value from the larger one. Therefore, $-30 + 110 = 110 - 30 = 80$.

5. Explain why the sum $-20 + 10$ is -10.

6. Add the following integers, using the observations just discussed.

a. $40 + (-10)$ **b.** $-60 + 85$

c. $-50 + 30$ **d.** $25 + (-35)$

The discussion above leads to the following rule.

Rule 2: To add two integers with different signs, determine their absolute values and then subtract the smaller from the larger. The sign of the result is the sign of the number with the larger absolute value.

7. Add the following integers.

a. $4 + 8$ **b.** $4 + (-8)$ **c.** $-3 + (-6)$

d. $10 + (-8)$ **e.** $(-7) + (4)$ **f.** $-3 + (+8)$

Subtracting Integers

The current balance in your checking account is \$195. You order merchandise for \$125 and record this amount as a debit. That is, you enter the amount as −\$125. The check was returned because the merchandise was no longer available. Adding the \$125 back to your checking account balance of \$70 brings the balance back to \$195.

$$70 + 125 = \$195$$

You also figured that if you had subtracted the debit from the \$70, you would get the same result. That means you reasoned that

$$70 - (-125) = 70 + 125 = \$195.$$

Of course, you prefer the simpler method of balancing your checkbook by adding the opposite of the debit, that is, adding the credit to your account. But subtracting the debit shows *an important mathematical fact.* The fact is that *subtracting a negative number is always equivalent to adding its opposite.* In the expression $70 - (-125)$, the term $-(-125)$ is replaced by $+125$ and $70 - (-125)$ becomes $70 + 125$.

8. Evaluate the following.

a. $7 - (-3)$ **b.** $-8 - (-12)$ **c.** $1 - (-7)$ **d.** $-5 - (-2)$

9. A friend had \$150 in her checking account and wrote a check for \$120. She calculated her checkbook balance as \$150 − (+ \$120) = \$30. You pointed out that she might as well have written \$150 − \$120 = \$30. Explain why.

10. Evaluate the following.

a. $11 - (+7)$ **b.** $5 - (+8)$ **c.** $-7 - (+9)$ **d.** $-3 - (+2)$

The discussion and problems above lead to the following rule.

Rule 3: To subtract an integer b from another integer a, change the subtraction to the addition of integer b's opposite. Then follow either rule 1 or rule 2 for adding integers.

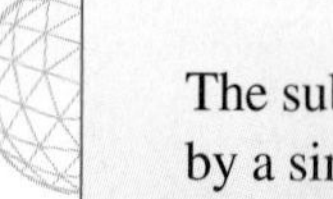

PROCEDURE

The subtraction sign and the sign of the number being subtracted can be replaced by a single sign.

$$a - (+b) = a - b$$

$$a - (-b) = a + b$$

For example, $-6 - (+2) = -6 - 2 = -8$ and $9 - (-4) = 9 + 4 = 13$.

11. Perform the following subtractions and state what rule(s) you used in each case.

a. $-8 - 2$ **b.** $10 - (-3)$

c. $-4 - (-6)$ **d.** $5 - 15$

e. $-5 - (+8)$ **f.** $+2 - (+10)$

Adding or Subtracting More Than Two Integers

When adding or subtracting more than two numbers, add or subtract from left to right.

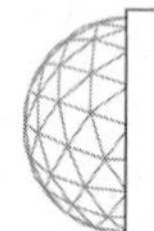

EXAMPLE 1 ***To compute 4 + 6 − 3, first add 4 + 6 = 10. Then subtract 3 from 10 to obtain the answer of 7.***

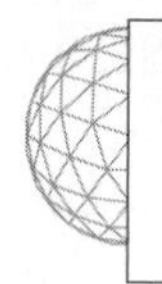

EXAMPLE 2 $\mathbf{-3 - 7 + 5 = -10 + 5 = -5}$

12. Perform the following calculations. Verify with your calculator.

a. $-3 + 7 - 5$ **b.** $5 + 3 + 10$ **c.** $-2 - 4 + 5$

d. $-4 - 2 - 4$ **e.** $3 - 7 - 6$ **f.** $-6 + 7 + 3 - 5$

Properties of Addition and Subtraction of Integers

13. a. Add: $9 + (-4)$

b. Add: $-4 + 9$

c. What property of addition is illustrated by comparing parts a and b?

14. a. Subtract: $5 - (-2)$

b. Subtract: $-2 - (+5)$

c. Is the operation of subtraction commutative? Explain.

15. a. Add the integers 5, -3, -9 by first adding 5 and -3, and then adding -9.

b. Add the integers given in part a: 5, -3, -9, but this time first add -3 and -9, and then add 5 to the sum.

c. Does it matter which two integers you add together first?

d. What property is being demonstrated in part c?

SUMMARY
ACTIVITY 3.2

1. Rules for adding integers.

To add two numbers with the same sign, add the absolute values of the numbers. The sign of the sum is the same as the sign of the numbers being added.

For example:

$$7 + 9 = 16$$
$$-6 + (-5) = -11$$

To add two integers with different signs, determine their absolute values and then subtract the smaller from the larger. The sign of the sum is the sign of the number with the larger absolute value.

For example:

$$-5 + 9 = 4$$
$$10 + (-13) = -3$$

2. Rules for subtracting integers.

To subtract an integer b from an integer a, change the subtraction to the addition of the opposite of b, then follow the rules for adding integers.

For example:

$$5 - (-7) = 5 + 7 = 12$$
$$-5 - (+3) = -5 + (-3) = -8$$

3. Procedure for simplifying the subtraction of an integer from another integer.

The subtraction sign and the sign of the number that follows it can be replaced by a single sign.

$$a - (+b) = a - b$$
$$a - (-b) = a + b$$

For example,

$$-3 - (+6) = -3 - 6 = -9 \quad \text{and} \quad 7 - (-8) = 7 + 8 = 15$$

4. Commutative property of addition.

For all integers a and b, $a + b = b + a$.

For example:

$$6 + (-9) = (-9) + 6, \quad \text{since } -3 = -3$$

Note: The commutative property is not true in general for subtraction.

For example:

$$3 - 4 \neq 4 - 3, \quad \text{because } -1 \neq 1$$

5. Associative property of addition.

For all integers a, b, and c, $(a + b) + c = a + (b + c)$.

For example:

$$(4 + 5) + (-3) = 4 + (5 + (-3))$$
$$9 - 3 = 4 + 2$$
$$6 = 6$$

Note: The associative property is not true in general for subtraction.

For example:

$$(7 - 5) - 1 \neq 7 - (5 - 1)$$
$$2 - 1 \neq 7 - 4$$
$$1 \neq 3$$

EXERCISES
ACTIVITY 3.2

Perform the indicated operation in Exercises 1–20.

1. $5 + (-7)$
2. $13 + (-17)$
3. $-32 + (+19)$
4. $-3 + (-2)$
5. $-21 - (-18)$
6. $-11 + 17$
7. $8 + (-8)$
8. $-54 - (+72)$
9. $3 + (-4) - 10$
10. $-5 - 6 + 3$
11. $-4 - 5 - 3$
12. $7 - (-3) - 7$
13. $4 + (-6) + 6$
14. $9 + 8 + (-3)$
15. $-4 - (-5) + (-6)$
16. The temperature increased by 8°F from −10°F. What is the new temperature?
17. The temperature was −7°F in the afternoon. That night, the temperature dropped by 5°F. What is the nighttime temperature?
18. The temperature increased 4°F from −1°F. What is the new temperature?
19. The temperature was 6°F in the morning. By noon, the temperature rose by 5°F. What was the noon temperature?

Exercise numbers appearing in color are answered in the Selected Answers appendix.

20. You began a diet on November 1 and lost 5 pounds. Then you gained 2 pounds over the Thanksgiving break. By how much did your weight change in November?

21. While descending into a valley in a hot air balloon, your elevation decreased by 500 feet from an elevation of -600 feet (600 feet below sea level). What is your new elevation?

22. a. You are a first-year lifeguard at Rockaway Beach and earn \$403 per week. On August 28 you deposited a week's pay into your checking account. The balance of your account before the deposit was \$39. What is the new balance?

b. Your \$460 monthly rent is due on September 1. So on September 1 you write and record a \$460 check. What is the new balance?

c. On September 4 you deposit your \$403 paycheck. What is the new balance?

d. You write a \$400 check on September 5 for your first installment on your college bill for the fall term. What is your new balance?

e. You write a check for groceries on September 10 in the amount of \$78. What is your new balance?

f. You get your last lifeguard paycheck on September 11. What is your new balance?

23. The highest point in Asia is at the top of Mount Everest, which is 8850 meters above sea level. The lowest point in Asia is the Dead Sea at 408 meters below sea level.

a. Write the elevation level of the Dead Sea as a signed number.

b. What is the difference between the highest and the lowest points in Asia?

24. Profits and losses in millions of dollars per quarter over a 2-year period for a telecommunications company are shown in the bar graph. Determine the total profit or loss for the company over the 2-year period.

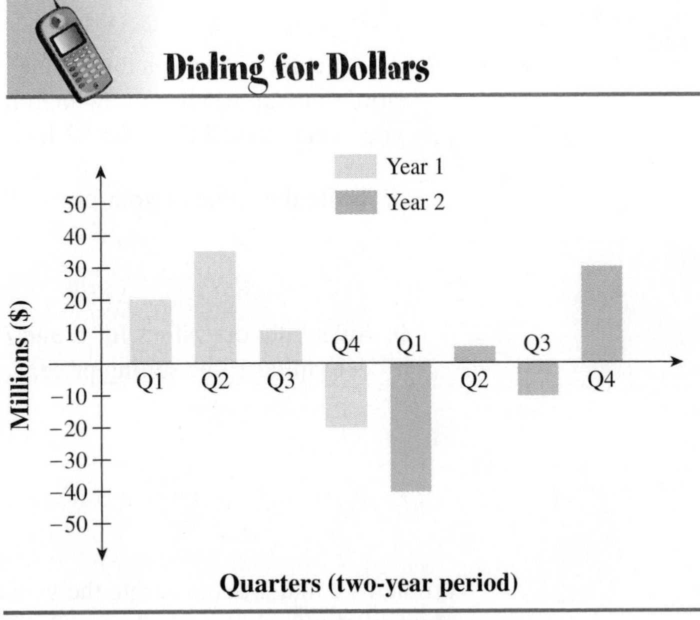

ACTIVITY 3.3

What's the Bottom Line?

OBJECTIVES

1. Write formulas from verbal statements.
2. Evaluate expressions in formulas.
3. Solve equations of the form $x + b = c$ and $b - x = c$.
4. Solve formulas for a given variable.

You run a retail shop in a tourist town. To determine the selling price of an item, you add the profit you want to what the item cost you. For example, if you want a profit of $4 on a decorated coffee mug that cost you $3, you would sell the mug for $7.

1. a. Write a formula for the selling price of an item in terms of your cost for the item and the profit you want. In your formula, let C represent your cost; P, the profit on the item; and S, its selling price.

   b. You make a profit of $8 on a wall plaque that costs you $12. Use your formula from part a to determine the selling price.

2. A particular style of picture frame has not sold well and you need space in the store to make room for new merchandise. The frames cost you $14 each and you decide to sell them for $2 less than your cost.

   a. Write the value of your profit, P, as an integer.

   b. Substitute the values for C and P in the equation $S = C + P$ and determine S, the selling price.

Problems 1 and 2 demonstrate the general method of determining an output from inputs that was shown in Chapter 2. In the equation $S = C + P$, *the inputs are* C and P in the expression on the right side of the equation, and S is the output variable. Replacing C and P with their values in the expression and evaluating their sum determines the value of S.

3. A wristwatch costs you $35 and you sell it for $60. Substitute these values of C and S in the equation $C + P = S$ and solve the equation for P, the profit.

In Problem 3, the unknown value, P, is part of the expression on the left side of the equation. The output value of the equation, S, is known. As shown in Chapter 2, determining the unknown input value in the equation is called *solving the equation*.

4. You make a profit of \$11 on a ring you sell for \$25. Replace P and S with these values in the equation $C + P = S$ and solve for C, the cost.

5. You sell a large candle for \$16 that cost you \$19.

a. Use the formula $C + P = S$ to determine your profit or loss.

b. Comment on your profit for this sale.

To solve the equations in Problems 3–5, you subtracted a given number from each side of the equation to isolate the unknown variable. This is the same as adding the opposite of the given number to each side.

6. Substitute the given values into the equation $c = a + b$ and then determine the value of the remaining unknown variable.

a. $a = 5, b = -17$

b. $a = 12, c = -30$

c. $b = -4, c = -9$

d. $b = 7, a = -23$

e. $c = 49, a = -81$

f. $a = -34, b = -55$

7. Substitute the given values into the equation $c = a - b$ and then determine the value of the remaining unknown variable.

a. $a = 12, b = -19$

b. $b = 7, c = -42$

c. $c = 36, b = 40$

d. $a = -17, b = 29$

Solving Formulas for a Given Variable

The process for solving formulas for a given variable is the same as the process for solving an equation. For example, to solve the formula $S = C + P$ for P, subtract C from both sides of the equation and simplify as demonstrated in Example 1.

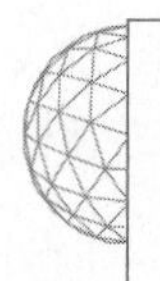

EXAMPLE 1

$$S = C + P$$
$$S - C = C - C + P$$
$$S - C = 0 + P$$
$$S - C = P$$
$$P = S - C$$

8. The formula $r_1 + r_2 = r$ is used in electronics. Solve for r_1.

9. The formula $p + q = 1$ is used in the study of probability. Solve for q.

10. The formula $P = R - C$ is used in business, where P represents profit; R, revenue or money earned; and C, cost of doing business. Solve for revenue, R.

To solve for cost, C, in the formula $P = R - C$ given in Problem 10, you add C to each side of the equation, simplify, and continue to solve, as demonstrated in Example 2.

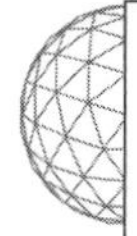

EXAMPLE 2 *Solve $P = R - C$ for C.*

SOLUTION

$$P + C = R - C + C$$
$$P + C = R$$
$$-P + P + C = R - P$$
$$C = R - P$$

11. Solve for x in the equation $y = b - x$.

SUMMARY
ACTIVITY 3.3

1. To solve for x in formulas or equations of the forms $c = x + b$ or $c = b + x$, add the opposite of b to both sides of the equation to obtain $x = c - b$.

2. To solve the equation $b - x = c$ for x, add x to both sides of the equation to obtain $b = c + x$. Then subtract c from each side to obtain $b - c = x$.

EXERCISES
ACTIVITY 3.3

1. When you began your diet, you weighed 154 pounds. After 2 weeks, you weighed 149 pounds. You can express the relationship between your initial weight, final weight, and the change between them by the equation

$$154 + x = 149,$$

where x represents your change in weight in the 2 weeks.

a. Solve the equation for x.

Exercise numbers appearing in color are answered in the Selected Answers appendix.

b. Was your change in weight a loss or a gain? Explain.

2. a. Another way to write an equation expressing your change in weight during the first 2 weeks of your diet is to subtract the initial weight, 154, from your final weight, 149. Then the equation is $x = 149 - 154$, where x represents the change in weight. Determine the value of x.

b. Write a formula to calculate the amount a quantity changes when you know the final value and the initial value. Use the words *final value*, *initial value*, and *change in quantity* to represent the variables in the formula. (*Hint:* Use the equation from part a as a guide.)

3. Suppose you lost 7 pounds during August. At the end of that month, you weighed 139 pounds. Let I represent your initial weight at the beginning of August.

a. Write an equation representing this situation. (*Hint:* Use your formula from Problem 2b as a guide.)

b. Solve this equation for I to determine your weight at the beginning of the month.

4. a. The following table gives information about your diet for a 6-week period. For each week, write an equation to solve for the unknown amount using the variables indicated. Enter the equations into the table.

Losing Proposition

Week	1	2	3	4	5	6
Initial Weight, *I*	154			150	146	144
Ending Weight, *E*		149	150	146		
Weight Change, *C*	−3	−2	1		−2	−5
Equation						

b. Solve each equation in part a and complete the table. State whether you are evaluating an expression or solving the equation.

5. The sale price, S, of an item is the difference between the regular price, P, and the discount, D.

a. Write a formula for S in terms of P and D.

b. If a video game is discounted \$8 and sells for \$35, use the formula in part a to determine the regular price.

6. You write a check for \$37 to pay for a laboratory manual that you need for chemistry class.

a. If the new balance in your checking account is \$314, write an equation to determine the original balance.

b. Solve the equation to determine the original balance.

7. Solve each of the following formulas for the given variable.

   a. $K = C + 273$, for C

   b. $S = P - D$, for D

   c. $A = B - C$, for B

8. In each of the following, let x represent an integer. Then translate the given verbal expression into an equation and solve for x.

   a. The sum of an integer and 10 is -12.

   b. 9 less than an integer is 16.

   c. The result of subtracting 17 from an integer is -8.

   d. If an integer is subtracted from 10, the result is 6.

ACTIVITY 3.4

Riding in the Wind

OBJECTIVES

1. Translate verbal rules into equations.
2. Determine an equation from a table of values.
3. Use a rectangular coordinate system to represent an equation graphically.

Windchill

Bicycling is enjoyable in New York State all year around, but it can get very cold in the winter. If there is a wind, it feels even colder than the actual temperature. This effect is called the windchill temperature, or windchill for short. The following table gives the actual temperature, T, and the windchill, W, for a 10 mph wind. Both are given in °F.

Blowin' in the Wind

ACTUAL TEMPERATURE, T (°F)	−15	−10	−5	0	5	10
WINDCHILL, W, FOR A 10 mph WIND (°F)	−20	−15	−10	−5	0	5

1. **a.** Write a verbal rule that describes how to determine the windchill for a 10 mph wind, if you know the actual temperature.

   b. Translate the verbal rule in part a into an equation where T represents the actual temperature and W represents windchill.

   c. Use the equation in part b to determine the windchill, W, for a temperature of −3 °F.

   d. Use the equation in part b to determine the actual temperature, T, for a windchill of −17°F.

Rectangular Coordinate System Revisited

To obtain a graphical view of the windchill data, you can plot the values in the table above on a rectangular coordinate grid. Notice that the table contains both positive and negative values for the actual temperature and the windchill.

Until now in this textbook, the horizontal (input) axis and the vertical (output) axis contained only nonnegative values. To plot the windchill data, you need to extend each axis in the negative direction.

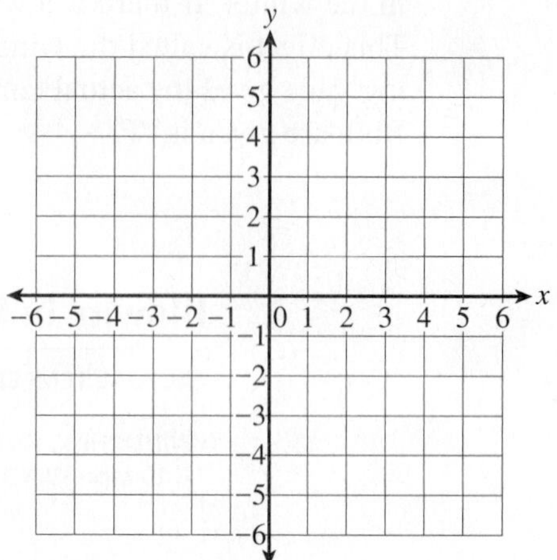

Now the grid has two complete number lines (axes) that intersect at a right (90°) angle at the **origin,** (0, 0). In this rectangular coordinate system, the two axes together divide the plane into four parts called **quadrants.** The quadrants are numbered as follows.

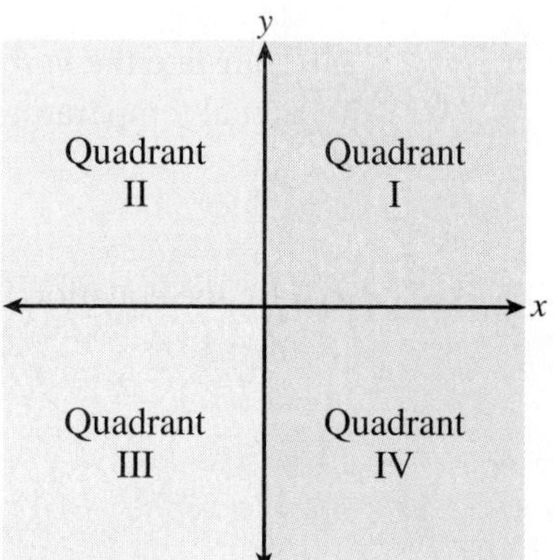

To locate the point with coordinates $(-5, -10)$, start at the origin (0, 0) and move 5 units left on the horizontal axis. Then, move 10 units down, parallel to the vertical axis. The point $(-5, -10)$ is located in quadrant III.

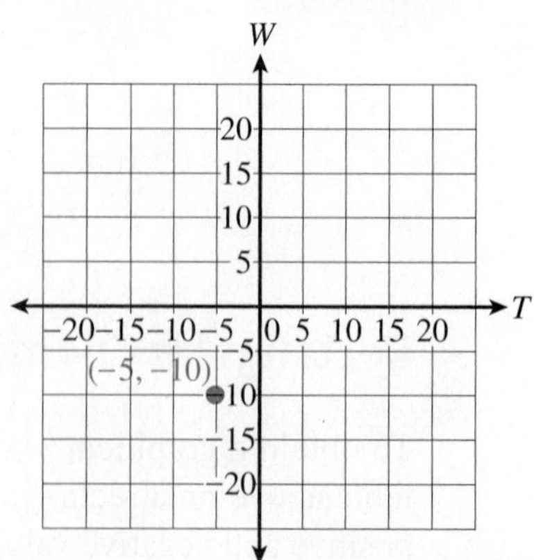

2. a. Plot the remaining values from the windchill table on page 175 on the preceding grid.

b. The points you graphed were determined by the equation $W = T - 5$. The points and their pattern are a graphical representation of the equation. What pattern do the graphed points suggest? Connect the points on the graph in the pattern you see.

3. Scale and label the following grid and plot the following points.

a. $(2, -5)$ **b.** $(-1, 3)$

c. $(4, 2)$ **d.** $(0, -4)$

e. $(5, 0)$ **f.** $(-2, -6)$

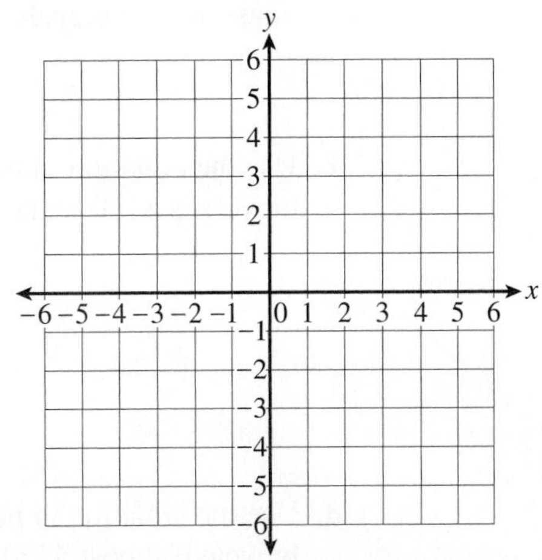

4. Without plotting the points, identify the quadrant in which each of the following points is located.

a. $(-12, 24)$ **b.** $(35, -26)$

c. $(56, 48)$ **d.** $(-28, -34)$

5. Identify the quadrant of each point whose coordinates have the given signs.

a. $(-, +)$ **b.** $(-, -)$

c. $(+, +)$ **d.** $(+, -)$

The Bicycle Shop

Your local sporting goods store sells a wide variety of bicycles priced from \$80 to \$500. The store sells bikes assembled or unassembled. The charge for assembly is \$20 regardless of the price of the bike.

6. a. What is the price of a \$100 bicycle with assembly?

b. What is the price of a $250 bicycle with assembly?

7. a. Write a verbal rule that describes how to determine the price of a bicycle with assembly.

b. Translate the verbal rule in part a into an equation where A represents the cost of the bicycle with assembly and U represents the price of the unassembled bicycle.

c. Use the equation in part b to determine the price, A, if the price of the bicycle is $160 without assembly.

d. Use the equation in part b to determine the price, U, of the unassembled bicycle if it cost $310 assembled.

8. Complete the following table.

PRICE WITHOUT ASSEMBLY, U	COST WITH ASSEMBLY, A
80	
	140
240	
	320
360	
460	480

9. a. Plot the values from the table in Problem 8 on an appropriately labeled and scaled coordinate system. Let P represent the input values on the horizontal axis. Let C represent the output values on the vertical axis.

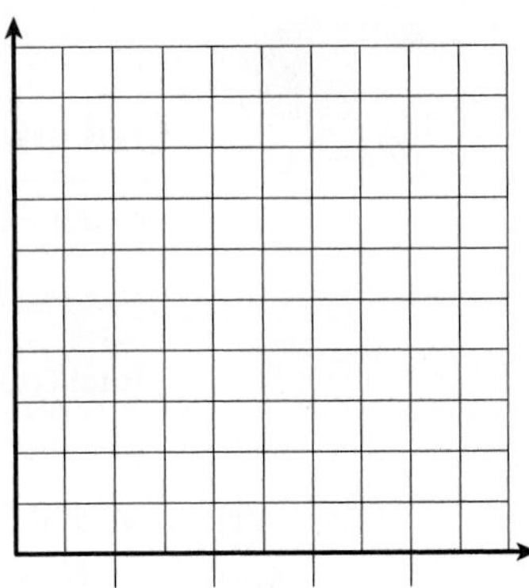

b. Connect the points on the graph to obtain a graphical representation of the equation in Problem 7b.

SUMMARY
ACTIVITY 3.4

Points on a rectangular coordinate system.

1. A point on a rectangular coordinate grid is written as an ordered pair in the form (x, y), where x is the input (horizontal axis) and y is the output (vertical axis).

2. Ordered pairs are plotted as points on a rectangular grid that is divided into four quadrants by a horizontal (input) axis and a vertical (output) axis.

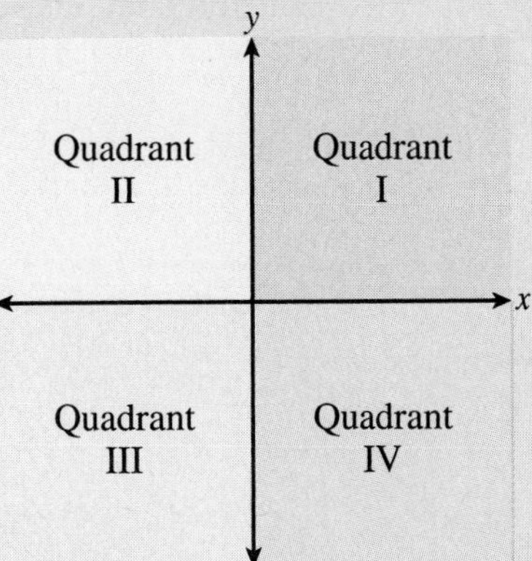

3. The sign of the coordinates of a point determine the quadrant in which the point lies.

x-COORDINATE	y-COORDINATE	QUADRANT
+	+	I
–	+	II
–	–	III
+	–	IV

EXERCISES
ACTIVITY 3.4

1. You make and sell birdhouses to earn some extra money. The following table lists the cost of materials and the total cost, both in dollars, for making the birdhouses.

Out on a Limb

Number of Birdhouses	1	2	3	4
Material Cost ($)	6	12	18	24
Total Cost ($)	8	14	20	26

a. Write a verbal rule that describes how to determine the total cost if you know the cost of the materials.

b. Translate the verbal rule in part a into an equation where T represents the total cost and M represents the cost of the materials.

c. Use the equation in part b to determine the total cost if materials cost $36.

d. Plot the values from the table above on an appropriately labeled and scaled rectangular coordinate system.

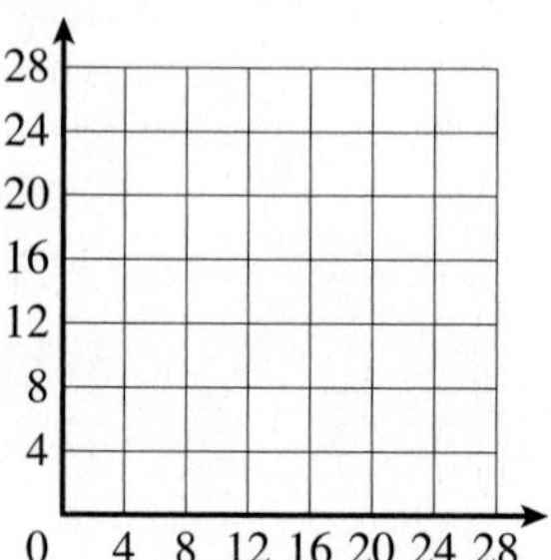

e. Connect the points on the graph to obtain a graphical representation of the equation in part d.

2. The sum of two integers is 3.

a. Translate this verbal rule into an equation where x represents one integer and y the other.

b. Use the equation in part a to complete the following table.

x	y
4	
	−3
−3	
	4
0	
	1

c. Plot the values from the table in part b on an appropriately labeled and scaled coordinate system.

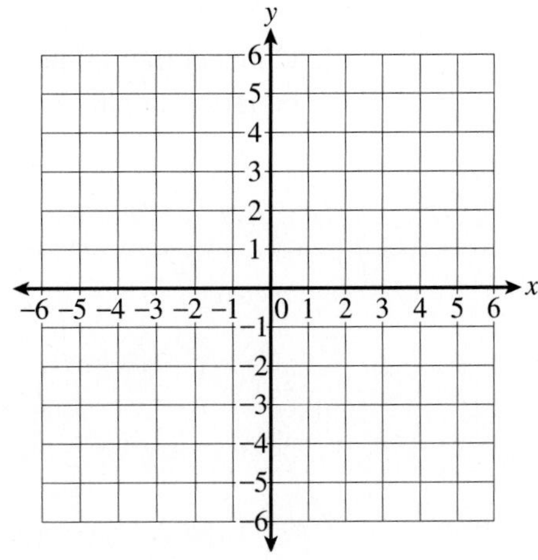

d. Connect the points on the graph to obtain a graphical representation of the equation in part a.

3. The difference of two integers is 5.

a. Translate this verbal rule into an equation where x represents the larger integer and y the smaller.

b. Use the equation in part a to complete the following table.

x	y
5	
	−6
−3	
	1
0	
	−1

c. Plot the values from the table in part b on an appropriately labeled and scaled coordinate system.

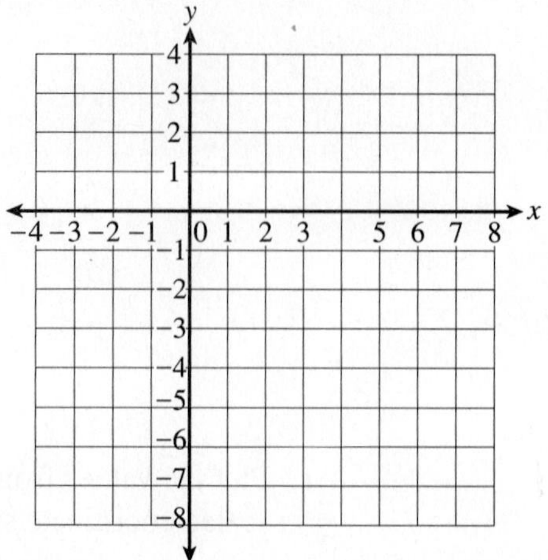

d. Connect the points on the graph to obtain graphical representation of the equation in part a.

CLUSTER 1 What Have I Learned?

1. Which number is always greater, a positive number or a negative number? Give a reason for your answer.

2. A number and its opposite are equal. What is the number?

3. Two numbers, x and y, are negative. If $|x| > |y|$, which number is smaller? Give an example to illustrate.

4. In each of the following, fill in the blank with a word that makes the statement true.

   a. When you add two positive numbers, the sign of the answer is always ________.

   b. When you add two negative numbers, the sign of the answer is always ________.

   c. The absolute value of zero is ________.

   d. When you subtract a negative number from a positive number, the answer is always ________.

   e. When you subtract a positive number from a negative number, the answer is always ________.

5. Describe in your own words how to add a positive number and a negative number. Use an example to help.

6. Describe in your own words how to add two negative numbers. Use an example to help.

Exercise numbers appearing in color are answered in the Selected Answers appendix.

7. Describe in your own words how to subtract two positive numbers. Use an example to help.

8. Describe in your own words how to subtract a positive number from a negative number. Use an example to help.

9. Describe in your own words how to subtract a negative number from a positive number. Use an example to help.

10. Describe in your own words how to subtract two negative numbers. Use an example to help.

11. Demonstrate how you would draw a rectangular coordinate grid to plot the following points: $(8, 90)$, $(6, -50)$, $(-4, 60)$, $(0, 0)$, $(-5, -50)$, $(-7, 0)$.

12. You process concert ticket orders for a civic center in your hometown. There is a \$3 processing fee for each order. Write a formula to represent the calculations you would do to determine the total cost of an order. State what the variables are in this situation and choose letters to represent them.

CLUSTER 1 How Can I Practice?

1. Determine the opposite of each of the following.

a. 3 b. -5 c. 0

2. Evaluate.

a. $|-13|$ b. $|15|$ c. $-|-39|$ d. $-|39|$

3. Calculate.

a. $15 + 36$

b. $-21 + 9$

c. $18 + (-35)$

d. $-5 + (-21)$

e. $-73 + 38 + (-49)$

f. $-21 + 45 + (-83) + (-5)$

g. $-18 + 12 + 69 + 32 + (-56)$

h. $14 + (-12) + (-25) + 7 + (-74)$

i. $23 + (-45) + (-51) + 82$

j. $-64 + (-49) + 28 + (-14) + 101$

4. Calculate.

a. $16 - 62$

b. $36 - 82$

c. $-14 - (-28)$

d. $-18 - (-6)$

e. $24 - (-48)$

f. $45 - 54$

g. $-39 - (-48) - 62$

h. $63 - (-72) - 101 - (-53) - 205$

i. $-81 - (-98) - 73 - (-49)$

Exercise numbers appearing in color are answered in the Selected Answers appendix.

5. Evaluate by performing the given operations.
   a. $-84 + 46 - (-98) - 108 - (-65)$
   b. $27 - 58 + (-72) - 85 - (-7)$
   c. $19 - 31 + (-42) + (-5)$
   d. $-51 + 27 - (-21) + 42$
   e. $-47 + 19 - (-12) + 71$
   f. $53 - 39 + (-88) - (-60)$
   g. $-89 - 11 + 45 - 61$

6. Evaluate the expression $x + y$, for each set of given values.
   a. $x = -18$ and $y = 7$
   b. $x = 13$ and $y = 8$
   c. $x = -21$ and $y = -12$
   d. $x = -17$ and $y = 5$

7. Evaluate the expression $x - y$, for each set of given values.
   a. $x = 14$ and $y = 6$
   b. $x = -16$ and $y = 9$
   c. $x = 23$ and $y = -62$
   d. $x = -27$ and $y = -32$

8. Solve the following equations and check your answer in the original equation.

a. $x + 21 = -63$

b. $x + 18 = 49$

c. $32 + x = 59$

d. $41 + x = -72$

e. $x - 17 = 19$

f. $x - 35 = 42$

g. $-22 + x = 16$

h. $-28 + x = 11$

9. Let x represent an integer. Translate the following verbal expression into an equation and solve for x.

a. The sum of an integer and 5 is 12.

b. 7 less than an integer is 13.

c. The difference of an integer and 8 is 6.

d. The result of subtracting 3 from an integer is 12.

10. The temperature in the morning was $-9°F$. The weather report indicated that by noon the temperature would be $4°F$ warmer. Determine the noontime temperature.

11. The temperature in the morning was $-13°F$. It was expected that the temperature would fall $5°F$ by midnight. What would be the midnight temperature?

12. The average temperature in July in town is $88°F$. The average temperature in January in the same town is $-5°F$. What is the change in average temperature from July to January?

13. The temperature on Monday was $-8°F$. By Tuesday the temperature was $-13°F$. What was the change in temperature?

14. The temperature at the beginning of the week was $-6°F$, and at the end of the week it was $7°F$. What was the change in the temperature?

15. **a.** Your favorite stock opened the day at \$29 per share and ended the day at \$23 per share. Let x represent the amount of your gain or loss per share. An equation that represents the situation is $29 + x = 23$. Solve for x.

b. The following table represents the beginning and ending values of your stock for 1 week. Complete the table indicating the amount of your gain or loss per share, each day.

Opening Value	23	21	21	18	20
Closing Value	21	21	18	20	23
Gain or Loss, x					

c. What is the total change in your stock's value for this week?

16. The following graph represents net exports in billions of dollars for six countries represented on the number line. (Net exports are obtained by subtracting total imports from total exports.)

The World of Exports

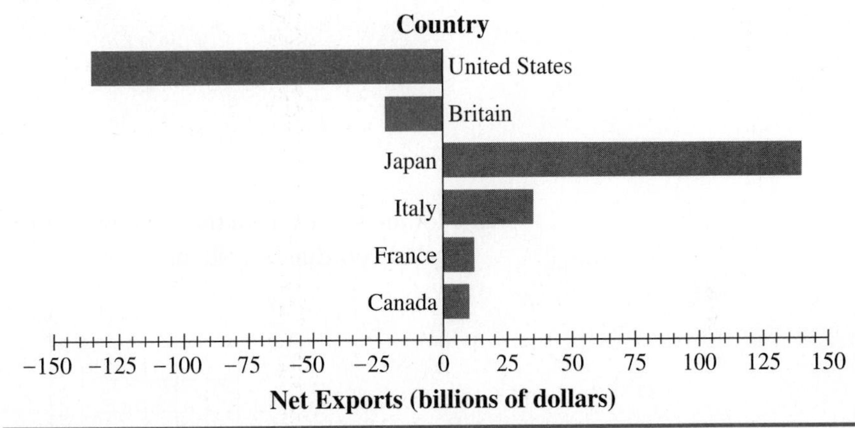

a. What is the difference between the net exports of Japan and Italy?

b. What is the difference between the net exports of Britain and France?

c. What is the total sum of net exports for the six countries listed?

17. A contestant on the TV quiz show *Jeopardy!* had \$1000 at the start of the second round of questions (double jeopardy). She rang in on the first two questions, incorrectly answering a \$1200 question but giving a correct answer to an \$800 question. What was her score after answering the two questions?

18. The difference of two integers is -3.

a. Translate this verbal rule into an equation, where x represents the smaller integer and y the larger.

b. Use the equation you wrote in part a to complete the following table.

x	y
0	
	0
5	
	−2
−2	
	6

c. Plot the values from the table in part b on an appropriately labeled and scaled coordinate system.

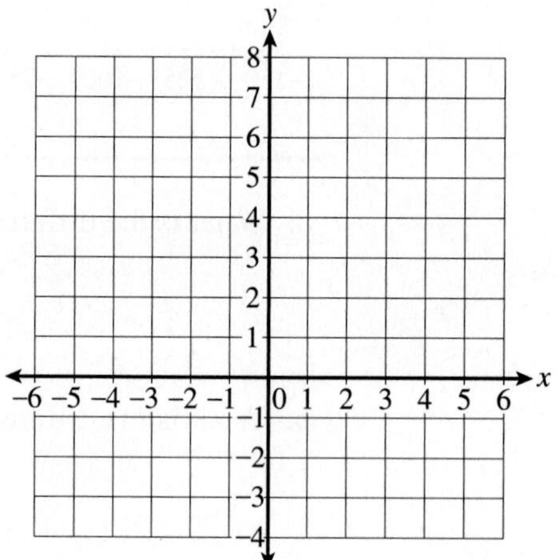

d. Connect the points on the graph to obtain the graphical representation of the equation in part a.

CLUSTER 2 Adding and Subtracting Fractions

ACTIVITY 3.5

Are You Hungry?

OBJECTIVES

1. Identify the numerator and the denominator of a fraction.
2. Determine the greatest common factor (GCF).
3. Determine equivalent fractions.
4. Reduce fractions to equivalent fractions in lowest terms.
5. Convert mixed numbers to improper fractions and improper fractions to mixed numbers.
6. Determine the least common denominator (LCD) of two or more fractions.
7. Compare fractions.

You decide to have some friends over to watch videos. After the first movie, you call the local sub shop to order three giant submarine sandwiches. When the subs are delivered, you pay the bill and everyone agrees to reimburse you, depending on how much they eat.

Because some friends are hungrier than others, you cut one sub into three equal (large) pieces, a second sub into six equal (medium) parts, and the third sub into twelve equal (small) parts.

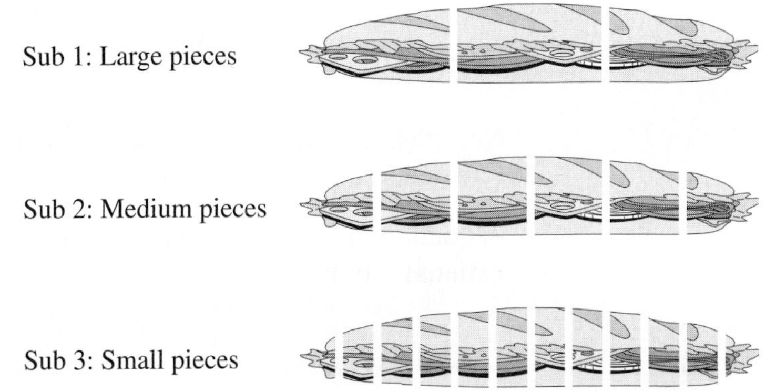

What Is a Fraction?

1. a. Because the first sub is divided into three large equal pieces, each piece represents a fractional part of the whole sub. Write a fraction that represents what part of the whole sub each large piece represents.

   b. What fraction of the sub does each medium piece represent? Explain.

   c. What fraction of the sub does each small piece represent? Explain.

DEFINITIONS

The bottom number of a fraction is called the **denominator.** The denominator indicates the total number of equal parts into which a whole unit is divided.

The top portion of a fraction is called the **numerator.** The numerator indicates the number of equal parts that the fraction represents out of the total number of parts.

2. a. If you eat five small pieces, what fractional part of the whole sub did you eat? Explain.

b. What is the denominator of this fraction? How many equal segments does the fraction indicate the sub is divided into?

c. What is the numerator of this fraction? What does this number represent?

Note that the fraction line, which separates the numerator and denominator, represents "out of" or "divided by."

A fraction that has an integer numerator and a nonzero integer denominator is a **rational number.**

DEFINITION

A **rational number** is a number that can be written in the form $\frac{a}{b}$, where a and b are integers and b is not zero. Every integer is also a rational number, since an integer can be written as $\frac{a}{1}$.

Note that the denominator of a fraction cannot be zero. It would indicate that the whole is divided into zero equal parts, which makes no sense.

3. a. Suppose a friend is very hungry and he eats all three large pieces of the sub. What fraction represents the amount he ate?

b. What is the value of the fraction in part a? Explain.

c. If $\frac{6}{6}$ represents how much of the medium sub that was eaten, what is the value of the fraction?

d. In general, what is the value of a fraction whose numerator and denominator are equal but non-zero?

4. a. If you did not eat any of the large pieces of the first sub, what fraction represents the amount of the first sub that you ate?

b. What is the value of the fraction in part a?

c. If $\frac{0}{12}$ represents the amount that you ate of the third sub, what does the numerator indicate?

d. What is the value of the fraction $\frac{0}{12}$?

e. In general, what is the value of a fraction whose numerator is zero?

Equivalent Fractions

5. As you are cutting the subs, you notice that two medium pieces placed end-to-end measure the same as one large piece. Therefore, $\frac{2}{6}$ of the sub represents the same portion as $\frac{1}{3}$.

a. How many small pieces represent the same portion as one large piece?

b. What fraction of the sub does the number of small pieces in part a represent?

In Problem 5, three different fractions were used to represent the same portion of a whole sub. These fractions, $\frac{1}{3}$, $\frac{2}{6}$, and $\frac{4}{12}$, are called **equivalent fractions.** They represent the same quantity.

> To obtain a fraction equivalent to a given fraction, multiply or divide the numerator and the denominator of the given fraction by the same non-zero number.

6. a. Divide the numerator and the denominator of $\frac{2}{6}$ by 2 to obtain an equivalent fraction.

b. Multiply the numerator and denominator of $\frac{1}{3}$ by 4 to obtain an equivalent fraction.

7. Write the given fraction as an equivalent fraction with the given denominator.

a. $\frac{3}{10} = \frac{?}{20}$

b. $\frac{4}{5} = \frac{?}{30}$

c. $\frac{7}{11} = \frac{?}{33}$

d. $\frac{8}{9} = \frac{?}{72}$

Reducing a Fraction to Lowest Terms

8. a. What whole number is a factor of both the numerator and denominator of $\frac{1}{3}$?

b. What is the largest whole number that is a factor of both the numerator and denominator of $\frac{8}{12}$?

c. What is the largest whole number that is a factor of both the numerator and denominator of $\frac{6}{8}$?

If the largest factor of both the numerator and denominator of a fraction is 1, the fraction is said to be in **lowest terms.** For example, $\frac{1}{3}$, $\frac{3}{4}$, and $\frac{4}{5}$ are written in lowest terms.

The **greatest common factor** (GCF) of two numbers is the largest factor common to both numbers. For example, the GCF of 8 and 12 is 4.

PROCEDURE

Reducing a Fraction to Lowest Terms Fractions that are not in lowest terms can be reduced to an equivalent fraction that is in lowest terms by dividing the numerator and denominator by their greatest common factor (GCF). For example, to reduce the fraction $\frac{8}{12}$ to an equivalent fraction in lowest terms, divide the numerator and denominator by 4.

$$\frac{8}{12} = \frac{8 \div 4}{12 \div 4} = \frac{2}{3}$$

9. Reduce the following fractions to equivalent fractions in lowest terms.

a. $\frac{6}{8}$

b. $\frac{6}{15}$

c. $\frac{3}{8}$

d. $\frac{15}{25}$

Mixed Numbers and Improper Fractions

10. a. All of the first (large pieces) and second (medium pieces) subs are eaten and only one small piece of the third sub (small pieces) is left. Represent the amount eaten as the sum of a whole number and a fraction.

b. Suppose all three subs were cut into twelve equal parts. How many twelfths were eaten if only one piece was left, as in part a? Write your answer in words and as a single fraction.

c. Is the sum in part a equivalent to the fraction in part b? Explain.

The sum $2 + \frac{11}{12}$ is more conveniently written as $2\frac{11}{12}$ and is called a **mixed number**.

The fraction $\frac{35}{12}$ is called an **improper fraction** because the numerator is greater than or equal to the denominator. An improper fraction and a mixed number that represent the same quantity are called **equivalent**.

DEFINITIONS

An **improper fraction** is a fraction where the absolute value of the numerator is greater than or equal to the absolute value of the denominator. A **mixed number** is a sum of an integer and a fraction.

It is often helpful to convert mixed numbers into improper fractions and vice versa.

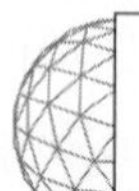

EXAMPLE 1 *Convert* $2\frac{11}{12}$ *to an improper fraction.*

SOLUTION

$$2\frac{11}{12} = 2 + \frac{11}{12} = \frac{2 \cdot 12}{1 \cdot 12} + \frac{11}{12} = \frac{24}{12} + \frac{11}{12} = \frac{35}{12}$$

Note in Example 1 that the integer 2 was rewritten as an equivalent fraction with denominator 12 so it could be added to the fractional part $\frac{11}{12}$. This observation leads to a shortcut method to convert a mixed number to an improper fraction:

$$2\frac{11}{12} = \frac{2 \cdot 12 + 11}{12} = \frac{35}{12}$$

PROCEDURE

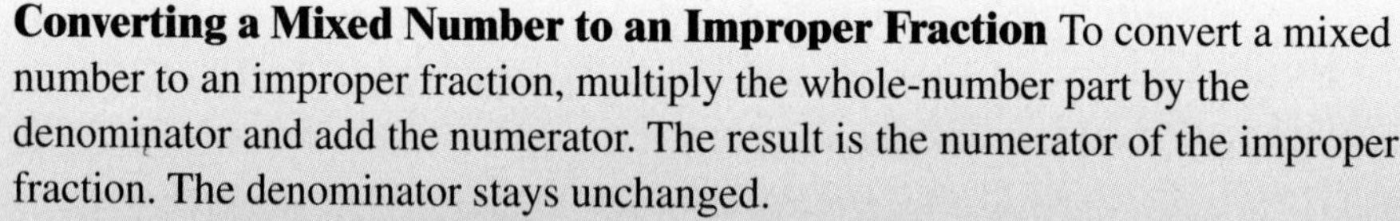

Converting a Mixed Number to an Improper Fraction To convert a mixed number to an improper fraction, multiply the whole-number part by the denominator and add the numerator. The result is the numerator of the improper fraction. The denominator stays unchanged.

EXAMPLE 2 *Convert* $\frac{35}{12}$ *to a mixed number, in lowest terms.*

SOLUTION

$$\begin{array}{r} 2 \\ 12\overline{)35} \\ -24 \\ \hline 11 \end{array}$$

$$\frac{35}{12} = 2 + \frac{11}{12} = 2\frac{11}{12}$$

PROCEDURE

Converting an Improper Fraction to a Mixed Number To convert an improper fraction to a mixed number, divide the numerator by the denominator. The quotient is the whole-number part of the mixed number, the remainder is the numerator of the fractional part, and the divisor is the denominator of the fractional part.

11. Write the following mixed numbers as improper fractions in lowest terms.

a. $1\frac{5}{8}$

b. $4\frac{9}{12}$

c. $5\frac{6}{10}$

12. Write the given improper fractions as mixed numbers in lowest terms.

a. $\frac{5}{2}$

b. $\frac{24}{10}$

c. $\frac{21}{7}$

d. $\frac{3}{3}$

Comparing Fractions: Determining Which Fraction Is Larger

13. Suppose one group of friends ate four medium pieces and another group ate ten small pieces.

a. What fractional part of the sub did the first group eat?

b. What fractional part of the sub did the second group eat?

c. Use the graphics display of the subs at the beginning of this activity to help determine which group ate more.

It is much easier to compare fractions that have the same denominator.

14. Write $\frac{4}{6}$ as an equivalent fraction with a denominator of 12.

It is easy to see that $\frac{10}{12}$ is larger than $\frac{8}{12}$ since they have the same denominator and $10 > 8$.

15. Which is the larger fraction, $\frac{5}{8}$ or $\frac{3}{4}$?

Problems 13, 14 and 15 show that two or more fractions are easily compared if they are expressed as equivalent fractions having the same (common) denominator. Then the fraction with the largest numerator is the largest fraction.

DEFINITIONS

A **common denominator** of two or more fractions is a number that is a multiple of each denominator. For example, $\frac{3}{4}$ and $\frac{5}{6}$ have common denominators of 12, 24, 36, ...

The **least common denominator (LCD)** of two fractions is the smallest common denominator. For example, the LCD of $\frac{3}{4}$ and $\frac{5}{6}$ is 12.

In general, to determine the LCD, identify the largest denominator of the fractions involved. Look at multiples of the largest denominator. The smallest multiple that is divisible by the smaller denominator(s) is the LCD.

EXAMPLE 3 ***Determine the least common denominator of $\frac{1}{6}$ and $\frac{3}{8}$.***

SOLUTION

In the case of $\frac{1}{6}$ and $\frac{3}{8}$, the larger denominator is 8.

The smallest multiple of the denominator 8 that is evenly divisible by the denominator 6 is the LCD, determined as follows:

$8 \cdot 1 = 8$ 8 is not a multiple of 6.

$8 \cdot 2 = 16$ 16 is not a multiple of 6.

$8 \cdot 3 = 24$ 24 is a multiple of 6, therefore 24 is the LCD.

16. a. Write $\frac{1}{6}$ and $\frac{3}{8}$ as equivalent fractions each with a denominator of 24.

b. Use the result from part a to compare $\frac{1}{6}$ and $\frac{3}{8}$.

17. Compare the following fractions and determine which is larger.

a. $\frac{3}{5}$ and $\frac{1}{2}$ **b.** $\frac{3}{8}$ and $\frac{1}{3}$ **c.** $\frac{1}{4}$ and $\frac{3}{16}$

SUMMARY ACTIVITY 3.5

1. The **greatest common factor (GCF)** of two or more numbers is the largest number that is a factor of each of the given numbers.
2. To write an **equivalent fraction,** multiply or divide both the numerator and denominator of a given fraction by the same number.
3. To reduce a fraction to an equivalent fraction in lowest terms, divide the numerator and the denominator by their GCD.
4. To convert a mixed number to an improper fraction, multiply the whole-number part by the denominator and add the numerator. The result is the numerator of the improper fraction; the denominator stays unchanged.
5. To convert an improper fraction to a mixed number, divide the numerator by the denominator. The quotient is the whole-number part of the mixed number and the remainder becomes the numerator of the fractional part; the denominator stays unchanged.
6. To find the **least common denominator (LCD)** of two or more fractions, identify the largest denominator and look at multiples of it. The smallest multiple that is divisible by the other denominator(s) is the LCD.
7. To compare two or more fractions, express them as equivalent fractions with a common positive denominator. The fraction with the largest numerator is the largest fraction.

EXERCISES ACTIVITY 3.5

Determine the missing number.

1. $\frac{2}{3} = \frac{?}{12}$ **2.** $\frac{3}{4} = \frac{?}{20}$

Reduce the given fraction to an equivalent fraction in lowest terms.

3. $\frac{9}{15}$ **4.** $\frac{8}{28}$

Exercise numbers appearing in color are answered in the Selected Answers appendix.

Use the inequality symbols < or > to compare the following fractions.

5. $\frac{5}{12}$ and $\frac{1}{3}$

6. $\frac{3}{7}$ and $\frac{2}{5}$

Reduce the following improper fractions to lowest terms and convert to mixed numbers.

7. $\frac{5}{3}$

8. $\frac{21}{14}$

Convert the following mixed numbers to improper fractions in lowest terms.

9. $3\frac{3}{8}$

10. $5\frac{6}{8}$

11. You and your sister ordered two individual pizzas. You ate $\frac{3}{4}$ of your pizza and your sister ate $\frac{5}{8}$ of her pizza. Who ate more pizza?

Your pizza

Your sister's pizza

12. You and your friend are painting a house. While you painted $\frac{2}{12}$ of the house, your friend painted $\frac{3}{18}$ of the house. Who painted more?

13. In a basketball game, you made 4 baskets out of 12 attempts. Your teammate made 3 baskets out of 8 attempts.

a. What fraction of your attempted shots did you make?

b. What fraction of your teammate's attempted shots did she make?

c. Who was the more accurate shooter?

14. An Ivy League college accepts 5 students for every 100 that apply. What fraction of applicants is accepted? Write your result in lowest terms.

15. You have two pieces of lumber. One piece measures 6 feet long and the second piece measures $\frac{25}{4}$ feet long.

a. Write $\frac{25}{4}$ as a mixed number.

b. Write 6 as an improper fraction with a denominator of 4.

c. Which piece is longer? Explain.

16. Measuring with rulers provides an opportunity to test your understanding of fractions. Use the following graphic of a ruler to answer the following questions about measuring the line segment drawn above the ruler.

a. Look at the ruler and determine the number of $\frac{1}{4}$-inch segments in one inch.

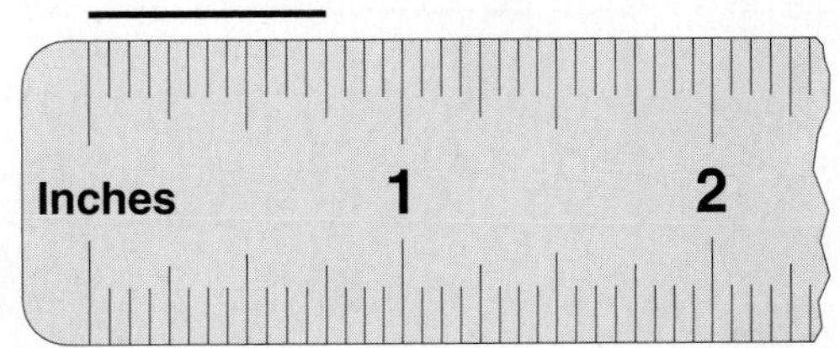

b. How many $\frac{1}{4}$-inch segments did the given line measure?

c. How long is the given line segment (in inches)?

d. Measure the given line segment in $\frac{1}{8}$-inch units.

e. Finally, measure the given line segment in $\frac{1}{16}$-inch units.

f. Explain why the three fractional answers in parts c, d, and e are equivalent.

g. Another line segment measures $\frac{2}{3}$ inch. Is it longer or shorter than the given line segment? Explain.

ACTIVITY 3.6

Food for Thought

OBJECTIVE

1. Add and subtract fractions and mixed numbers with the same denominators.

Many recipes include ingredients that are measured in terms of fractions and mixed numbers. For example, a cake recipe may call for $2\frac{1}{2}$ cups of flour, $\frac{1}{2}$ teaspoon of baking powder, etc. Sometimes you may want to change a recipe and will have to do some basic calculations with the fractions and mixed numbers.

Adding and Subtracting Fractions with Same Denominators

EXAMPLE 1 ***You are preparing a meal to celebrate your mother's birthday. You use her favorite recipes to make a zucchini bread for dinner and a chocolate birthday cake. The recipes call for $\frac{1}{4}$ teaspoon of baking powder for the bread and $\frac{3}{4}$ teaspoon of baking powder for the cake. How much baking powder do you need?***

SOLUTION

Add the amounts of baking powder to obtain the total amount you will need.

$$\frac{1}{4} + \frac{3}{4} = \frac{4}{4} = 1.$$

You will need 1 teaspoon of baking powder.

Example 1 illustrates a procedure to add or subtract fractions with the same denominator. Note that fractions with the same denominators are sometimes referred to as **like fractions**.

To add two like fractions, add the numerators and keep the common denominator:

$$\frac{a}{c} + \frac{b}{c} = \frac{a+b}{c}$$

To subtract one like fraction from another, subtract one numerator from the other and keep the common denominator:

$$\frac{a}{c} - \frac{b}{c} = \frac{a-b}{c}$$

1. For each of the following, perform the indicated operation. Convert improper fractions to mixed numbers. Write the result in lowest terms.

a. $\frac{1}{3} + \frac{1}{3}$ **b.** $\frac{3}{5} + \frac{4}{5}$ **c.** $\frac{3}{10} + \frac{9}{10}$ **d.** $\frac{3}{8} + \frac{5}{8}$

e. $\frac{6}{7} - \frac{4}{7}$ f. $\frac{7}{8} - \frac{3}{8}$ g. $\frac{1}{4} - \frac{3}{4}$ h. $\frac{3}{18} - \frac{5}{18}$

Note that the results of the subtraction in parts g and h are negative. A fraction that represents a negative number is written with the negative sign in front of the fraction or in the numerator. That is, $-\frac{a}{b}$ and $\frac{-a}{b}$ are two ways of writing the same fraction.

For example, $\frac{-2}{3}$ and $-\frac{2}{3}$ are two ways to write the same number.

Adding Mixed Numbers with Like Fraction Parts

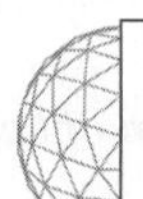

EXAMPLE 2 *A Scandinavian rye bread recipe calls for $3\frac{1}{4}$ cups of white flour and $2\frac{1}{4}$ cups of rye flour. What is the total amount of flour in the bread?*

SOLUTION

$3\frac{1}{4} = 3 + \frac{1}{4}$ Definition of mixed number

$+\ 2\frac{1}{4} = 2 + \frac{1}{4}$

$= 5 + \frac{2}{4}$ Sum of integer part and fraction part

$= 5\frac{1}{2}$ Answer in reduced form

The total amount of flour in the bread is $5\frac{1}{2}$ cups.

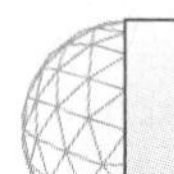

PROCEDURE

Adding Mixed Numbers To add mixed numbers with the same denominators, add the integer parts and the fraction parts separately. If the sum of the fraction part is an improper fraction, convert it to a mixed number and add the integer parts. Reduce the fraction part to lowest terms if necessary.

2. For each of the following, perform the indicated operation. Write the fractional part of the mixed number in lowest terms.

a. $10\frac{3}{7} + 15\frac{2}{7}$ b. $1\frac{1}{6} + 9\frac{5}{6}$

c. $12\frac{1}{8} + 48\frac{5}{8}$

d. $1\frac{13}{16} + 4\frac{5}{16}$

Subtracting Mixed Numbers with Like Fraction Parts

EXAMPLE 3 ***Your bread machine holds a maximum of $5\frac{1}{4}$ cups of flour. You want to use a favorite recipe that calls for $6\frac{3}{4}$ cups of flour. By how much must your recipe be reduced so you will be able to use the bread machine?***

SOLUTION

Subtract $5\frac{1}{4}$ from $6\frac{3}{4}$ to determine if there is more flour than can be held by your machine.

$$6\frac{3}{4} = +6 + \frac{3}{4} \quad \text{Positive mixed number definition}$$

$$-5\frac{1}{4} = -5 - \frac{1}{4} \quad \text{Subtraction of a mixed number}$$

$$= 1 + \frac{2}{4} \quad \text{Result of subtraction}$$

$$= 1 + \frac{1}{2}$$

$$= 1\frac{1}{2} \quad \text{Answer in final form}$$

There are $1\frac{1}{2}$ cups more than the machine can hold, so it can't be used unless the recipe is reduced by $1\frac{1}{2}$ cups.

> To subtract mixed numbers, both the integer parts and the fraction parts must be subtracted.

3. Perform each subtraction. Write the fractional part of the mixed number in lowest terms.

   a. $13\frac{5}{7} - 9\frac{4}{7}$

   b. $9\frac{5}{6} - 8\frac{1}{6}$

c. $45\frac{9}{13} - 17\frac{5}{13}$

d. $14\frac{11}{16} - 4\frac{5}{16}$

4. You need $5\frac{1}{3}$ tablespoons of butter to bake a small apple tart and have one stick of butter, which is 8 tablespoons. How much butter will you have left after you make the tart?

5. The purpose of baseboard molding in a room is to finish the area where the wall meets the floor. A carpenter got a good deal on the molding he wanted because he was willing to buy the last $112\frac{3}{10}$ feet of it that the store had. If he used $96\frac{9}{10}$ feet to finish two rooms, how much did he have left over?

6. Adding and subtracting like fractions and mixed numbers are used in a variety of situations. For example, consider the ingredient amounts in the following beef stew recipe.

Basic Beef Stew	
$2\frac{1}{8}$ *cups of chunked, cooked beef*	$1\frac{5}{8}$ *cups chopped onions*
$3\frac{3}{8}$ *cups of broth*	*2 cups chunked carrots*
$\frac{1}{8}$ *cup of garlic salt*	$1\frac{7}{8}$ *cups chopped potatoes*

a. You need to mix all of these ingredients together in a large mixing bowl. To determine what size mixing bowl to use, add all ingredient amounts listed in the recipe. What is the total number of cups of ingredients in the stew?

b. Your mixing bowls come in 10-cup, 15-cup, and 20-cup sizes. Which mixing bowl should you use? Explain.

c. How much space do you have for mixing the ingredients in the bowl you have chosen?

d. A friend who does not like potatoes is coming to dinner. So you decide to take the potatoes out of the recipe. Use subtraction to determine the new total number of cups of ingredients in the potato-free version of the stew.

7. A negative mixed number such as $-4\frac{3}{7}$ is defined as $-4 - \frac{3}{7}$. Use this information to perform the indicated operations for each of the following.

a. $-4\frac{3}{7} + 2\frac{2}{7}$

b. $-2\frac{3}{8} - \left(-5\frac{2}{8}\right)$

c. $-4\frac{2}{9} - 7\frac{8}{9}$

SUMMARY
ACTIVITY 3.6

Add or Subtract Fractions or Mixed Numbers with the Same Denominator

1. To add or subtract fractions with the same denominator, add or subtract the numerators and write the sum or difference over the given denominator.
2. To add or subtract mixed numbers that have the same denominator, add or subtract the integer parts and the fraction parts separately. Then combine the parts into the form $a\,\frac{b}{c}$.
3. A fraction or the fractional part of a mixed number in an answer should always be written in lowest terms.

EXERCISES
ACTIVITY 3.6

Perform the indicated operation. Write all results in lowest terms.

1. $\frac{3}{8} + \frac{3}{8}$

2. $\frac{4}{5} - \frac{1}{5}$

3. $\frac{5}{12} + \frac{1}{12}$

4. $\frac{5}{8} - \frac{3}{8}$

5. $\frac{2}{3} + \frac{2}{3}$

6. $\frac{9}{4} - \frac{6}{4}$

7. $\frac{1}{4} + \frac{5}{4}$

8. $\frac{20}{12} - \frac{4}{12}$

9. $2\frac{2}{5} + 5\frac{1}{5}$

10. $4\frac{5}{7} - 3\frac{2}{7}$

11. $3\frac{1}{4} + 2\frac{1}{4}$

12. $8\frac{2}{5} - 5\frac{4}{5}$

13. $-7\frac{4}{9} + 5\frac{2}{9}$

14. $6\frac{11}{17} - 8\frac{15}{17}$

15. $8\frac{7}{11} + \left(-13\frac{5}{11}\right)$

16. $-14\frac{7}{8} - \left(-3\frac{3}{8}\right)$

17. $-12\frac{5}{13} - 8\frac{9}{13}$

18. A pizza is cut into 8 equal slices. You eat 2 slices and your friend eats 3 slices.

a. What fraction of the pizza did you and your friend eat?

b. What fraction of the pizza is left?

19. You take home $400 a month from your part-time job as a cashier. Each month you budget $120 for car expenses, $160 for food, and the rest for entertainment.

a. What fraction of your take-home pay is budgeted for car expenses?

b. What fraction of your take-home pay is budgeted for food?

c. What fraction of your take-home pay is budgeted for entertainment?

20. Your bedroom measures $12\frac{1}{8}$ feet by $10\frac{3}{8}$ feet. You want to put a wallpaper border around the perimeter of the room. How much wallpaper border do you need? Remember, the perimeter of a rectangle is the sum of the lengths of the four sides of the rectangle.

ACTIVITY 3.7

Math Is a Trip

OBJECTIVES

1. Determine the least common denominator (LCD) for two or more fractions.
2. Add and subtract fractions with different denominators.
3. Solve equations in the form $x + b = c$ and $x - b = c$ that involve fractions.

The highlight of your summer was a cross-country trip to see your college friends. You drove a total of 2400 miles to get there and made stops along the way to do some sightseeing.

1. The first day, you drove 600 miles and stopped to see statues of famous historical figures in a Wax Museum. What fraction of the 2400 miles did you complete on the first day? Write the fraction in lowest terms.

2. The second day, you drove only 400 miles so you could visit an art gallery that featured the works of M. C. Escher. What fraction of the 2400 miles did you complete on the second day? Write the fraction in lowest terms.

3. a. You completed $\frac{1}{4}$ of the total 2400-mile mileage on the first day and $\frac{1}{6}$ of the total mileage on the second day. Write a numerical expression that can be used to determine what fraction of the 2400-mile trip you completed on the first two days.

   b. How is this addition problem different from those of the previous activity?

Recall (Activity 3.6) that to add or subtract two or more fractions, they all must have the same denominator. Also recall (Activity 3.5) that a common denominator for a set of fractions can be determined by finding a number that is a multiple of each of the denominators in the fractions to be added or to be subtracted.

4. In Problem 3 you can choose any number that is divisible by both 4 and 6 as the common denominator. List at least four numbers divisible by both 4 and 6.

Is 12 one of the numbers that you listed in Problem 4? Notice that 12 is divisible by both 4 and 6 and that it is the smallest such number. Therefore, 12 is the **least common denominator (LCD)** of $\frac{1}{4}$ and $\frac{1}{6}$.

5. a. Write $\frac{1}{4}$ as an equivalent fraction with a denominator of 12.

b. Write $\frac{1}{6}$ as an equivalent fraction with a denominator of 12.

c. Add the two like fractions.

6. On the third day, you drove 800 miles and stopped to see a flower garden where the leaf and petal arrangements form mathematical patterns. What fraction of the 2400 miles did you complete on the third day? Write the fraction in lowest terms.

7. You completed $\frac{5}{12}$ of the trip on the first two days and $\frac{1}{3}$ of the trip on the third day. What fraction of the trip did you complete on the first three days?

8. During the fourth day, you drove another 200 miles and stayed at the Motel 8, where the windows were all regular octagons. What fraction of the 2400 miles did you complete on the fourth day? Write the fraction in lowest terms.

9. You completed $\frac{3}{4}$ of the trip on the first three days and $\frac{1}{12}$ of the trip on the fourth day. What fraction of the trip did you finish in the first four days?

10. What fraction of the trip do you have to complete to get to your friends on the fifth day?

Solving Equations

Problem 10 can also be solved using algebra. If x represents the fractional part of the trip to be completed on the fifth day, then

$$x + \frac{5}{6} = 1.$$

Recall that this equation may be solved for x by *subtracting* $\frac{5}{6}$ from both sides of the equation.

$$x + \frac{5}{6} - \frac{5}{6} = 1 - \frac{5}{6}$$

$$x = \frac{6}{6} - \frac{5}{6} = \frac{1}{6}$$

So, you have $\frac{1}{6}$ of the trip to complete on the fifth day.

11. Once you arrive, you and your friends compare college experiences, grades and the different methods professors use to determine grades. For example, the final grade in one of your courses is determined by quizzes, exams, a project, and class participation. Quizzes count for $\frac{1}{4}$ of the final grade, exams $\frac{1}{3}$, and the project $\frac{1}{4}$ of the final grade.

a. If x represents the fractional part of the final grade for class participation, write an equation relating x, $\frac{1}{4}$, $\frac{1}{3}$, $\frac{1}{4}$, and 1.

b. Solve the equation for x.

12. You decide to make lunch and discover that you all like salad. Your favorite recipe for a simple salad dressing is made of oil and vinegar. You need a total of $2\frac{1}{2}$ cups of salad dressing, of which $1\frac{7}{8}$ cups is oil.

a. If x represents the amount of vinegar to be used in the salad dressing, write an equation relating the amounts of oil and vinegar for the $2\frac{1}{2}$ cups of salad dressing.

b. Solve the equation for x.

13. You are still hungry after eating the salad and order a submarine sandwich. The total weight of a sub is the sum of the weights of its ingredients. A formula for total weight is give by $W = M + C + R + V$, where M is the weight of the meat, C is the weight of the cheese, R is the weight of the roll, and V is the weight of all the other ingredients, such as lettuce, tomato, and so on. Use this formula to calculate the weight of a regular submarine sandwich that includes $\frac{1}{4}$ pound of meat, $\frac{1}{8}$ pound of cheese, a $\frac{1}{2}$ pound roll and other ingredients weighing $\frac{1}{4}$ pound.

SUMMARY
ACTIVITY 3.7

1. To determine the **least common denominator (LCD)** of fractions,

a. Identify the largest denominator of the fractions involved.

b. Look at multiples of the largest denominator. The smallest multiple that is divisible by the smaller denominator(s) is the LCD.

2. To add or subtract fractions with different denominators,

a. Find the LCD and convert each fraction to an equivalent fraction, that has the LCD you found.

b. Add or subtract the numerators of the equivalent fractions to obtain the new numerator, leaving the LCD in the denominator.

c. If necessary, reduce the resulting fraction to lowest terms.

d. If the result is an improper fraction, convert it to a mixed number where appropriate.

EXERCISES
ACTIVITY 3.7

In Exercises 1–17, perform the indicated operation.

1. $\frac{1}{6} + \frac{1}{2}$

2. $\frac{3}{8} - \frac{1}{4}$

3. $-\frac{2}{5} + \frac{9}{10}$

4. $3\frac{1}{3} + 1\frac{1}{6}$

5. $5\frac{7}{12} + 8\frac{13}{16}$

Exercise numbers appearing in color are answered in the Selected answers appendix.

6. $12\frac{3}{4} + 6\frac{2}{5}$

7. $5\frac{2}{12} + 3\frac{7}{18}$

8. $12\frac{1}{4} - 7\frac{1}{8}$

9. $14\frac{3}{4} - 6\frac{5}{12}$

10. $11 - 6\frac{3}{7}$

11. $12\frac{1}{4} - 5$

12. $8\frac{5}{6} - 3\frac{11}{12}$

13. $4\frac{3}{11} - 2\frac{9}{22}$

14. $-7\frac{5}{8} + 2\frac{1}{6} = -7 + 2 - \frac{5}{8} + \frac{1}{6}$

15. $-9\frac{3}{5} + \left(-3\frac{4}{15}\right)$

16. $2\frac{2}{7} - \left(-3\frac{3}{8}\right)$

17. $-5\frac{2}{5} + \left(-6\frac{4}{9}\right)$

18. You are considering learning to play the piano and check with a friend about practice time. For 3 consecutive days before a recital, she practiced $1\frac{1}{4}$ hours, $2\frac{1}{2}$ hours, and $3\frac{2}{3}$ hours. What was her total practice time for 3 days?

19. A favorite muffin recipe calls for $2\frac{2}{3}$ cups of flour, 1 cup of sugar, $\frac{1}{2}$ cup of crushed cashews, and $\frac{5}{8}$ cup of milk, plus assorted spices. How many cups of batter does the recipe make?

20. A student spends $\frac{1}{3}$ of a typical day sleeping, $\frac{1}{6}$ of the day in classes, and $\frac{1}{8}$ of the day watching TV.

   a. If x represents the fraction of the rest of the day available for study, write an equation describing the situation.

b. Solve the equation for x.

21. The final grade in one of your friend's courses is determined by a term paper, exams, quizzes, and class participation. The term paper is worth $\frac{1}{5}$, the exams $\frac{1}{3}$, and the quizzes $\frac{1}{4}$ of the final grade.

a. If x represents the fractional part of the final grade for class participation, write an equation describing the situation.

b. Solve the equation for x.

22. A cake recipe calls for $3\frac{1}{2}$ cups of flour, $1\frac{1}{4}$ cups of brown sugar, and $\frac{5}{8}$ cup of white sugar. What is the total amount of dry ingredients?

23. To create a table for a report on your word processor, you need two columns, each $1\frac{1}{2}$ inches wide, and five columns, each $\frac{3}{4}$ inch wide. Will your table fit on a piece of paper $8\frac{1}{2}$ inches wide?

24. Your living room wall is 14 feet long. You want to buy a couch and center it on the wall. You have two end tables, each $2\frac{1}{4}$ feet long, that will be on each side of the couch. You plan to leave $1\frac{1}{2}$ feet next to each end table for floor plants. Determine how long a couch you can buy.

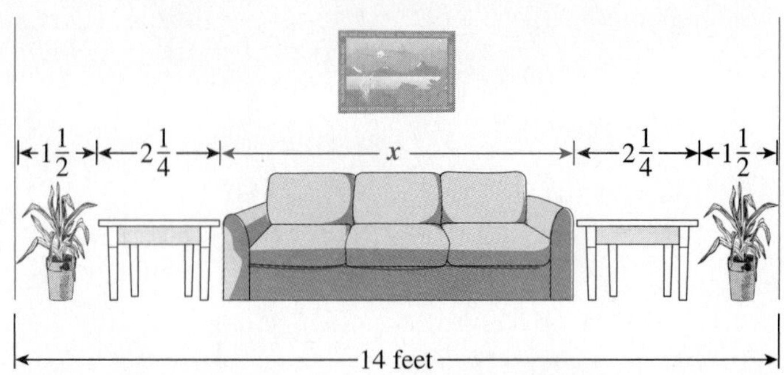

a. If x represents the length of the couch, write an equation describing the situation.

b. Solve the equation for x.

In Exercises 25–30, solve each equation for the unknown quantity. Check your answers.

25. $\frac{3}{10} = c - \frac{1}{5}$

26. $x - 3\frac{1}{2} = 6\frac{3}{4}$

$10\frac{1}{4}$

27. $x + 4 = 2\frac{1}{2}$

$-1\frac{1}{2}$

28. $\frac{1}{2} + b = \frac{2}{3}$

29. $2\frac{1}{3} + x = 5\frac{5}{6}$

$3\frac{1}{2}$

30. $x + 5\frac{1}{4} = -7\frac{1}{3}$

CLUSTER 2 What Have I Learned?

1. a. How do you convert a mixed number to an improper fraction?

b. Give an example.

2. a. How do you convert an improper fraction to a mixed number?

b. Give an example.

3. Determine whether the following statements are true or false. Give a reason for each answer.

a. For a fraction to be called improper, the absolute value of the numerator has to be greater than the absolute value of the denominator.

b. The LCD of two denominators must be greater than or equal to either of those denominators.

c. $-6\frac{1}{4} = -6 + \frac{1}{4}$

d. $7\frac{2}{9} = \frac{65}{9}$

e. $\frac{2}{7} + \frac{4}{7} = \frac{6}{14}$

4. a. Outline the general steps in adding mixed numbers. Be sure to allow for differences when the denominators are the same or different.

b. Give an example of two or more mixed numbers with same denominators. Then determine their sum.

c. Give an example of two or more mixed numbers with different denominators. Then determine their sum.

5. a. Outline the general steps in subtracting mixed numbers. Be sure to indicate how you make a decision to regroup from the integer part.

b. Give an example of subtracting two mixed numbers with the same denominator. Then determine their difference.

c. Give an example of subtraction in which two mixed numbers have different denominators and you do *not* need to regroup. Then determine the difference.

d. Give an example of subtraction in which two mixed numbers have different denominators and you *do* need to regroup. Then determine the difference.

CLUSTER 2 How Can I Practice?

1. Determine the missing numerators.

a. $\frac{3}{11} = \frac{?}{33}$

b. $\frac{5}{7} = \frac{?}{42}$

c. $\frac{8}{9} = \frac{?}{81}$

d. $\frac{7}{12} = \frac{?}{36}$

e. $\frac{13}{18} = \frac{?}{72}$

2. Convert the following mixed numbers into improper fractions in lowest terms.

a. $3\frac{7}{9}$

b. $5\frac{5}{8}$

c. $4\frac{3}{12}$

d. $10\frac{8}{13}$

e. $1\frac{27}{31}$

3. Convert the following improper fractions to mixed numbers in lowest terms.

a. $\frac{71}{32}$

b. $\frac{28}{13}$

c. $\frac{89}{12}$

d. $\frac{45}{5}$

e. $\frac{92}{14}$

4. Reduce the following fractions to lowest terms.

a. $\frac{12}{18}$

b. $\frac{8}{72}$

c. $\frac{36}{39}$

d. $\frac{42}{48}$

e. $\frac{54}{82}$

5. Rearrange the following fractions in order from smallest to largest.

$$\frac{3}{2}, \frac{1}{6}, \frac{3}{4}, \frac{5}{8}, \frac{11}{12}, \frac{3}{24}$$

6. When fog hit the New York City area, visibility was reduced to $\frac{1}{16}$ mile at JFK Airport, $\frac{1}{8}$ mile at LaGuardia Airport, and $\frac{1}{2}$ mile at Newark Airport.

a. Which airport had the best visibility?

b. Which airport had the worst visibility?

7. Add the following and write the result in lowest terms.

a. $\frac{4}{9} + \frac{7}{9}$

b. $\frac{13}{17} + \frac{5}{17}$

c. $1\frac{5}{12} + 3\frac{3}{4}$

d. $8\frac{6}{7} + 4\frac{5}{9}$

e. $5\frac{5}{12} + 3\frac{11}{18}$

f. $13 + 4\frac{9}{13}$

g. $-8\frac{4}{6} + \left(-25\frac{7}{24}\right)$

h. $-6\frac{17}{24} + \left(-12\frac{31}{48}\right)$

i. $-7\frac{6}{21} + \left(-8\frac{15}{39}\right)$

j. $-9\frac{3}{14} + \left(-11\frac{6}{35}\right)$

8. In each case, do the subtraction and write the result in lowest terms.

a. $\frac{9}{13} - \frac{2}{13}$

b. $13\frac{28}{54} - 2\frac{11}{54}$

c. $8\frac{15}{48} - 7\frac{5}{24}$

d. $11\frac{4}{9} - 7\frac{5}{24}$

e. $22\frac{8}{11} - 17\frac{15}{44}$

f. $46\frac{13}{28} - 34\frac{10}{56}$

g. $74\frac{3}{8} - 61$

h. $28\frac{7}{12} - 15$

i. $48 - 21\frac{5}{9}$

j. $72 - 13\frac{7}{12}$

k. $13\frac{2}{15} - 8\frac{4}{5}$

l. $14\frac{1}{36} - 7\frac{1}{6}$

m. $29\frac{7}{12} - 21\frac{15}{18}$

n. $23\frac{13}{42} - 18\frac{5}{7}$

o. $17\frac{9}{14} - 9\frac{8}{21}$

p. $16\frac{5}{8} - (-3\frac{7}{12})$

q. $-6\frac{4}{5} - (-7\frac{11}{15})$

r. $-13\frac{5}{6} - 8\frac{6}{7}$

9. While testing a new drug, doctors found that $\frac{1}{2}$ of the patients given the drug improved, $\frac{2}{5}$ showed no change in their condition, and the remaining patients got worse. What fraction of the patients taking the new drug got worse?

10. You purchased a roll of wallpaper that is $30\frac{1}{2}$ yards long. Your contractor used $26\frac{7}{8}$ yards to wallpaper one room. Is there enough wallpaper left on the roll for a job that requires 4 yards of wallpaper?

11. Solve each of the following equations for x.

a. $14\frac{3}{7} + x = 21$

b. $-17\frac{3}{5} + x = -12\frac{4}{5}$

c. $x - 13\frac{5}{6} = 17\frac{4}{7}$

d. $x + 8\frac{4}{9} = 19\frac{2}{3}$

e. $-18\frac{2}{11} + x = -23\frac{1}{3}$

CLUSTER 3 Adding and Subtracting Decimals

ACTIVITY 3.8

What Are You Made of?

OBJECTIVES

1. Identify place values of numbers written in decimal form.
2. Convert a decimal to a fraction or a mixed number.
3. Compare decimals.
4. Read and write decimals.
5. Round decimals.

Have you ever wondered about the chemical elements that are contained in your body? According to the *Universal Almanac*, a 150-pound person is made up of the following elements.

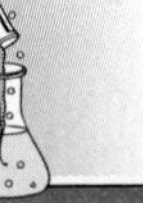

It's Elementary

ELEMENT	WEIGHT IN POUNDS	ELEMENT	WEIGHT IN POUNDS
Oxygen	97.5	Cobalt	0.00024
Carbon	27.0	Copper	0.00023
Hydrogen	15.0	Manganese	0.00020
Nitrogen	4.5	Iodine	0.00006
Calcium	3.0	Zinc	Trace
Phosphorus	1.8	Boron	Trace
Potassium	0.3	Aluminum	Trace
Sulfur	0.3	Vanadium	Trace
Chlorine	0.3	Molybdenum	Trace
Sodium	0.165	Silicon	Trace
Magnesium	0.06	Fluorine	Trace
Iron	0.006	Chromium	Trace
		Selenium	Trace

Source: *Universal Almanac*

Note that the weights are expressed as **decimal numbers**. A number written as a decimal has an integer part to the left of the decimal point, and a fractional part that is to the right of the decimal point. Decimal numbers extend the place value system for whole numbers to include fractional parts. Also, decimal numbers, like fractions, may be used to express portions of a whole, that is, numbers less than 1.

Comparing Decimals

As you recall from Chapter 1, the place values for the integer part, to the left of the decimal point, are powers of 10: 1, 10, 100, 1000, etc. The place values for the fractional part of a decimal number, to the *right* of the decimal point, are powers of 10 in the denominator: $\frac{1}{10}, \frac{1}{100}, \frac{1}{1000}$, etc.

1. Does a 150-lb person have more sodium or potassium in their body? Explain why.

EXAMPLE 1 ***The place values of the decimal number 79.653 are as follows.***

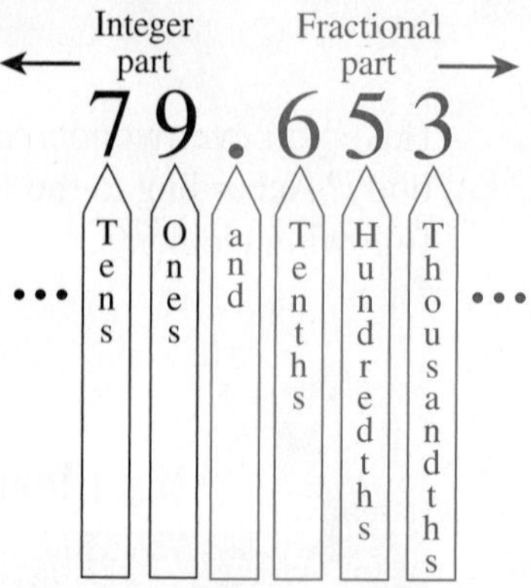

Therefore, $79.653 = 70 + 9 + \frac{6}{10} + \frac{5}{100} + \frac{3}{1000}$.

The fractional part of the number 79.653 is $\frac{6}{10} + \frac{5}{100} + \frac{3}{1000}$. Notice that the least common denominator of these three fractions is 1000. So,

$$\frac{6}{10} + \frac{5}{100} + \frac{3}{1000} = \frac{600}{1000} + \frac{50}{1000} + \frac{3}{1000} = \frac{653}{1000}.$$

Therefore, 79.653 can be written as $79\frac{653}{1000}$, which is read as "79 and 653 thousandths." Note that when reading decimal numbers, the word "and" separates the integer part from the fractional part, replacing the decimal point.

When a decimal number is written as a fraction or mixed number, the denominator corresponds to the place value of the rightmost digit in the decimal number. This leads to the following procedure to convert a decimal number to an equivalent fraction or mixed number.

PROCEDURE

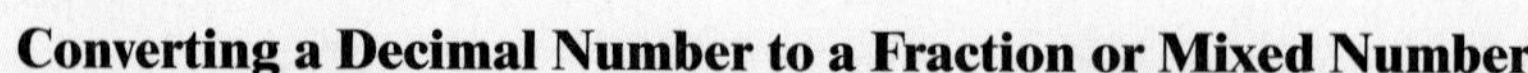

Converting a Decimal Number to a Fraction or Mixed Number

1. Write the nonzero integer part of the number and drop the decimal point.
2. Write the fractional part of the number as the numerator of a fraction. The denominator is determined by the place value of the rightmost digit.
3. Reduce the fraction if necessary.

2. a. Write the weight of sodium given in the table as a fraction with denominator 1000.

b. Write the weight of potassium given in the table as a fraction with denominator 1000.

c. Compare the weights of potassium and sodium in the fraction form that you determined in parts a and b. Which one is larger?

There is a faster and easier method for comparing decimal numbers. Suppose you want to compare 2.657 and 2.68. The first step is to line up the decimal points as follows.

$$\begin{array}{l} 2.657 \\ 2.680 \end{array}$$

Zero since no digit is present.

Now, moving left to right, compare digits that have the same place value. The digit in the ones and tenths places is the same in both numbers. In the hundredths place, 8 is greater than 5, and it follows that $2.68 > 2.657$.

PROCEDURE

Comparing Decimal Numbers

1. First compare the integer parts (to the left of the decimal point). The number with the larger integer part is the larger number.
2. If the integer parts are the same, compare the fractional parts (to the right of the decimal point).
   a. Write the decimal parts as fractions with the same denominator and compare the fractions.

   or

   b. Line up the decimal points and compare the digits that have the same place value, moving from left to right. The first number with the larger digit is the larger number.

3. a. Use the method just described to determine if an average person has more sodium or potassium in his body.

   b. Does the average person have more cobalt or iron in their body?

Reading and Writing Decimals

It is important to know how to name, read, and write decimal numbers as fractions. For example, 0.165 is named as "one hundred sixty-five thousandths." More generally, an extension of the place value system to the right of the decimal point is found in the following chart.

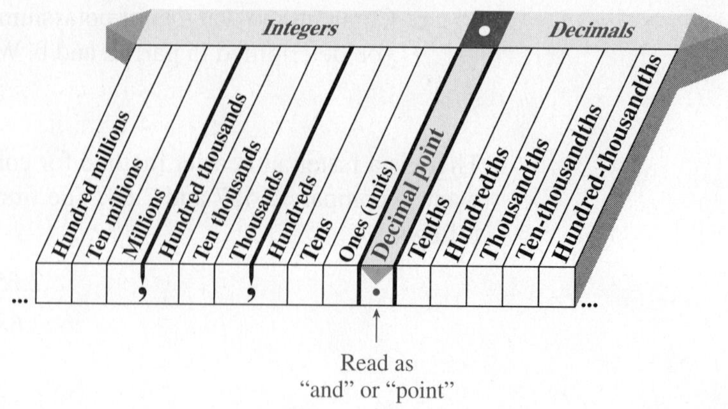

To read a decimal number, read the digits to the right of the decimal point as though they were *not* preceded by a decimal point, and then attach the place value of its rightmost digit. Use the word "and" to separate the integer part from the decimal or fractional part.

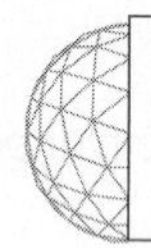

EXAMPLE 2 ***The decimal number* 251.737 *is read two hundred fifty-one and seven hundred thirty-seven thousandths.***

4. a. Write in words the weight of potassium in a 150-pound body.

b. Write in words the weight of the amount of copper in a 150-pound body.

c. Which element would have a weight of "six hundred-thousandths" of a pound?

5. a. Write the number 9,467.00624 in words.

b. Write the number 35,454,666.007 in words.

c. Write the following number using decimal notation:
four million sixty-four and seventy-two ten-thousandths.

d. Write the following number using decimal notation:
seven and forty-three thousand fifty-two millionths.

Rounding Decimal Numbers

In Chapter 1 you estimated whole numbers by a technique called rounding-off, or rounding for short. Decimal numbers can also be rounded to a specific place value.

EXAMPLE 3 ***Round the weight of sodium in a 150-pound body by rounding to the tenths place.***

SOLUTION

Observe that 0.165 has a 1 in the tenths place and that $0.165 > 0.1$. So the task is to determine whether 0.165 is closer to 0.1 or 0.2.

0.15 is exactly halfway between 0.1 and 0.2.

$0.165 > 0.15$, so 0.165 must be closer to 0.2.

There is approximately 0.2 pound of sodium in a 150-pound body.

PROCEDURE:

Rounding a Decimal Fraction to a Specified Place Value

1. If the specified place value is 10 or more, drop the fractional part and round the resulting whole number.
2. Otherwise, locate the digit with the specified place value.
3. If the digit directly to the right is less than 5, keep the digit in step 2 and delete all the digits to the right.
4. If the digit directly to the right is greater than or equal to 5, increase the digit in step 2 by 1 and delete all the digits to the right.

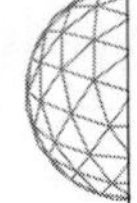

EXAMPLE 4 ***Round 279.583 to the ones place.***

SOLUTION

The digit 9 in the ones place will either stay the same or increase by 1. Basically, you want to know whether the number is closer to 279 or 280. Since the tenths digit is 5, the number rounds to 280 (following step 4 in the procedure).

6. Copper's weight of 0.00023 lb, rounded to the nearest ten-thousandths place, is 0.0002 lb. Explain how this answer is obtained.

7. a. What is the weight of sodium, rounded to the nearest hundredth?

   b. What is cobalt's weight, rounded to the nearest ten-thousandth?

SUMMARY
ACTIVITY 3.8

- **Converting a Decimal to a Fraction or a Mixed Number**

  1. Write the nonzero integer part of the number and drop the decimal point.
  2. Write the fractional part of the number as the numerator of a new fraction, whose denominator is the same as the denominator of the place value of the rightmost digit.
  3. Reduce the fraction if necessary.

- **Comparing Decimal Numbers**

  Align the decimal points vertically and compare the digits that have the same place value, moving from left to right. The first number with the larger digit in the same value place is the larger number.

- **Reading and Writing Decimal Numbers**

  To name a decimal number, read the digits to the right of the decimal point as though they were not preceded by a decimal point, and then attach the place value of its rightmost digit. Use the word *and* to separate the whole-number part from the decimal or fractional part.

- **Rounding a Decimal Number to a Specified Place Value**

  1. If the specified place value is 10 or more, drop the fractional part and round the resulting whole number.
  2. Otherwise, locate the digit with the specified place value.
  3. If the digit directly to the right is less than 5, keep the digit in step 2 and delete all the digits to the right.
  4. If the digit directly to the right is greater than or equal to 5, increase the digit in step 2 by 1 and delete all the digits to the right.

EXERCISES
ACTIVITY 3.8

1. Using the table on page 223, list the chemical substances chlorine, calcium, iron, magnesium, iodine, cobalt, and manganese in the order in which their amounts are in your body from largest to smallest.

2. a. The euro, the basic unit of money in the European Union, was recently worth \$1.28817 American dollars. Round to estimate the worth of 1 euro in cents.

   b. To what place did you round in part a?

3. a. The Canadian dollar was recently worth \$0.889597. Round this value to the nearest cent.

   b. Round the value of the Canadian dollar to the nearest dime.

4. a. The Mexican peso was recently worth \$0.0917. Round this value to the nearest cent.

   b. Round the value of the peso to the nearest dime.

5. a. Write 0.052 in words.

   b. Write 0.00256 in words.

   c. Write 3402.05891 in words.

   d. Write 64 ten thousandths as a decimal.

   e. Write one hundred twenty-five thousandths as a decimal.

   f. Write two thousand forty-one and six hundred seventy-three ten-thousandths as a decimal.

6. The *New York Times Almanac* reported that U.S. alcohol consumption from 1940–1990 (in gallons of ethanol, per person) was as follows.

	GALLONS PER PERSON			
YEAR	BEER	WINE	SPIRITS	ALL BEVERAGES
1940	0.73	0.16	0.67	1.56
1950	1.04	0.23	0.77	2.04
1960	0.99	0.22	0.86	2.07
1970	1.14	0.27	1.11	2.52
1980	1.38	0.34	1.04	2.76
1990	1.34	0.33	0.78	2.46

a. In which year was beer consumption the highest?

b. Round the amount of all beverage consumption in 1990 to the nearest tenth and write your answer as a mixed number.

c. In which year was the least amount of spirits consumed?

d. Write in words the amount of spirits consumption in 1970.

e. Round the amount of wine consumption to the nearest tenth in the years 1970, 1980, and 1990. Can you use the rounded values to determine when the most wine was consumed? Explain.

7. Circle either true or false for each of the following statements and explain your choice.

a. True or False: 456.77892 < 456.778902

b. True or False: 0.000501 > 0.000510

c. True or False: 7832.00375 rounded to the nearest thousandth is 7832.0038

d. True or False: Seventy-three and four hundred thousandths is written 73.0004.

ACTIVITY 3.9

Think Metric

OBJECTIVES

1. Know the metric prefixes and their decimal values.
2. Convert measurements between metric quantities.

The shop where you have a new job has just received a big contract from overseas. A significant part of your job will involve converting measurements in the English system to the metric system, used by most of the world. To find out more about changing measurements to the metric system, visit *www.metrication.com.*

The Old English system of measurement (for example, feet for length, pounds for weight) used throughout most of the United States was not designed for its computational efficiency. It evolved over time before the decimal system was adopted by Western European cultures. The metric system of measurement, which you may already know something about, was devised to take advantage of our decimal numeration system.

EXAMPLE 1 ***A foot is divided into 12 equal parts, 12 inches. But as a fraction of a foot, $\frac{1}{12}$ is not easily expressed as a decimal number. The best you can do is approximate. One inch equals $\frac{1}{12}$ of a foot, or approximately $1 \div 12 \approx 0.08333$ feet. In the metric system, each unit of measurement is divided into 10 equal parts. So 1 meter (a little longer than 3 feet) is divided into 10 decimeters, each decimeter equaling $\frac{1}{10} = 0.1$ meter. In this case, the fraction can be expressed exactly as a simple decimal.***

1. How many decimeters are in 1 meter?

2. If the specifications for a certain part is 8 decimeters long, what is the length in meters?

In general, the metric system of measurement relies upon each unit being divided into 10 equal subunits, which in turn are divided into 10 equal subunits, and so on. The prefix added to the base unit signifies how far the unit has been divided. Starting with the meter, as the basic unit for measuring distance, the following smaller units are defined by Latin prefixes.

UNIT OF LENGTH	FRACTIONAL PART OF A METER
meter	1 meter
decimeter	0.1 meter
centimeter	0.01 meter
millimeter	0.001 meter

3. Complete each statement to show how many of each unit are in 1 meter.

_______ decimeters equal 1 meter.

_______ centimeters equal 1 meter.

_______ millimeters equal 1 meter.

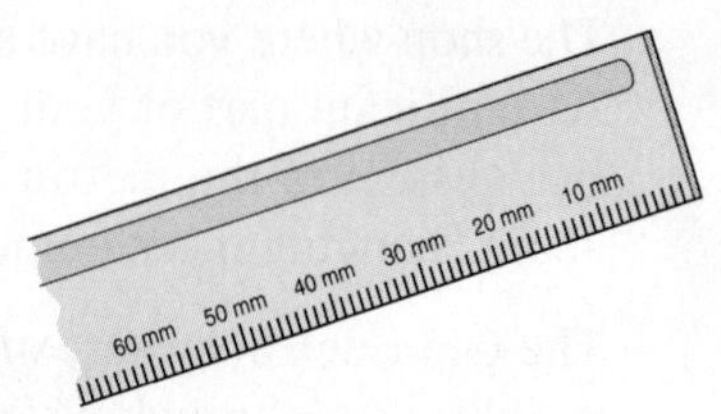

So, in Latin, the prefixes mean exactly what you found in Problem 3: deci- is 10, centi- is 100, and milli- is 1000.

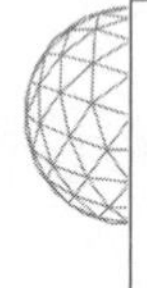

EXAMPLE 2 ***Any measurement in one metric unit can be easily converted to another metric unit by simply relocating the decimal point.***
235 centimeters = 2.35 meters since 200 centimeters is the same as 2 meters and 35 centimeters is $\frac{35}{100}$ of 1 meter.

4. Express a distance of 480 centimeters in meters.

Abbreviations are commonly used for all metric units:

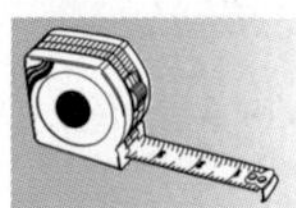

Going the Distance

METRIC UNIT OF DISTANCE	ABBREVIATION
meter	m
decimeter	dm
centimeter	cm
millimeter	mm

5. How would you express 2.5 m in centimeters?

6. Since there are 1000 mm in 1 m, how many meters are there in 4500 mm?

7. How many millimeters are there in 1 cm?

8. Which distance is greater, 345 cm or 3400 mm?

In the metric system, each place value is equivalent to the unit of measurement. Each digit can be interpreted as a number of units.

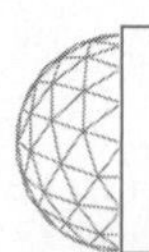

EXAMPLE 3 ***The distance 6.358 meters can be expanded to 6 meters + 3 decimeters + 5 centimeters + 8 millimeters.***

9. Expand 7.25 meters, as in Example 3.

Larger metric units are defined using Greek prefixes.

UNIT OF LENGTH	ABBREVIATION	DISTANCE IN METERS
meter	m	1 meter
decameter	dam	10 meters
hectometer	hm	100 meters
kilometer	km	1000 meters

In practice, the decameter and hectometer units are rarely used.

10. How many meters are in a 10-kilometer race?

11. Convert 3450 meters into kilometers.

The metric system is also used to measure the mass of an object and volume. Grams are the basic unit for measuring the mass of an object (basically, how heavy it is). Liters are the basic unit for measuring volume (usually of a liquid). The same prefixes are used to indicate larger and smaller units of measurement.

12. Complete the following tables.

UNIT OF MASS	ABBREVIATION	MASS IN GRAMS
milligram	mg	
centigram	cg	
decigram	dg	
gram	g	1 g
decagram	dag	
hectogram	hg	
kilogram	kg	

UNIT OF VOLUME	ABBREVIATION	VOLUME IN LITERS
milliliter	$m\ell$	
centiliter	$c\ell$	
deciliter	$d\ell$	
liter	ℓ	
decaliter	$da\ell$	
hectoliter	$h\ell$	
kiloliter	$k\ell$	

SUMMARY
ACTIVITY 3.9

1. The basic metric units for measuring distance, mass, and volume are meter, gram, and liter respectively.

2. Larger and smaller units in the metric system are based on the decimal system. The prefix before the basic unit determines the size of the unit, as shown in the table.

UNIT PREFIX (BEFORE METER, GRAM, OR LITER)	ABBREVIATION	SIZE IN BASIC UNITS (METERS, GRAMS, OR LITERS)
milli-	mm, mg, mℓ	0.001
centi-	cm, cg, cℓ	0.01
deci-	dm, dg, dℓ	0.1
deca-	dam, dag, daℓ	10
hecto-	hm, hg, hℓ	100
kilo-	km, kg, kℓ	1000

EXERCISES
ACTIVITY 3.9

1. Convert 3.6 m into cm.

2. Convert 287 cm into m.

3. Convert 1.9 m into mm.

4. Convert 4705 mm into m.

5. Convert 4675 m into km.

6. Convert 3.25 km into m.

7. Convert 42.55 g into cg.

8. Convert 236 g into kg.

9. Convert 83.2 ℓ into mℓ.

10. Convert 742 mℓ into ℓ.

11. Which is greater, 23.67 cm or 237 mm?

12. Which is smaller, 124 mg or 0.15 g?

13. Which is greater, 14.5 mℓ or 0.015 ℓ?

Exercise numbers appearing in color are answered in the Selected Answers appendix.

ACTIVITY 3.10

Dive into Decimals

OBJECTIVES

1. Add and subtract decimals.
2. Compare and interpret decimal numbers.

In Olympic diving, a panel of judges awards a score for each dive. There is a preliminary round and a final round of dives. The preliminary round total and the scores of the five dives in the final round are given below for the top six divers in the women's 10-meter platform event at the 2004 Athens Olympics. The total score determines who wins the gold, silver, and bronze medals for first, second, and third place.

Going for Gold

DIVER	SCORE FOR PRELIMINARY ROUND	SCORES FOR FINAL ROUND					TOTAL SCORE
		DIVE 1	DIVE 2	DIVE 3	DIVE 4	DIVE 5	
Chantelle Newbery (Australia)	198.3	77.4	82.62	72.96	70.29	88.74	
Lishi Lao (China)	203.04	74.7	72.0	80.64	65.34	80.58	
Loudy Tourky (Australia)	192.87	80.1	85.44	55.68	78.21	69.36	
Emilie Heymans (Canada)	187.05	84.48	77.22	61.44	86.7	58.14	
Laura Wilkinson (USA)	194.02	64.32	75.84	79.2	61.38	74.46	
Ting Li (China)	198.33	69.3	83.52	52.2	51.33	91.8	

1. Based *only* on the preliminary round, which divers would have won the gold, silver, and bronze medals?

Adding Decimals

When adding whole numbers, add the digits in each place value position and regroup where needed. Add decimal numbers in a similar way.

2. a. Determine the total of the five dives in the final round for Chantelle Newbery.

b. Describe in words how to line up the digits when adding decimal numbers.

c. How do you locate the decimal point in your answer?

d. Add the preliminary round score for Chantelle Newbery to the total for the five dives in part a. This sum represents the total score for Chantelle in the competition.

3. Determine the total score for each of the six divers listed in the table and record your answer in the appropriate place in the table.

4. a. Which diver won the gold medal?

b. Which diver won the silver medal?

c. Which diver won the bronze medal?

5. Explain why Ting Li did not win the silver medal as predicted in Problem 1.

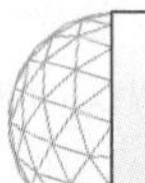

PROCEDURE

Adding Decimal Numbers

1. Line up the place values vertically with decimal points as a guide.
2. Add as usual.
3. Insert the decimal point in the answer directly in line with the decimal points of the numbers being added.

Subtracting Decimals

6. By how many points did the gold medalist win?

The answer to Problem 6 involves the subtraction of decimals. The procedure for subtracting decimal numbers is very similar to the procedure for adding decimal numbers.

7. Use the procedure for adding decimal numbers as a model to write a three-step procedure for subtracting decimal numbers.

8. For each diver, determine the difference between her highest and lowest scores in the final round for each diver.

   a. Chantelle Newbery b. Lishi Lao c. Loudy Tourky

   d. Emilie Heymans e. Laura Wilkinson f. Ting Li

9. Based on the results in Problem 8, which diver was most consistent? Explain.

10. Based *only* on the final round, which diver should have won the gold, silver, and bronze medals?

SUMMARY
ACTIVITY 3.10

Adding or Subtracting Decimals

1. Line up the place values vertically with decimal points as a guide.
2. Add or subtract as usual.
3. Insert the decimal point in the answer directly in line with the decimal points of the numbers being added.

EXERCISES
ACTIVITY 3.10

1. Perform the indicated operation. Verify your calculations with a calculator.

   a. 0.023 + 0.45 b. 5.25 + 0.069 + 12 c. 3.408 + 1.733

   d. 5.008 − 0.079 e. 0.0044 − 0.056 f. 2.034 − 5.246

Exercise numbers appearing in color are answered in the Selected Answers appendix.

2. You had $149.73 in your bank account on Monday. You wrote a check for $83.69 on Tuesday and two checks for $12.96 and $26.48 on Wednesday. You deposited your paycheck for $212.05 on Thursday and withdrew $50.00 in cash on Friday. What is the balance in your account?

3. You have $20 for groceries. Your shopping list has the following items: bagels ($2.99), cream cheese ($2.99), hummus ($2.69), olives ($4.49), cookies ($3.29), turkey ($2.09), and mustard ($2.49). Is $20 enough for everything on the list? If not, which item should you eliminate so that you will be within your budget?

4. Preparing a meal, you use several cookbooks, some with metric units and others with English units. To make sure you do not make a mistake in measuring, you check a book of math tables for the following conversions.

Measure for Measure

ENGLISH SYSTEM	METRIC SYSTEM
1 fluid ounce	0.02957 liters
1 cup	0.236588 liters
1 pint	0.473176 liters
1 quart	0.946352 liters
1 gallon	3.78541 liters

a. One recipe calls for mixing 1 pint of broth, 1 cup of cream, and 1 ounce of vinegar. What is the total amount of liquid, measured in liters?

b. Another recipe requires 1 gallon of water, 1 quart of milk, 1 one pint of cream. You have a 6-liter pan to cook the liquid ingredients. How much space (in liters) is left in the pan after you have added the three ingredients?

c. You also notice in the book of tables that there are 2 cups in a pint. Is this fact consistent with the information in the table above?

d. Based on the table, how many pints are there in 1 quart? Explain.

5. The results for the top eight countries in the women's gymnastics competition at the 2004 Athens Olympics are summarized in the following table. Each team's final score is the total of the scores for each of the four events: the vault, uneven bars, balance beam, and the floor exercise.

A Balancing Act

COUNTRY	VAULT	UNEVEN BARS	BEAM	FLOOR	TOTAL
Australia	27.449	27.1	26.974	27.324	
China	28.149	27.549	27.244	27.086	
France	28.162	27.011	27.537	27.449	
Romania	28.437	28.136	28.961	28.749	
Russia	28.325	28.137	28.511	28.262	
Spain	27.599	28.412	27.724	27.837	
Ukraine	28.099	28.199	28.024	27.987	
USA	28.387	28.524	28.499	28.174	

a. In each event, determine the order of the top three finishers.

b. Determine the totals for each country. What countries won the gold, silver, and bronze medals?

c. Consider the four event scores for each country. Which team had the largest difference between their lowest- and highest-scoring events? Which team had the smallest difference?

ACTIVITY 3.11

Boiling, Freezing, and Financial Aid

OBJECTIVES

1. Add and subtract positive and negative decimal numbers.
2. Interpret and compare decimal numbers.
3. Solve equations of the type $x + b = c$ and $x - b = c$ involving decimal numbers.

Just as water can freeze to become a solid, or boil to become a gas, the elements of the earth can exist in solid, liquid, or gaseous form. For example, the metal mercury is a liquid at room temperature. Mercury boils at 673.9°F and freezes at −38.0°F.

1. What is the difference between the freezing point and the boiling point of mercury?

2. You have some mercury stored at a temperature of 75°F.

   a. By how many degrees Fahrenheit must the mercury be heated for it to boil?

   b. By how many degrees Fahrenheit must the mercury be cooled to freeze it?

3. The use of the element krypton in energy-efficient windows helps meet new energy conservation guidelines. Krypton has a boiling point of −242.1°F. This means that krypton boils and becomes a gas at temperatures above −242.1°F. Is krypton a solid, liquid, or gas at room temperature?

4. You have some krypton stored at −250°F in a special container in your laboratory.

   a. If you heat the krypton by 7.5°F, does it boil? If not, how much more must it be heated to become a gas?

   b. Krypton freezes and becomes a solid below −249.9°F. What is the difference between krypton's boiling and freezing points?

5. The melting points of various elements are given in parts a–d. In each part, circle the element with the higher melting point, then determine the difference between the two melting points.

   a. radon: −96°F, xenon: −169.4°F

   b. fluorine: −363.3°F, nitrogen: −345.8°F

c. bromine: −7.2°C, cesium: 28.4 °C

d. argon: −308.6°F, hydrogen: 434.6°F

6. Complete the table to indicate whether each element is solid (below its freezing point), liquid (between its freezing and boiling points), or gas (above its boiling point) at 82°F.

How Low Can It Go?

ELEMENT	FREEZING POINT	BOILING POINT	STATE AT 82°F
Neon	−416.7°F	−411°F	
Lithium	356.9°F	2248°F	
Francium	80.6°F	1256°F	
Bromine	19.0°F	137.8°F	

Solving Equations Involving Decimal Numbers

In preparing the chemicals for a lab experiment in chemistry class, precise measurements are required. In Problems 7, 8 and 9, represent the unknown quantity with the variable x, and write an equation that accurately describes the situation. Solve the equation to get your answer.

7. You need to have precisely 1.250 grams of sulfur for your experiment. Your lab partner has measured 0.975 gram.

   a. How much more sulfur is needed?

   b. Convert your answer to milligrams.

8. You also need 2.120 grams of copper. This time your lab partner measures 2.314 grams of copper.

   a. How much of this copper needs to be removed for your experiment?

   b. Convert your answer to milligrams.

9. The acidic solution needed for your experiment must measure precisely 0.685 liter.

a. How much water must be added to 0.325 liter of sulfuric acid and 0.13 liter of hydrochloric acid to result in the desired volume?

b. Convert your answer to milliliters.

10. Solve each of the following equations for the unknown variable.

a. $x + 51.78 = 548.2$

b. $y - 14.5 = 31.75$

c. $1.637 - z = 9.002$

d. $0.1013 + t = 0$

SUMMARY
ACTIVITY 3.11

- Adding or subtracting decimals
  1. Line up the numbers vertically on their decimal points.
  2. Add or subtract as usual.
  3. Insert the decimal point in the answer directly in line with the decimal points of the numbers being added.
- To solve equations of the form $x + b = c$, add the opposite of b to both sides of the equation to obtain $x = c - b$.
- To solve equations of the form $x - b = c$, add b to both sides of the equation to obtain $x = b + c$.

EXERCISES
ACTIVITY 3.11

1. Use the table on page 241 to answer the following questions.

   a. Determine the difference between the boiling point and freezing point of neon.

   b. Determine the difference between the freezing points of lithium and neon.

   c. Determine the difference between the boiling points of lithium and neon.

2. Perform the indicated operations. Verify by using your calculator.

   a. $30.6 + 1.19$ b. $5.205 + 5.971$ c. $1.78 - 1.865$

   d. $0.298 + 0.113$ e. $2.242 - 1.015$ f. $-18.27 + 13.45$

   g. $-20.11 - 0.294$ h. $-10.078 - 46.2$ i. $15.59 - 0.374$

   j. $-876.1 - 73.63$ k. $0.0023 - 0.00658$ l. $-0.372 + 0.1701$

3. You have a $95.13 balance in your checking account. You will deposit your paycheck for $125.14 as soon as you get it 3 days from now. You need to write some checks now to pay some bills. You write checks for $25.48, $69.11, and $33.15. Unfortunately, the checks are cashed before you deposit your paycheck. The bank pays the $33.15 check but charges a $15.00 fee. What is your account balance after you deposit your paycheck?

4. One milligram is one thousandth of a gram.

   a. Express 1 milligram as a part of a gram by writing one thousandth as a decimal.

   b. Your father takes 40 milligrams (mg) per day of Zocor, a cholesterol-reducing medication. Express 40 milligrams as a decimal part of a gram.

   c. Today, your father has taken 0.010 gram of Zocor. Let x represent the number of grams of Zocor he still needs today. Write an equation and solve for x.

5. A full bottle of Joy detergent contains 1.75 pints. Your son empties a partially used bottle of Joy into a half-pint container. The amount fits perfectly! Choose a variable to represent the amount of detergent already used, in pints. Write an equation for this situation, and solve for your variable, expressing the answer as a decimal.

6. A tube of Colgate Whitening Toothpaste and Mouthwash contains 4.6 ounces. Last month, your family used up a 7.5-ounce tube of toothpaste. Let t be the number of additional ounces you would need for your family this month. Write an equation and solve for t.

7. A dental floss package you have contains 91.4 meters of cinnamon-flavored dental floss. You also have a small package of dental floss that you notice contains only 11 meters of floss. Let x be the difference in length for the two kinds. Write an equation and solve for x.

8. Arm & Hammer baking soda deodorant weighs 56.7 grams. Sure Clear Dry deodorant weighs 45.0 grams. Let x be the difference in weight. Write an equation and solve for x.

9. Solve each equation for the given variable.

   a. $1.04 + 0.293 + x = 5.3$

b. $0.01003 - x = 0.0091$

c. $y - 7.623 = 84.212$

d. $1.626 + b = 14.503$

e. $x + 8.28 = 3.4$

f. $0.0114 + z = 0.005 + 0.0101$

10. You need precisely 1.25 grams of mercury for a lab experiment. If you have 0.76 gram of mercury in your test tube, how much more mercury do you need? Write an equation to solve for this unknown amount.

11. The human body requires many minerals and vitamins. The minimum daily requirement of magnesium has been set at 400 milligrams. If your intake today has been 0.15 gram, how many milligrams of magnesium do you still need to meet the minimum daily requirement? Write an equation to solve for this unknown amount.

CLUSTER 3 What Have I Learned?

1. How do you determine which of two decimal numbers is the larger?

2. Why can zeros written to the right of the decimal point and after the rightmost nonzero digit be omitted without changing its value?

3. a. Round 0.31 to the nearest tenth.

   b. Determine the difference between 0.31 and 0.3

   c. Determine the difference between 0.4 and 0.31

   d. Which of the two differences you calculated is smaller, the one in part b or in part c?

   e. In the light of your answer to part d, give a justification for the rounding off procedure.

4. Which step in the process of adding or subtracting decimals must be done first and very carefully?

5. Are the rules for addition or subtraction of decimals different from those of whole numbers?

CLUSTER 3 How Can I Practice?

1. Write the value of the given decimal number.

   a. 3495.065

   b. 71,008.0049

   c. 13,053.3891

   d. 6487.08649

2. Write the following in decimal notation.

   a. Eighty-two thousand, seventy-six and four hundred eight ten-thousandths.

   b. Six hundred eleven thousand, seven hundred twelve and sixty-eight hundred-thousandths.

   c. Three million, five thousand, ninety-two and forty-one thousandths.

   d. Forty-nine thousand, eight hundred nine and six thousand four hundred thirteen hundred-thousandths.

   e. Nine billion, five and twenty-eight hundredths.

   f. Ninety-three million, five hundred sixty-four thousand, seven hundred and forty-nine hundredths.

3. Rearrange the following decimals from smallest to the largest.

   a. 0.9 0.909 0.099 0.0099 0.9900

   b. 0.384 0.0987 0.392 0.561 0.0983

   c. 1.49 1.23 1.795 0.842 0.01423

   d. 0.0838 0.8383 0.3883 0.00888 0.3838

Exercise numbers appearing in color are answered in the Selected Answers appendix.

4. a. Round 639.438 to the nearest hundredth.

b. Round 31.2695 to the nearest thousandth.

c. Round 182.09998 to the nearest ten-thousandth.

d. Round 59.999 to the nearest tenth.

5. Add the following decimals. Verify by using your calculator.

a. $3.28 + 17.063 + 0.084$

b. $628.13 + 271.78 + 68.456$

c. $9628.13 + 271.78 + 6814.56 + 13.91$

d. $171.004 + 18.028 + 51.68 + 4.5$

e. $-69.73 + (-198.32)$

f. $-131.02 + (-7.689)$

g. $18.62 - 29.999$

h. $-13.58 + 6.729$

i. $258.1204 + 49.0683 + 71.099$

j. $8171.004 + 4318.028 + 51.683 + 4.495$

6. Subtract the following decimals. Verify by using your calculator.

a. $100 - 71.998$

b. $212.085 - 63.0498$

c. $329.79 - 84.6591$

d. $8000.49 - 6583.725$

e. $678.146 - 39.08$

f. $-72.43 - 8.058$

g. $49.62 - (-6.088)$

h. $-18.173 - 12.59$

7. Evaluate.

a. $x + y$, where $x = -7.29$ and $y = 4.8$

b. $h + k$, where $h = -6.13$ and $k = -5.017$

c. $x - y$, where $x = -13.478$ and $y = 7.399$

d. $p - r$, where $p = -4.802$ and $r = -19.99$

8. Solve each of the equations for x.

a. $x + 23.49 = -71.11$

b. $x - 13.14 = 69.17$

c. $-28.99 + x = -55.03$

d. $35.17 + x = 12.19$

9. Your hourly salary was increased from $6.75 to $7.04. How much was your raise?

10. Convert 3872 mg to grams.

11. Convert 7.34 km to meters.

12. Convert 392 mℓ to liters.

13. Convert 0.64 kg to grams.

14. Convert 4.5 km to mm.

15. Convert 0.014 liter to mℓ.

16. The boundaries around a parcel of land measure 378 m, 725 m, 489 m, and 821 m. What is the perimeter, measured in kilometers?

17. You have 670 grams of carbon, but need precisely 2.25 kg for your experiment. How much more carbon do you need?

18. You start with 1.275 liters of solution in a beaker. Into four separate test tubes you pour 45 milliliters, 75 milliliters, 90 milliliters, and 110 milliliters of the solution. How much solution is left in your original beaker?

19. To prepare for you new job, you spent $25.37 for a shirt, $39.41 for pants, and $52.04 for shoes. How much did you spend?

20. The gross monthly income for your new job is $2105.96. You have the following monthly deductions: $311.93 for federal income tax, $64.72 for state income tax, and $161.11 for Social Security. What is your take-home pay?

CHAPTER 3 SUMMARY

The bracketed numbers following each concept indicate the activity in which the concept is discussed.

CONCEPT / SKILL	DESCRIPTION	EXAMPLE
Comparing integers [3.1]	The numbers increase from left to right on a number line. The further to the left a number is, the smaller it is.	$-18 < -3$ $-32 < 1$
Absolute Value [3.1]	The distance of a number from zero on the number line represents its absolute value. The absolute value of a number is always nonnegative.	$\|0\| = 0$ $\|-17\| = 17$ $\|13\| = 13$
Adding/Subtracting Integers [3.2]	1. Rules for adding integers:	
	a. When adding two numbers with the same sign, add the absolute values of the numbers. The sign of the sum is the same as the sign of the numbers being added.	$4 + (+7) = 4 + 7 = 11$ $-3 + (-5) = -8$
	b. When adding two numbers with different signs, find their absolute values and then subtract the smaller from the larger. The sign of the sum is the sign of the number with the larger absolute value.	$-5 + (+7) = 2$ $+10 + (-13) = -3$
	2. Rules for subtracting integers: To subtract two integers, change the operation of subtraction to addition and change the number being subtracted to its opposite; then follow the rules for adding integers.	$-13 - (-5) =$ $-13 + 5 = -8$ $-11 - (+7) =$ $-11 + (-7) = -18$
Evaluating expressions [3.3]	To evaluate an expression, substitute the given number for the letter. Perform the arithmetic, using the order of operations.	Evaluate $a - b$, where $a = 15$, $b = -7$. $15 - (-7) = 15 + 7 = 22$
Solving equations of the form $x + b = c$ and $x - b = c$ [3.3], [3.4], [3.7], [3.11]	• For $x + b = c$, add the opposite of b to both sides of the equation to obtain $x = c - b$. • For $x - b = c$, add b to both sides of the equation to obtain $x = c + b$.	$x + 3 = 19$ $-3 \quad -3$ $x = 16$ $x - 4 = -2$ $x - 4 + 4 = -2 + 4 = 2$ $x = 2$

CONCEPT / SKILL	DESCRIPTION	EXAMPLE
Points plotted on a rectangular coordinate system [3.4]	An ordered pair is always given in the form of (x, y), where x is the input and y is the output. The x-axis (horizontal) and the y-axis (vertical) determine a coordinate plane, divided into four quadrants numbered counterclockwise from the upper right.	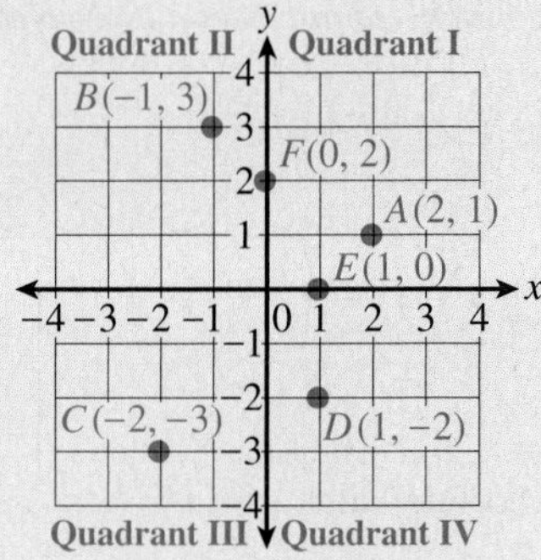 A: (2, 1) B: (−1, 3) C: (−2, −3) D: (1, −2) E: (1, 0) F: (0, 2)
Greatest common factor (GCF) [3.5]	The greatest common factor of two numbers is the largest factor of both numbers.	GCF of 18 and 12 is 6.
Equivalent fractions [3.5]	To write an equivalent fraction, multiply or divide both the numerator and the denominator of a given fraction by the same number.	$\frac{4}{5} = \frac{?}{25}$ $\frac{4}{5} \cdot 1 = \frac{4}{5} \cdot \frac{5}{5}$ $= \frac{20}{25}$
Reducing fractions [3.5]	To reduce a fraction to lowest terms, divide the numerator and denominator by their greatest common factor (GCF).	$\frac{6}{12} = \frac{6}{12} \div 1$ $= \frac{6}{12} \div \frac{6}{6}$ $= \frac{6 \div 6}{12 \div 6} = \frac{1}{2}$
Converting mixed numbers into improper fractions [3.5]	To convert a mixed number into an improper fraction, multiply the whole-number part by the denominator and add the numerator. The result is the numerator of the improper fraction. The denominator remains unchanged.	$6\frac{2}{3} = \frac{6 \cdot 3 + 2}{3}$ $= \frac{20}{3}$
Converting improper fractions to a mixed number [3.5]	To convert an improper fraction to a mixed number, divide the numerator by the denominator. The quotient is the integer part of the mixed number. The remainder becomes the numerator and the denominator is unchanged.	$\frac{27}{4}$; $4\overline{)27}$ = 6, -24, remainder 3 $\frac{27}{4} = 6\frac{3}{4}$

QUADRANT	X COORDINATE	Y COORDINATE
I	+	+
II	−	+
III	−	−
IV	+	−

CONCEPT / SKILL	DESCRIPTION	EXAMPLE
Least common denominator (LCD) [3.5], [3.7]	To determine the LCD, identify the largest denominator of the fractions involved. Look at multiples of the larger denominator. The smallest multiple that is divisible by all the denominators involved is the LCD.	Find the LCD of $\frac{1}{12}, \frac{5}{18}$: The largest denominator is 18. Multiples of 18 are 18, 36, 54, ... So the LCD is 36.
Comparing fractions [3.5]	Write the fractions as equivalent fractions with a common denominator. The larger fraction is the one with the larger numerator.	Compare $\frac{6}{7}, \frac{8}{9}$. $\frac{6}{7} = \frac{54}{63}; \frac{8}{9} = \frac{56}{63}$ $56 > 54$ $\frac{8}{9} > \frac{6}{7}$
Adding and subtracting fractional numbers with the same denominator [3.6]	• If you have two fractions with common denominators, you add and subtract them by adding or subtracting the numerators, leaving the denominator the same. • If you have an improper fraction as a result, convert it to a mixed number. • If the fraction in your answer is not in lowest terms, reduce it. • If you have a mixed number, you add or subtract the whole part separately.	$2\frac{13}{15} + 1\frac{7}{15}$ $= 3\frac{13+7}{15}$ $= 3\frac{20}{15}$ $15\overline{)20}$ = 1; -15; 5 $3 + 1\frac{5}{15} = 4\frac{5}{15} = 4\frac{1}{3}$
Adding and subtracting fractions with different denominators [3.7]	First find the LCD, then convert each fraction to an equivalent fraction, using the LCD you found. Add or subtract the numerators, leaving the LCD in the denominator. If you get an improper fraction, convert to a mixed number. If the fraction is not in lowest terms, reduce it.	Add: $\frac{3}{4} + \frac{5}{7}$ $\frac{3}{4} = \frac{21}{28}$ $+\frac{5}{7} = \frac{20}{28}$ $\frac{41}{28} = 1\frac{13}{28}$
Converting a decimal to a fraction or a mixed number [3.8]	1. Write the nonzero integer part of the number and drop the decimal point. 2. Write the fractional part of the number as the numerator of a new fraction, whose denominator is the same as the place value of the rightmost digit. If possible, reduce the fraction.	3.0025 $= 3\frac{25}{10{,}000}$ $= 3\frac{1}{400}$
Comparing decimals [3.8], [3.10], [3.11]	To compare two or more decimals, line up the decimal points and compare the digits that have the same place value, moving from left to right. The first number with the largest digit in the same "value place" is the largest number.	Compare 0.035 and 0.038 0 · 0 3 5 0 · 0 3 8 same same same $8 > 5$ $0.038 > 0.035$

CONCEPT / SKILL	DESCRIPTION	EXAMPLE
Reading and writing decimals [3.8]	To name a decimal number, read the digits to the right of the decimal point as an integer, and then attach the place value of its rightmost digit. Use "and" to separate the integer part from the decimal or fractional part.	• Write 3.42987 in words. three and forty-two thousand nine hundred eighty-seven hundred thousandths • Write "seven and sixty-nine thousandths" in standard form. 7.069
Rounding decimals [3.8]	1. Locate the digit with the specified place value. 2. If the digit directly to its right is less than 5, keep the digit that is in the specified place and delete all the digits to the right. 3. If the digit directly to its right is 5 or greater, increase this digit by 1 and delete all the digits to the right. 4. If the specified place value is located in the integer part, proceed as instructed in step 2 or step 3. Then insert trailing zeros to the right and up to the decimal point as place holders. Drop the decimal point.	Round 0.985 to the nearest hundredth. *Answer:* 0.99 Round 0.653 to the nearest hundredth. *Answer:* 0.65 Round 424.6 to the nearest ten. *Answer:* 420
The metric system [3.9]	Basic metric units for measuring: distance : meter (m) mass : gram (g) volume : liter (ℓ) Larger and smaller units in the metric system are based on the decimal system. The prefix before the basic unit determines the size of the unit, as shown in the table.	1000 mm = 1 m 100 cm = 1 m 10 dm = 1 m 0.001 m = 1 mm 0.01 m = 1 cm 0.1 m = 1 dm 1000 g = 1 kg 0.001 kg = 1 g 100 mℓ = 0.1 ℓ 243 mℓ = 0.243 ℓ 6.8 ℓ = 6800 mℓ 735 cm = 0.735 m 3.92 km = 3920 m

UNIT PREFIX	ABBR.	SIZE IN BASIC UNITS
milli-	m-	0.001
centi-	c-	0.01
deci-	d-	0.1
deca-	da-	10
hecto-	h-	100
kilo-	k-	1000

CONCEPT / SKILL	DESCRIPTION	EXAMPLE
Adding and subtracting decimals [3.10], [3.11]	1. Rewrite the numbers vertically, lining up the decimal points. 2. Add or subtract as usual. 3. Insert the decimal point in the answer directly in line with the decimal points of the numbers being added.	Add: 3.689 + 41 + 12.07 3.689 41.000 Assume zeros for any +12.070 numbers not shown. 56.759 Decimal point carries down.

CHAPTER 3 GATEWAY REVIEW

1. Determine the absolute value for each of the following.

a. $|15|$ **b.** $|23|$ **c.** $|0|$ **d.** $|-42|$ **e.** $|67|$

2. Perform the indicated operations.

a. $4 - (+13)$ **b.** $11 - (+17)$ **c.** $5 - (-11)$

d. $-2 + (-12)$ **e.** $27 + (-15)$ **f.** $15 + (-23)$

g. $-18 + (-35)$ **h.** $-27 + (+21)$

3. Evaluate.

a. $x - y$, where $x = 3$ and $y = -3$

b. $-x + y$, where $x = 5$ and $y = -2$

c. $-x - y$, where $x = -3$ and $y = -7$

d. $x + y$, where $x = -2$ and $y = -5$

4. Solve for x.

a. $x - 15 = -17$ **b.** $x + 7 = -12$ **c.** $x + 18 = 3$

d. $-11 + x = 32$ **e.** $-23 + x = -61$

5. Translate the following into equations where x represents the number. Then solve for x.

a. A number increased by eighteen is negative seven.

b. The sum of a number and eleven is twenty-nine.

Answers to all Gateway exercises are included in the Selected Answers appendix.

c. Fifteen increased by a number is negative twenty-eight.

d. The difference of a number and twenty is thirty-nine.

e. A number subtracted from eight is twelve.

f. Seventeen subtracted from a number is forty-two.

g. Thirteen less than a number is negative thirty-four.

6. Convert each to a mixed number.

a. $\frac{63}{5}$ **b.** $\frac{63}{7}$

c. $\frac{77}{15}$ **d.** $\frac{77}{13}$

7. Convert each to an improper fraction.

a. $4\frac{3}{5}$ **b.** $2\frac{5}{7}$

c. $5\frac{2}{11}$ **d.** $10\frac{3}{8}$

8. Compare the following fractions.

a. $\frac{3}{7} \square \frac{2}{5}$ **b.** $\frac{4}{11} \square \frac{1}{3}$

c. $\frac{4}{7} \square \frac{5}{8}$ **d.** $\frac{6}{11} \square \frac{1}{12}$

e. $\frac{7}{10} \square \frac{5}{8}$ **f.** $\frac{8}{11} \square \frac{3}{4}$

CHAPTER 4

MULTIPLICATION AND DIVISION OF RATIONAL NUMBERS

Have you ever wondered how your college grade point average is computed? The skills involved are the same as those necessary to answer financial questions, to tile a bathroom, or to calculate a baseball player's batting average.

In Chapter 4, you will learn the skills needed for solving problems involving multiplying and dividing integers, fractions, and decimals.

CLUSTER 1 Multiplying and Dividing Integers

ACTIVITY 4.1

Are You Physically Fit?

OBJECTIVES

1. Multiply and divide integers.
2. Perform calculations involving a sequence of operations.
3. Apply exponents to integers.
4. Identify properties of calculations involving multiplication and division with zero.

As a member of a health-and-fitness club, you have a special diet and exercise program developed by the club's registered dietitian and your personal trainer. Your weight gain (positive value) or weight loss (negative value) over the first 6 weeks of the program is recorded in the following table.

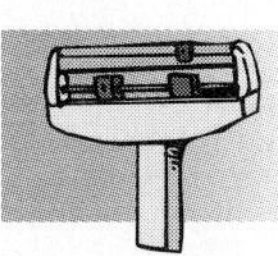

Weight and See

NUMBER OF WEEKS	1	2	3	4	5	6
CHANGE IN WEIGHT (LB.)	−3	−3	4	−3	−3	4

1. a. Counting only those weeks in which you gained weight, what was your weight gain during the 6-week period? Write your answer as an integer (signed number) and in words.

   b. Counting only those weeks in which you lost weight, what was your weight loss during the 6-week period? Write your answer as an integer (signed number) and in words.

c. Explain how you calculated the answers to parts a and b.

d. At the end of the first six weeks, what is the total change in your weight?

There are two ways to determine the answers to parts a and b of Problem 1. One way to determine the increase in weight in part a is by repeated addition:

$$4 + 4 = 8 \text{ lb.}$$

A second way uses the fact that multiplication is repeated addition. So, the increase in weight is also determined by the product:

$$4(2) = 8 \text{ lb.}$$

Similarly in part b, the weight loss is determined by repeated addition:

$$-3 + (-3) + (-3) + (-3) = -12.$$

Again similar to part a, multiplication can be used to determine the weight loss:

$$(-3)(4) = -12.$$

In general, **the product of a negative number and a positive number is negative.** This is true when a negative number is multiplied by a positive number, as Problem 1 showed. It is also true when a positive number is multiplied by a negative number. For example, the product, $3(-2)$ may be interpreted as adding two instances of 3 **but in the negative direction,** resulting in a negative product. That means $3(-2) = -6$.

2. Multiply each of the following; then check by using your calculator.

a. $5(-2)$ b. $(-2)(5)$ c. $7(-8)$

d. $-8(7)$ e. $6(-4)$ f. $-4(6)$

3. Compare parts a and b in Problem 2 as follows:

a. Compare the order of the factors. Are they the same or different?

b. Now compare the products. Do they have the same sign?

c. What conclusion can you make about multiplying a negative and positive number?

Problem 3 illustrates the fact that multiplication of integers is commutative. For example,

$$3(-2) = (-2)(3) = -6.$$

4. Compare parts c and d in Problem 2 and write what you observe. Do the same for parts e and f.

So far in this activity, you have seen that the product of two positive integers is positive and that the product of a positive integer and a negative integer is negative. One more case has to be considered to answer the question: What is the sign of the product of two negative integers? The following problem suggests a pattern that leads to the answer.

5. a. Fill in the blanks in the following table to complete the pattern begun in the first three lines:

4(−2)	−8
3(−2)	−6
2(−2)	−4
1(−2)	
0(−2)	
−1(−2)	
−2(−2)	
−3(−2)	
−4(−2)	
−5(−2)	

b. What does the pattern you completed suggest about the sign of the product of two negative integers? Illustrate with a specific example.

6. In each of the following, multiply the two integers. Then check your answer using your calculator.

a. $(-2)(-4)$ **b.** $-6(-7)$ **c.** $(-1)(4)$

d. $5(-8)$ **e.** $(-3)(-9)$ **f.** $(-12)(3)$

Product of Two Integers

Rule 1: The product of two integers with the same sign is positive.

$$(+)(+) = (+) \rightarrow (4)(5) = 20$$
$$(-)(-) = (+) \rightarrow (-4)(-5) = 20$$

Rule 2: The product of two integers with opposite signs is negative.

$$(-)(+) = (-) \rightarrow (-4)(5) = -20$$
$$(+)(-) = (-) \rightarrow (4)(-5) = -20$$

Division of Integers

Your friend gained 15 pounds over a 5-week period. If his weight gain was the same each week, then the calculation $15 \div 5 = 3$ or $\frac{15}{5} = 3$ shows that he gained 3 pounds each week. You can check that a gain of 3 pounds per week is correct by multiplying 5 by 3 to obtain 15. In other words, the quotient $\frac{15}{5}$ is 3 because $5 \cdot 3 = 15$ by a multiplication check.

Another friend lost 15 pounds over the same 5-week period, losing the same amount each week. The calculation $-15 \div 5 = -3$ (or $\frac{-15}{5} = -3$) shows that she lost 3 pounds each week. The quotient $\frac{-15}{5}$ is -3. The product $5(-3) = -15$ shows the quotient $\frac{-15}{3}$ has to be -3.

7. Evaluate each of the following. Use the multiplication check to verify your answers.

a. $27 \div 9$ **b.** $-32 \div 8$ **c.** $18 \div (-2)$

d. $\frac{-10}{5}$ **e.** $\frac{21}{-7}$ **f.** $\frac{38}{2}$

8. a. Use the multiplication check for division to obtain the only reasonable answer for $(-22) \div (-11)$.

b. What rule does part a suggest for dividing a negative integer by a negative integer?

c. Evaluate each of the following and verify your answer by the multiplication check.

i. $(-42) \div (-7)$ **ii.** $\frac{-9}{-3}$ **iii.** $(-7) \div (-1)$

9. a. The sign of the quotient of two integers with the same sign is

______________.

b. The sign of the quotient of two integers with opposite signs is

______________.

Quotient of Two Integers

Rule 1: The quotient of two integers with the same sign is positive.

$$\frac{(+)}{(+)} = (+) \longrightarrow \frac{12}{4} = 3$$

$$\frac{(-)}{(-)} = (+) \longrightarrow \frac{-12}{-4} = 3$$

Rule 2: The quotient of two integers with opposite signs is negative.

$$\frac{(-)}{(+)} = (-) \longrightarrow \frac{-12}{4} = -3$$

$$\frac{(+)}{(-)} = (-) \longrightarrow \frac{12}{-4} = -3$$

Calculations Involving a Combination of Operations

10. A friend joins you on your health-and-fitness program. The following table lists his weight loss and gain during a 5-week period.

WEEK	1	2	3	4	5
WEIGHT CHANGE (LB.)	−4	−3	2	−3	−2

Your friend computes his average weight change by adding the weekly changes in weight and then dividing the result by the number of weeks. You volunteer to check his results.

a. What is his change in weight for the 5 weeks?

b. What is your friend's average weight change for the 5-week period?

c. Did your friend have an average gain or an average loss of weight during the 5-week period? Explain.

11. The following table gives weight gains and losses for 16 persons during a given week.

NUMBER OF PERSONS	WEIGHT CHANGE
2	−5
3	−4
4	−3
2	−2
1	0
3	1
1	3

a. What is the weight change of the group during the given week?

b. What is the average weight change for the group of 16 persons?

c. Does this average change represent a weight gain or loss?

12. a. What is the sign of the product $(-3)(-5)(-7)(-2)$? Explain.

b. What is the sign of the product of $(-1)(-1)(-1)(-1)(-1)(-1)(-1)$? Explain.

c. If you multiplied 16 factors of -1, what would be the sign of the product?

d. What can you conclude about the sign of a product with an odd number of factors with negative signs?

e. What can you conclude about the sign of a product with an even number of factors with negative signs?

13. Perform the indicated operations. Check the results using your calculator.

a. $-5 \cdot (-4)$

b. $-50 \div (-10)$

c. $2 \cdot 6$

d. $6 \div 3$

e. $-3(-6)(-2)$

f. $5(-3)(-4)$

g. $-15 \div -3$

h. $2(-3)(5)(-1)$

i. $(-2)(6)$

j. $60 \div (-5)$

k. $(-1)(-1)(-1)(-1)(-1)(-1)$

l. $(-2)(-3)(-4)(5)(-6)$

Exponents and Negative Integers

The product $(-3)(-3)$ can be written as $(-3)^2$. The integer -3 is the factor that appears two times in the product and is called the **base.** The integer 2 represents the number of factors of -3 in the product and is called the **exponent.**

14. a. Evaluate $(-3)(-3)$.

b. Evaluate $(-3)^2$ using the exponent key on your calculator. Be sure to key in the parentheses when you do the calculation.

c. Suppose, when using your calculator, you omit the parentheses in $(-3)^2$ and instead you calculate the expression -3^2. Does -3^2 have the same value as $(-3)^2$?

You should think of -3^2 as the opposite of 3^2. Therefore,

$$-3^2 = -(3^2) = -(3)(3) = -9.$$

15. Evaluate the following expressions by hand. Use your calculator to check the results.

a. -5^2

b. $(-5)^2$

c. $(-3)^3$

d. -1^4 e. $2 - 4^2$ f. $(2 - 4)^2$

g. $-5^2 - (-5)^2$ h. $(-1)^8$ i. $-5^2 + (-5)^2$

Multiplication and Division Involving 0

16. **a.** Recall that multiplication is repeated addition. For example, $4 \cdot 3 = 4 + 4 + 4$. Use this fact to write $0 \cdot 6$ as an addition problem.

b. Compute the answer in part a.

c. What is 0 times any integer? Explain.

17. **a.** Recall the multiplication check for division. For example, $8 \div 2 = 4$ because $2 \cdot 4 = 8$. Use the multiplication check to compute the answer to the division problem $0 \div 7$.

b. What is 0 divided by any nonzero integer? Explain.

c. Use the multiplication check to explain why $7 \div 0$ has no answer.

d. Can any integer be divided by 0? Explain.

18. Evaluate each of the following, if possible. If not possible, explain why not.

a. $0 \cdot (-8)$ **b.** $12 \cdot 0$ **c.** $0 \div 4$

d. $-6 \div 0$ **e.** $0 \div (-6)$ **f.** $0 \div 0$

When multiplying any integer by 0, the answer is 0. When dividing 0 by any nonzero integer, the answer is 0. No integer can be divided by 0.

SUMMARY ACTIVITY 4.1

1. Product of Two Integers

Rule 1: The product of two integers with the same sign is positive.

$$(+)(+) = (+) \rightarrow (4)(5) = 20$$

$$(-)(-) = (+) \rightarrow (-4)(-5) = 20$$

Rule 2: The product of two integers with opposite signs is negative.

$$(-)(+) = (-) \rightarrow (-4)(5) = -20$$

$$(+)(-) = (-) \rightarrow (4)(-5) = -20$$

2. Quotient of Two Integers

Rule 1: The quotient of two integers with the same sign is positive.

$$\frac{(+)}{(+)} = (+) \rightarrow \frac{12}{4} = 3$$

$$\frac{(-)}{(-)} = (+) \rightarrow \frac{-12}{-4} = 3$$

Rule 2: The quotient of two integers with opposite signs is negative.

$$\frac{(-)}{(+)} = (-) \rightarrow \frac{-12}{4} = -3$$

$$\frac{(+)}{(-)} = (-) \rightarrow \frac{12}{-4} = -3$$

3. **a.** The base for $(-2)^4$ is -2; $(-2)^4 = (-2)(-2)(-2)(-2) = 16$.
   b. The base for -2^4 is 2; $-2^4 = -(2)(2)(2)(2) = -16$ because the expression -2^4 represents the opposite of 2^4.

4. **a.** The product of an integer and 0 is always 0.
   b. 0 divided by any nonzero integer is 0.
   c. No integer may be divided by 0.

EXERCISES ACTIVITY 4.1

1. Determine the product or quotient. Check your results using your calculator.

a. $-6 \cdot 7$

b. $(3)(-8)$

c. $8(-200)$

d. $(-1)(7)$

e. $8 \cdot 0$

f. $(-3)(-268)$

g. $-6(-40)$

h. $-30 \div 6$

i. $(-45) \div (-9)$

Exercise numbers appearing in color are answered in the Selected Answers appendix.

j. $-3 \div 0$ k. $-25 \div (-1)$ l. $0 \div (-6)$

m. $\frac{-64}{16}$ n. $\frac{-56}{-8}$ o. $\frac{0}{-167}$

2. Perform the following calculations. Check the results using your calculator.

a. $(-2)(-4)(1)(-5)$ b. $(3)(-4)(2)(-4)(2)$ c. $(-1)(-1)(-1)(-1)$

d. $(-1)(-5)(-11)$ e. -6^2 f. $(-6)^2$

g. $(-4)^3$ h. -4^3

3. The daily low temperature in Lake Placid, New York, dropped 5 degrees Celsius per day for 6 consecutive days. Use a negative integer to represent the drop in temperature each day and calculate the total drop over the 6-day period.

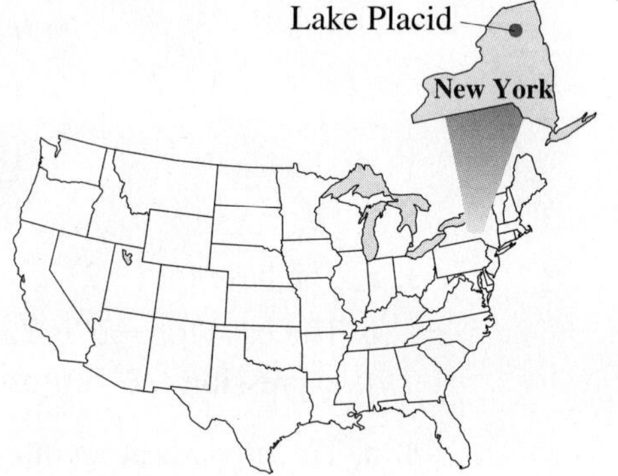

4. You recently bought a piece of property in a rural area and need a well for your water supply. The well-drilling company you hire says they can drill down about 25 feet per day. It takes the company approximately 5 days to reach water. Represent the company's daily drilling rate by a negative integer and calculate the approximate depth of the water supply. Report your result as a negative integer.

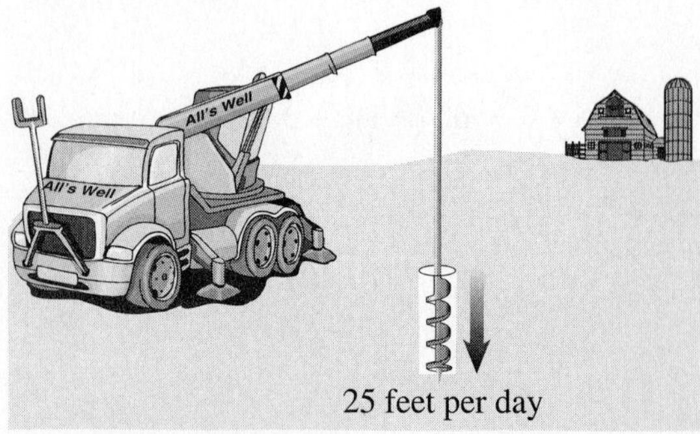

5. You plan to dive to a shipwreck located 168 feet below sea level. You can dive at the rate of approximately 2 feet per second. Use negative integers to represent the distance below sea level in the following calculations.

   a. Calculate the depth you dove in 1 minute.

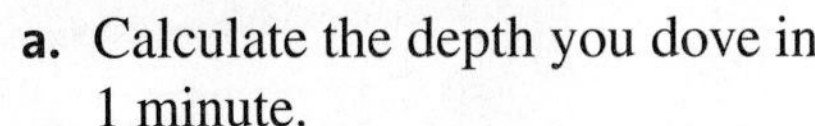

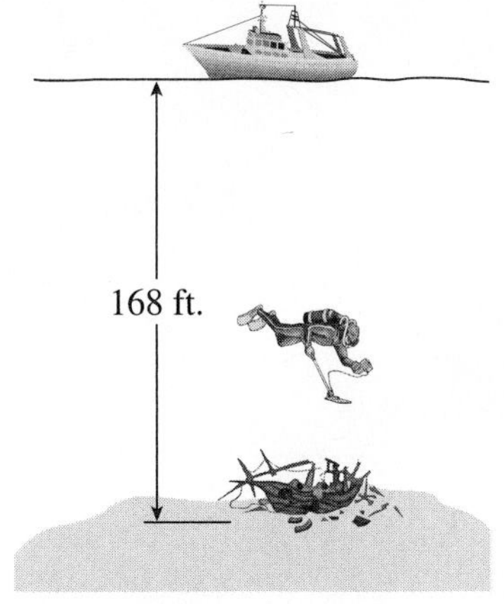

   b. Calculate where you are in relationship to the shipwreck after 1 minute.

6. You are one of three equal partners in a company. Your company experienced a loss of \$150,000 in 2002. What was your share of the loss? Write the answer in words and as a signed number.

7. The running back of your favorite football team lost 6 yards per carry for a total of a 30-yard loss. Use negative integers to represent a loss in yardage and calculate the number of plays that were involved in this yardage loss.

8. To discourage random guessing on a multiple-choice exam, the instructor assigns 5 points for a correct response, −2 points for an incorrect answer, and 0 points for leaving the question blank. What is the score for a student who had 22 correct answers, 7 incorrect answers, and left 6 questions blank?

9. You are interning at the weather station. This week the low temperatures were 4°C, −8°C, 4°C, −1°C, −2°C, −2°C, −2°C. Determine the average low temperature for the week.

10. It has not rained in northern New York State for a month. You are worried about the depth of the water in your boathouse. You need 3 feet of water so that your boat motor does not hit bottom and get stuck in the mud. Each week you measure the water depth. The initial measurement was 42 inches. The following table tells the story of the change in depth each week.

WEEK	1	2	3	4
CHANGE IN WATER LEVEL (INCHES)	−2	−2	−2	−3

a. What was the depth of the water in your boathouse after week 1?

b. Determine the total change in the water level for the first 3 weeks.

c. Include the fourth week in part b. Determine the total change in the water level.

d. Determine the depth of the water in your boathouse after the fourth week.

e. Was your motor in the mud? Explain.

11. Each individual account in your small business has a current dollar balance, which can be either positive (a credit) or negative (a debit). Your records show the following balances.

−$230	−$230	$350	−$230	−$230	$350

What is the total balance for these accounts?

12. You wrote four checks, each in the amount of $23, for your daughters to play summer league soccer. You had $82 in your checking account and thought you had just enough to cover the checks.

a. Write an arithmetic expression and simplify to determine if you will have enough money.

b. What possible mathematical error made you believe you could write the four checks?

ACTIVITY 4.2

Integers and Tiger Woods

OBJECTIVES

1. Use order of operations with expressions involving integers.
2. Apply the distributive property.
3. Evaluate algebraic expressions and formulas using integers.
4. Solve equations of the form $ax = b$, where $a \neq 0$, involving integers.

The popularity of golf has increased greatly in recent years, due in large part to the achievements of Tiger Woods. In the 1997 Masters Tournament, played every year in Augusta, Georgia, Woods won by an amazing 12 strokes, the widest margin of victory the tournament has ever seen. At 21 years of age, he became the youngest golfer to win the Masters Tournament, and the first of African or Asian descent. By 2005 he added three more Masters victories to his record.

In the game of golf, the object is to hit a ball with a club into a hole in the ground with as few strokes (or swings of the club) as possible. The golfer with the lowest total number of strokes is the winner. A standard game consists of 18 different holes laid out in a parklike setting. Each hole, depending upon its length and difficulty, is assigned a number of strokes that represents the average number of strokes a very good player should expect to need to get the ball into the hole. This number is called *par* (hence the cliché you may have heard, "that's par for the course").

Most golfers refer to how their score relates to par; they give a score as the number of strokes above or below par. In this way, par acts like zero. Scoring at par can be represented by zero, scoring below par by a negative integer, and scoring above par by a positive integer. Golfers also use special terminology for the number of strokes above or below par, as summarized in the following table.

Getting the Swing of It . . .

TERM	MEANING	DIFFERS FROM PAR
Double eagle	3 strokes less than par	−3
Eagle	2 strokes less than par	−2
Birdie	1 stroke less than par	−1
Par	Expected number of strokes	0
Bogey	1 stroke more than par	1
Double bogey	2 strokes more than par	2

In the Masters, the expected number of strokes to complete the 18 holes is 72. In the last round of the 2000 Masters Tournament, Tiger Woods had a score of 68. This meant that he completed the course in 4 fewer strokes than was expected. That is, he was 4 under par.

Rather than keeping a tally of the actual number of strokes, you can record the number of birdies, pars, bogeys, etc. For example, if you parred every hole, your score would be 72. You would not add or subtract anything to par.

1. Use the preceding table to answer the following questions.

   a. If you birdied each of the 18 holes at the Masters, what is your score?

   b. If you bogeyed every hole, what is your score?

c. If you birdied every hole on the first 9 holes, and bogeyed every hole on the last 9 holes, what is your score?

2. The Corning Country Club is the site of the LPGA Corning Classic Golf Tournament for Lady Professionals. The course consists of 18 holes and par is 70. One golfer summarized her scores for the first round by the following table.

Join the Club

SCORE ON A HOLE	NUMBER OF TIMES SCORE OCCURRED
Eagle: −2	1
Birdie: −1	5
Par: 0	9
Bogey: +1	2
Double bogey: +2	1

a. Describe the process you would use to determine the total number of strokes it took her to complete her first round (all 18 holes).

b. Calculate her score for the first round.

The golfer's first-round score can be determined by the arithmetic expression

$$70 + (-2) \cdot 1 + (-1) \cdot 5 + 0 \cdot 9 + 1 \cdot 2 + 2 \cdot 1.$$

The order of operations discussed in Chapter 1 is valid for all numbers, positive and negative. Therefore, in the given expression, multiplication is performed first (left to right) *before* addition or subtraction.

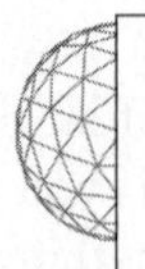

EXAMPLE 1 *Evaluate the expression* $7 + (-3) \cdot 5$ *using order of operations.*

$7 + (-3) \cdot 5$

$= 7 + (-15)$ Multiplication before addition

$= -8$ Addition of integers with opposite signs

3. Evaluate the following expressions using the order of operations. Then check the answer with your calculator.

a. $6 + 4 \cdot (-2)$

b. $-6 + 2 - 3$

c. $(6 - 10) \div 2$

d. $-3 + (2 - 5)$

e. $(2 - 7) \cdot 5$

f. $-3 \cdot (-3 + 4)$

g. $-2 \cdot 3^2 - 15$

h. $(2 + 3)^2 - 10$

i. $-7 + 8 \div (5 - 7)$

4. a. Evaluate the expression $-2(-3 + 7)$ by first performing the operation within the parentheses.

b. Evaluate the expression $-2(-3 + 7)$ by using the distributive property. Recall that $a(b + c) = ab + ac$

c. Compare the results from parts a and b.

Evaluating Algebraic Expressions

The International Golf Championship on the Men's PGA tour, played annually at Castle Pines Golf Club, Castle Rock, Colorado, uses a rather unique scoring system. Rather than counting strokes, players are awarded points on each hole depending on how well they played the hole. The points are summarized in the following table.

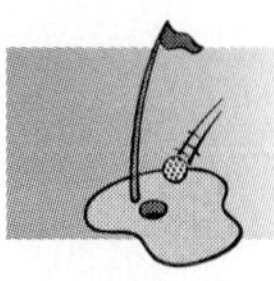

Stroke of Luck

SCORE	POINTS
Double eagle	8
Eagle	5
Birdie	2
Par	0
Bogey	−1
Double bogey or worse	−3

A player's score could be determined by the following expression:

$$8 \cdot a + 5 \cdot b + 2 \cdot c + 0 \cdot d + (-1) \cdot e + (-3) \cdot f,$$

where a = number of double eagles; b = number of eagles; c = number of birdies; d = number of pars; e = number of bogeys; and f = number of double bogeys or worse.

5. Determine the score of golfer Corey Pavin if his round included no double eagles, 1 eagle, 5 birdies, 9 pars, 2 bogeys, and 1 double bogey.

The process in Problem 5 is an example of evaluating an expression involving integers. Recall from Chapter 2 that you can evaluate an algebraic expression when you know the value of each variable. To evaluate, replace each variable with its given value, then calculate using the order of operations.

EXAMPLE 2 ***Evaluate the expression $3cd - 4d^2$ for $c = -2$ and $d = -5$.***

$3cd - 4d^2 = 3(-2)(-5) - 4(-5)^2$	Substituting values for c and d
$= 30 - 4(25)$	Simplifying using the order of operations
$= 30 - 100$	Simplifying using the order of operations
$= -70$	

6. Evaluate each of the following expressions for the given values.

 a. $6a + 3b - a^2$ for $a = -5$ and $b = 4$

 b. $-4xy + 3x - 6y$ for $x = -4$, $y = -5$

 c. $\dfrac{7c - 2d^2}{-3w}$ for $c = -10$, $d = -4$ and $w = 2$

Solving Equations of the Form $ax = b$ that Involve Integers

7. a. You join the fitness center and lose weight at the rate of 2 pounds per week for 10 weeks. Represent the rate of weight loss as a negative number and use it to determine your total weight loss.

 b. Write a verbal rule that expresses your total weight loss in terms of the rate of weight loss per week and the number of weeks.

 c. Let t represent the total weight loss, r the rate of weight loss per week, and w the number of weeks. Translate the verbal statement from part b into an equation.

d. One of the members of the Fitness Center lost a total of 42 pounds, represented by -42, at the rate of 3 pounds per week, represented by -3. Use the formula in part c to write an equation to determine the number of weeks it took the member to lose the weight.

e. Solve the equation in part d.

8. Solve the following equations for the given variables.

a. $-4x = 36$ **b.** $6s = -48$ **c.** $-32 = -4y$

SUMMARY
ACTIVITY 4.2

1. The order of operations first introduced in Chapter 1 for whole numbers also applies to integers.
   a. Perform all operations *inside* parentheses.
   b. Apply *all exponents* as you read the expression from left to right.
   c. Perform all *multiplications and divisions* from left to right.
   d. Perform all *additions and subtractions* from left to right.
2. To evaluate formulas or expressions involving integers, substitute for the variables and evaluate using the order of operations.
3. To solve the equation $ax = c$ for x, where $a \neq 0$, divide each side of the equation by a to obtain $x = \frac{c}{a}$.

EXERCISES
ACTIVITY 4.2

1. In 2001, Tiger Woods again won the Masters Tournament. His performance in the last round (all 18 holes) is given in the following table.

Mastering the Game

SCORE ON A HOLE	NUMBER OF TIMES SCORE OCCURRED
Birdie: -1	6
Par: 0	10
Bogey: $+1$	2

If par for the 18 holes is 72, determine Tiger's score in the last round of the tournament.

Exercise numbers appearing in color are answered in the Selected Answers appendix.

2. You own 24 shares of stock in a high-tech company and have been following this stock over the past 4 weeks.

WEEK	GAIN OR LOSS PER SHARE ($)
1	2
2	−3
3	4
4	−5

a. What is the net gain or loss in total value of the stock over the 4-week period?

b. If your stock was worth $18 per share at the beginning of the first week, how much was your stock worth at the end of the fourth week?

c. Let the variable x represent the cost per share of your stock at the beginning of the first week. Write an expression to determine the cost per share at the end of the fourth week.

d. Use the expression in part c to determine the value of your stock at the end of the fourth week if it was worth $27 per share at the beginning of the first week.

3. Evaluate the following. Check your results by using your calculator.

a. $3 + 2 \cdot 4$

b. $3 - 2(-4)$

c. $-2 + (2 \cdot 4 - 3 \cdot 5)$

d. $4 \cdot 3^2 - 50$

e. -6^2

f. $(-6)^2$

g. $(-2)^2 + 10 \div (-5)$

h. $12 - 3^2$

i. $-12 \div 3 - 4^2$

j. $(2 \cdot 3 - 4 \cdot 5) \div (-2)$

k. $(-13 - 5) \div 3 \cdot 2 \cdot 1$

l. $-5 + 9 \div (8 - 11)$

m. $14 - (-4)^2$

n. $3 - 2^3$

o. $3^2(-5)^2 \div (-1)$

4. Use the distributive property to evaluate the following.

a. $5(-3 + 4)$

b. $-2(5 - 7)$

c. $4(-6 - 2)$

5. Evaluate the following expressions using the given values.

a. $ab^2 + 4c$ for $a = 3, b = -5, c = -3$

b. $(5x + 3y)(2x - y)$ for $x = -4, y = 6$

c. $-5c(14cd - 3d^2)$ for $c = 3, d = -2$

d. $\dfrac{3ax - 4c}{4ac}$ for $a = -2, c = 4, x = -8$

6. You are on a diet to lose 15 pounds in 8 weeks. The first week you lost 3 pounds, and then you lost 2 pounds per week for each of the next 3 weeks. The fifth week showed a gain of 2 pounds, but the sixth and seventh each had a loss of 2 pounds. You are beginning your eighth week. How many pounds do you need to lose in the eighth week in order to meet your goal?

7. Translate each of the following into an equation and solve. Let x represent the number.

a. The product of a number and -6 is 54.

b. -30 times a number is -150.

c. -88 is the product of a number and 8.

d. Twice a number is -28.

8. Solve each equation.

a. $3x = 27$

b. $9s = -81$

c. $-24 = -3y$

d. $-5x = 45$

e. $-6x = -18$

For Exercises 9–11, write an equation to represent the situation and solve.

9. On average, you can lose 5 pounds per month while dieting. Your goal is to lose 65 pounds. Represent each weight loss by a negative number. How many months will it take you to do this?

10. As winter approached in Alaska, the temperature dropped each day for 19 consecutive days. The total decrease in temperature was 57°F, represented by -57. What was the average drop in temperature per day?

11. Over a 10-day period the low temperature for International Falls, Minnesota, was recorded as -18°F on 3 days, $-15°$ on 2 days, $-10°$ on 2 days, and $-8°$, $-5°$, and $-3°$ on the remaining 3 days. What was the average low temperature over the 10-day period?

CLUSTER 1 What Have I Learned?

1. When you multiply a positive integer, n, by a negative integer, is the result greater or less than n? Give a reason for your answer.

2. a. When you multiply a negative integer, n, by a second negative integer, is the result greater or less than n? Explain your answer.

b. Does your answer to part a depend on the absolute value of the second negative integer? Why or why not?

3. a. Complete the following table to list the rules for the multiplication and division of two integers.

MULTIPLICATION	DIVISION
$(+)\cdot(+)=$	$(+)\div(+)=$
$(+)\cdot(-)=$	$(+)\div(-)=$
$(-)\cdot(+)=$	$(-)\div(+)=$
$(-)\cdot(-)=$	$(-)\div(-)=$

b. Use the table you completed above to show how the rules for division are related to the rules for multiplication of two integers.

4. How will you remember the rules for multiplying and dividing two integers?

5. When you multiply several signed numbers, some positive and some negative, how do you determine the sign of the product?

6. When a negative number is raised to a power, is the result negative? Illustrate your answer with examples.

7. Explain why -7^2 is not the same as $(-7)^2$.

CLUSTER 1 How Can I Practice?

1. Calculate the following products and quotients. Use your calculator to check the results.

a. $-2 \cdot 8$

b. $(5)(-6)$

c. $-8(-20)$

d. $-9 \cdot 0$

e. $-36 \div 9$

f. $(-54) \div (-6)$

g. $-8 \div 0$

h. $\dfrac{72}{-12}$

i. $\dfrac{-24}{-3}$

j. $(-2)(-1)(3)(-2)(-1)$

k. $(5)(-1)(2)(-3)(-1)$

2. Evaluate each expression. Check the results using your calculator.

a. $-6 + 4(-3)$

b. $10 - 5(-2)$

c. $(6 - 26 + 5 - 10) \div -5^2$

d. $40 + 8 \div (-4) \cdot 2$

e. $-14 + 8 \div (10 - 12)$

f. $-27 \div 3 - 6^2$

g. $18 - 6 + 3 \cdot 2$

h. $-7^2 - 7^2$

i. $(-7)^2 - 7 \cdot 2 + 5$

3. Evaluate each expression using the given values of the variables.

a. $4x - 3(x - 2)$, where $x = -5$

b. $x^2 - xy$, where $x = -6$ and $y = 4$

c. $2(x + 3) - 5y$, where $x = -2$ and $y = -4$

4. In hockey, a defenseman's plus/minus record (+/− total) determines his success. If he is on the ice for a goal by the opposing team, he has a −1. If he is on the ice for a goal by his team, he gets a +1. A defenseman for a professional hockey team has the following on-ice record for five games.

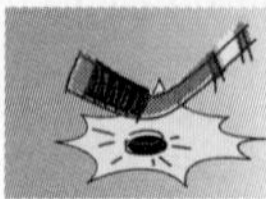

Shot...Score!

GAME	YOUR TEAM'S GOALS	OPPOSITION GOALS	+/− TOTAL
1	4	1	$1(4) + (-1)(1) = +3$
2	3	5	
3	0	3	
4	3	0	
5	1	6	

For example, an expression that represents his +/− total for game 1 is

$$1(4) + (-1)(1).$$

Evaluating this expression, he was a +3 for game 1.

a. Write an expression for each game and use it to determine the +/− total for that game.

b. Write an expression and use it to determine the defenseman's +/− total for the five games.

5. At the end of a hockey season a defenseman summarized his +/− total in the following chart. For example: In the first column, in three games his +/− total was +6. His total +/− for the 3 games was +18.

NO. OF GAMES	3	4	12	8	6	11	3	15	9	6	3
+/− TOTAL	+6	+4	+3	+2	+1	0	−1	−2	−3	−4	−5

Write an expression that will determine the +/− total for the season. What is his +/− total for the season?

6. The following table contains the daily midnight temperatures in Buffalo for a week in January.

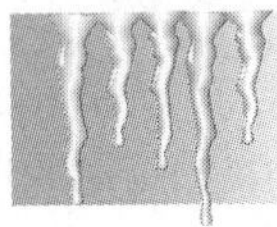

A Balmy −12°

DAY	Sun.	Mon.	Tues.	Wed.	Thurs.	Fri.	Sat.
TEMP. (°F)	−3°	6°	−3°	−12°	6°	−3°	−12°

a. Write an expression to determine the average daily temperature for the week.

b. What is the average daily temperature for the week?

7. Translate each of the following into an equation and solve. Let x represent the number.

a. Three times a number is -15.

b. -54 is the product of -3 and what number?

c. The product of a number and -4 is 48.

d. 90 is the product of -3, -6, and a number.

CLUSTER 2 Multiplying and Dividing Fractions

ACTIVITY 4.3

Get Your Homestead Land

OBJECTIVES

1. Multiply and divide fractions.
2. Recognize the sign of a fraction.
3. Determine the reciprocal of a fraction.
4. Solve equations of the form $ax = b$, $a \neq 0$, that involve fractions.

In 1862, Congress passed the Homestead Act. It provided for the transfer of 160 acres of unoccupied public land to each homesteader after paying a nominal fee and living on the land for 5 years. As part of an assignment, you are asked to determine the size of 160 acres, in square miles. Your research reveals that there are 640 acres in 1 square mile, so you draw the following diagram, dividing the square mile into four equal portions (quarters).

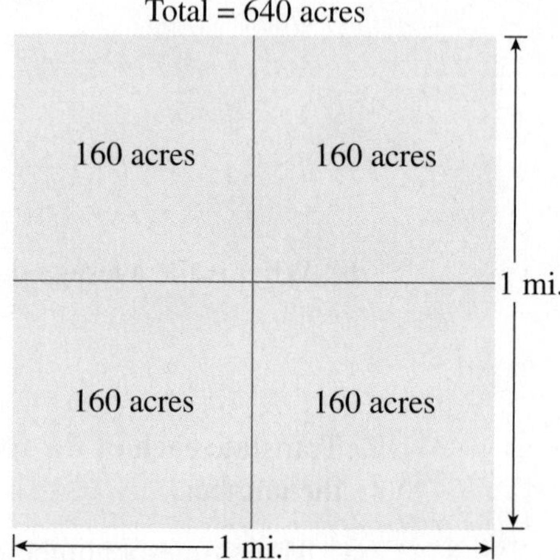

1. Use the diagram to determine the fractional part of the original square mile that the 160 acres represents.

You decide to investigate further. Suppose each 160-acre homestead is further divided into four equal squares. Each *new square* represents 40 acres, as illustrated in the diagram.

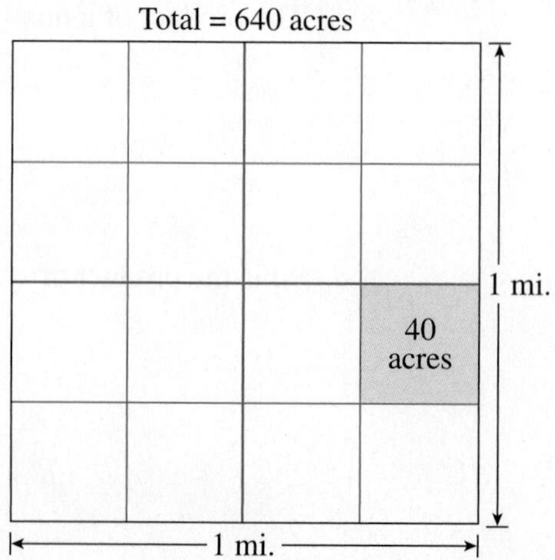

2. Use the diagram to determine the fractional part of the original square mile that 40 acres represents.

3. a. Use the diagram on the previous page to write the dimensions of a 40-acre square in terms of miles.

b. What is the area of a 40-acre square in square miles?

4. Since the area of a square is determined by multiplying the lengths of two sides, you can combine the results of Problem 3a and b as follows:

$$\left(\frac{1}{4}\text{ mile}\right) \cdot \left(\frac{1}{4}\text{ mile}\right) = \frac{1}{16}\text{ sq. mi.}$$

a. Suppose someone is able to acquire three adjacent lots of 40 acres along one side of the square mile. What fractional part of the square mile did he acquire?

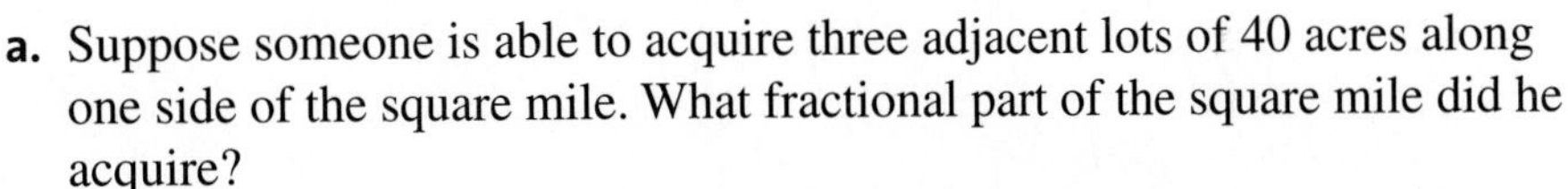

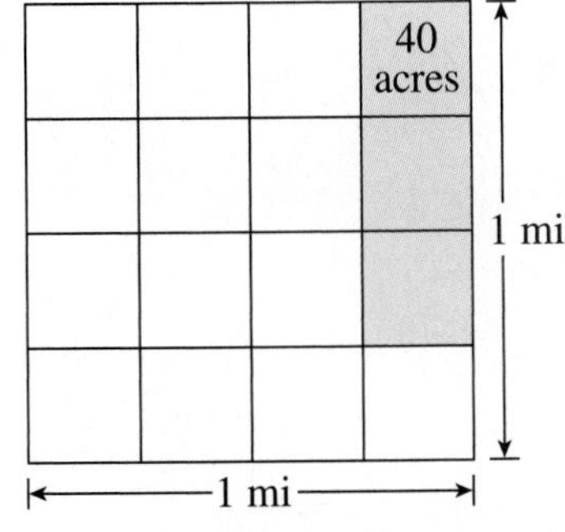

b. What are the overall dimensions of this combined lot, in miles?

Because he acquired 3 of the 16 lots, the area in Problem 4a is $\frac{3}{16}$ square mile. The dimensions of his lot are $\frac{1}{4}$ mile by $\frac{3}{4}$ mile. So the area of his lot is

$$\frac{1}{4} \cdot \frac{3}{4} = \frac{3}{16}\text{ sq. mi.}$$

5. In Problems 3 and 4, you determined that $\frac{1}{4} \cdot \frac{1}{4} = \frac{1}{16}$ and $\frac{1}{4} \cdot \frac{3}{4} = \frac{3}{16}$. Describe in your own words a rule for multiplying fractions.

PROCEDURE

Multiplying Fractions To multiply fractions, multiply the numerators to obtain the new numerator and multiply the denominators to obtain the new denominator. Stated symbolically:

$$\frac{a}{b} \cdot \frac{c}{d} = \frac{ac}{bd}$$

EXAMPLE 1 *Multiply* $\frac{1}{2} \cdot \frac{3}{5}$.

SOLUTION

$$\frac{1}{2} \cdot \frac{3}{5} = \frac{1 \cdot 3}{2 \cdot 5} = \frac{3}{10}$$

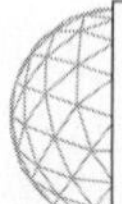

EXAMPLE 2 *Multiply* $-\frac{2}{3} \cdot \frac{5}{7}$.

The rules for multiplying negative integers also apply to fractions.

SOLUTION

$$-\frac{2}{3} \cdot \frac{5}{7} = -\frac{2 \cdot 5}{3 \cdot 7} = -\frac{10}{21}$$

EXAMPLE 3 *Multiply* $28 \cdot \frac{3}{4}$.

Think of the whole number 28 as the fraction $\frac{28}{1}$, then multiply: $\frac{28}{1} \cdot \frac{3}{4} = \frac{84}{4} = 21$.

EXAMPLE 4 *Multiply* $\frac{3}{8} \cdot \frac{1}{6}$.

SOLUTION

Multiply the fractions, $\frac{3}{8} \cdot \frac{1}{6} = \frac{3 \cdot 1}{8 \cdot 6} = \frac{3}{48}$.

Then reduce to lowest terms, $\frac{3}{48} = \frac{3 \cdot 1}{3 \cdot 16} = \frac{3}{3} \cdot \frac{1}{16} = 1 \cdot \frac{1}{16} = \frac{1}{16}$.

In Example 4, the common factor 3 in the numerator and denominator can be divided out, because, in effect, it amounts to multiplication by 1. It is actually preferable to remove (divide out) the common factor of 3 from both numerator and denominator before the final multiplication.

$$\frac{3}{8} \cdot \frac{1}{6} = \frac{\overset{1}{\not{3}} \cdot 1}{8 \cdot \underset{1}{\not{3}} \cdot 2} = \frac{1}{8 \cdot 2} = \frac{1}{16}$$

This will guarantee that your final answer will be written in lowest terms.

EXAMPLE 5 *Multiply* $-\frac{12}{25} \cdot -\frac{10}{9}$.

SOLUTION

$$-\frac{12}{25} \cdot -\frac{10}{9} = \frac{\overset{1}{\not{3}} \cdot 4 \cdot 2 \cdot \overset{1}{\not{5}}}{\underset{1}{\not{5}} \cdot 5 \cdot \underset{1}{\not{3}} \cdot 3} = \frac{4 \cdot 2}{5 \cdot 3} = \frac{8}{15}$$

Note that the common factors of 3 and 5 were divided out. This procedure is sometimes called *canceling common factors.*

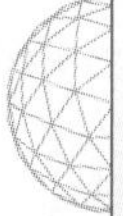

PROCEDURE

Determining the Square Root of a Fraction or a Mixed Number

1. If finding the square root of a mixed number, first change it to an improper fraction.
2. Determine the square root of the numerator.
3. Determine the square root of the denominator.
4. The square root of the fraction is the quotient $\frac{\text{square root of the numerator}}{\text{square root of the denominator}}$.
5. Write the result as a mixed number if necessary.

16. Calculate the following square roots.

a. $\sqrt{\frac{9}{100}}$

b. $\sqrt{\frac{64}{49}}$

c. $\sqrt{\frac{81}{16}}$

SUMMARY
ACTIVITY 4.4

1. To multiply or divide mixed numbers:
   a. Change the mixed numbers to improper fractions.
   b. Multiply or divide as you would proper fractions.
   c. If the product is an improper fraction, convert it to a mixed number, then reduce if possible.
2. The rules for order of operation when evaluating expressions involving fractions and/or mixed numbers are the same as for integers and whole numbers.
3. To determine the square root of a fraction or a mixed number:
   a. Convert a mixed number to an improper fraction.
   b. Determine the square root of the numerator.
   c. Determine the square root of the denominator.
   d. Divide the square root of the numerator by the square root of the denominator.
   e. Write the result as a mixed number if necessary. Symbolically,

$$\sqrt{\frac{a}{b}} = \frac{\sqrt{a}}{\sqrt{b}}, \; b \neq 0.$$

EXERCISES
ACTIVITY 4.4

1. Multiply and write your answer as a mixed number, if necessary.

   a. $3\frac{2}{3} \cdot 1\frac{4}{5}$

   b. $-2\frac{2}{15} \cdot 4\frac{1}{2}$

   c. $2\frac{4}{11} \cdot 4\frac{2}{5}$

   d. $\left(-5\frac{3}{8}\right) \cdot (-24)$

2. Divide and write your answer as a mixed number, if necessary.

   a. $3\frac{2}{7} \div 1\frac{9}{14}$

   b. $-1\frac{7}{8} \div \frac{3}{4}$

   c. $4\frac{2}{9} \div 8\frac{1}{3}$

   d. $\dfrac{-12\frac{2}{3}}{-3\frac{1}{6}}$

3. Simplify.

   a. $\left(\frac{2}{9}\right)^2$

   b. $\left(-\frac{3}{5}\right)^3$

   c. $\left(-\frac{4}{7}\right)^2$

   d. $\left(\frac{2}{3}\right)^5$

   e. $\sqrt{\frac{36}{81}}$

   f. $\sqrt{\frac{64}{100}}$

   g. $\sqrt{\frac{144}{49}}$

   h. $\sqrt{\frac{375}{15}}$

4. Evaluate each expression.

   a. $2x - 4(x - 3)$, where $x = 1\frac{1}{2}$

b. $a(a - 2) + 3a(a + 4)$, where $a = \frac{2}{3}$

c. $2x + 6(x + 2)$, where $x = 2\frac{1}{4}$

d. $6xy - y^2$, where $x = 1\frac{5}{16}$ and $y = \frac{1}{8}$

e. $4x^3 + x^2$, where $x = \frac{1}{2}$

5. The measurement of one side of a square rug is $5\frac{3}{4}$ feet. Determine the area of the rug in square feet (use $A = s^2$).

6. As an apprentice at your uncle's cabinet shop, you are cutting out pieces of wood for a dresser. You need pieces that are $5\frac{1}{4}$ inches long. Each saw cut wastes $\frac{1}{8}$ inch of wood. The board you have is 8 feet long. How many pieces can you make from this board?

7. Solve the following equations.

a. $2\frac{3}{4}x = 22$

b. $6\frac{1}{8}w = -24\frac{1}{2}$

c. $-1\frac{4}{5}t = -9\frac{9}{10}$

d. $5\frac{1}{3}x = \frac{-5}{6}$

8. The maximum distance, d, in kilometers that a person can see from a height h meters above the ground is given by the formula

$$d = \frac{7}{2}\sqrt{h}.$$

a. Use the formula to determine the maximum distance a person can see from a height of 64 meters.

b. Use the formula to determine the maximum distance a person can see from a height of $20\frac{1}{4}$ meters.

9. An object dropped from a height falls a distance of d feet in t seconds. The formula that describes this relationship is $d = 16t^2$. A math book is dropped from the top of a building that is 400 feet high.

a. How far has the book fallen in $\frac{1}{2}$ second?

b. After $3\frac{1}{2}$ seconds, how far above the ground is the book?

c. Another book was dropped from the top of a building across the street. It took $6\frac{1}{2}$ seconds to hit the ground. How tall is that building?

CLUSTER 2 What Have I Learned?

1. What would be the advantage of dividing both the numerator and denominator of a fraction by a common factor before multiplying fractions?

2. Will a fraction that has a negative numerator and a negative denominator be positive or negative?

3. What is the sign of the product when you multiply a positive fraction by a negative fraction?

4. What is the sign of the quotient when you divide a negative fraction by a negative fraction?

5. Explain, by using an example, how to divide a positive fraction by a negative fraction.

6. Explain, by using an example, how to multiply a negative fraction by a negative fraction.

7. Why do you have to know the meaning of the word *reciprocal* in this cluster?

8. When multiplying or dividing mixed numbers, what must be done first?

9. Explain how you would solve the following equation:

   one-fifth times a number equals negative four.

10. Explain the difference between squaring $\frac{1}{4}$ and taking the square root of $\frac{1}{4}$.

CLUSTER 2 How Can I Practice?

1. Multiply and write each answer in simplest form. Use a calculator to check answers.

a. $\frac{1}{9} \cdot \frac{2}{9}$
b. $\frac{10}{11} \cdot \frac{3}{13}$
c. $\frac{14}{9} \cdot \frac{-3}{28}$
d. $\frac{30}{35} \cdot \frac{10}{25}$

e. $\frac{-9}{10} \cdot \frac{-4}{3}$
f. $-20 \cdot \frac{2}{35}$
g. $\frac{10}{9} \cdot \frac{87}{100}$
h. $\frac{3}{16} \cdot \frac{14}{-3}$

2. Divide and write each answer in simplest form. Use a calculator to check answers.

a. $\frac{-3}{7} \div \frac{9}{7}$
b. $\frac{8}{11} \div \frac{12}{33}$
c. $-\frac{4}{9} \div -20$

d. $\frac{24}{45} \div \frac{15}{18}$
e. $\dfrac{\frac{6}{7}}{-\frac{9}{14}}$
f. $\dfrac{-\frac{35}{10}}{-3}$

3. Multiply or divide and write each answer as a mixed number in simplest form. Use a calculator to check answers.

a. $\frac{-1}{2} \cdot 5\frac{5}{6}$
b. $-2\frac{5}{6} \div \frac{3}{7}$
c. $-37\frac{1}{2} \cdot -1\frac{3}{5}$

d. $7\frac{1}{5} \div 2\frac{2}{5}$
e. $2\frac{1}{3} \cdot 1\frac{1}{2}$
f. $-8\frac{1}{4} \div -1\frac{1}{2}$

4. Simplify.

a. $-\left(\frac{4}{9}\right)^2$
b. $\sqrt{\frac{81}{121}}$
c. $\left(-\frac{3}{5}\right)^3$
d. $\sqrt{\frac{64}{49}}$

Exercise numbers appearing in color are answered in the Selected Answers appendix.

5. Evaluate each expression.

a. $3x - 5(x - 6)$, where $x = -1\frac{2}{3}$

b. $x^2 - xy$, where $x = \frac{1}{4}$ and $y = -1\frac{1}{3}$

6. Solve to determine the value of the unknown in each of the following.

a. $-6x = 25$

b. $-42 = 5x$

c. $-63x = 0$

d. $\frac{-1}{6}x = -12$

e. $\frac{2}{5} = -8x$

f. $\frac{-3}{4}x = \frac{-9}{16}$

7. According to an ad on television, a man lost 30 pounds over $6\frac{1}{4}$ months. How much weight did he lose per month?

8. On your last math test, you answered $\frac{7}{8}$ of the questions. Of the questions that you answered, you got $\frac{4}{5}$ of them correct. What fraction of the questions on the test did you get right?

9. A large-quantity recipe for pea soup calls for $64\frac{1}{2}$ ounces of split peas. You decide to make $\frac{1}{3}$ of the recipe. How many ounces of split peas will you need?

10. A large corporation recently reported that 16,936 employees were paid at an hourly rate. Approximately 1 out of every 5 of these workers earned $10 or more per hour. How many workers earned $10 or more per hour?

11. Your town has purchased a rectangular piece of land to develop a park. Three-fourths of the property will be used for the public. The rest of the land will be for park administration. The land will be divided so that $\frac{5}{6}$ of the public land will be used for picnicking and the remainder will be a swimming area.

Picnics Public area

a. Divide the rectangle into fourths and separate the public part from the park administration.

b. Shade the area that will be used for picnics.

c. Determine from the picture what fractional part of the whole piece of property will be used for picnics?

d. Write an expression and evaluate to show that part c is correct.

e. What fractional part of the property will be used by the park administration?

f. What fractional part of the property will be used for swimming?

g. Show that the fractional parts add up to the whole.

12. You drop your sunglasses out of a boat. Your depth finder on the boat measures 27 feet of water. You jump overboard with your goggles to see if you can rescue the glasses before they land on the bottom. You return for a gulp of air with your sunglasses in hand. You think they were about $\frac{2}{3}$ of the way down.

a. Estimate at what depth from the surface you found your glasses.

b. Write an expression that will show the distance from the surface of the water.

c. Evaluate the expression.

$$\frac{2}{\cancel{3}_1} \cdot \frac{\overset{9}{\cancel{27}}}{1} = 18.$$

13. a. Your swimming pool is 15 feet by 30 feet and will be filled to a height of $5\frac{1}{2}$ feet. Use $V = lwh$ (volume equals length times width times height) to determine the volume of the water in the pool.

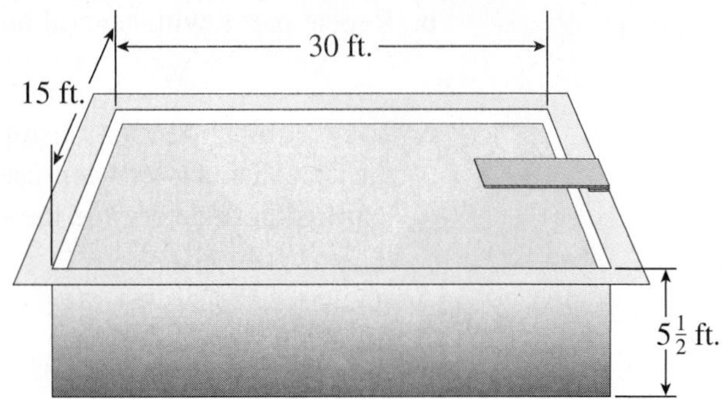

b. If 1 cubic foot of water weighs $62\frac{2}{5}$ pounds, determine the weight of the water.

14. You decide to drop a coin into the gorge at Niagara Falls. The distance the coin will fall, in feet s, and in time t, in seconds, is given by the formula $s = 16t^2$.

a. After $\frac{1}{2}$ second how far has the coin fallen?

b. How far is the coin from the top of the gorge after $2\frac{1}{2}$ seconds?

15. As a magician, you ask a friend to do this problem with you.

a. Ask your friend to select a number.

Then he should add three to the number.

Now have your friend multiply the result by $\frac{1}{3}$.

Next have him subtract 4.

Then, multiply by 6.

Finally, multiply by $\frac{1}{2}$, and ask your friend to tell you the answer.

Add 9 to the number your friend gives you. It will be the original number your friend selected.

b. Repeat part a with several numbers.

c. To determine the relationship between the original number selected and the final answer, let x represent the original number and write an expression to determine the final answer.

d. Simplify the expression.

CLUSTER 3 Multiplying and Dividing Decimals

ACTIVITY 4.5

Quality Points and GPA: Tracking Academic Standing

OBJECTIVES

1. Multiply and divide decimals.
2. Estimate products and quotients involving decimals.

Your college tracks your academic standing by an average called a *grade point average* (GPA). Each semester the college determines two grade point averages for you. One average is for the semester and the other includes all your courses up to the end of your current semester. The second grade point average is called a *cumulative GPA*. At graduation, your cumulative GPA represents your final academic standing.

At most colleges the GPA ranges from 0 to 4.0. A minimum GPA of 2.0 is required for graduation. You can track your own academic standing by calculating your GPA yourself. You may find this useful if you apply for internships or jobs at your college and elsewhere where a minimum GPA is required.

Quality Points and Multiplication

To calculate a GPA, you must first determine the number of points (known as quality points) that you earn for each course you take. The following problems will show you why.

1. Suppose that your schedule this term includes a four-credit literature class and a two-credit photography class. Is it fair to say that the literature course should be worth more towards your academic standing than the photography course? Explain.

The GPA gives more weight to courses with a higher number of credits than to ones with lower credit values. For example, a B in literature and an A in photography means that you have four credits each worth a B but only two credits that are each worth an A. To determine how much more the literature course is worth in the calculation of your GPA, you must first change the letter grades into quality point numerical equivalents. You then multiply the number of credits by the numerical equivalents of the letter grade to obtain the quality points for the course.

The following table lists the numerical equivalents of a typical set of letter grades.

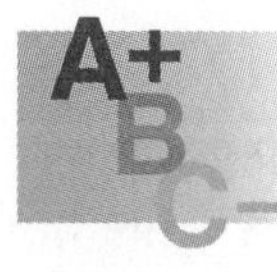

Numerical Equivalents for Letter Grades

LETTER GRADES	A	A−	B+	B	B−	C+	C	C−	D+	D	D−	F
QUALITY POINTS (NUMERICAL EQUIVALENTS)	4.00	3.67	3.33	3.00	2.67	2.33	2.00	1.67	1.33	1.00	0.67	0.00

2. a. The table shows that each B credit is worth 3 quality points. How many quality points is your B worth in the four-credit literature course?

b. How many quality points did you earn in your two-credit photography course if your grade was an A?

3. a. Your grade was a B+ in a three-credit psychology course. How many quality points did you earn?

b. Complete the following table for each pair of grades and credits. For example, in the row labeled 4 credits and the column labeled A, the number of quality points is 4 × 4.00 or 16.00. Similarly, the number of quality points for a grade of C− in a two-credit course is 3.34.

QUALITY POINTS FOR GRADE AND CREDIT PAIRS												
LETTER GRADES	A	A−	B+	B	B−	C+	C	C−	D+	D	D−	F
NUMERICAL EQUIVALENTS	4.00	3.67	3.33	3.00	2.67	2.33	2.00	1.67	1.33	1.00	0.67	0.00
COURSE CREDITS												
4 CREDITS	16.00											0.00
3 CREDITS			9.99									
2 CREDITS								3.34				
1 CREDIT										1.00		

4. A community college currently offers an internship course on geriatric health issues. The course, which includes one lecture hour and four fieldwork hours, is worth 1.5 credits.

a. Calculate the number of quality points for a grade of C in this course.

b. Estimate the number of quality points for a grade of C−.

c. Now, determine the exact number of quality points for a grade of C− by first multiplying 15 times 167, basically ignoring the decimal point, for the moment. Then, use your estimate in part b to determine where the decimal point should be placed.

Notice in Problem 4c that there are three digits to the right of the decimal point in your answer. This is the sum total of digits to the right of the decimal point in both factors. Example 1 illustrates how this pattern occurs.

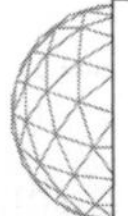

EXAMPLE 1 ***Why does* 0.3 · 0.5 = 0.15?**

SOLUTION

Since $0.3 = \frac{3}{10}$ and $0.5 = \frac{5}{10}$, you have

$$\begin{aligned} 0.3 \cdot 0.5 &= \frac{3}{10} \cdot \frac{5}{10} && \text{decimals to fractions} \\ &= \frac{15}{100} && \text{multiplying fractions} \\ &= 0.15 && \text{fractions to decimals} \end{aligned}$$

Example 1 shows that the number of decimal places to the right of the decimal point in the product 0.15 has to be 2, one from each of the factors.

PROCEDURE

Multiplying Two Decimals

1. Multiply two numbers as if they were whole numbers, ignoring decimal points for the moment.
2. Add the number of digits to the right of the decimal point in each factor, to obtain the number of digits that must be to the right of the decimal point in the product.
3. Place the decimal point in the product by counting digits from the right.

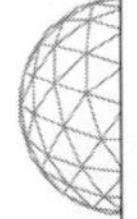

EXAMPLE 2 ***Multiply* 8.731 *by* 0.25.**

SOLUTION

$$\begin{array}{rl} 8.731 & \text{3 decimal places} \\ \times\ \ 0.25 & \text{2 decimal places} \\ \hline 43655 & \\ 17462 & \\ \hline 2.18275 & \text{5 decimal places} \end{array}$$

In Example 2, since 0.25 is $\frac{1}{4}$, and $\frac{1}{4}$ of 8.731 is approximately 2, you see that the general rule for multiplying two decimals placed the decimal point in the correct position.

5. Calculate the following products. Check the results using your calculator.

   a. $0.008 \cdot 57.2$

   b. $0.0201 \cdot -27.8$

   c. $-0.45 \cdot -3.12$

GPA and Division

Once you calculate the quality points you earned in a semester, you are ready to determine your GPA for that semester.

6. Suppose that in addition to the credit courses in literature, photography, and psychology that you took this term, you also enrolled in a one-credit weight training class. Using the following table to record your results, calculate the quality points for each course. Then determine the total number of credits and the total number of quality points. The calculation for literature has been done for you.

	CREDITS	GRADE	NUMERICAL EQUIVALENT	QUALITY POINTS
LITERATURE	4	B	3.00	12.00
PHOTOGRAPHY	2	A		
PSYCHOLOGY	3	B+		
WEIGHT TRAINING	1	B−		
TOTALS		n/a		

You earned 32.66 quality points for 10 credits this semester. To calculate the average number of quality points per credit, divide 32.66 by 10 to obtain 3.266. Note that dividing by 10 moves the decimal point in 32.66 one place to the left.

GPAs are usually rounded to the nearest hundredth, so your GPA is 3.27.

PROCEDURE

Determining a GPA

1. Calculate the quality points for each course.
2. Determine the sum of the quality points.
3. Determine the total number of credits.
4. Divide the total quality points by the total number of credits.

7. Your cousin enrolled in five courses this term and earned the grades listed in the following table. Determine her GPA to the nearest hundredth.

	CREDITS	GRADE	NUMERICAL EQUIVALENT	QUALITY POINTS
POETRY	3	C+		
CHEMISTRY	4	B		
HISTORY	3	A−		
MATH	3	B+		
KARATE	1	B−		
TOTALS		n/a	n/a	
			GPA	

In Problem 7, note that $42.66 \div 14$ is approximately $45 \div 15 = 3$. Therefore, the decimal point in 3.05 is correctly placed.

EXAMPLE 3 ***In Problem 4, you read that a community college offers a 1.5 credit course to investigate issues in geriatric health. A co-worker told you he earned 4.5 quality points in that course. What grade did he earn?***

SOLUTION

To determine your co-worker's grade, divide the 4.5 quality points by 1.5 credits, written as $4.5 \div 1.5$. The division can be written as a fraction, $\frac{4.5}{1.5}$, where the numerator is *divided* by the denominator.

$$4.5 \div 1.5 = \frac{4.5}{1.5} \quad \text{division as a fraction}$$

$$= \frac{4.5}{1.5} \cdot \frac{10}{10} \quad \text{multiplying by } \tfrac{10}{10} = 1 \text{ to obtain equivalent fraction}$$

$$= \frac{45}{15} \quad \text{multiplying fractions to obtain whole-number denominator}$$

$$= 3 \quad \text{dividing numerator by denominator}$$

He earned a grade of B.

Example 3 leads to a general method for dividing decimals.

$$1.\overline{)\ \ 4.5}\ \text{ becomes } 15\overline{)45.}\quad 3.$$

PROCEDURE

Dividing Decimals

1. Write the division in long division format.
2. Move the decimal point the same number of places to the right in both divisor and dividend so that the divisor becomes a whole number.
3. Place the decimal point in the quotient directly above the decimal point in the dividend and divide as usual.

EXAMPLE 4 **a.** ***Divide 92.4 by 0.25.*** (92.4: dividend; 0.25: divisor) **b.** ***Divide 0.00052 ÷ 0.004.*** (0.00052: dividend; 0.004: divisor)

SOLUTION

a. $0.25\overline{)92.40}$ becomes $25\overline{)9240.}$

$$\begin{array}{r} 369.6 \\ 25\overline{)9240.0} \\ -75 \\ \hline 174 \\ -150 \\ \hline 240 \\ -225 \\ \hline 150 \\ -150 \\ \hline 0 \end{array}$$

b. $0.004\overline{)0.00052}$ becomes $4\overline{)0.52}$.

$$\begin{array}{r} 0.13 \\ 4\overline{)0.52} \\ -4 \\ \hline 12 \\ -12 \\ \hline 0 \end{array}$$

8. Determine the following quotients. Check the results using your calculator.

a. $11.525 \div 2.5$

b. $-45.525 \div 0.0004$

c. Calculate $-3.912 \div (-0.13)$ and round the result to the nearest ten thousandths.

9. You need to calculate your GPA to determine if you qualify for a summer internship that requires a cumulative GPA of at least 3.3. You had 14 credits and 44.5 quality points in the fall term. In the spring you earned 16 credits and 55.2 quality points. Did you qualify for the internship?

Estimating Products and Quotients

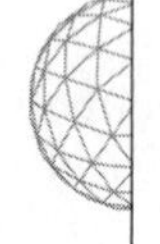

EXAMPLE 5 ***You need to drive to campus, but your gas tank is just about empty. The tank holds about 18.5 gallons of gas. Today, gas is selling for $2.489 per gallon in your town. Estimate how much it will cost to fill the tank.***

SOLUTION

Note that 18.5 gallons is about 20 gallons and $2.489 is about $2.50. Therefore, $20 \cdot 2.5 =$ $50. You will need approximately $50 to fill the tank.

PROCEDURE

Estimating Products of Decimals

1. Round each factor to one or two nonzero digits (the number you can handle easily in your head).
2. Multiply the results of step 1. This is your estimate.

10. Estimate the following products.

a. $57.3 \cdot 615.3$

b. $-0.031 \cdot 322.76$

c. $7893.65 \cdot 0.0016$

d. If you multiply a positive number by a decimal between 0 and 1, is the product greater or smaller than the original number?

11. Your uncle teaches at a college in San Francisco where gas costs \$3.259 a gallon. His car holds 17.1 gallons of gasoline.

a. Estimate how much he would spend to fill an empty tank.

b. Calculate the amount he would actually spend to fill an empty tank.

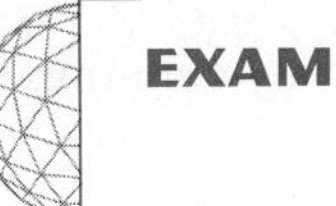

EXAMPLE 6 ***In 2002, a Japanese research laboratory built the world's fastest supercomputer to study weather and other environmental conditions around Earth. The speed of the NEC Earth Simulator was reported at 35.6 teraflops, or trillions of mathematical operations per second. The next fastest supercomputer was listed as the IBM ASCII White-Pacific with a speed of 7.226 teraflops. Estimate how many times faster the Earth Simulator is than the White-Pacific.***

SOLUTION

35.6 is approximately 35 and 7.226 is about 7. Therefore,

$$\frac{35.6}{7.226} \approx \frac{35}{7} = 5.$$

The Earth Simulator is about 5 times faster than the White-Pacific.

PROCEDURE

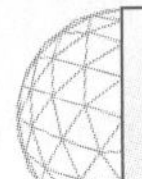

Estimating Quotients of Decimals

1. Think of the division as a fraction with the dividend as the numerator and the divisor as the denominator.
2. Round the numerator and denominator to one or two nonzero digits (a number you can handle easily in your head).
3. Divide the results of step 2. This is your estimate.

12. Estimate the following quotients.

a. $857.3 \div 61.53$

b. $-0.0315 \div 32.2$

c. $8934.65 \div 0.0018$

d. If you divide a positive number by a decimal fraction between 0 and 1, is the quotient greater than or smaller than the original number?

SUMMARY ACTIVITY 4.5

1. Multiplying two decimals:
   a. Multiply two numbers as if they were whole numbers, ignoring decimal points for the moment.
   b. Add the number of digits to the right of the decimal point in each factor, to obtain the number of digits that must be to the right of the decimal point in the product.
   c. Place the decimal point in the product by counting digits from the right.
2. Dividing decimals:
   a. Write the division in long division format.
   b. Move the decimal point the same number of places to the right in both divisor and dividend so that the divisor becomes a whole number.
   c. Place the decimal point in the quotient directly above the decimal point in the dividend and divide as usual.
3. Estimating products of decimals:
   a. Round each factor to one or two nonzero digits (the number you can handle easily in your head).
   b. Multiply the results of step a. This is your estimate.
4. Estimating quotients of decimals:
   a. Think of the division as a fraction, with the dividend as the numerator and the divisor as the denominator.
   b. Round the numerator and denominator to one or two nonzero digits (the number you can handle easily in your head).
   c. Divide the results of step b. This is your estimate.

EXERCISES ACTIVITY 4.5

1. You drive your own car while delivering pizzas for a local pizzeria. Your boss said she would reimburse you $0.32 per mile for the wear and tear on your car. The first week of work you put 178 miles on your car.

   a. Estimate how much you will receive for using your own car.

   b. Explain how you determined your estimate.

   c. How much did you receive for using your own car?

Exercise numbers appearing in color are answered in the Selected Answers appendix.

2. Perform the following calculations. Use your calculator to check your answer.

   a. $-12.53 \cdot -8.2$ b. $115.3 \cdot -0.003$ c. $14.62 \cdot -0.75$

   d. Divide 12.05 by 2.5.

   e. Divide 18.99729 by 78.

   f. Divide 14.05 by 0.0002.

   g. $150 \div 0.03$

   h. $0.00442 \div 0.017$

   i. $69.115 \div 0.0023$

3. You and some college friends go out to lunch to celebrate the end of midterm exams. There are six of you, and you decide to split the $47.98 bill evenly.

   a. Estimate how much each person owed for lunch.

   b. Explain how you determined your estimate.

   c. How much did each person actually pay? Round your answer to the nearest cent.

4. You have just purchased a new home. The town's assessed value of your home (which is usually much *lower* than the purchase price) is $68,700. For every $1000 of assessed value, you will pay $7.48 in taxes. How much do you pay in taxes?

5. Alex Rodriguez of the New York Yankees led the American League in home runs with 48 in the 2005 season. He also made 194 hits in 605 times at bat. What was his batting average? (Batting average = number of hits ÷ number of at bats.) Use decimal notation and round to the nearest thousandth.

6. The top-grossing North American concert tour between 1985 and 1999 was the Rolling Stones tour in 1994. The tour included 60 shows and sold $121.2 million worth of tickets.

   a. Estimate the average amount of ticket sales per show.

   b. What was the actual average amount of ticket sales per show?

c. The Stones 2005 tour broke their previous record, selling $162 million worth of tickets in 42 performances. What was the average amount of ticket sales per show in 2005, rounded to the nearest $1000?

7. In 1992, the NEC SX-3/44 supercomputer had a speed of 0.020 teraflops (trillions of mathematical operations per second).

a. How many times faster is the NEC Earth Simulator supercomputer whose speed is 35.6 teraflops?

b. In 1993, the speed of the Fujitsu NWT supercomputer was 0.124 teraflops. How many times faster was it than the NEC SX–3/44?

c. In 2005, a new supercomputer, designed specifically to simulate proteins for biology and drug research, the IBM Blue Gene/L (Watson), posted a speed of 91.3 teraflops. How many times faster was it than the NEC SX-3/44?

8. Your friend tracked his GPAs for his first three semesters as shown in the following table.

a. Calculate your friend's GPA for each of the semesters listed in the table. Record the results in the table.

	CREDITS	QUALITY POINTS	GPA
SEMESTER 1	14	36.54	
SEMESTER 2	15	49.05	
SEMESTER 3	16	45.92	

b. Determine his cumulative GPA for the first two semesters.

c. Determine his cumulative GPA for all three semesters.

ACTIVITY 4.6

Tracking Temperature

OBJECTIVES

1. Use the order of operations to evaluate expressions that include decimals.
2. Use the distributive property in calculations involving decimals.
3. Evaluate formulas that include decimals.
4. Solve equations of the form $ax = b$ that include decimals.

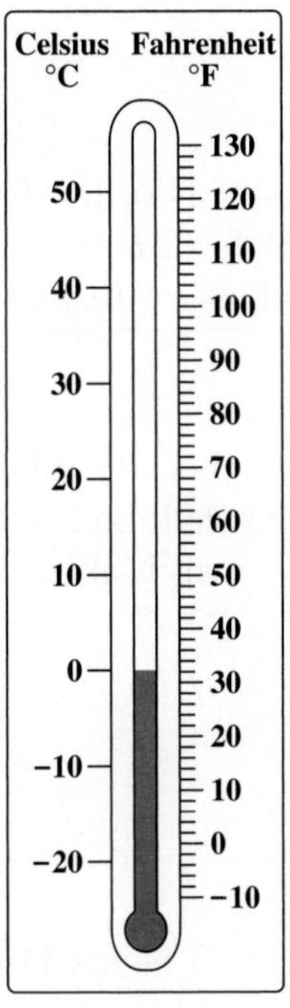

Temperature measures warmth or coolness of everything around us—in lake water, in the human body, in the air we breathe. The thermometer is Alaska's favorite scientific instrument. During the winter, says Ned Rozell at the Geophysical Institute of the University of Alaska at Fairbanks, "we check our beloved thermometers thousands of times a day."

Thermometers have scales whose units are called *degrees*. The two commonly used scales are Fahrenheit and Celsius. The Celsius scale is the choice in science. Most English-speaking countries began changing to Celsius in the late 1960s, but the United States is still the outstanding exception to this changeover.

Each scale was determined by considering two reference temperatures, the freezing point and the boiling point of water. On the Fahrenheit scale, the freezing point of water is taken as 32°F and the boiling point is 212°F. On the Celsius scale, the freezing point is taken as 0°C and the boiling point is 100°C. This leads to a formula that converts Celsius degrees to Fahrenheit degrees,

$$y = 1.8x + 32,$$

where y represents degrees Fahrenheit and x represents degrees Celsius.

1. Each year, the now-famous Iditarod Trail Sled Dog Race in Alaska starts on the first Saturday in March. Temperatures on the trail can range from about 5°C down to about −50°C.

   a. Use the formula $y = 1.8x + 32$ to convert 5°C to degrees Fahrenheit.

   b. Convert −50°C to degrees Fahrenheit.

   c. In a sentence, describe the temperature range in degrees Fahrenheit.

2. The Tour de France bicycle race is held in July in France. The race ends in Paris where July's temperature ranges from 57°F to 77°F. Use the formula $x = (y - 32) \div 1.8$ to convert these temperatures to degrees Celsius.

The previous problems show that expressions that involve decimals are evaluated by the usual order of operations procedures.

3. Evaluate the following expressions using order of operations. Use your calculator to check your answer.

   a. $100 + 100(0.075)(2.5)$

   b. $2(8.4 + 11.7)$

   c. $0.5(9.16 + 6.38) \cdot 4.21$

Solving Equations of the Form $ax = b$ that Involve Decimals

4. The Tour de France is held in 21 stages of varying distances. One stage runs from Orleans to Evry, a distance of 149.5 km. The best time for completing this stage was 3.2075 hours.

   a. Use the formula $d = rt$ to determine the average speed of the cyclists with the best time for this stage. Recall that d represents distance, t, time, and r, average speed.

   b. The distance in miles for this stage is 93.5 miles. Determine the average speed in terms of miles per hour (mph) for this stage.

The formula $d = rt$ is an example of an equation that is of the form $b = ax$. Problem 4 illustrates the fact that equations of the form $b = ax$ (or $ax = b$) that involve decimals are solved in the same way as those with whole numbers, integers, or fractions.

5. a. In the 2006 Iditarod Trail Sled Dog Race, the winner, Jeff King of Denali, Alaska, and his dog team ran the last leg of the race from Safety to Nome in 3.31 hours. The distance between these two towns is about 22.0 miles. What was the average speed of King's team?

   b. Jeff King finished the entire 2006 race in 9 days, 11 hours, 11 minutes, and 36 seconds. In terms of hours, this is 227.193 hours. The race followed the northern route of about 1112 miles. What was his team's average speed over the entire race?

Distributive Property

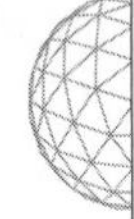

EXAMPLE 1 ***The vegetable garden along the side of your house measures 20.2 feet long by 12.6 feet wide. You want to enclose it with a picket fence. How many feet of fencing will you need?***

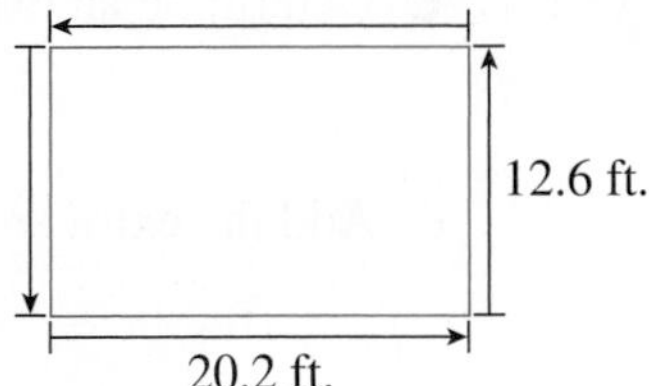

The diagram indicates that you can add the length to the width and then double it to determine the amount of fencing. Numerically, you write

$$2(20.2 + 12.6) = 2 \cdot 32.8 = 65.6 \text{ ft.}$$

Note that another way to solve the problem is to add *twice* the length to *twice* the width. Numerically,

$$2 \cdot 20.2 + 2 \cdot 12.6 = 40.4 + 25.2 = 65.6 \text{ ft.}$$

In either case, the amount of fencing is 65.6 feet. This means that the expression $2(20.2 + 12.6)$ is equal to the expression $2 \cdot 20.2 + 2 \cdot 12.6$. This example illustrates that the distributive property you learned for whole numbers, integers, and fractions also holds for decimal numbers.

As discussed in Chapter 2, the amount of fencing can be determined by using a formula for the perimeter of a rectangle. If P represents the perimeter, l, the length, and w, the width, then

$$P = 2(l + w) \quad \text{or} \quad P = 2l + 2w.$$

6. Your neighbor also has a rectangular garden, but its length is 10.5 feet by 13.8 feet wide. Use a perimeter formula to determine the distance around your neighbor's garden.

7. Next year you plan to make your garden longer and keep the width the same to gain more area for growing vegetables. You have not decided how much you want to increase the length. Call the extra length x. As you look at the diagram, note that a formula for the area of the new garden can be determined in two ways.

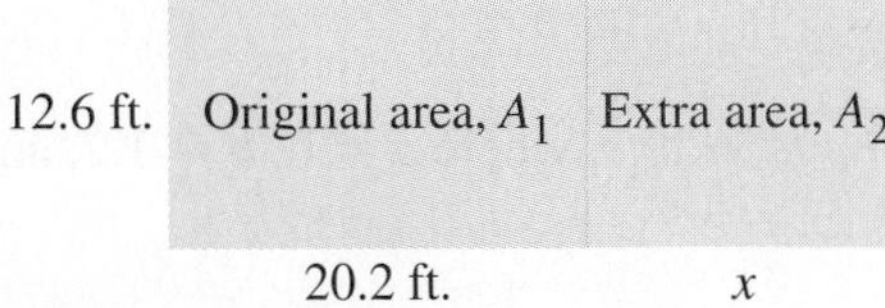

Method 1

a. Determine an expression for the new length in terms of x

b. Multiply the expression in part a by 12.6 to obtain the new area, A. Recall that area equals the product of the length and width, or $A = lw$.

Method 2

c. Determine an expression for the extra area, A_2.

d. Add the extra area, A_2, to the original area, A_1, to obtain the new area A.

8. Explain how you may obtain the expression in Problem 7d directly from the expression in Problem 8b by the distributive property.

9. Use the distributive property to simplify the following.

a. $0.2(4x - 0.15) =$

b. $3.4(2.05x - 0.1)$

10. Simplify each of the following expressions. Check the results using your calculator.

a. $0.2 - 0.42(5.4 - 6)^2$

b. $0.24 \div 0.06 \cdot 0.2$

c. $4.96 - 6 + 5.3 \cdot 0.2 - (0.007)^2$

11. Evaluate the following.

a. $2x - 4(y - 3)$, where $x = 0.5$ and $y = 1.2$

b. $b^2 - 4ac$, where $a = 6$, $b = 1.2$, and $c = -0.03$

c. $\dfrac{2.6x - 0.13y}{x - y}$, where $x = 0.3$ and $y = 6.8$

12. Evaluate the following using the order of operations. Check the results using your calculator.

a. $(-0.5)^2 + 3$

b. $(2.6 - 1.08)^2$

c. $(-2.1)^2 - (0.07)^2$

SUMMARY
ACTIVITY 4.6

1. The rules that apply to whole numbers, integers, and fractions, *also apply to decimals.* They are:

- the order of operations procedure,
- the distributive property, and
- the evaluation of formulas

EXERCISES
ACTIVITY 4.6

In Exercises 1–4, calculate using the order of operations. Check the results using your calculator.

1. $48.5 - 10.34 + 4.66$

2. $0.56 \div 0.08 \cdot 0.7$

3. $10.31 + 8.05 \cdot 0.4 - (0.08)^2$

4. $-2.6 \cdot .01 + (-.01) \cdot (-8.3)$

In Exercises 5–7, evaluate the expression for the given values.

5. $\frac{1}{2}h(a + b)$, where $h = 0.3$, $a = 1.24$, and $b = 2.006$

6. $3.14\, r^2$, where $r = 0.25$ inches

7. $A \div B - C \cdot D$, where $A = 10.8$, $B = 0.12$, $C = 2.4$, and $D = 6$

8. You earn \$7.25 per hour and are paid once a month. You record the number of hours worked each week in your table.

WEEK	1	2	3	4
HOURS WORKED	35	20.5	12	17.75

a. Write a numerical expression using parentheses to show how you will determine your gross salary for the month.

b. Simplify the expression in part a.

c. You think you are going to get a raise but don't know how much it might be. Write an expression to show your new hourly rate. Represent the amount of the raise by x.

d. Calculate the total number of hours worked for the month. Use it with your result in part c to write an expression to represent your new gross salary after the raise.

e. Use the distributive property to simplify the expression in part d.

f. Your boss tells you that your raise will be \$0.50 an hour. How much did you make this month?

9. You have been following your stock investment over the past 4 months. Your friend tells you that his stock's value has doubled in the past 4 months. You know that your stock did better than that. Without telling your friend how much your original investment was, you would like to do a little boasting. The following table shows your stock's activity each month for 4 months.

Up with the Dow-Jones

MONTH	1	2	3	4
STOCK VALUE	decreased \$75	doubled	increased \$150	tripled

a. Represent your original investment by x dollars and write an algebraic expression to represent the value of your stock after the first month.

b. Use the result from part a to determine an expression for the value at the end of the second month. Simplify the expression if possible. Continue this procedure until you determine the value of your stock at the end of the fourth month.

c. Explain to your friend what has happened to the value of your stock over 4 months.

d. Instead of simplifying after each step, write a single algebraic expression that represents the changes over the 4-month period.

e. Simplify the expression in part d. How does the simplified algebraic expression compare with the result in part b?

f. Your stock's value was $4678.75 four months ago. What is your stock's value today?

10. Translate each of the following verbal statements into equations and solve. Let x represent the unknown number.

a. The quotient of a number and 5.3 is -6.7.

b. A number times 3.6 is 23.4.

c. 42.75 is the product of a number and -7.5.

d. A number divided by 2.3 is 13.2.

e. -9.4 times a number is 47.

11. Solve each of the following equations. Round the result to the nearest hundredth if necessary.

a. $3x = 15.3$ **b.** $-2.3x = 10.35$ **c.** $-5.2a = -44.2$

d. $15.2 = 15.2y$ **e.** $98.8 = -4y$ **f.** $4.2x = -\sqrt{64}$

12. You and your best friend are avid cyclists. You are planning a bike trip from Corning, New York, to West Point. On the Internet, you find that the distance from Corning to West Point is approximately 236.4 miles. On a recent similar trip, you both averaged about 35.4 miles per hour. You are interested in estimating the riding time for this trip to the nearest tenth of an hour.

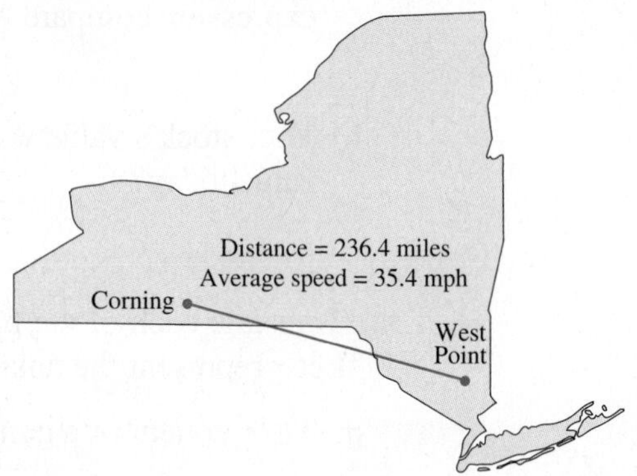

a. Reread the problem and list all the numerical information you have been given.

b. What are you trying to determine? Give it a letter name.

c. What is the relationship between the length of your intended trip, the rate at which you travel, and what you want to determine?

d. Write an equation for the relationship in part c.

e. Solve the equation in part d.

f. Is your solution reasonable? Explain.

CLUSTER 3 What Have I Learned?

1. When multiplying two decimals, how do you decide where to place the decimal point in the product? Illustrate with an example.

2. a. When faced with a division problem such as $2.35\overline{)1950.5}$, what would you do first?

   b. Explain how you would estimate this quotient.

3. How do the rules for order of operations and the distributive property apply to decimals?

4. Explain the steps you would take to determine the value for I from the formula $I = P + Prt$ if $P = 1500$, $r = 0.095$, and $t = 2.8$.

5. a. Each day, you estimate the average speed at which you are traveling on a car trip from Massachusetts to North Carolina. You keep track of the hours and mileage. Explain how you would use the formula $d = rt$ to determine your average speed.

b. Explain how the two formulas, $d = rt$ and $I = Pr$ are mathematically similar.

6. a. Give at least one reason why estimating a product or quotient would be beneficial.

b. Give a reason(s) for the steps you would take to estimate $15{,}776 \div 425$.

7. If you divide a negative number by a number between 0 and 1, is the quotient greater than or less than the original number? Give an example to illustrate your answer.

CLUSTER 3 How Can I Practice?

In Exercises 1–4, estimate the products and quotients.

1. $36.42 \cdot (-4.2)$

2. $-226.90 \cdot (-0.005)$

3. $14.62 \div (-0.75)$

4. Divide 43,096.48 by 78.4.

In Exercises 5–8, calculate the products and quotients. Check your result with a calculator.

5. $36.42 \cdot (-4.2)$

6. $-226.90 \cdot (-0.005)$

7. $14.625 \div (-0.75)$

8. $43{,}096.48 \div 78.4$

9. $0.63 \div 0.09 \cdot 0.5$

10. $30.38 + 4.05 \cdot 0.6 - (0.03)^2$

11. $-6.2 \cdot .03 + (-.01) \cdot (-6.3)$

12. Evaluate the expression $\frac{1}{2}h(a + b)$, where $h = 0.7$, $a = 3.45$, and $b = 5.00$.

13. Evaluate the expression πr^2, where $r = 0.84$ inches. Use 3.14 for π and round to the hundredths place.

14. Evaluate the expression $A \div B - C \cdot D$, where $A = 9.9$, $B = 0.33$, $C = 2.7$, and $D = 4$.

Translate each of the following verbal statements into equations and solve. Let x represent the unknown number.

15. The quotient of a number and -7.4 is 13.5.

Exercise numbers appearing in color are answered in the Selected Answers appendix.

16. A number times -9.76 is 678.32.

17. A number divided by -9.5 is 78.3.

Solve each of the following equations. Round to the nearest hundredths if necessary.

18. $7x = 5.81$

19. $-2.3x = 58.65$

20. $-5.2a = -57.72$

21. You buy a DVD player at a discount and pay \$136.15. This discounted price represents seven-tenths of the original price. Answer the following questions to determine the original price.

a. List all of the information you have been given.

b. What quantity are you trying to determine?

c. Write a verbal statement to express the relationship between the price you paid, the discount rate, and what quantity you want to determine.

d. Select a letter to represent the original price. Then write an equation based on the verbal statement in part c.

e. Solve the equation in part d to obtain the original price.

f. Is your solution reasonable? Explain.

22. George Herman "Babe" Ruth was baseball's first great batter and is one of the most famous players of all time.

a. In 1921, Babe Ruth played for the New York Yankees, hitting 59 home runs and making 204 hits out of 540 at bats. What was his batting average? Recall that batting averages are recorded to the nearest thousandth.

b. In 2001, Barry Bonds of the San Francisco Giants broke the single-season record for most home runs (73). During that season, he made 156 hits out of 476 at bats. What was his batting average?

c. Which of the two batting averages is the highest?

23. A student majoring in science earned the credits and quality points listed in the following table. Complete the table by determining the student's GPA for each semester and his cumulative GPA at the end of the fourth semester.

Majoring in Science

	CREDITS	QUALITY POINTS	GPA
SEMESTER 1	14	35.42	
SEMESTER 2	15	44.85	
SEMESTER 3	16	55.20	
SEMESTER 4	16	51.68	
CUMULATIVE GPA			

24. A sophomore at your college took the courses and earned the grades that are listed in the following table. Complete the table and determine the GPA to the nearest hundredth.

	CREDITS	GRADE	NUMERICAL EQUIVALENT	QUALITY POINTS
ENGLISH	3	C-	1.67	
PHYSICS	4	B	3.00	
HISTORY	3	C+	2.33	
MATH	4	A-	3.67	
MODERN DANCE	2	B+	3.33	
TOTALS		n/a	n/a	
GPA	n/a	n/a	n/a	

25. Every May since 1977, Bike New York offers cyclists a 42-mile tour of the five boroughs (counties) that make up New York City. In 2005, a week before the tour, the temperature was predicted to range from a low of 54°F to a high of 71°F. Determine these temperatures in degrees Celsius for your friend coming from Canada to do the tour. Recall the formula $x = (y - 32)1.8$ where x is the temperature in Celsius and y is the temperature in Farenheit.

26. In your chemistry class, you need to weigh a sample of pure carbon. You will weigh the amount that you have 3 times and take the average (called the *mean*) of the three weighings. The average weight is the one you will report as the actual weight of the carbon. The three weights you obtain are 12.17 grams, 12.14 grams, and 12.20 grams.

a. Explain how you will calculate the average.

b. Determine the average weight of the carbon to the nearest hundredth of a gram.

CHAPTER 4 SUMMARY

The bracketed numbers following each concept indicate the activity in which the concept is discussed.

CONCEPT / SKILL	DESCRIPTION	EXAMPLE
Multiply/divide integers with the same sign [4.1]	To multiply or divide two integers with the same sign: 1. Multiply or divide their absolute values. 2. The product or quotient will always be positive.	$(-6) \cdot (-7) = 42$ $4 \cdot 3 = 12$ $8 \div 2 = 4$ $(-20) \div (-4) = 5$
Multiply/divide integers with opposite signs [4.1]	To multiply or divide two integers with opposite signs: 1. Multiply or divide their absolute values. 2. The product or quotient will always be negative.	$(-42) \div 14 = -3$ $11 \cdot (-4) = -44$
Product of an even number of negative factors [4.1]	The product will be positive if you are multiplying an even number of negative integers.	$(-3) \cdot (-4) \cdot (-2) \cdot (-1)$ $= 24$
Product of an odd number of negative factors [4.1]	The product will be negative if you are multiplying an odd number of negative integers.	$(-3) \cdot (-4) \cdot (-2) = -24$
The distributive properties [4.2], [4.4], [4.6]	The distributive properties, $a(b + c) = ab + ac$ $a(b - c) = ab - ac,$ hold for rational numbers.	$\frac{3}{4}\left(\frac{1}{2} + \frac{1}{2}\right) = \frac{3}{4} \cdot \frac{1}{2} + \frac{3}{4} \cdot \frac{1}{2}$ $= \frac{3}{8} + \frac{3}{8} = \frac{6}{8} = \frac{3}{4}$ $\frac{1}{2}\left(\frac{3}{5} - \frac{2}{5}\right) = \frac{1}{2} \cdot \frac{3}{5} - \frac{1}{2} \cdot \frac{2}{5}$ $= \frac{3}{10} - \frac{2}{10} = \frac{1}{10}$
Order of operations [4.2], [4.4], [4.6]	The order of operation rules for rational numbers are the same as those applied to whole numbers.	$30.38 + 4.5 \cdot 0.6 - (0.3)^2$ $= 30.38 + 4.5 \cdot 0.6 - 0.09$ $= 30.38 + 2.7 - 0.09$ $= 33.08 - 0.09$ $= 32.99$
Negation of a squared number [4.2]	Negation, or determining the opposite, always follows exponentiation in the order of operations.	$-5^2 = -(5^2)$ $= -1 \cdot 25$ $= -25$
Evaluate expressions or formulas that involve integers, fractions, and decimals [4.2], [4.4], [4.6]	To evaluate formulas or expressions involving rational numbers, substitute for the variables and evaluate using the order of operations.	Evaluate $6xy - y^2$, where $x = \frac{1}{3}$ and $y = 3$. $6xy - y^2 = 6(\frac{1}{3})(3) - 3^2$ $= 6 - 9 = -3$

CONCEPT / SKILL	DESCRIPTION	EXAMPLE
Solve $ax = b$ for x that involve integers, fractions, and decimals [4.2], [4.4], [4.6]	To solve the equation $$ax = b, a \neq 0,$$ you need to divide each side of the equation by a to obtain the value for x.	Solve $-3x = -216$. $\frac{-3x}{-3} = \frac{-216}{-3}$ $x = 72$
General problem-solving strategy [4.2], [4.4], [4.6]	To solve verbal problems: 1. Understand the problem. 2. Compile the information. 3. Solve the problem. 4. Check the equation for reasonableness.	
Multiply fractions [4.3]	To multiply fractions, multiply the numerators and the denominators. $$\frac{a}{b} \cdot \frac{c}{d} = \frac{ac}{bd}$$	$\frac{3}{5} \cdot \frac{2}{7} = \frac{3 \cdot 2}{5 \cdot 7} = \frac{6}{35}$
The negative sign for a negative fraction [4.3]	There are three different ways to place the sign for a negative fraction.	$-\frac{2}{5} = \frac{-2}{5} = \frac{2}{-5}$ are all the same number.
Reciprocals [4.3]	The reciprocal of a nonzero number, written in fraction form, is the fraction obtained by interchanging the numerator and denominator of the original fraction. The product of a number and its reciprocal is always 1.	The reciprocal of $\frac{5}{7}$ is $\frac{7}{5}$, because $\frac{5}{7} \cdot \frac{7}{5} = 1$.
Divide fractions [4.3]	To divide fractions: 1. Multiply the dividend by the reciprocal of the divisor. 2. Reduce to lowest terms. $$\frac{a}{b} \div \frac{c}{d} = \frac{a}{b} \cdot \frac{d}{c}$$	$\frac{3}{8} \div \frac{5}{7} = \frac{3}{8} \cdot \frac{7}{5} = \frac{21}{40}$ $\frac{4}{5} \div \frac{2}{3} = \frac{4}{5} \cdot \frac{3}{2} = \frac{12}{10}$ $= \frac{6}{5}$ or $1\frac{1}{5}$
Multiplications and divisions involving 0 [4.1]	1. Any number times 0 is 0. 2. 0 divided by any nonzero integer is 0. 3. No integer may be divided by 0.	$3 \cdot 0 = 0$ $0 \div 6 = 0$ $6 \div 0$ is not possible.
Multiply mixed numbers [4.4]	To multiply mixed numbers: 1. Change the mixed numbers to improper fractions. 2. Multiply as you would proper fractions. 3. If the product is an improper fraction, convert it to a mixed number.	$3\frac{1}{2} \cdot 4\frac{3}{8} = \frac{7}{2} \cdot \frac{35}{8}$ $= \frac{245}{16} = 15\frac{5}{16}$

CONCEPT / SKILL	DESCRIPTION	EXAMPLE
Divide mixed numbers [4.4]	To divide mixed numbers: 1. Change the mixed numbers to improper fractions. 2. Divide as you would proper fractions. 3. If the quotient is an improper fraction, convert it to a mixed number.	$4\frac{5}{8} \div 2\frac{1}{3} = \frac{37}{8} \div \frac{7}{3}$ $= \frac{37}{8} \cdot \frac{3}{7} = \frac{111}{56}$ $= 1\frac{55}{56}$
Squaring fractions [4.4]	To determine the square of a fraction: 1. Determine the square of the numerator. 2. Determine the square of the denominator. 3. The square of the fraction is the quotient of the two squares. 4. Write the result as a mixed number if necessary.	$\left(\frac{3}{4}\right)^2 = \frac{3^2}{4^2} = \frac{9}{16}$
Square roots of fractions [4.4]	To determine the square root of a fraction: 1. Determine the square root of the numerator. 2. Determine the square root of the denominator. 3. The square root of the fraction is the quotient of the two square roots. 4. Write the result as a mixed number if necessary.	$\sqrt{\frac{25}{144}} = \frac{\sqrt{25}}{\sqrt{144}} = \frac{5}{12}$
Multiply decimals [4.5]	To multiply decimals: 1. Multiply two numbers as if they were whole numbers, ignoring decimal points for the moment. 2. Add the number of digits to the right of the decimal point in each factor, to obtain the number of digits that must be to the right of the decimal point in the product. 3. Place the decimal point in the product by counting digits from the right.	23.4 × 2.45 1170 936 468 57.330

CONCEPT / SKILL	DESCRIPTION	EXAMPLE
Divide decimal fractions [4.5]	To divide decimals: 1. Write the division in long division format. 2. Move the decimal point the same number of places to the right in both divisor and dividend so that the divisor becomes a whole number. 3. Place the decimal point in the quotient directly above the decimal point in the dividend and divide as usual.	$2.3\overline{)130.41}$ = 56.7 115 154 138 161 161 0
Estimate products [4.5]	To estimate products: 1. Round each factor to one or two nonzero digits. 2. Multiply the rounded factors. This is your estimate.	Estimate 279×52. $300 \times 50 = 15{,}000$
Estimate quotients [4.5]	To estimate quotients: 1. Think of the division as a fraction. 2. Round the numerator and denominator to one or two nonzero digits. 3. Divide the results of step 2. This is your estimate.	Estimate $93.4\overline{)345.26}$. $\frac{345.26}{93.4} \approx \frac{350}{100} = 3.5$

CHAPTER 4 GATEWAY REVIEW

1. Determine the product or quotient.

a. $-4 \cdot 38$

b. $\frac{4}{9} \cdot -\frac{15}{14}$

c. $-42 \div -6$

d. $30.8 \div .002$

e. $2\frac{4}{5} \div 2\frac{1}{10}$

f. $-4.89(-.0008)$

2. Perform the indicated operations.

a. $\left(-\frac{8}{13}\right)^2$

b. -4^2

c. $\sqrt{\frac{121}{144}}$

3. Determine the product of 13 factors of -1.

4. Explain the difference between -5^2 and $(-5)^2$.

5. Evaluate each expression.

a. $-7.2 \cdot 0.06 + .81 \div (-0.9)$

b. $\dfrac{10 - 6 \cdot 4 - 3^2}{6 - (2)^2}$

6. Evaluate each expression.

a. $2x - 6(x - 3)$, where $x = -1\frac{1}{2}$

b. $-3xy - x^2$, where $x = -2$ and $y = -6$

Answers to all Gateway exercises are included in the Selected Answers appendix.

c. $\pi r^2 h$, where $r = 2.4$ and $h = 0.9$

7. Solve for the unknown in each of the following.

a. $-12x = -72$

b. $\dfrac{-x}{6} = 14$

c. $-\dfrac{4}{9} = -18x$

d. $8n = 73.84$

e. $-\dfrac{5}{9} = \dfrac{20}{33}s$

f. $12.9 = 0.0387x$

8. Translate each of the following verbal statements into an equation and solve for the unknown value. Let x represent the unknown number.

a. The quotient of a number and -15 is -7.

b. $-\dfrac{11}{12}$ times a number is $2\dfrac{1}{16}$.

c. 1.08 is the product of a number and 0.2.

9. You have prepared $9\frac{1}{2}$ gallons of fruit punch for your friend's bridal shower. How many $\frac{3}{4}$-cup servings will you have for the guests?

d. I saved \$10.

e. I saved 40%.

f. 4 out of 5 students work to help pay tuition.

g. Derek Jeter's batting average is 0.296.

h. Barry Bonds hit 73 home runs, a major league record, in 2001.

i. In 2005, 16 million Americans owned a Palm personal digital assistant.

j. By 2008, 40% of all physicians will rely on a personal digital assistant.

5. What mathematical notation or verbal phrases in Problem 4 indicated a relative measure?

DEFINITIONS

Relative measure is a term used to describe the comparison of two similar quantities.

Ratio is the term used to describe relative measure as a quotient of **two similar quantities,** often a "part" and a "total," where "part" is *divided by* the "total."

6. Use the free-throw statistics from the 2005 championship games to express each player's relative performance as a ratio in verbal, fraction, and decimal form. The ratios for Chauncey Billups are completed for you. Round decimals to the nearest thousandth.

Nothing but Net

PLAYER	FREE THROWS MADE	FREE THROW ATTEMPTS	EXPRESSED VERBALLY	EXPRESSED AS A FRACTION	EXPRESSED AS A DECIMAL
Billups	40	44	40 out of 44	$\frac{40}{44}$	0.909
Duncan	36	54			
Hamilton	18	24			
Parker	7	15			

7. a. Three friends, shooting baskets in the schoolyard, kept track of their performance. Andy made 9 out of 15 shots, Pat made 28 out of 40, and Val made 15 out of 24. Rank their relative performance.

   b. Which ratio form (fraction, decimal, or other) did you use to determine the ranking?

8. Describe how to determine if the two ratios "12 out of 20" and "21 out of 35" are equivalent?

9. a. Match each ratio from column A with the equivalent ratio in column B.

Column A	Column B
15 out of 25	84 out of 100
42 out of 60	65 out of 100
63 out of 75	60 out of 100
52 out of 80	70 out of 100

   b. Which set of ratios, those in column A or those in column B, is more useful in comparing and ranking the ratios? Why?

Percents

Relative measure based on 100 is familiar and natural. There are 100 cents in a dollar and 100 points on many tests. You have probably been using a ranking scale from 0 to 100 since childhood. You most likely have an instinctive understanding of ratios relative to 100. A ratio such as 40 out of 100 can be expressed as 40 **per** 100, or, more commonly, as 40 **percent**, 40%. Percent *always* indicates a ratio "out of 100."

10. Express each ratio in column B of Problem 9 as a percent, using the symbol "%."

Since each ratio in Problem 10 is already a ratio "out of 100," you replaced the phrase "out of 100" with the % symbol. But suppose you need to write a ratio such as 21 out of 25 in percent format. You may recognize that the denominator, 25, is a factor of 100 $(25 \cdot 4 = 100)$. Then the fraction $\frac{21}{25}$ can be written equivalently as

$$\frac{21 \cdot 4}{25 \cdot 4} = \frac{84}{100}, \text{ which is precisely } 84\%.$$

A more general method to convert the ratio 21 out of 25 into percent format is to first calculate the quotient, 21 divided by 25, by calculator or by long division.

On a calculator:

Key in [2] [1] [÷] [2] [5] [=] to obtain 0.84.

Using long division:

$$\begin{array}{r} 0.84 \\ 25\overline{)21.00} \\ \underline{20.0} \\ 100 \\ \underline{100} \\ 0 \end{array}$$

Next, you convert the decimal form of the ratio into percent format by multiplying the decimal by $\frac{100}{100} = 1$.

$$0.84 = 0.84 \cdot \frac{100}{100} = (0.84 \cdot 100) \cdot \frac{1}{100} = \frac{84}{100} = 84\%$$

Note that multiplying the decimal by 100 moves the decimal point in 0.84 two places to the right. This leads to a shortcut for converting a decimal to a percent.

PROCEDURE

Converting a Fraction or Decimal to a Percent

1. Convert the fraction to decimal form by dividing the numerator by the denominator.
2. Move the decimal point in the quotient from step 1 two places to the right, inserting placeholder zeros if necessary.
3. Then attach the % symbol to the right of the number.

Example: $\frac{4}{5} = 5\overline{)4} = .80 = 80\%$

11. Rewrite the following ratios in percent format.

 a. 35 out of 100

 b. 16 out of 50

 c. 8 out of 20

 d. 7 out of 8

In many applications, you will need to convert a percent into decimal format. The following examples demonstrate the process.

EXAMPLE 1 *Convert 73% to a decimal.*

SOLUTION

First locate the decimal point.

$$73.\%$$

Next, since percent means "out of 100,"

$$73.\% = \frac{73}{100}.$$

Finally, since division by 100 means moving the decimal point two places to the left,

$$\frac{73.}{100} = 0.73.$$

Therefore, 73% = 0.73.

EXAMPLE 2 *Convert 4.5% to a decimal.*

SOLUTION

$$4.5\% = \frac{4.5}{100} = 0.045$$

Note that it was necessary to insert a placeholding zero.

PROCEDURE

Converting a Percent to a Decimal

1. Locate the decimal point in the number attached to the % symbol.
2. Move the decimal point two places to the left, inserting place-holding zeros if needed, and write the decimal number without the % symbol.

12. Write the following percents in decimal format.

a. 75% **b.** 3.5% **c.** 200% **d.** 0.75%

13. Use the three-point shot statistics from the 2005 NBA championship series to express each player's relative performance as a ratio in verbal, fraction, decimal, and percent form. Round the decimal to the nearest hundredth. The ratios for the Piston's Chauncey Billups are completed for you.

PLAYER	THREE-POINT SHOTS MADE	THREE-POINT SHOTS ATTEMPTED	EXPRESSED			
			VERBALLY	AS A FRACTION	AS A DECIMAL	AS A PERCENT
C. Billups (Det)	11	34	11 out of 34	$\frac{11}{34}$	0.32	32%
M. Ginobili (SA)	12	31				
R. Horry (SA)	15	31				
R. Wallace (Det)	5	17				

SUMMARY ACTIVITY 5.1

1. **Relative measure** is a term used to describe the comparison of two similar quantities.
2. **Ratio** is the term used to describe relative measure as a quotient of **two similar quantities,** often a "part" and a "total," where "part" is *divided by* the "total."
3. Ratios can be expressed in several forms: **verbally** (4 out of 5), as a **fraction** $\left(\frac{4}{5}\right)$, as a **decimal** (0.8), or as a **percent** (80%).
4. **Percent** always indicates a ratio out of 100.
5. **Converting a fraction or decimal to a percent**
   i. Convert the fraction to decimal form by dividing the numerator by the denominator.
   ii. Move the decimal point two places to the right, and then attach the % symbol.
6. **Converting a percent to a decimal**
   i. Locate the decimal point in the number attached to the % symbol.
   ii. Move the decimal point two places to the left, inserting place-holding zeros if needed, and write the decimal number without the % symbol.
7. Two ratios are **equivalent** if their decimal or reduced fraction forms are equal.

EXERCISES
ACTIVITY 5.1

1. Complete the following table by representing each ratio in all four formats. Round decimals to the thousandths place and percents to the tenths place.

3.14
6.02 2
50% 0.007
99 ½

Numerically Speaking...

VERBAL	REDUCED FRACTION	DECIMAL	PERCENT
1 out of 3			
2 out of 5			
18 out of 25			
8 out of 9			
3 out of 8			
25 out of 45			
120 out of 40			
3 out of 4			
27 out of 40			
3 out of 5			
2 out of 3			
4 out of 5			
1 out of 200			
2 out of 1			

2. a. Match each ratio from column 1 with the equivalent ratio in column 2.

Column 1	Column 2
12 out of 27	60 out of 75
28 out of 36	25 out of 40
45 out of 75	21 out of 27
64 out of 80	42 out of 70
35 out of 56	20 out of 45

b. Write each matched pair of equivalent ratios as a percent. If necessary, round to the nearest tenth of a percent.

3. Your biology instructor returned three quizzes today. On which quiz did you perform best? Explain how you determined the best score.

Quiz 1: 18 out of 25 **Quiz 2:** 32 out of 40 **Quiz 3:** 35 out of 50

$\frac{18}{25} = 72\%$ $\frac{32}{40} = 80\%$ $\frac{35}{50} = 70\%$

4. A baseball batting average is the ratio of hits to the number of times at bat. It is reported as a three-digit decimal. Determine the batting averages of three players with the given records.

   a. 16 hits out of 54 at bats

   b. 25 hits out of 80 at bats

   c. 32 hits out of 98 at bats

5. There are 1720 females among the 3200 students at the local community college. Express this ratio in each of the following forms.

   a. fraction b. reduced fraction c. decimal d. percent

6. At the state university campus near the community college in Exercise 5, there are 2304 females and 2196 males enrolled. In which school, the community college or university, is the relative number of females greater? Explain your reasoning.

7. In the 2005 Major League Baseball season, the world champion Chicago White Sox ended their regular season with a 99–63 win-loss record. During their playoff season, their win-loss record was 11–1. Did the White Sox play better in the regular season or in the playoff season? Justify your answer mathematically.

8. A random check of 150 Southwest Air flights last month identified that 113 of them arrived on time. What "on-time" percent does this represent?

9. A consumer magazine reported that of the 13,350 subscribers who owned brand A dishwasher, 2940 required a service call. Only 730 of the 1860 owners of brand B needed repairs. Which brand of dishwasher has the better repair record?

10. The admissions office in a local college has organized data in the following table listing the number of men and women who are currently enrolled. Admissions will use the data to help recruit students for the next academic year. Note that a student is matriculated if he or she is enrolled in a program that leads to a college degree.

School-Bound

	FULL-TIME MATRICULATED STUDENTS ($\geq$ 12 CREDITS)	PART-TIME MATRICULATED STUDENTS ($<$ 12 CREDITS)	PART-TIME NONMATRICULATED STUDENTS ($<$ 12 CREDITS)
MEN	214	174	65
WOMEN	262	87	29

a. How many men attend the college?

b. What percent of the men are full-time students?

c. How many women attend the college?

d. What percent of the women are full-time students?

e. How many students are enrolled full-time?

f. How many students are enrolled part-time?

g. What percent of the full-time students are women?

h. What percent of the part-time students are women?

i. How many students are nonmatriculated?

j. What percent of the student body is nonmatriculated?

ACTIVITY 5.2

The Devastation of AIDS in Africa

OBJECTIVES

1. Use proportional reasoning to apply a known ratio to a given piece of information.
2. Write an equation using the relationship "ratio · total = part" and then solve the resulting equation.

International public health experts are desperately seeking to stem the spread of the AIDS epidemic in sub-Saharan Africa, especially in the small nation of Botswana. According to United Nations officials, over 35% of Botswana's 800,000 adults (ages 15–49) are currently infected with the AIDS virus.

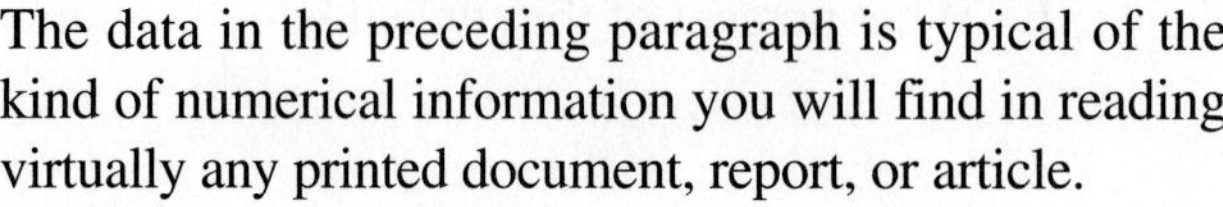

The data in the preceding paragraph is typical of the kind of numerical information you will find in reading virtually any printed document, report, or article.

1. What relative data (ratio) appears in the opening paragraph? What phrase or symbol identifies it as a relative measure?

2. Express the ratio in Problem 1 in fraction, decimal, and verbal form.

Proportional Reasoning

From the information given and your answers to Problems 1 and 2, you know that the adult population of Botswana is 800,000 and that 35% is the percentage of the adult population that has AIDS. The unknown quantity in this case is the number of adults that have AIDS. To determine the unknown quantity, take 35% of 800,000 as follows:

$$0.35 \cdot 800{,}000 = 280{,}000$$

Therefore, 280,000 adults in Botswana are infected with AIDS. Because the ratio $\frac{280{,}000}{800{,}000} = 0.35$, or 35%, you know that the calculation is correct.

This example illustrates a thought process called **proportional reasoning.** In this case, proportional reasoning is used to apply a known ratio to one piece of information (the total) to determine a related, but yet unknown, second piece of information (the part). The formula that was used,

$$ratio \cdot total = part$$

is equivalent to the formula

$$ratio = \frac{part}{total}.$$

A common procedure to apply a known ratio to a given piece of information is described as follows.

PROCEDURE

Setting Up and Solving ratio · total = part, When the Ratio Is Known

Step 1. Identify whether the given piece of information represents a *part* or a *total*.

Step 2. Substitute all the known information into the formula *ratio* · *total* = *part* to form an equation. The unknown may be represented by a letter.

Step 3. Solve the equation.

3. You do further research about this disturbing health problem and discover that by 2005 there were approximately 25,800,000 adults and children living in sub-Saharan Africa with HIV/AIDS. This represents 7.2% of the total population there.

a. The ratio of the total population that is infected with AIDS is given in the preceding information. What is it? In what form is the ratio expressed?

b. Identify the other given information.

c. Identify the unknown piece of information.

d. Use the formula $ratio \cdot total = part$ to estimate the total population.

4. It was estimated that 330,000 adults from Botswana were infected with HIV/AIDS in 2003, and that 190,000 of them were women. What percent of those infected were women? Use the formula $ratio = \frac{part}{total}$.

5. Africa has buried three-fourths of the 25 million people worldwide who have died from AIDS since the beginning of the AIDS epidemic. Calculate how many people from Africa have died from AIDS.

6. In the year 2005, 1.9 million sub-Saharan African children under the age of 15 were living with AIDS. This represents 80% of all children worldwide who were infected in 2005. Use the following steps to estimate the total number of all children worldwide who were infected.

a. Identify the known ratio.

b. Identify the given piece of information. Does this number represent a part or a total?

c. Identify the unknown piece of information. Does this number represent a part or a total?

d. Use the formula *ratio* · *total* = *part* to determine your estimate. Let the unknown quantity be represented by *x*.

7. In 2005, 4.9 million adults and children became newly infected with HIV. Sixty-five percent of these new cases occurred in sub-Saharan Africa and 20% of the new cases occurred in South and Southeast Asia.

a. Determine how many of these new cases of AIDS occurred in sub-Saharan Africa.

b. How many of these new cases of AIDS occurred in South and Southeast Asia?

8. Of the world's 6.5 billion people, 3 billion live on less than $2 per day. Calculate the percent of the world's population who live on less than $2 per day.

SUMMARY
ACTIVITY 5.2

Proportional reasoning is the process that uses the equation *ratio* · *total* = *part* to determine the unknown quantity when two of the quantities in the equation are known. This relationship is equivalently written as $ratio = \frac{part}{total}$.

Direct approach to solve for an unknown part or total:

Step 1. Identify whether the given piece of information represents a part or a total.

Step 2. Substitute all the known information into the formula *ratio* · *total* = *part* to form an equation. The unknown may be represented by a letter.

Step 3. Solve the equation.

Note that sometimes the unknown piece of information is the ratio. In this case, use the formula, $ratio = \frac{part}{total}$ to calculate the ratio directly.

EXERCISES
ACTIVITY 5.2

1. Determine the value of each expression.

a. $12 \cdot \frac{2}{3}$

b. $18 \div \frac{2}{3}$

c. $\frac{3}{5}$ of 25

d. $24 \div \frac{3}{4}$

e. 30% of 60

f. $150 \div 0.60$

g. $45 \cdot \frac{4}{9}$

h. 25% of 960

i. $5280 \div 75\%$

j. $2000 \div \frac{4}{5}$

k. 80% of 225

l. $30{,}000 \div 50\%$

2. Sub-Saharan Africa accounted for 77% of the 3.1 million adults and children who died of AIDS in the year 2005. How many sub-Saharan children died of AIDS that year?

a. Identify the known ratio.

b. Identify the given piece of information. Does this number represent a part or a total?

c. How many sub-Saharan adults and children died of AIDS in 2005?

3. The 25.8 million sub-Saharan people (adults and children) currently living with AIDS account for 64% of the entire world population infected with the disease. How many people worldwide are infected with AIDS?

4. The customary tip on waiter service in New York City restaurants is approximately 20% of the food and beverage cost. This means that your server is counting on a tip of \$0.20 for each dollar you spend on your meal. What is the expected tip on a dinner for two costing \$65?

5. Your new car cost \$22,500. The state sales tax rate is 8%. How much sales tax will you be paying on the car purchase?

6. You happened to notice that the 8% sales tax that your uncle paid on his new luxury car came to \$3800. How much did the car itself cost him?

7. In your recent school board elections only 45% of the registered voters went to the polls. If 22,000 votes were cast, approximately how many voters are registered in your school district?

8. In 2005, new car and truck sales totaled approximately 16.9 million in the United States. The breakdown among the leading automobile manufacturers was

General Motors 26.7% Honda 8.6%
Ford 11.8% Toyota 13.4%
Chrysler 13.6%

Approximately how many new cars did each company sell that year?

9. In a very disappointing season, your softball team won only 40% of the games it played this year. If you won 8 games, how many games did you play? How many did you lose?

10. In a typical telemarketing campaign, 5% of the people contacted purchase the product. If your quota is to sign up 50 people, approximately how many phone calls do you anticipate making?

11. Approximately three-fourths of the teacher-education students in your college pass the licensing exam on their first attempt. This year, 92 students will sit for the exam. How many are expected to pass?

12. You paid $90 for your economics textbook. At the end of the semester, the bookstore will buy back your book for 20% of the original purchase price. The book sells used for $65. How much money does the bookstore earn on the resale?

ACTIVITY 5.3

Who Really Did Better?

OBJECTIVES

1. Define actual and relative change.
2. Distinguish between actual and relative change.

After researching several companies, you and your cousin decided to purchase your first shares of stock. You bought \$200 worth of stock in one company and your cousin invested \$400 in another. At the end of a year, your stock was worth \$280 and your cousin's stock was valued at \$500.

1. How much money did you earn from your purchase?

2. How much did your cousin earn?

3. Who earned more money?

Your answers to Problems 1 and 2 represent **actual change**, the actual numerical value by which a quantity has changed. When a quantity increases in value, such as your stock, the actual change is positive. When a quantity decreases in value, the actual change is negative.

4. Keeping in mind the amount each of you originally invested, whose investment do you believe did better? Explain your answer.

In answering Problem 4, you may have considered the actual change *relative* to the original amount invested. This comparison leads to the mathematical idea of relative change.

DEFINITION

The ratio formed by comparing the actual change to the original value is called the **relative change**. Relative change is written in fraction form with the actual change placed in the numerator, and the original (or earlier) amount placed in the denominator.

$$\text{relative change} = \frac{\text{actual change}}{\text{original value}}$$

Since *relative change* (increase or decrease) is frequently reported as a percent, it is often called *percent change*. The two terms are interchangeable.

5. Determine the relative change in your \$200 investment. Write your answer as a fraction and as a percent.

6. Determine the relative change in your cousin's \$400 investment. Write your answer as a fraction and as a percent.

7. In relative terms, whose stock performed better?

8. Determine the actual change and the percent change of the following quantities.

   a. You paid \$20 for a rare baseball card that is now worth \$30.

   b. Last year's graduating class had 180 students; this year's class size is 225.

   c. Ann invested \$300 in an energy stock, and now it is worth \$540.

   d. Bob invested \$300 in a technology stock, but now it is worth only \$225.

   e. Pat invested \$300 in a risky dot-com venture, and his stock is now worth \$60.

When speaking about percent change, a percent greater than 100 makes sense. For example, suppose a \$50 investment tripled to \$150. The actual increase in the investment is

$$\$150 - \$50 = \$100.$$

Then the relative change is $\dfrac{\text{actual increase}}{\text{original value}} = \dfrac{\$100}{\$50} = 2.00$. Therefore, the investment increased by 200%.

9. Determine the actual change and the percent change of the following quantities.

a. Your parents purchased their home in 1970 for $40,000. They sold it recently for $200,000.

b. After building a new convention center, the number of hotel rooms in a city increased from 1500 to 6000.

Actual Change vs. Relative Change

Actual change and relative (percent) change provide two perspectives to understanding change. Actual change looks only at the *change itself,* ignoring the original value. Relative change indicates the significance or importance of the change by *comparing it to the original value.*

The following problem will show you how relative change is different from actual change.

10. You and your colleague are bragging about your respective stock performance. Each of your stocks earned $1000 last year. Then it slipped out that your original investment was $2000, while your colleague had invested $20,000 in her stock.

a. What was the actual increase in each investment?

b. Calculate the relative increase (in percent format) for each investment.

c. Whose investment performance was more impressive? Why?

11. During the first week of classes, campus clubs actively recruit members. The Ultimate Frisbee Club signed up 8 new members, reaching an all-time high membership of 48 students. The Yoga Club also attracted 8 new members and now boasts 18 members.

a. What was the membership of the ultimate frisbee club before recruitment week?

b. By what percent did the frisbee club increase its membership?

c. What was the membership of the yoga club before recruitment week?

d. By what percent did the yoga club increase its membership?

SUMMARY
ACTIVITY 5.3

1. The ratio formed by comparing the actual change to the original value is called the **relative change.** Relative change is written in fraction form with the actual change placed in the numerator and the original (or earlier) amount placed in the denominator.

$$\text{relative change} = \frac{\text{actual change}}{\text{original value}}$$

Since *relative change* (increase or decrease) is frequently reported as a percent, it is often called *percent change.* The two terms are interchangeable.

EXERCISES
ACTIVITY 5.3

1. The full-time enrollment in your college last year was 3200 students. This year there are 3560 full-time students on campus.

a. Determine the actual increase in full-time enrollment.

b. Determine the relative increase (as a percent) in full-time enrollment.

2. In the spring of 2004, the price of gasoline seemed to be rising every day. The average price for regular gas began the year at \$2.20 per gallon. By July, the price had risen to \$2.65 per gallon.

a. Determine the actual increase in gasoline price.

b. Determine the relative increase (as a percent) in gasoline price.

Exercise numbers appearing in color are answered in the Selected Answers appendix.

3. This year, the hourly wage from your part-time job is \$6.50, up from last year's \$6.25. What percent increase does this represent?

4. Last month, Acme and Arco corporations were forced to lay off 500 workers each. Acme's workforce is now 1000. Arco currently has 4500 employees. What percent of their workforce did each company lay off?

5. You are now totally committed to changing your eating habits for better health. You have decided to cut back your daily caloric intake from 2400 to 1800 calories. By what percent are you cutting back on calories?

6. **a.** You paid \$10 per share for a promising technology stock. Within a year, the stock price skyrocketed to \$50 per share. By what percent did your stock value increase?

   b. You held on to the stock for a bit too long. The price has just fallen from its \$50 high back to \$10. By what percent did your stock decrease from \$50?

7. Perform the following calculations.

   a. Start with a value of 10, double it, and then calculate the percent increase.

   b. Start with a value of 25, double it, and then calculate the percent increase.

   c. Start with a value of 40, double it, and then calculate the percent increase.

   d. When any quantity doubles in size, by what percent has it increased?

8. Perform a similar set of calculations to the ones you did in Exercise 7 to help determine the percent increase of any quantity that triples in size.

ACTIVITY 5.4

Don't Forget the Sales Tax

OBJECTIVES

1. Define and determine growth factors.
2. Use growth factors in problems involving percent increases.

In order to earn revenue (income), many state and local governments require merchants to collect sales tax on the items they sell. In a number of localities in the United States, the sales tax is as much as 8% of the selling price, or even higher. The purchaser pays both the selling price and the sales tax.

1. Determine the total cost (including the 8% sales tax) to the customer of the following items. Include each step of your calculation in your answer.

   a. A greeting card selling for $1.50.

   b. A shirt selling for $35.

   c. A vacuum cleaner priced at $300.

   d. A car priced at $25,000.

Many people will correctly determine the total costs for Problem 1 in two steps.

- They will first compute the sales tax on the item.
- Then they will add the sales tax to the selling price to obtain the total cost.

It is also possible to compute the total cost in one step by using the idea of a **growth factor**. If you use the growth factor, you won't need to calculate the sales tax. Later in the activity you will be asked to determine the selling price of an item for which you know the total checkout price. You will see that the *only* direct way to answer this "reverse" question is through use of a growth factor. The problems that follow will introduce you to the growth factor concept and how to use it.

Growth Factors

2. If a quantity increases by 50%, how does its new value compare to its original value? That is, what is the ratio of new value to original value? Complete the second and third rows of the table to confirm/discover your answer. The first row is completed for you.

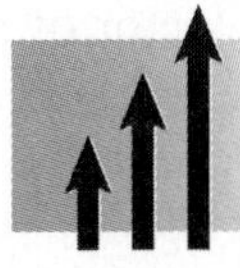

On the Up and Up

ORIGINAL VALUE	NEW VALUE (INCREASED BY 50%)	RATIO OF NEW VALUE TO ORIGINAL VALUE		
		FRACTION FORM	DECIMAL FORM	PERCENT FORM
10	15	$\frac{15}{10} = \frac{3}{2}$	1.50	150%
50				
100				

3. Use the results from the table in Problem 2 to answer the question, "What is the ratio of new value to original value of any quantity that increases by 50%?" Write the ratio as a reduced fraction, as a decimal, and as a percent.

The ratio, $\frac{3}{2} = \frac{\text{new value}}{\text{original value}}$, that you calculated in the table and observed in Problem 3 is called the **growth factor** for a 50% increase in any quantity. The definition of a growth factor for any percent increase is given in the following.

DEFINITION

For a specified percent increase, no matter what the original value may be, the ratio of the new value to the original value is always the same. This ratio is called the **growth factor** associated with the specified percent increase. The formula is

$$\text{growth factor} = \frac{\text{new value}}{\text{original value}}.$$

The growth factor is most often written in decimal form. As you determined in Problems 2 and 3, the growth factor, in decimal form, associated with a 50% increase is 1.50. Since 150% is the same as 100% + 50%, the growth factor is the sum of 100% and 50%. Since 150% = 1.50 in decimal form, the growth factor is also the sum of 1.00 and 0.50. This leads to a second way to determine a growth factor.

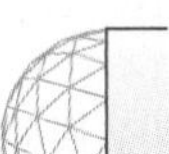

PROCEDURE

Determining a Growth Factor from a Percent Increase The growth factor is determined by adding the specified percent increase to 100% and then changing this percent into its decimal form.

4. In each of the following, determine the growth factor of any quantity that increases by the given percent. Write your result as a percent and then as a decimal.

   a. 30%

   b. 75%

   c. 15%

   d. 7.5%

5. A growth factor can always be written in decimal form. Explain why this decimal will always be greater than 1.

Using Growth Factors in Percent Increase Problems

In Problems 2 and 3, you saw that a growth factor is given by the formula,

$$\text{growth factor} = \frac{\text{new value}}{\text{original value}}.$$

This formula is equivalent to the formula

$$\text{original value} \cdot \text{growth factor} = \text{new value}.$$

For example, $1.5 = \frac{75}{50}$ is the same as $50 \cdot 1.5 = 75$. This observation leads to a procedure for determining a new value using growth factors.

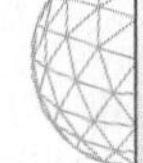

PROCEDURE

Determining the New Value Using Growth Factors When a quantity increases by a specified percent, its new value can be obtained by multiplying the original value by the corresponding growth factor. That is,

$$\text{original value} \cdot \text{growth factor} = \text{new value}.$$

6. a. Use the growth factor 1.50 to determine the new price of a stock portfolio that increased by 50% from its original value of $400.

   b. Use the growth factor 1.50 to determine the population of a town that has grown by 50% from its previous population of 120,000 residents.

7. a. Determine the growth factor for any quantity that increases by 20%.

b. Use the growth factor to determine this year's budget, which has increased by 20% over last year's budget of $75,000.

c. Use the growth factor to determine the enrollment of an organization that has grown by 20% from its original size of 250 members.

8. a. Determine the growth factor for any quantity that increases by 8%.

b. Use the growth factor to determine the total cost of each item from Problem 1.

In the previous problems you were given an original (earlier) value and a percent increase and were asked to determine the new value. Suppose you are asked a "reverse" question: Sales in your small retail business have increased by 20% over last year. This year you had gross sales receipts of $75,000. How much did you gross last year?

Using the growth factor 1.20, you can write

$$\text{original value} \cdot 1.20 = 75{,}000.$$

Let x represent the original value (last year's gross sales). This equation can now be written as

$$x \cdot 1.20 = 75{,}000.$$

To solve for x, divide both sides by 1.20.

$$x = \frac{75{,}000}{1.20} = 62{,}500$$

So, last year's gross sales were $62,500. This example leads to a general procedure for determining an original value from a new value and the growth factor.

PROCEDURE

Determining the Original Value Using Growth Factors When a quantity has *already* increased by a specified percent, its original value is obtained by dividing the new value by the corresponding growth factor.

$$\text{new value} \div \text{growth factor} = \text{original value}$$

9. You determined a growth factor for an 8% increase in Problem 8. Use it to answer the following questions.

a. The cash register receipt for your new coat, which includes the 8% sales tax, totals $243. What was the ticketed price of the coat?

b. The credit-card receipt for your new sofa, which includes the 8% sales tax, came to $1620. What was the ticketed price of the sofa?

SUMMARY
ACTIVITY 5.4

1. When a quantity increases by a specific percent, the ratio of the new value to the original value is called the **growth factor** associated with the specified percent increase. The growth factor is also formed by *adding* the specified percent increase to 100%, and then changing the resulting percent into decimal form.

2. The equation that defines the growth factor as the ratio of the new value to the original value is

$$\text{growth factor} = \frac{\text{new value}}{\text{original value}}$$

3. When the growth factor equation is multiplied on both sides of the equal sign by the original value, an equivalent equation is obtained:

$$\text{original value} \cdot \text{growth factor} = \frac{\text{new value}}{\cancel{\text{original value}}} \cdot \cancel{\text{original value}}$$

Therefore, when a quantity increases by a specified percent, its new value can be obtained by multiplying the original value by the corresponding growth factor.

$$\text{new value} = \text{original value} \cdot \text{growth factor}$$

4. When the equation, new value = original value · growth factor is divided on both sides by the growth factor, another equivalent equation is obtained

$$\frac{\text{new value}}{\text{growth factor}} = \frac{\text{original value} \cdot \cancel{\text{growth factor}}}{\cancel{\text{growth factor}}}$$

Therefore, when a quantity has *already* increased by a specified percent, its original value can be obtained by dividing the new value by the corresponding growth factor.

$$\text{original value} = \text{new value} \div \text{growth factor}$$

EXERCISES
ACTIVITY 5.4

1. Complete the following table.

% INCREASE	5%		15%		100%	300%	
GROWTH FACTOR		1.35		1.045			11

2. You are purchasing a new car. The price you negotiated with a car dealer in Staten Island, New York, was $17,944 excluding sales tax. The sales tax rate for Staten Island is 8.375%.

   a. What growth factor is associated with the sales tax rate?

   b. Use the growth factor to determine the total cost of the car.

Exercise numbers appearing in color are answered in the Selected Answers appendix.

3. The cost of first-class postage in the United States increased by approximately 5% on January 8, 2006. If the cost is now 39 cents, how much did first-class postage cost in 2005? Round your answer to the nearest penny.

4. A state university in the western United States reported that its enrollment increased from 11,952 students in 2003–2004 to 12,144 students in 2004–2005.

a. What is the ratio of the number of students in 2004–2005 to the number of students in 2003–2004?

b. From your result in part a, determine the growth factor. Write it in decimal form to the nearest hundredth.

c. Write the growth factor in percent form.

5. The 2000 United States Census showed that from 1990 to 2000, Florida's population increased by approximately 23.5%. The population in 2000 was 15,982,378. What was Florida's population in 1990?

6. Your grandparents bought a house in 1967 for $26,000. The United States Bureau of Labor Statistics reports that due to inflation, the cost of the same house in 2006 would be $158,413.

a. What was the inflation growth factor for housing from 1967 to 2006?

b. What was the inflation rate (to the nearest whole-number percent) for housing from 1967 to 2006?

c. Your grandparents sold their house in 2006 for $465,000. What is the profit they made in terms of 2006 dollars?

7. Your friend plans to move from Newark, New Jersey, to Anchorage, Alaska. She earns $50,000 a year in Newark. How much must she earn in Anchorage to maintain the same standard of living if the cost of living increases from Newark to Anchorage by 35%?

8. You decide to invest $3000 in a 1-year certificate of deposit (CD) that earns an annual percentage yield (APY) of 5.05%. How much will your investment be worth in a year?

9. Your company is moving you from the United States to one of the other countries listed in the following table. For each country, the table lists the cost of a basket of goods and services that would cost $100 in the United States.

In this problem, you will compare the cost of the same basket of goods in each of the countries listed to the $100 cost in the United States. The comparison ratio in each case will be a growth factor relative to the cost in the United States.

Living Abroad

COUNTRY	COST OF BASKET	GROWTH FACTOR IN COST OF LIVING	PERCENT INCREASE IN SALARY
United States	100		
Argentina	140		
Australia	114		
Belgium	127		
Canada	105		
Saudi Arabia	131		
Singapore	126		

a. Determine the growth factor for each cost in the table and list it in the third column.

b. Determine the percent increase in your salary that would allow you the same cost of living as you have in the United States. List the percent increase in the fourth column of the table.

c. You earn a salary of $45,000 in the United States. What would your salary have to be in Argentina to maintain the same standard of living?

10. In 1965, Gordon Moore, cofounder of Intel, predicted that the number of transistors that can be placed on a computer chip would double approximately every 18 months. The press dubbed this prediction Moore's law and the name has stuck.

a. Moore's law can be restated in terms of a growth factor. The number of transistors that can be placed on a computer chip *will grow by a factor of 1.59 each year.* One chip that Intel produced in 1999 has 24 million transistors and another has 28 million transistors. Use this data and the growth factor 1.59 to predict a range for the number of transistors on one chip that Intel produced in 2000.

b. The Intel chip produced in 2000 actually had 42 million transistors. If an upgraded chip is made about a year later and Moore's law still applies, predict the number of transistors on the new chip.

ACTIVITY 5.5

It's All on Sale!

OBJECTIVES

1. Define and determine decay factors.
2. Use decay factors in problems involving percent decreases.

It's the sale you have been waiting for all season. Everything in the store is marked 40% off the original ticket price.

1. Determine the discounted price of the following items. Include each step of your calculation in your answer.

   a. A pair of sunglasses originally selling for $20

   b. A shirt originally selling for $35

   c. A bathing suit originally priced at $80

   d. A dress originally priced at $250

Many people will correctly determine the discounted prices for Problem 1 in two steps.

- They will first compute the actual dollar discount on the item.
- Then they will subtract the dollar discount from the original price to obtain the sale price.

It is also possible to compute the sale price in one step by using the idea of a **decay factor**. If you use a decay factor, you will not have to calculate the actual dollar discount separately. Later in the activity you will determine the original price of an item for which you know the discounted sale price. You will see that the *only* direct way to answer this "reverse" question is through use of a decay factor. The problems that follow will introduce you to the decay factor concept and how to use it.

Decay Factors

2. If a quantity decreases by 20%, how does its new value compare to its original value? That is, what is the ratio of new value to original value? Complete the second and third rows in the following table. The first row has been done for you. Notice that the new values are less than the originals, so the ratios of new values to old will be less than 1.

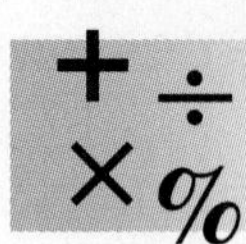

A Matter of Values

ORIGINAL VALUE	NEW VALUE (DECREASED BY 20%)	RATIO OF NEW VALUE TO ORIGINAL VALUE		
		FRACTION FORM	DECIMAL FORM	PERCENT FORM
20	16	$\frac{16}{20} = \frac{4}{5}$	0.80	80%
50	40			
100				

3. What is the ratio of the new value to the original value of any quantity that decreases by 20%? Use the results from the table in Problem 2 to answer this question. Express the ratio as a reduced fraction, a decimal, and as a percent.

The ratio, $\frac{4}{5} = \frac{\text{new value}}{\text{original value}}$, that you calculated in the table and observed in Problem 3 is called the **decay factor** for a 20% decrease in any quantity. The definition of a decay factor for any percent decrease is given in the following.

DEFINITION

For any specified percent *decrease,* no matter what the original value may be, the ratio of the new value to the original value is always the same. This ratio is called the **decay factor** associated with the specific percent decrease. The equation is

$$\text{decay factor} = \frac{\text{new value}}{\text{original value}}.$$

It is important to remember that a percent decrease describes the percent that has been *removed* (such as the discount taken off an original price). Therefore, the corresponding decay factor always represents the percent remaining (the portion that you still must pay).

The decay factor is most often written in decimal form. As you determined in Problems 2 and 3, the decay factor associated with a 20% decrease is 80%. Since 80% = 100% − 20%, the decay factor is the difference of 100% and 20%. Since 80% is 0.80 in decimal form, the decay factor is also the difference of 1.00 and 0.20. This leads to a second way to determine a decay factor.

PROCEDURE

Determining a Decay Factor from a Percent Decrease The decay factor is determined by *subtracting* the specified percent decrease from 100% and changing the resulting percent into decimal form.

4. In each of the following, determine the decay factor of a quantity that decreases by the given percent. Write your result as a percent and then as a decimal.

   a. 30%

   b. 75%

   c. 15%

   d. 7.5%

   e. 5%

5. Explain why every decay factor will have a decimal format that is less than 1.

Using Decay Factors in Percent Decrease Problems

In Problems 2 and 3, you saw that a decay factor is given by the formula

$$\text{decay factor} = \frac{\text{new value}}{\text{original value}}.$$

This formula is equivalent to the formula

$$\text{original value} \cdot \text{decay factor} = \text{new value}.$$

For example, $0.80 = \frac{16}{20}$ is the same as $20 \cdot 0.8 = 16$. This observation leads to a procedure for determining a new value using decay factors.

PROCEDURE

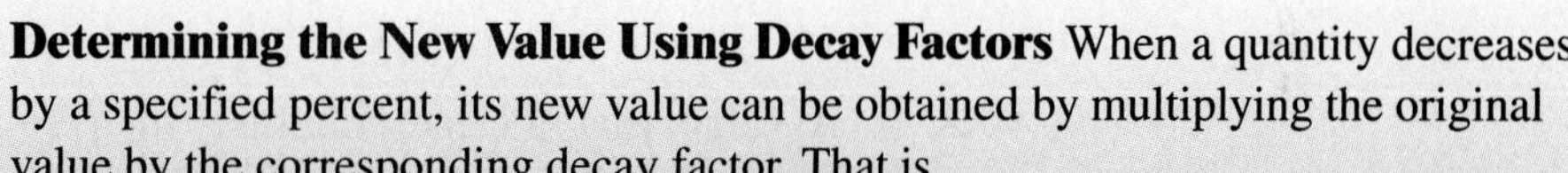

Determining the New Value Using Decay Factors When a quantity decreases by a specified percent, its new value can be obtained by multiplying the original value by the corresponding decay factor. That is,

$$\text{original value} \cdot \text{decay factor} = \text{new value}.$$

6. a. Use the decay factor 0.80 to determine the new value of a stock portfolio that has decreased by 20% from its original value of $400.

   b. Use the decay factor 0.80 to determine the population of a town that has decreased by 20% from its previous size of 300,000 residents.

In the previous questions you were given an original value and a percent decrease and were asked to determine the new (smaller) value. Suppose you are asked a "reverse" question: Voter turnout in the local school board elections declined by 40% from last year. This year only 3600 eligible voters went to the polls. How many people voted in last year's school board elections?

The decay factor is $100\% - 40\% = 60\%$. Using the decay factor 0.60 and the new value 3600, you can write

$$\text{original value} \cdot 0.60 = 3600.$$

If you let x represent the original value (last year's voter turnout), you can rewrite the equation as

$$x \cdot 0.60 = 3600.$$

Now it should be easy to see that you have to divide this year's turnout by the decay factor to obtain the original value. In the equation, this means divide both sides by 0.60. So,

$$x = \frac{3600}{0.60} = 6000.$$

Therefore, last year 6000 people voted in the school board elections.

The example just given leads to a general procedure for determining an original value from a new value and the decay factor.

PROCEDURE

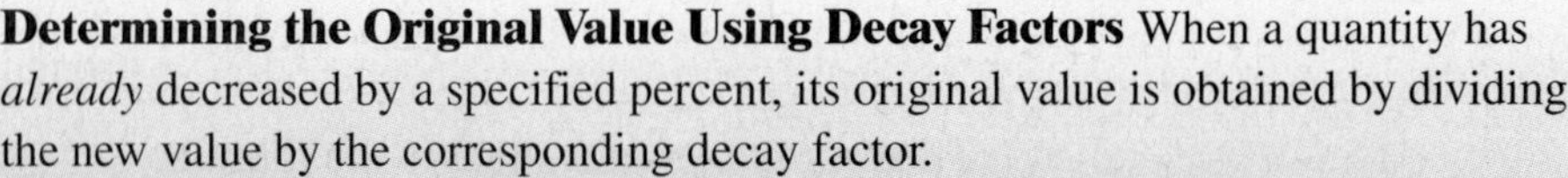

Determining the Original Value Using Decay Factors When a quantity has *already* decreased by a specified percent, its original value is obtained by dividing the new value by the corresponding decay factor.

$$\text{new value} \div \text{decay factor} = \text{original value}$$

7. You purchased a two-line answering machine and telephone on sale for \$150. The discount was 40%. What was the original price?

8. a. Assume that all merchandise in a store has been reduced by 25%. Determine the decay factor for any price that decreases by 25%.

Use the decay factor to determine the prices requested in parts b through e.

b. The sale price of a blouse that originally sold for \$80

c. The sale price of a suit that originally sold for \$360

d. The original price of a fur jacket that is on sale for \$1620

e. The original price of a Nordic Track treadmill that is on sale for \$1140

SUMMARY
ACTIVITY 5.5

1. When a quantity decreases by a specific percent, the ratio of the new value to the original value is called the decay factor associated with the specified percent decrease. The decay factor is formed by *subtracting* the specified percent decrease from 100% and changing the resulting percent into decimal form.

2. The formula that defines the decay factor as the ratio of the new value to the original value is

$$\text{decay factor} = \frac{\text{new value}}{\text{original value}}.$$

3. When the decay factor equation is multiplied on both sides of the equals sign by the original value, an equivalent formula is obtained.

$$\text{original value} \cdot \text{decay factor} = \frac{\text{new value}}{\cancel{\text{original value}}} \cdot \cancel{\text{original value}}$$

Therefore, when a quantity decreases by a specified percent, its new value can be obtained by multiplying the original value by the corresponding decay factor.

$$\text{new value} = \text{original value} \cdot \text{decay factor}$$

4. When the formula new value = original value · decay factor is divided on both sides by the decay factor, another equivalent formula is obtained.

$$\frac{\text{new value}}{\text{decay factor}} = \frac{\text{original value} \cdot \cancel{\text{decay factor}}}{\cancel{\text{decay factor}}}$$

Therefore, when a quantity has *already* decreased by a specified percent, its original value can be obtained by dividing the new value by the corresponding decay factor.

$$\text{original value} = \text{new value} \div \text{decay factor}$$

EXERCISES
ACTIVITY 5.5

1. Complete the following table.

% DECREASE	5%		15%		12.5%
DECAY FACTOR		0.45		0.94	

2. You wrote an 8-page article that will be published in a journal. The editor asked you to revise the article and to reduce the number of pages to 6. By what percent must you reduce the length of your article?

3. In 1995, there were 551,226 live births recorded in the state of California. In 2000, the number of live births dropped to 531,285. What was the decay factor for live births for the 5-year period from 1995 to 2000? What was the percent decrease?

4. A car dealer will sell you the car you want for $18,194, which is just $200 over the dealer's invoice price (the price the dealer pays the manufacturer for the car). You tell him that you will think about it. The dealer is anxious to meet his monthly quota of sales, so he calls the next day to offer you the car for $17,994 if you agree to buy it tomorrow, October 31. You decide to accept the deal.

   a. What is the decay factor associated with the decrease in the price to you? Write your answer in decimal form to the nearest thousandth.

   b. What is the percent decrease of the price, to the nearest tenth of a percent?

   c. The sales tax is 6.5%. How much did you save on sales tax by taking the dealer's second offer?

5. Your company is moving you from the United States to one of the other countries listed in the following table. For each country, the table lists the cost of a basket of goods and services that would cost $100 in the United States.

   In this problem, you will compare the cost of the same basket of goods in each of the countries listed to the $100 cost in the United States. The comparison ratio in each case will be a decay factor relative to the cost in the United States.

Over There

COUNTRY	COST OF GOODS BASKET	DECAY FACTOR IN COST OF LIVING	PERCENT DECREASE IN COST OF LIVING
United States	100		
Colombia	83		
India	93		
Indonesia	98		
Hungary	80		
Poland	100		
South Africa	87		

   a. Determine the decay factor for each cost in the table and list it in the third column.

b. Determine the percent decrease in the cost of living for each country compared to living in the United States. List the percent decrease in the fourth column of the table.

c. You earn a salary of $45,000 in the United States. If you move to South Africa, how much of your salary will you need for the cost of living there?

d. How much of your salary can you save if you move to Hungary?

6. You needed a fax machine but planned to wait for a sale. Yesterday, the model you wanted was reduced by 30%. It originally cost $129.95. Use the decay factor to determine the sale price of the fax machine.

7. Your doctor advises you that losing 10% of your body weight will significantly improve your health.

a. Before you start dieting, you weigh 175 pounds. Determine and use the decay factor to calculate your goal weight to the nearest pound.

b. After 6 months, you reach your goal weight. However, you still have more weight to lose to reach your ideal weight range, which is between 125 and 150 pounds. If you lose 10% of your new weight, from part a, will you be in your ideal weight range? Explain.

8. The World Wildlife Fund (WWF) estimated that in 2001 the world tiger population was between 5000 and 7000. Accurate data is difficult to obtain but one estimate is that this is a 5% reduction from the tiger population in 1900.

a. What is a lower-range estimate for the tiger population in 1900?

b. What is an upper-range estimate for the tiger population in 1900?

9. Airlines often encourage their customers to book online by offering a 5% discount on their ticket. You are traveling from Kansas City to Denver. A fully refundable fare is $558.60. A restricted nonrefundable fare is $273.60. Determine and use the decay factor to calculate the cost of each fare if you book online.

ACTIVITY 5.6

Take an Additional 20% Off

OBJECTIVE

1. Apply consecutive growth and/or decay factors to problems involving two or more percent changes.

Your friend Maureen arrives at your house after shopping for a jacket. She explains that today's newspaper contained a 20% off coupon at Old Navy. The $100 jacket she had been eyeing all season was already reduced by 40%. She clipped the coupon, drove to the store, selected her jacket and walked up to the register. The cashier brought up a price of $48; Maureen thought that the price should have been only $40. The store manager arrived, reentered the transaction and, again, the register displayed $48. Maureen decided not to purchase the jacket and drove to your house to get your advice.

1. How do you think Maureen calculated a price of $40?

2. You grab a pencil and start your own calculation. You determine the decay factor and use it to calculate the ticketed price that reflects the 40% reduction. Explain how to calculate this price.

3. To what price does the 20% off coupon apply?

4. Use the decay factor associated with the 20% discount to determine the final price of the jacket.

You are now curious if you could justify a better price by applying the discounts in the reverse order. You start a new set of calculations, again using the decay factor approach.

5. Starting with the list price, determine the sale price after taking the 20% reduction.

6. Apply the 40% discount to the intermediate sale price.

7. Which sequence of discounts gives a better sale price?

Consecutive Decay Factors

The important point to understand from Problems 1–7 is that multiple discounts are always applied sequentially, one after the other. That is, discounts are *never* added together first before calculating the discounted price. With this in mind, you can determine the sale price simply and quickly by using the associated decay factors.

The following example shows how applying multiple discounts as decay factors can be done in a single chain of multiplications.

EXAMPLE 1 ***A stunning $2000 gold-and-diamond necklace was far too expensive to consider. However, over the next several weeks you tracked the following successive discounts:***

20% off list; 30% off marked price; an additional 40% off every item

Determine the selling price after each of the discounts is taken.

SOLUTION

Step 1. Calculate the first reduced price by applying the decay factor corresponding to the 20% discount to the original price.

Decay factor: 100% − 20% = 80% = 0.80

First sale price: $2000 · 0.80 = $1600

Step 2. Next, determine the decay factor corresponding to the 30% discount and apply it to the first sale price to obtain the second sale price.

Decay factor: 100% − 30% = 70% = 0.70

Second sale price: $1600 · 0.70 = $1120

Step 3. Finally, determine the decay factor corresponding to the 40% discount and apply it to the second sale price to get the final sale price.

Decay factor: 100% − 40% = 60% = 0.60

Final sale price: $1120 · 0.60 = $672

The final cost after the three consecutive discounts is $672.

Note that the three steps can be combined into a single step once you know each decay factor. Then the final price is calculated from the original price by using a single "chain" of multiplications by the decay factors.

Final sale price: 2000 · 0.80 · 0.70 · 0.60 = $672

Check this calculation using your calculator.

8. a. Use the chain of multiplications approach shown in Example 1 to determine the final sale price of the necklace if the discounts had been taken in the reverse order (40%, 30%, and 20%).

b. Did you obtain the same final price in Problem 8a as in Example 1? Why?

Forming a Single Decay Factor for Consecutive Percent Decreases

The single chain of multiplication approach in Example 1 can be used to produce a single decay factor to represent two or more consecutive percent decreases. The single decay factor is the *product* of the consecutive decay factors.

In Example 1, the single decay factor is given by the product 0.80 · 0.70 · 0.6, which equals .336 (or 33.6%). The associated single discount is calculated by subtracting the decay factor (in percent form) from 100%, to obtain 66.4%.

Therefore, the effect of applying 20%, 30%, and 40% consecutive discounts is identical to a single discount of 66.4%. The single discount is called the *effective discount* and the associated single decay factor is called the *effective decay factor.*

9. a. Determine the effective decay factor that represents the cumulative effect of consecutively applying Old Navy's 40% and 20% discounts.

b. Use this decay factor to determine the effective discount on Maureen's jacket.

10. a. Determine the effective decay factor that represents the cumulative effect of consecutively applying discounts of 40% and 50%.

b. Use this decay factor to determine the effective discount.

Consecutive Growth Factors

You can also use the single chain of multiplications approach to apply multiple percent increases. In this case, you will multiply by growth factors. The following problem will guide you in using this method for such growth.

11. Your computer repair business is growing faster than you had ever imagined. Last year, you had 100 employees statewide. This year, you opened several additional locations and increased the number of workers by 30%. With demand so high, next year you will be opening new stores nationwide and plan to increase your employee roll by an additional 50%.

To determine the projected number of employees next year, you can use growth factors to simplify the calculations.

a. Determine the growth factor corresponding to a 30% increase.

b. Apply this growth factor to calculate your current workforce.

c. Determine the growth factor corresponding to an additional 50% increase.

d. Apply this growth factor to your current workforce (130 employees) to determine the projected number of employees next year.

e. Write a single chain of multiplications to calculate the projected number of employees from last year's workforce of 100. Compare your answer to the result in part d.

You can form a single growth factor to represent two or more consecutive percent increases. The single growth factor is the product of the associated growth factors. In Problem 11, the single growth factor is given by the product 1.30 · 1.50, which equals 1.95. This shows that the number of employees nearly doubled since last year. Note also that the associated percent increase is calculated by subtracting 100% from the single growth factor 195% to obtain 95%. Therefore, the effect of applying consecutive 30% and 50% increases is identical to a single increase of 95%. The single percent increase is called the *effective increase* and the associated single growth factor is called the *effective growth factor.*

12. A college laboratory technician started his job at $30,000 and expects a 3% increase in his salary each year.

a. What is the growth factor associated with the percent increase expected?

b. What is the technician's salary after 3 years?

Forming Single Factors for Consecutive Percent Increases and Decreases

The next problem will now guide you in applying consecutive growths and decays by using growth and decay factors together.

13. You purchased $1000 in a recommended stock last year and gleefully watched as it rose quickly by 30%. Unfortunately, the economy turned downward and you discover that your stock has recently fallen 30% from last year's high. The question is, have you made or lost money on your investment? You might be surprised by the answers obtained in the following.

a. What is the growth factor corresponding to a 30% increase?

b. What is the stock worth after the 30% increase?

c. Form the decay factor corresponding to a 30% decrease.

d. What was the stock worth after the 30% decrease?

Problem 13 shows that you can form a single factor that represents the cumulative effect of applying the consecutive percent increase and decrease—the single factor is the *product* of the growth and decay factors. In Problem 13, the effective factor is given by the product 1.30 · 0.70, which equals 0.91.

14. a. Does 0.91 in Problem 13 represent a growth factor or a decay factor? How can you tell?

b. Use the single factor from part a to determine the current value of your stock. What is the cumulative effect (as a percent change) of applying a 30% increase followed by a 30% decrease?

c. What is the cumulative effect if the 30% decrease had been applied first, followed by the 30% increase?

SUMMARY
ACTIVITY 5.6

1. The cumulative effect of a sequence of percent changes is the *product* of the associated growth or decay factors. For example,

a. To calculate the effect of consecutively applying 20% and 50% increases, determine the respective growth factors and multiply. The effective growth factor is

$$1.20 \cdot 1.50 = 1.80,$$

representing an effective increase of 80%.

b. To calculate the effect of applying a 30% *decrease* followed by a 40% *increase,* determine the respective growth and decay factors and then multiply. The effective factor is

$$0.70 \cdot 1.40 = 0.98,$$

representing an effective *decrease* of 2%.

c. To calculate the effect of applying a 25% *increase* followed by a 20% *decrease*, determine the respective growth and decay factors and then multiply. The effective factor is

$$1.25 \cdot 0.80 = 1.00,$$

indicating neither growth nor decay. That is, the quantity has returned to its original value.

2. The cumulative effect of a sequence of percent changes is the same regardless of the order in which the changes are applied. That is, the cumulative effect of applying a 20% *increase* followed by a 50% *decrease* is equivalent to first having applied the 50% *decrease* followed by the 20% *increase*.

3. It is important to remember that you do not add the individual percent changes when applying the percent changes consecutively. That is, a 20% increase followed by a 50% increase is *not* equivalent to a 70% increase.

EXERCISES
ACTIVITY 5.6

1. A $300 suit is on sale for 30% off. You present a coupon at the cash register for an additional 20% off.

   a. Form the decay factor corresponding to each percent decrease.

   b. Use these decay factors to determine the price you paid for the suit.

2. Your union has just negotiated a 3-year contract containing annual raises of 3%, 4%, and 5% during the term of contract. Your current salary is $42,000. What will you be earning in 3 years?

3. You had anticipated a large demand for a popular toy and increased your inventory of 1600 by 25%. You were able to sell 75% of your inventory. How many toys remain?

4. You deposit $2000 in a 5-year certificate of deposit that pays 4% interest compounded annually. To the nearest dollar, what balance will the account show when your certificate comes due?

5. Budget cuts have severely crippled your department over the last few years. Your operating budget of $600,000 has decreased by 5% each of the last 3 years. What is your current operating budget?

6. When you became a manager, your $60,000 annual salary increased by 25%. You found the new job too stressful and requested a return to your original job. You resumed your former duties at a 20% reduction in salary. How much are you making at your old job due to the transfer and return?

Exercise numbers appearing in color are answered in the Selected Answers appendix.

7. A coat with an original price tag of $400 was marked down by 40%. You have a coupon good for an additional 25% off.

 a. What is its final cost?

 b. What is the equivalent percent discount?

8. A digital camera with an original price tag of $500 was marked down by 40%. You have a coupon good for an additional 30% off.

 a. What is the decay factor for the first discount?

 b. What is the decay factor for the additional discount?

 c. What is the effective decay factor?

 d. What is the final cost of the camera?

 e. What is the equivalent percent discount?

9. Your friend took a job 3 years ago that started at $30,000 a year. Last year, she got a 5% raise. She stayed at the job for another year but decided to make a career change and took another job that paid 5% less than her current salary. Was the starting salary at the new job more, or less, or the same as her starting salary at her previous job?

10. You wait for the price to drop on a diamond-studded watch at Macy's. Originally, it cost $2500. The first discount was 20% and the second discount is 50%.

 a. What are the decay factors for each of the discounts?

 b. What is the effective decay factor for the two discounts?

 c. Use the effective decay factor in part b to determine how much you will pay for the watch.

ACTIVITY 5.7

Fuel Economy

OBJECTIVES

1. Apply rates directly to solve problems.
2. Use unit analysis or dimensional analysis to solve problems that involve consecutive rates.

You are excited about purchasing a new car for commuting to college. Concerned about the cost of driving, you did some research on the Internet and came across the Web site, *http://www.fueleconomy.gov*. You found that this Web site listed fuel efficiency, in miles per gallons (mpg), for five cars that you are considering. You recorded the mpg for city and highway driving in the table after Problem 1.

1. **a.** For each of the cars listed in the table, how many city miles can you travel per week on 5 gallons of gasoline? Explain the calculation you will do to obtain the answers. Record your answers in the third column of the table.

   b. The daily round-trip drive to your college is 38 city miles, which you do 4 days per week. Which of the cars would get you to school each week on 5 gallons of gas?

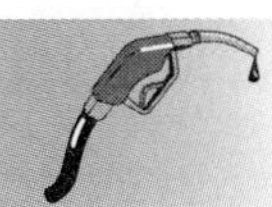

Every Drop Counts

FUEL ECONOMY GUIDE: MODEL YEAR 2006

MAKE/MODEL	CITY MPG	CITY MILES ON 5 GALLONS OF GAS	HIGHWAY MPG	GALLONS NEEDED TO DRIVE 304 HIGHWAY MILES	FUEL TANK CAPACITY IN GALLONS
Chevrolet Aveo5	26		35		12.0
Ford Focus	26		34		14.0
Honda Civic	30		38		13.2
Hyundai Accent	32		35		11.9
Toyota Corolla	32		41		13.2

2. Suppose you plan to move and your round-trip commute to the college will be 304 highway miles each week. How many gallons of gas would each of the cars require? Explain the calculation you will do to obtain the answers. Record your answers to the nearest tenth in the fifth column of the table.

Miles per gallon (mpg) is an example of a rate. Mathematically, a rate is a comparison by division of two quantities that have different units of measurement. A rate such as 32 mpg expresses the number of miles that can be traveled per 1 gallon of gas. The word *per* signifies division.

Solving problems involving rates is similar to solving problems using ratios. The difference is that to solve rate problems you will use the units of measurement as a guide to set up the appropriate calculations. The problems that follow will guide you to do this efficiently.

Applying a Known Rate Directly by Multiplication/Division to Solve a Problem

In Problem 1, the known rate is miles per gallon for several cars. You can write the mpg in fraction form, $\frac{\text{number of miles}}{1 \text{ gallon}}$. In the case of the Chevrolet Aveo5, the mpg is $\frac{26 \text{ miles}}{1 \text{ gal}}$. To determine how many miles the Aveo5 can travel on 5 gallons, you notice that the units in the answer should be miles. So, you multiply the rate by 5 gallons.

$$5 \cancel{\text{gal.}} \cdot \frac{26 \text{ mi.}}{1 \cancel{\text{gal.}}} = 130 \text{ mi.}$$

Notice that gallon occurs in both a numerator and a denominator. You divide out common measurement units in the same way that you reduce a fraction by dividing out common numerical factors. Therefore, you can drive for 130 miles in the Chevrolet Aveo5 on 5 gallons of gas.

PROCEDURE

Multiplying and Dividing Directly to Solve Problems Involving Rates

- Identify the measurement unit of the answer to the problem.
- Set up the calculation so the appropriate units will divide out, leaving the measurement unit of the answer.
- Multiply or divide the numbers as usual to obtain the numerical part of the answer.
- Divide out the common units to obtain the measurement unit of the answer.

3. Solve each of the following problems involving rates.

a. The gas tank of a Honda Civic holds 13.2 gallons. How many highway miles can you travel on a full tank of gas?

b. The Hyundai Accent gas tank holds 11.9 gallons. Is it possible to travel as far in this car as in the Honda Civic on the highway?

4. After you purchase your new car, you would like to take a trip to see a good friend in another state. The highway distance is approximately 560 miles.

a. If you bought the Hyundai Accent, how many gallons of gas would you need to make the round-trip?

b. How many tanks of gas would you need for the trip?

Solving Rate Problems Using Measurement Units as a Guide

In Problem 2, you may have directly divided the total miles by the mpg. One way to decide that the calculation involves division is to consider the measurement units. From reading the problem, you identify that the measurement unit of the answer should be gallons. In the case of the Honda Civic, dividing 304 miles by 38 miles per gallon produces the correct unit for the answer, as the following calculation shows.

$$304 \text{ mi.} \div \frac{38 \text{ mi.}}{1 \text{ gal.}} = 304 \cancel{\text{mi.}} \cdot \frac{1 \text{ gal.}}{38 \cancel{\text{mi.}}} = \frac{304}{38} \text{ gal.} \approx 8 \text{ gal.}$$

Notice that the unit, miles, occurs in both the numerator and the denominator. You divide out miles to obtain gallons as the unit in the result. This is known as *unit analysis* or *dimensional analysis*.

5. You decide to read a 15-page article that provides advice for negotiating the lowest price for a car. You read at the rate of 10 pages an hour. How long will it take you to read the article?

Unit Conversion

In many countries, distance is measured in kilometers (km) and gasoline in liters (ℓ). A kilometer is equivalent to 0.6214 miles and 1 liter is equivalent to 0.264 gallons. In general, the equivalence of two units can be treated as a rate written in fraction form. For example, the fact that a kilometer is equivalent to 0.6214 miles is written as

$$\frac{1 \text{ km}}{0.6214 \text{ mi.}} \quad \text{or} \quad \frac{0.6214 \text{ mi.}}{1 \text{ km}}$$

Using the fraction form, you can convert one unit to another by applying multiplication directly.

6. Your friend joins you on a trip through Canada where gasoline is measured in liters and distance in kilometers.

a. Write the equivalence of liters and gallons in fraction form.

b. If you bought 20 liters of gas, how many gallons did you buy?

Note that conversion equivalents are given on the inside front cover of this textbook.

7. To keep track of mileage and fuel needs in Canada your friend suggests that you convert your car's mpg into kilometers per liter. Your car's highway fuel efficiency is 45 mpg.

a. Convert your car's fuel efficiency to *miles per liter.*

b. Use the result you obtained in part a to determine your car's fuel efficiency in *kilometers per liter.*

Using Unit Analysis to Solve a Problem Involving Consecutive Rates

Problem 6 involved a one-step rate problem. In Problem 7, to convert the car's fuel efficiency from miles per gallon (mpg) to kilometers per liter (kpl) you were guided to do two one-step conversions. You may have observed that the two calculations could be done in a single chain of multiplications.

$$\frac{45 \cancel{\text{mi.}}}{1 \cancel{\text{gal}}} \cdot \frac{0.264 \cancel{\text{gal}}}{1 \ \ell} \cdot \frac{1 \text{ km}}{0.6214 \cancel{\text{mi.}}} = 19.1 \text{ km per liter (kpl)}$$

Note how the units given in the problem guide you to choose the appropriate equivalent fractions so the method becomes a single chain of multiplications.

8. You plan to buy 3 gallons of juice to serve guests at a breakfast rally that your local political group is hosting. How many cups would you have?

9. You are on a 1500-mile trip where gas stations are far apart. Your car is averaging 40 mpg and you are traveling at 60 miles per hour (mph). The fuel tank holds 12 gallons of gas and you just filled the tank. Use a chain of multiplications to determine how long it will be before you have to refill the tank.

PROCEDURE

Applying Consecutive Rates

- Identify the measurement unit of the answer.
- Set up the sequence of multiplications and/or divisions so the appropriate units divide out, leaving the appropriate measurement unit.
- Multiply and divide the numbers as usual to obtain the numerical part of the answer.
- Check that the appropriate measurement units divide out, leaving the expected measurement unit for the answer.

10. You have been driving for several hours and notice that your car's 13.2-gallon fuel tank registers half empty. How many more miles can you travel if your car is averaging 30 mpg?

SUMMARY
ACTIVITY 5.7

1. Multiplying/dividing directly by rates to solve problems:

a. Identify the unit of the result.

b. Set up the calculation so the appropriate units will divide out, leaving the unit of the result.

c. Multiply or divide the numbers as usual to obtain the numerical part of the result.

d. Divide out the common units to obtain the unit of the answer.

2. Applying several rates consecutively to solve problems:

a. Identify the unit of the result.

b. Set up the sequence of multiplications and/or divisions so the appropriate units divide out, leaving the unit of the result.

c. Multiply and divide the numbers as usual to obtain the numerical part of the result.

d. Check that the appropriate units divide out, leaving the expected unit of the result.

EXERCISES
ACTIVITY 5.7

Use the conversion tables or formulas on the inside front cover of the textbook for conversion equivalencies.

1. The length of a football playing field is 100 yards between the opposing goal lines. What is the length of the football field in feet?

2. The distance between New York City, NY, and Los Angeles, CA, is approximately 4485 kilometers. What is the distance between these two major U.S. cities in miles?

3. As part of your job as a quality control worker in a factory you can check 16 parts in 3 minutes. How long will it take you to check 80 parts?

4. Your car averages about 27 miles per gallon on highways. With gasoline priced at $2.74 per gallon, how much will you expect to spend on gasoline during your 500-mile trip?

5. You currently earn $11.50 per hour. Assuming that you work fifty-two 40-hour weeks per year with no raises, what total gross salary will you earn over the next 5 years?

6. The aorta is the largest artery in the human body. The aorta attains a maximum diameter of about 1.18 inches where it adjoins the heart in the average adult. What is the maximum diameter of the aorta in terms of centimeters? In millimeters?

7. Mount Everest in the Himalaya mountain range along the border of Tibet and Nepal reaches upward to a record height of 29,035 feet. How high is Mt. Everest in miles? In kilometers? In meters?

8. The average weight of a mature human brain is approximately 1400 grams. What is the equivalent weight in kilograms? In pounds?

9. Approximately 4.5 liters of blood circulates in the body of the average human adult. How many quarts of blood does the average person have? How many pints?

10. How many seconds are in a day? In a week? In a non-leap year?

11. Lava flowing out of shield volcanoes has reached incredibly hot temperatures of about 1200°C. What is the Fahrenheit temperature of these lava flows?

12. The following places are three of the wettest locations on Earth. Determine the rainfall for each site in centimeters per year.

LOCATION	ANNUAL RAINFALL (INCHES)	ANNUAL RAINFALL (CENTIMETERS)
Mawsynram, Meghala, India	467 inches	
Tutenendo, Columbia	463.5 inches	
Mt. Waialeali, Kauai	410 inches	

13. The mass of diamonds is commonly measured in carats. Five carats is equivalent to 1 gram. How many grams are in a 24-carat diamond? How many ounces?

14. The tissue of Earth organisms contains carbon molecules known as proteins, carbohydrates, and fats. A healthy human body is approximately 18% carbon by weight. Determine how many pounds and kilograms of carbon your own body contains.

ACTIVITY 5.8

Four out of Five Dentists Prefer the Brooklyn Dodgers?

OBJECTIVES

1. Recognize that equivalent fractions lead to proportions.
2. Use proportions to solve problems involving ratios and rates.

Manufacturers of retail products often conduct surveys to see how well their products are selling compared to competing products. In some cases, they use the results in advertising campaigns.

One company, Proctor and Gamble, ran a TV ad in the 1970s claiming that "four out of five dentists surveyed preferred Crest toothpaste over other leading brands." The ad became a classic, and the phrase, *four out of five prefer*, has become a popular cliché in advertising, in appeals, and in one-line quips.

What does "four out of five prefer" mean in a survey? How does that lead to solving an equation called a proportion? Continue in this activity to find out.

1. Suppose Proctor and Gamble asked 250 dentists what brand toothpaste he or she preferred and 200 dentists responded that they preferred Crest.

   a. What is the ratio of the number of dentists who preferred Crest to the number who were asked the question? Write your answer in words, and as a fraction.

   b. Reduce the ratio in part a to lowest terms and write the result in words.

Note that part b of Problem 1 shows that the ratio $\frac{200}{250}$ is equivalent to the ratio $\frac{4}{5}$, or $\frac{200}{250} = \frac{4}{5}$. This is an example of a **proportion**.

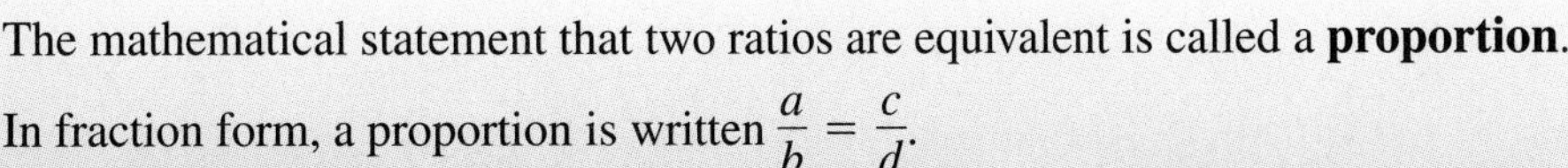

DEFINITION

The mathematical statement that two ratios are equivalent is called a **proportion**. In fraction form, a proportion is written $\frac{a}{b} = \frac{c}{d}$.

If one of the component numbers, a, b, c, or d in a proportion is unknown and the other three are known, the equation can be solved for the unknown component. The next example shows how.

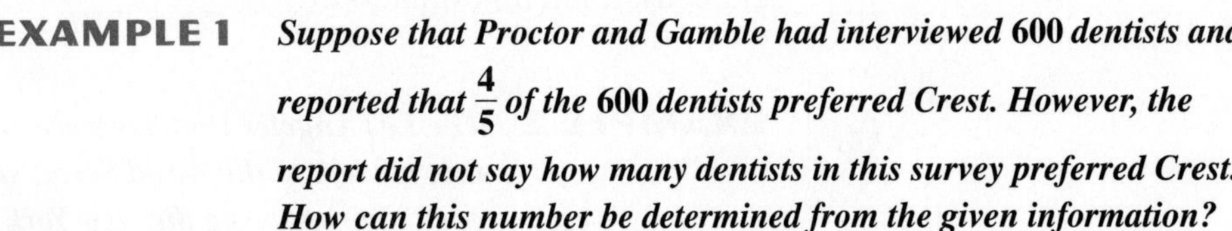

EXAMPLE 1 ***Suppose that Proctor and Gamble had interviewed 600 dentists and reported that $\frac{4}{5}$ of the 600 dentists preferred Crest. However, the report did not say how many dentists in this survey preferred Crest. How can this number be determined from the given information?***

SOLUTION

At this point, the comparison ratio for all the dentists interviewed is $\frac{x}{600}$, where the variable x represents the unknown number of dentists who preferred Crest and 600 is the total number of dentists surveyed. According to the report, this ratio has to be equal to $\frac{4}{5}$, that is,

$$\frac{x}{600} = \frac{4}{5}.$$

The equation is solved for x by multiplying each fraction by 600 and simplifying:

$$600 \cdot \frac{x}{600} = \frac{4}{5} \cdot 600$$

$$\frac{\cancel{600}}{1} \cdot \frac{x}{\cancel{600}} = \frac{4}{\cancel{5}} \cdot \frac{\overset{120}{\cancel{600}}}{1}$$

$$x = 4 \cdot 120 = 480$$

Therefore, 480 dentists in the survey preferred Crest.

2. Suppose Proctor and Gamble surveyed another group of 85 dentists in Idaho. If $\frac{4}{5}$ of the group preferred Crest, how many of this group preferred Crest? Let x represent the number of dentists who preferred Crest.

3. The best season for the New York Mets baseball team was 1986 when they won $\frac{2}{3}$ of the games they played and won the World Series, beating the Boston Red Sox. If the Mets played 162 games, how many did they win?

Sometimes, the unknown in a proportion is in the denominator of one of the fractions. The next example shows a method of solving a proportion that is especially useful in this situation.

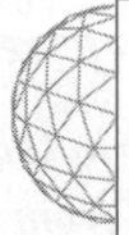

EXAMPLE 2 ***The Los Angeles Dodgers were once the Brooklyn Dodgers team that made it to the World Series seven times from 1941 to 1956, each time opposing the New York Yankees. The Brooklyn Dodgers won only one World Series Championship against the Yankees and that was in 1955.***

In the 1955 season, the Dodgers won about 16 out of every 25 games that they played. They won a total of 98 games. How many games did they play during the season?

SOLUTION

Let x represent the total number of games that the Dodgers played in the 1955 season. To write a proportion, the units of the numerators in the proportion must be the same. Similarly, the units of the denominators must be the same. In this case, the proportion is

$$\frac{16 \text{ games won}}{25 \text{ total games played}} = \frac{98 \text{ games won}}{x \text{ total games played}} \quad \text{or} \quad \frac{16}{25} = \frac{98}{x}$$

This equation can be solved in three steps that will lead to a shortcut method. First follow the three steps.

Step 1. Multiply each fraction by x and simplify.

$$\frac{16}{25} \cdot x = \frac{98}{\cancel{x}} \cdot \cancel{x}$$

$$\frac{16x}{25} = 98$$

Step 2. Multiply each term of the equation by 25.

$$\frac{16x}{\cancel{25}} \cdot \cancel{25} = 98 \cdot 25$$

$$16x = 98 \cdot 25$$

Step 3. Divide each term of the equation by 16 and calculate the value for x.

$$\frac{\cancel{16}x}{\cancel{16}} = \frac{98 \cdot 25}{16}$$

$$x = \frac{98 \cdot 25}{16} \approx 153.1$$

The Dodgers played 153 games in the 1955 season.

A Shortcut Method for Solving Proportions

In steps 1 and 2 of the solution in Example 2, the proportion $\frac{16}{25} = \frac{98}{x}$ was rewritten as $16x = 98 \cdot 25$. Observe that the denominator of each fraction is multiplied by the numerator of the other.

$$\frac{16}{25} = \frac{98}{x}$$

$$16x = 98 \cdot 25$$

Divide both terms in the equation by 16 to obtain the value for x.

$$x = \frac{98 \cdot 25}{16} \approx 153$$

The shortcut process just described is called *cross multiplication* because the numerator of the first fraction multiplies the denominator of the second, and vice versa.

PROCEDURE

Solving Proportions by Cross Multiplication Rewrite the proportion $\frac{a}{b} = \frac{c}{d}$ as $a \cdot d = b \cdot c$ and solve for the unknown quantity.

4. a. Solve the proportion $\frac{7}{10} = \frac{x}{24}$.

b. Solve the proportion $\frac{9}{x} = \frac{6}{50}$.

5. As a volunteer for a charity, you were given a job stuffing envelopes for an appeal for donations. After stuffing 240 envelopes, you were informed that you are two-thirds done. How many envelopes in total are you expected to stuff? Let x represent the number of envelopes to be stuffed.

6. You are feeding a group of 6 friends who are joining you to watch a figure skating competition and everyone wants soup. The information on a can of soup states that one can of soup contains about 2.5 servings. Use a proportion to determine the number of cans of soup to open so that all seven of you have one serving of soup. Let x represent the number of cans to be opened.

7. Solve each proportion for x.

a. $\frac{2}{3} = \frac{x}{48}$ **b.** $\frac{5}{8} = \frac{120}{x}$ **c.** $\frac{3}{20} = \frac{x}{3500}$

SUMMARY
ACTIVITY 5.8

1. The mathematical statement that two ratios are equivalent is called a **proportion.** In fraction form, a proportion is written as $\frac{a}{b} = \frac{c}{d}$.

2. Shortcut (cross multiplication) procedure for solving proportions:

   Rewrite the proportion $\frac{a}{b} = \frac{c}{d}$ as $a \cdot d = b \cdot c$ and solve for the unknown quantity.

EXERCISES
ACTIVITY 5.8

1. A company that manufactures optical products estimated that three out of five people in North America wore eyeglasses in 2006. The population estimate for North America in 2006 was about 334 million. Use a proportion to determine how many people in this part of the world were estimated to wear eyeglasses in 2006?

2. Solve each proportion for x.

   a. $\frac{2}{9} = \frac{x}{108}$ **b.** $\frac{8}{7} = \frac{120}{x}$ c. $\frac{x}{20} = \frac{70}{100}$

3. The Center for Education Reform is an organization founded in 1993 to help foster better education opportunities in American communities. It spends 4 cents out of every dollar in its spending budget for administrative expenses. In 2004, the center's spending budget was approximately 2.8 million dollars. Use a proportion to estimate the administrative expenses for 2004.

Exercise numbers appearing in color are answered in the Selected Answers appendix.

4. Powdered skim milk sells for $6.99 a box in the supermarket. The amount in the box is enough to make 8 quarts of skim milk. What is the price for 12 quarts of skim milk prepared in this way?

5. A person who weighs 120 pounds on Earth would weigh 42.5 pounds on Mars. What would be a person's weight on Mars if she weighs 150 pounds on Earth?

6. You want to make up a saline (salt) solution in chemistry lab that has 12 grams of salt per 100 milliliters of water. How many grams of salt would you use if you needed 15 milliliters of solution for your experiment?

7. Tealeaf, a company that offers management solutions to companies that sell online, announced in a 2005 consumer survey that 9 out of 10 customers reported problems with transactions online. The survey sampled 1859 adults in the United States, 18 years and older, who had conducted an online transaction in the past year. How many of those adults reported problems with transactions online?

8. In 2005, the birth rate in the United States was estimated to be 14.14 births per 1000 persons. If the population estimate for the United States was 296.7 million, how many births were expected that year?

What Have I Learned?

1. On a 40-question practice test for this course, you answered 32 questions correctly. On the test itself, you correctly answered 16 out of 20 questions. Does this mean that you did better on the practice test than you did on the test itself? Explain your answer using the concepts of actual and relative comparison.

2. a. Ratios are often written as a fraction in the form $\frac{a}{b}$. List the other ways you can express a ratio.

   b. Thirty-three of the 108 colleges and universities in Michigan are 2-year institutions. Write this ratio in each of the ways you listed in part a.

3. Florida's total population, reported in the 2000 census, was 15,982,378 persons and 183 out of every 1000 residents were 65 years and older. Show how you would determine the actual number of Florida residents who are 65 years or older in 2000.

4. Explain how you would determine the percent increase in the cost of a cup of coffee at your local diner if the price increased from 75 cents to 85 cents last week.

Exercise numbers appearing in color are answered in the Selected Answers appendix.

5. a. Think of an example and use it to show how to determine the growth factor associated with a percent increase. For instance, you might refer to growth factors associated with yearly interest rates for a savings account.

b. Show how you would use the growth factor to apply the percent increase twice, then three times.

6. Current health research shows that losing just 10% of body weight produces significant health benefits, including a reduced risk for a heart attack. Suppose a relative who weighs 199 pounds begins a diet to reach a goal weight of 145 pounds.

a. Use the idea of a decay factor to show your relative how much he will weigh after losing the first 10% of his body weight.

b. Explain to your relative how he can use the decay factor to determine how many times he has to lose 10% of his body weight to reach his goal weight of 145 pounds.

7. a. In conversion tables, conversions between two measurement units are usually given in an equation format. For example, 1 quart = 0.946 liters. Write this conversion in two different fraction formats.

b. Write the conversion 1 liter = 1.057 quarts in two different fraction formats.

c. Are the two conversions given in parts a and b equivalent? Explain.

How Can I Practice?

1. Write the following percents in decimal format.

a. 25%
b. 87.5%
c. 6%
d. 3.5%
e. 250%
f. 0.3%

2. An error in a measurement is the difference between the measured value and the true value. The relative error in measurement is the ratio of the absolute value of the error to the true value. That is,

$$\text{relative error} = \frac{|\text{error}|}{\text{true value}}.$$

Determine the actual and relative errors in the following measurements.

MEASUREMENT	TRUE VALUE	ERROR	RELATIVE ERROR (AS A PERCENT)
107 inches	100 inches		
5.7 ounces	5.0 ounces		
4.3 grams	5.0 grams		
11.5 cm	12.5 cm		

3. You must interpret a study of cocaine-addiction relapse after three different treatment programs. The subjects were treated with the standard therapy, with an antidepressant, or with a placebo. They were tracked for 3 years.

TREATMENT	SUBJECTS WHO RELAPSED	SUBJECTS STILL SUBSTANCE FREE
Standard therapy	36	20
Antidepressant therapy	27	18
Placebo	30	9

Which therapy has proved most effective? Explain.

Exercise numbers appearing in color are answered in the Selected Answers appendix.

4. The 2619 female students on campus comprise 54% of the entire student body. What is the total enrollment of the college?

5. The New York City Board of Education recently announced plans to eliminate 1500 administrative staff jobs over the next several years, reducing the number of employees at its central headquarters by 40%. What is the current size of its administrative staff (prior to any layoffs)?

6. According to the 2000 U.S. Census, the Cajun community in Louisiana has dwindled from 407,000 to 44,000 in the last 10 years. What percent decrease does this represent?

7. A suit with an original price of $400 was marked down by 30%. You have a coupon good for an additional 20% off.

   a. What is the suit's final cost?

   b. What is the equivalent percent discount?

8. In recent contract negotiations between the Transit Workers Union and the New York City Metropolitan Transit Authority, the workers agreed to a 3-year contract, with 5%, 4%, and 3% wage increases, respectively, over each year of the contract. At the end of this 3-year period, what will be the salary of a motorman who had been earning $48,000 at the start of the contract?

9. You are moving into a rent-controlled apartment and the landlord is raising the rent by the maximum allowable, 20%. You will be paying $900. What was the former tenant's rent?

10. Gasoline prices increased from $2.25 per gallon to a price of $2.75. What percent increase does this represent?

11. During October 2005, the Ford Motor Company sold a total of 28,000 new Explorers, a decrease of approximately 14% from the 2004 model year's October sales period. Approximately how many Explorers were sold during October 2004?

12. The Mariana Trench near the Mariana Islands in the South Pacific Ocean extends down to a maximum depth of 35,827 feet. How deep is this deepest point in miles? In kilometers? In meters?

13. Under the current union contract, you earn $22.50 per hour as a college laboratory technician. The contract is valid for the next 2 years. Assuming that you stay in this position and work 40 hours per week, 52 weeks a year, what will your total gross earnings be over the next 2 years?

14. As a nurse in a rehabilitation center, you have received an order to administer 60 milligrams of a drug. The drug is available at a strength of 12 milligrams per milliliter. How many milliliters would you administer?

15. Solve each proportion for x.

a. $\frac{x}{8} = \frac{120}{320}$ **b.** $\frac{6}{13} = \frac{42}{x}$ **c.** $\frac{24}{x} = \frac{6}{5}$

16. A three-pound bag of rice cost $2.67. At that rate, what would a 20-pound bag of rice cost?

CHAPTER 5 SUMMARY

The bracketed numbers following each concept indicate the activity in which the concept is discussed.

CONCEPT / SKILL	DESCRIPTION	EXAMPLE
Ratio [5.1]	A quotient that compares two similar numerical quantities, such as part to total.	4 out of 5 is a ratio. It can be expressed as a fraction, $\frac{4}{5}$; a decimal, 0.8; and a percent, 80%.
Convert from decimal to percent format. [5.1]	Move the decimal place two places to the right and then attach the % symbol.	0.125 becomes 12.5%. 0.04 becomes 4%. 2.50 becomes 250%.
Convert from percent to decimal format. [5.1]	Drop the percent symbol and move the decimal point two places to the left.	35% becomes 0.35. 6% becomes 0.06. 200% becomes 2.00.
Apply a known ratio to a given piece of information. [5.2]	total · known ratio = unknown part	40% of the 350 children play an instrument. 40% is the known ratio; 350 is the total. $350 \cdot 0.40 = 140$ children play an instrument.
	part ÷ known ratio = unknown total	24 children, comprising 30% of the marching band, play the saxophone. 30% is the known ratio, 24 is the part. $24 \div 0.30 = 80$, the total number of children in the marching band.
Relative change [5.3]	$\text{relative change} = \frac{\text{actual change}}{\text{original value}}$	A quantity changes from 25 to 35. The actual change is 10; the relative change is $\frac{10}{25} = \frac{2}{5}$, or 40%.
Growth factor [5.4]	When a quantity increases by a specified percent, the ratio of its new value to its original value is called the growth factor associated with the specified percent increase. The growth factor also is formed by adding the specified percent increase to 100% and then changing the percent into decimal form.	To determine the growth factor associated with a 25% increase, add 25% to 100% to obtain 125%. Change 125% into a decimal to obtain the growth factor, 1.25.
Apply a growth factor to an original value. [5.4]	original value · growth factor = new value	A $120 item increases by 25%. Its new value is $\$120 \cdot 1.25 = \$150$.
Apply a growth factor to a new value. [5.4]	new value ÷ growth factor = original value	An item has already increased by 25% and is now worth $200. Its original value was $\$200 \div 1.25 = \$160$.

CONCEPT / SKILL	DESCRIPTION	EXAMPLE
Decay factor [5.5]	When a quantity decreases by a specified percent, the ratio of its new value to the original value is called the decay factor associated with the specified percent decrease. The decay factor is also formed by *subtracting* the specified percent decrease from 100% and then changing the percent into decimal form.	To determine the decay factor associated with a 25% decrease, subtract 25% from 100%, 100% − 25%, to obtain 75%. Change 75% into a decimal to obtain the decay factor of 0.75.
Apply a decay factor to an original value. [5.5]	original value · decay factor = new value	A $120 item decreases by 25%. Its new value is $120 · 0.75 = $90.
Apply a decay factor to a new value. [5.5]	new value ÷ decay factor = original value	An item has already decreased by 25% and is now worth $180. Its original value was $180 ÷ 0.75 = $240.
Apply a sequence of percent changes. [5.6]	The cumulative effect of a sequence of percent changes is the *product* of the associated growth or decay factors.	The effect of a 25% increase followed by a 25% decrease is $1.25 \times 0.75 = 0.9375$. That is, the item is now worth only 93.75% of its original value. The item's value has decreased by 6.25%.
Use unit or dimensional analysis to solve conversion problems. [5.7]	1. Identify the measurement unit of the result. 2. Set up the sequence of multiplications and/or divisions so the appropriate measurement units cancel, leaving the measurement unit of the result. 3. Multiply and divide the numbers as usual to obtain the numerical part of the result. 4. Check that the appropriate measurement units divide out, leaving the expected measurement unit of the result.	To convert your height of 70 inches to centimeters, $70 \text{ in.} \cdot \frac{2.54 \text{ cm}}{1 \text{ in.}}$ $= 177.8 \text{ cm}$ Reading a 512-page book at the rate of 16 pages per hour will take how many 8-hour days? $512 \text{ pages} \div \frac{16 \text{ pages}}{1 \text{ hr.}} \cdot \frac{1 \text{ day}}{8 \text{ hr.}}$ $= 512 \cancel{\text{pages}} \cdot \frac{1 \cancel{\text{hr.}}}{16 \cancel{\text{pages}}} \cdot \frac{1 \text{ day}}{8 \cancel{\text{hr}}}$ $= 4 \text{ days}$
Proportion [5.8]	A mathematical statement that two ratios are equivalent. In fraction form a proportion is written as $\frac{a}{b} = \frac{c}{d}$	$\frac{2}{3} = \frac{50}{75}$ is a proportion because $\frac{50}{75}$ in reduced form equals $\frac{2}{3}$.
Solve a proportion [5.8]	Rewrite $\frac{a}{b} = \frac{c}{d}$ as $a \cdot d = b \cdot c$ and solve for the unknown quantity.	$\frac{7}{30} = \frac{x}{4140}$ $30 \cdot x = 7 \cdot 4140$ $x = \frac{7 \cdot 4140}{30} = 966$

CHAPTER 5

GATEWAY REVIEW

1. Determine the value of each expression.

a. $20 \cdot \frac{7}{4}$

b. $\frac{4}{5} \div 2$

c. 27% of 44

d. $1300 \div 20\%$

e. $45.7 \div 0.0012$

f. $\frac{1}{6} \cdot 37.34$

2. Solve the proportion for x.

a. $\frac{4}{9} = \frac{x}{45}$

b. $\frac{x}{4} = \frac{5}{4}$

c. $\frac{1}{x} = \frac{2}{3}$

d. $\frac{2.3}{1.7} = \frac{x}{4}$

e. $\frac{\frac{1}{2}}{7} = \frac{x}{6}$

f. $\frac{x}{3} = 4$

3. In a recent survey, 70% of the 1400 female students and 30% of the 1000 male students on campus indicated that shopping was their favorite leisure activity. How many students placed shopping at the top of their list? What percent of the entire student body does this represent?

4. The least-favorite responsibility in your summer job is to stuff and seal preaddressed envelopes. After sealing 600 envelopes you discover that you have only completed two-thirds of the job. How large is the entire mailing list?

Answers to all Gateway exercises are included in the Selected Answers appendix.

5. In November 2001, Alcatel, one of Europe's leading makers of telecommunications equipment, announced that it was slashing its worldwide workforce from 110,000 to 77,000. What percent of its workforce was being laid off?

6. In 1985, a total of 470,816 live births were recorded in the state of California. In 1990, the total number of live births rose to 611,666. What is the growth factor for live births for the 5-year period from 1985 to 1990? What is the percent increase (also called the growth rate)?

7. Aspirin is typically absorbed into the bloodstream from the duodenum. In general, 75% of a dose of aspirin is eliminated from the bloodstream in an hour. A patient is given a dose of 650 mg. How much aspirin is expected to remain in the patient after 1 hour?

8. In 2004, your company's sales revenues were $400,000. A vigorous advertising campaign increased sales by 20% in 2005, and then by another 30% in 2006. What were your sales revenues in 2006?

9. During a 1-month period during the 2005 model year, Chevrolet sold a total of 53,800 new Silverado pickup trucks, an increase of approximately 13.5% over the same month in 2004. Approximately how many Silverados were sold during the same 1-month period in 2004?

10. The 2001 economic slowdown and consequent decline in state aid forced public colleges in several states to cut costs and raise tuition. Determine the new tuition at the following institutions.

a. University of Minnesota, which raised its $4900 tuition by 12%

b. Clemson University, which raised its tuition of $3590 by 25%

c. Clemson's trustees anticipated a continued slowdown. They voted to raise the 2002 tuition by another 13%. What was the new tuition amount?

11. The size of an optical telescope is given by the diameter of its mirror, which is circular in shape. Currently, the world's largest optical telescope is the 10-meter Keck Telescope atop Mauna Kea, Hawaii. What is the diameter of the mirror in feet?

12. In the middle of a playing season, NBA players Kevin Garnet (Minn), Kobe Bryant (LA), and Shaquille O'Neal (MIA) had these respective field goal statistics:

Garnet 87 field goals out of 167 attempts

Bryant 170 field goals out of 342 attempts

O'Neal 229 field goals out of 408 attempts

Rank them according to their relative (%) performance.

13. Hybrid electric/gas cars seem to be friendlier to the environment than all-gas vehicles, and more models are being introduced each year in the United States. One model reports a 56 mpg for highway driving. How far can you travel on one tank of gas if the tank holds 10.6 gallons?

14. You run at the rate of 6 mph. Convert your rate to feet per second.

15. Solve each proportion for x.

a. $\frac{x}{5} = \frac{120}{300}$ **b.** $\frac{8}{15} = \frac{48}{x}$ **c.** $\frac{18}{x} = \frac{3}{2}$

16. An eight-pound bag of flour cost \$5.28. At that rate, what would a 100-pound bag of flour cost?

CHAPTER 6

PROBLEM SOLVING WITH GEOMETRY

Things around us come in different shapes and sizes and can be measured in a number of ways. A garden can be circular, a room is often rectangular, and a pool of water may be one of many interesting shapes. A fence around a garden is usually measured in terms of its perimeter, a carpet on the floor of a room in terms of its area, and a pool of water in terms of its volume. These ideas are all part of a field of mathematics called geometry. In this chapter you will determine the perimeters and areas of rectangles, triangles, circles, and figures composed of these shapes. You will also calculate the surface areas and volumes of prisms, cones, and spheres.

CLUSTER 1 The Geometry of Two-Dimensional Plane Figures

ACTIVITY 6.1

Walking around Bases, Gardens, Trusses, and Other Figures

OBJECTIVES

1. Recognize perimeter as a geometric property of plane figures.
2. Write formulas for, and calculate perimeters of, squares, rectangles, triangles, parallelograms, trapezoids, and polygons.
3. Use unit analysis to solve problems involving perimeter.

In this first cluster of activities, you will explore the properties of geometric figures or shapes that are two-dimensional. This means they exist in a **plane**—a surface like the floor at your feet or the walls in your classroom. To make sure you understand these various shapes, some preliminary definitions are needed.

DEFINITION

Parallel lines are lines in a plane that never intersect. No matter how far you extend the lines, in either direction, they will never meet.

Example:

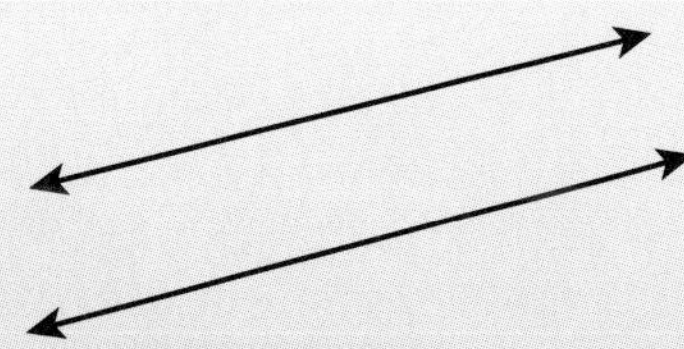

A **ray** is a portion of a line that starts from a point and continues indefinitely in one direction, much like a ray of light coming from the Sun.

Example:

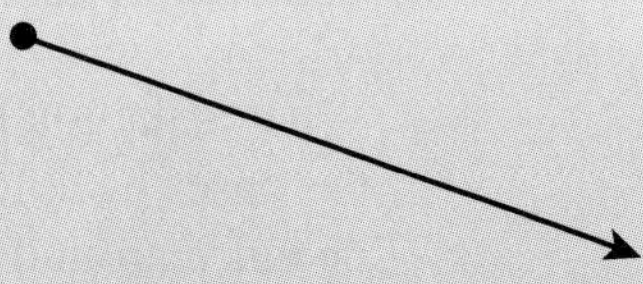

An **angle** is formed by two rays that have a common starting point. The common starting point is called the **vertex** of the angle.

Example:

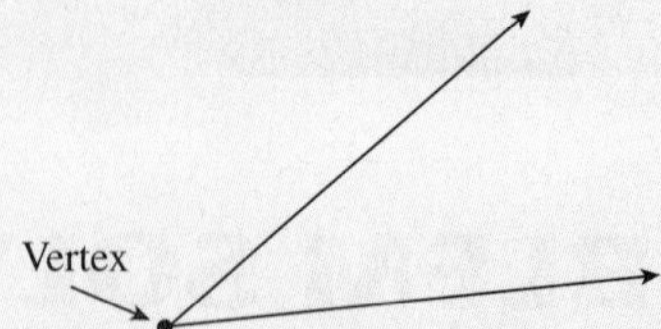

Perpendicular lines are intersecting lines that form four angles of equal size. The resulting angles are defined to be **right angles**. Each right angle measures 90 degrees. In a diagram, right angles are usually designated by little squares at the point of intersection.

Example:

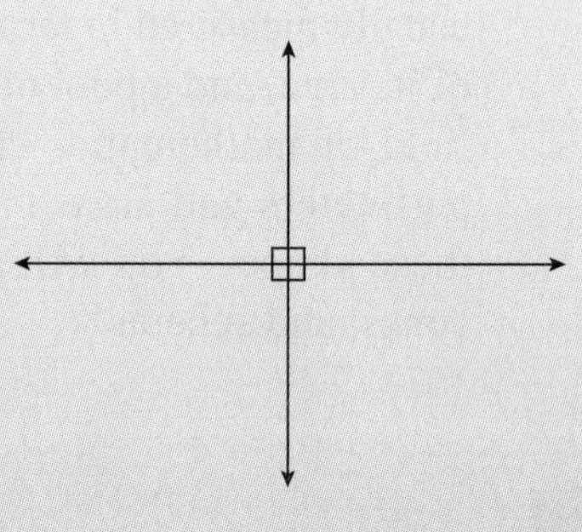

Squares

DEFINITION

A **square** is a closed plane figure whose four sides have equal length and are at right angles to each other.

Examples:

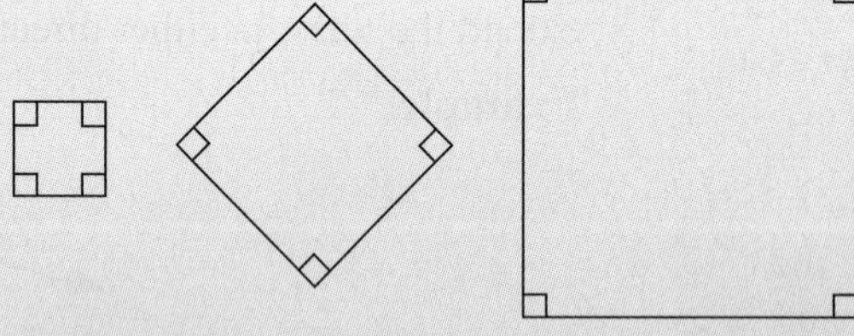

1. You are at bat in the middle of an exciting baseball game. You are a fairly good hitter and can run the bases at a speed of 15 feet per second. You know that the baseball diamond has the shape of a square, measuring 90 feet on each side.

   a. What is the total distance in feet you must run from home plate through the bases back to home? (See figure on next page.) Recall that the total distance around the square is called the *perimeter* of the square.

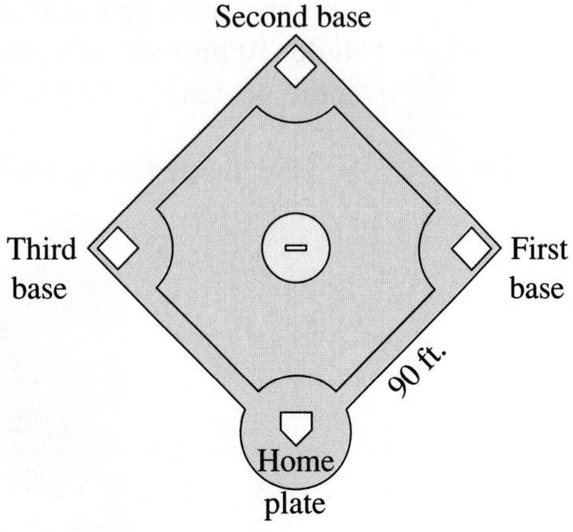

b. How many seconds will it take you to run this total distance?

2. Your niece plays Little League baseball. The square baseball diamond for Little League is 60 feet on each side. What is the total distance if she ran all of the bases?

DEFINITION

The **perimeter of a square** is the total distance around all its edges or sides.

PROCEDURE

Calculating the Perimeter of a Square The formula for the perimeter, P, of a square whose sides have length s has two equivalent forms.

$$P = s + s + s + s \quad \text{and} \quad P = 4s$$

Rectangles

DEFINITION

A **rectangle** is a closed plane figure whose four sides are at right angles to each other.

Examples:

3. You are interested in planting a rectangular garden, 10 feet long by 15 feet wide. To protect your plants, you decide to purchase a fence to enclose your entire garden.

   a. How many feet of fencing must you buy to enclose your garden?

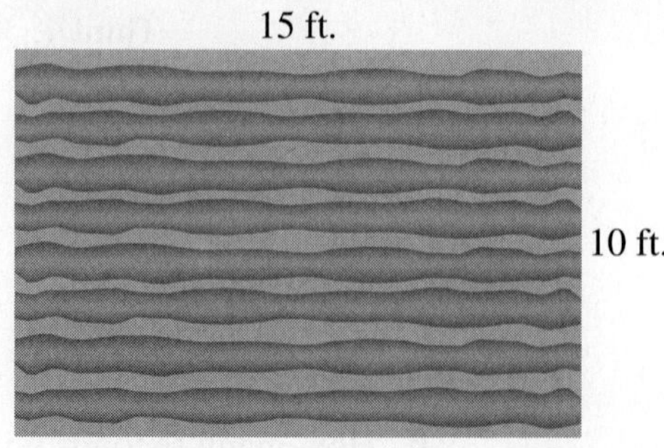

   b. A friend suggests surrounding the garden with a 2-foot-wide path and then enclosing the garden and path with fencing. If you do this, how many feet of fencing must you buy? Explain by including a labeled sketch of the garden and path.

DEFINITION

The **perimeter of a rectangle** is the total distance around all its edges or sides.

PROCEDURE

Calculating the Perimeter of a Rectangle The formula for the perimeter, P, of a rectangle with length l and width w has three equivalent forms.

$$P = l + w + l + w$$

$$P = 2l + 2w \quad \text{and}$$

$$P = 2(l + w)$$

Triangles

DEFINITION

A **triangle** is a closed plane figure with three sides.

Examples:

Many houses and garages have roofs supported by trusses, triangular structures usually made of wood. Trusses provide the greatest strength in building design. See the accompanying figure.

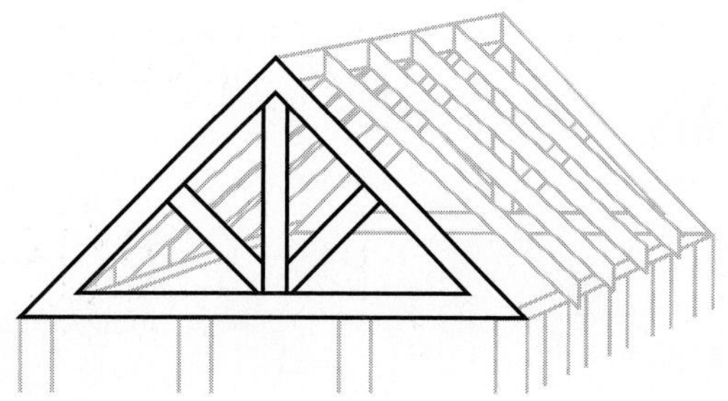

4. Why would a builder need to know about the perimeter of a triangular truss?

5. The dimensions of the three sides of a triangular truss are 13 feet, 13 feet, and 20 feet. What is the perimeter of the truss?

6. If the sides of a triangle measure a, b, and c, write a formula for the perimeter, P, of the following triangle.

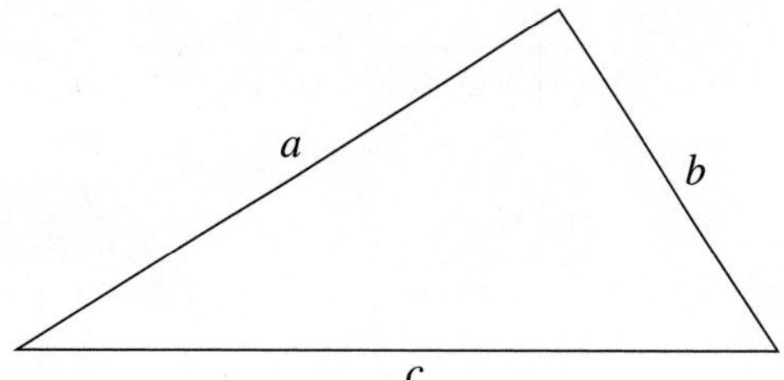

Parallelograms

DEFINITION

A **parallelogram** is a closed four-sided plane figure whose opposite sides are parallel and the same length.

Examples:

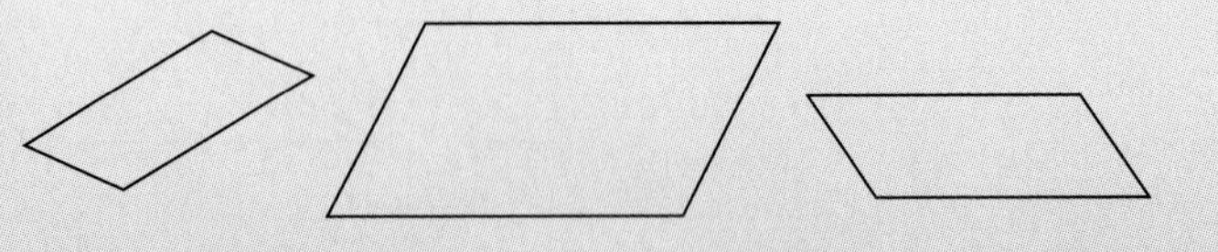

7. Where do you see parallelograms in the real world?

8. You decide to install a walkway diagonally from the street to your front steps. The width of your steps is 3 feet, and you determine that the length of the walkway is 23 feet. What is the perimeter of your walkway?

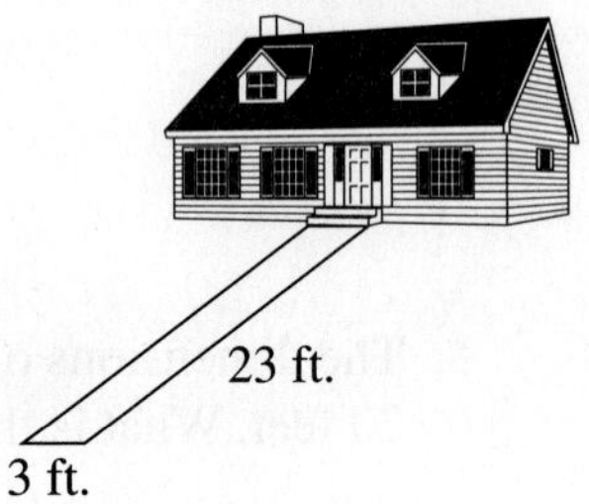

9. If the sides of a parallelogram measure a and b, write the formula for the perimeter, P, of a parallelogram.

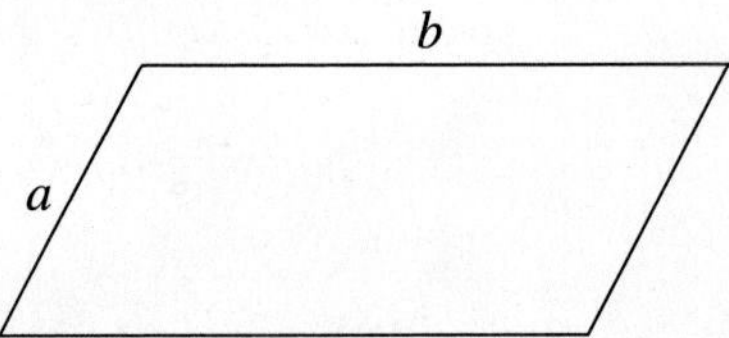

Trapezoids

DEFINITION

A **trapezoid** is a closed four-sided plane figure with two opposite sides that are parallel and two opposite sides that are not parallel.

Examples:

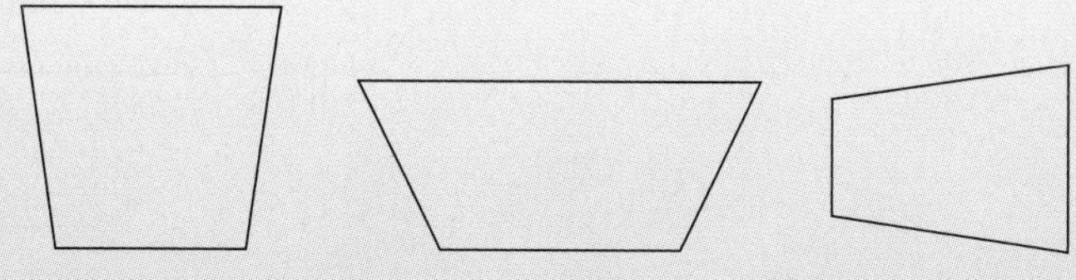

10. The sides of a trapezoid measure *a*, *b*, *c*, and *d*. Write a formula for the perimeter, *P*, of a trapezoid.

11. You buy a piece of land in the shape of a trapezoid with sides measuring 100 feet, 40 feet, 130 feet, and 50 feet. The parallel sides are 100 feet and 130 feet. Draw the trapezoid and calculate its perimeter.

Polygons

DEFINITION

A **polygon** is a closed many-sided plane figure.

Examples:

DEFINITION

A **regular polygon** is a polygon whose sides are all equal and whose angles are all equal.

Examples:

12. Calculate the perimeter of the following polygon.

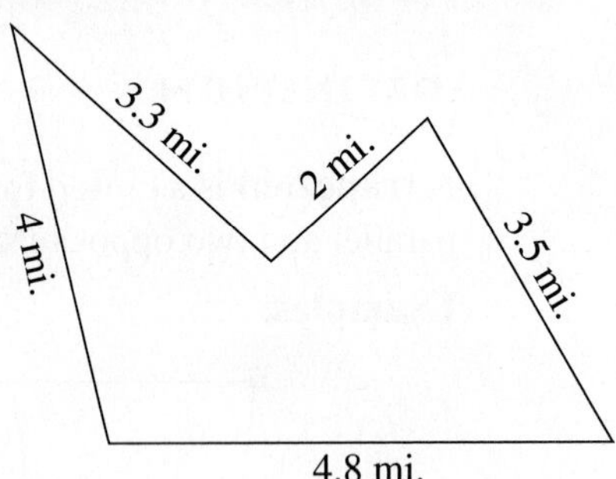

DEFINITION

The **perimeter of a polygon** is the total distance around all its edges or sides.

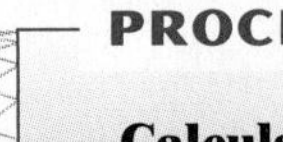

PROCEDURE

Calculating the Perimeter of a Polygon The perimeter of a polygon is calculated by adding all of the lengths of the sides that make up the figure.

SUMMARY
ACTIVITY 6.1

Two-Dimensional Figure	Labeled Sketch	Perimeter Formula
Square	s, s, s, s	$P = 4s$
Rectangle	l, w, w, l	$P = 2l + 2w$
Triangle	a, b, c	$P = a + b + c$
Parallelogram	b, a, a, b	$P = 2a + 2b$
Trapezoid	b, a, c, d	$P = a + b + c + d$

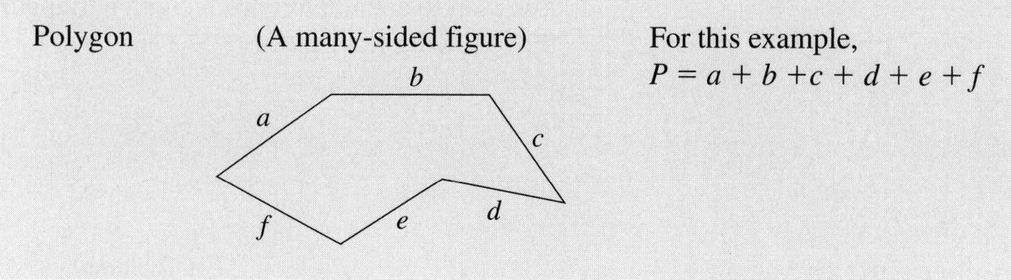

EXERCISES
ACTIVITY 6.1

1. You are interested in buying an older home. The first thing you learn about older homes is that they frequently have had additions over the years, and the floor plans often have the shape of a polygon.

   a. When the house you want to buy was built, 100 years ago, it had a simple rectangular floor plan.

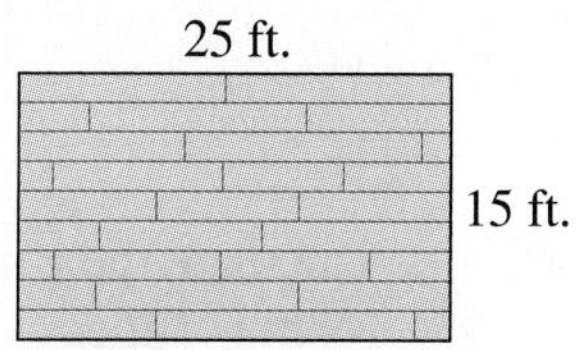

   Calculate the perimeter of the floor plan.

   b. Sixty years ago, a rectangular 10-foot by 25-foot garage was added to the original structure.

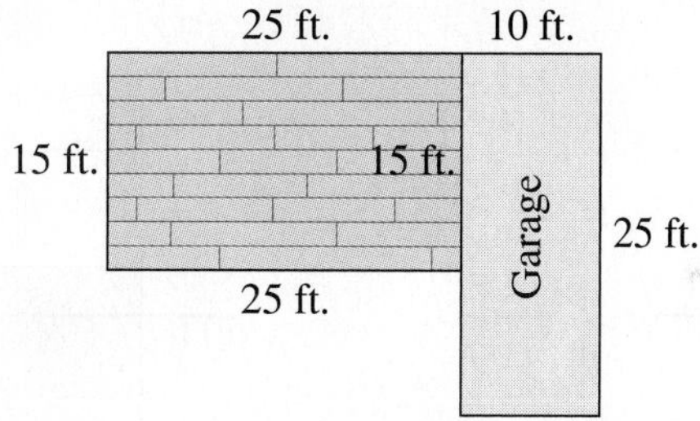

   Calculate the perimeter of the floor plan.

   c. Twenty-five years ago, a new master bedroom, with the same size and shape as the garage, was added onto the other side of the original floor plan.

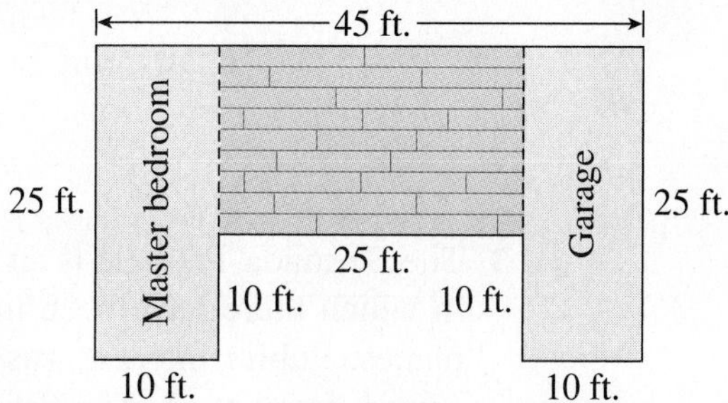

   Calculate the perimeter of the remodeled floor plan.

d. You plan to add a family room with a triangular floor plan, as shown in the following floor plan.

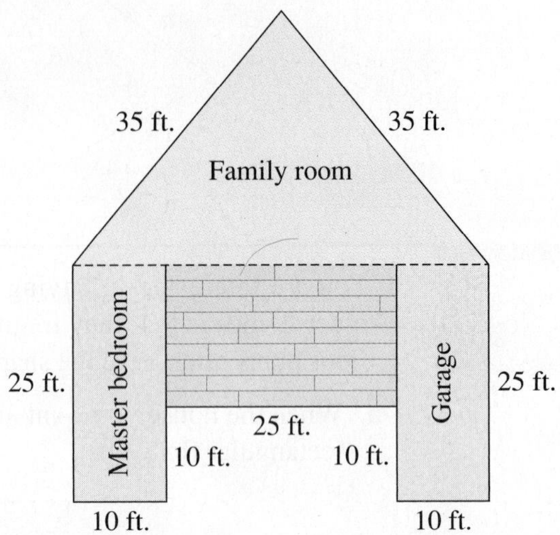

Calculate the perimeter of the floor plan of the house after the family room is added.

2. A standard basketball court has the following dimensions:

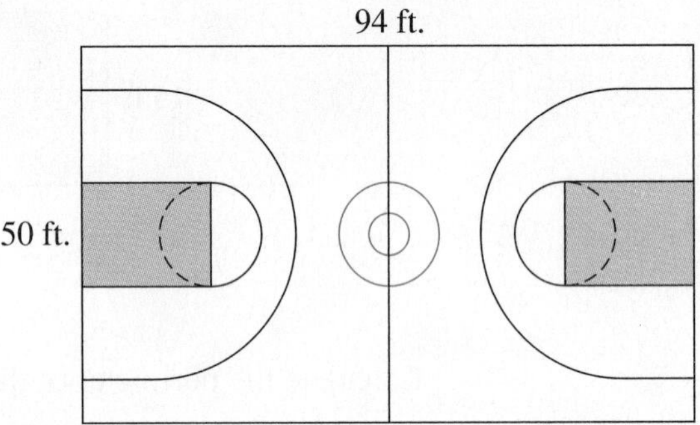

a. Calculate the perimeter of the court.

b. If you play a half-court game, calculate the perimeter of the half-court.

3. The Bermuda Triangle is an imaginary triangular area in the Atlantic Ocean in which there have been many unexplained disappearances of boats and planes. Public interest was aroused by the publication of a popular and controversial book, *The Bermuda Triangle*, by Charles Berlitz in 1974. The triangle starts at Miami, Florida, goes to San Juan, Puerto Rico (1038 miles), then to Bermuda (965 miles), and back to Miami (1042 miles).

a. What is the perimeter of this triangle?

EXERCISES
ACTIVITY 6.2

For calculations in Exercises 1–5, use π on your calculator and round your answers to the nearest hundredths unless otherwise indicated.

1. You order a pizza in the shape of a circle with diameter 14 inches. Calculate the "length" of the crust (that is, find the circumference of the pizza).

2. You enjoy playing darts. You decide to make your own dartboard consisting of four concentric circles (that is, four circles with the same center). The smallest circle (the "bull's-eye") has radius 1 cm, the next largest circle has radius 3 cm, the third circle has radius 6 cm, and the largest circle has radius 10 cm. You decide to compare the circumferences of the circles. What are these circumferences?

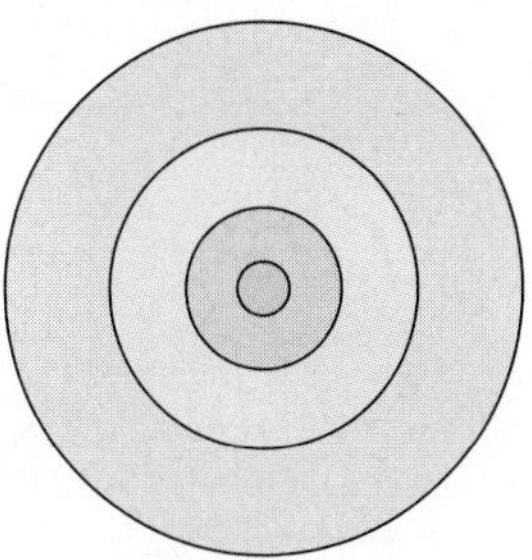

3. United States coins are circular. Choose a quarter, dime, nickel, and penny.

a. Use a tape measure (or ruler) to estimate the diameter of each coin in terms of centimeters. Record your results to the nearest tenth of a centimeter.

b. Use a tape measure (or string and ruler) to estimate the circumference of each coin in terms of centimeters. Record your results to the nearest tenth of a centimeter.

Exercise numbers appearing in color are answered in the Selected Answers appendix.

c. Check your estimates by using the formulas derived in this activity. Record your results to the nearest tenth of a centimeter.

4. Use the appropriate geometric formulas to calculate the circumference, or fraction thereof, for each of the following circles. Round to the nearest tenth.

a.

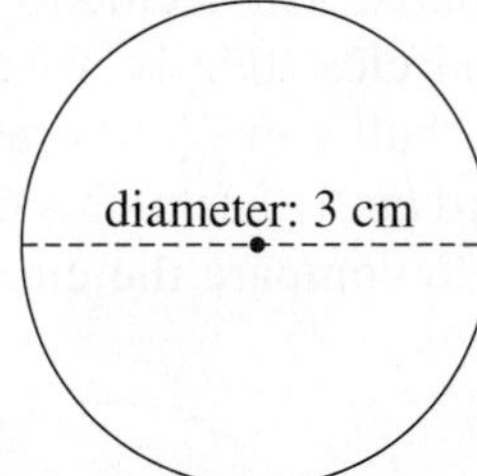

b.

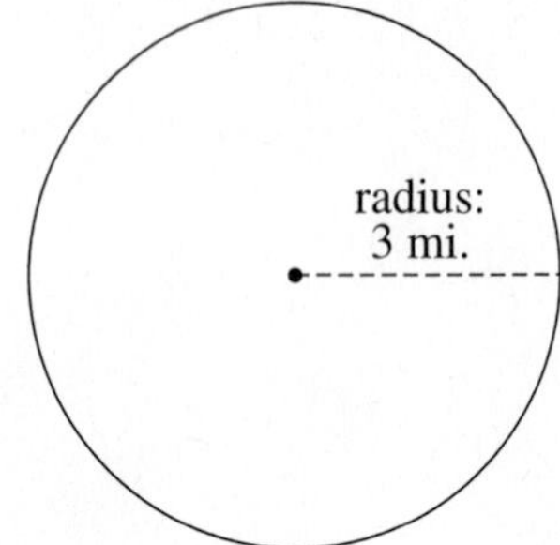

c.

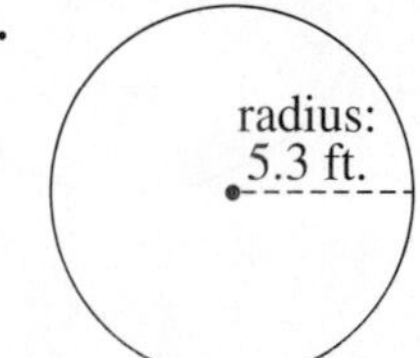

d.

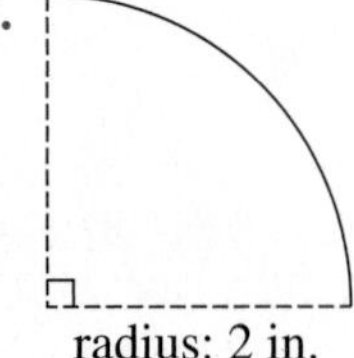

5. If a circle has circumference 63 inches, approximate its radius.

ACTIVITY 6.3

Lance Armstrong and You

OBJECTIVES

1. Calculate perimeters of many-sided plane figures using formulas and combinations of formulas.
2. Use unit analysis to solve problems involving perimeters.

Inspired by Lance Armstrong's remarkable performance in the Tour de France, you decide to experiment with long-distance biking. You choose the route shown by the solid line path in the following figure, made up of rectangles and a quarter circle.

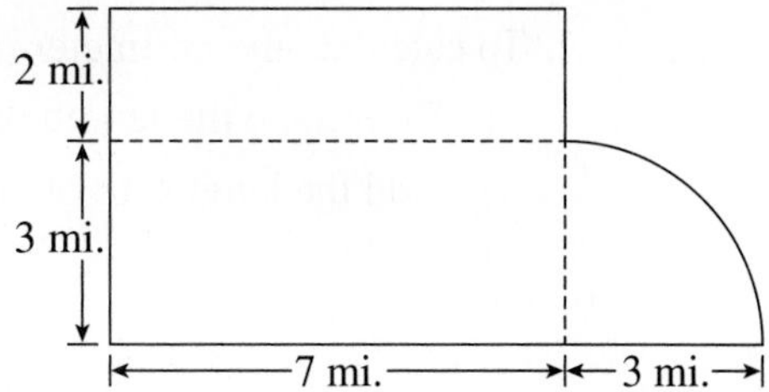

1. Calculate the total length of your bike trip in miles (that is, determine the perimeter of the figure). Use π on your calculator and round to the nearest thousandth.

2. If you can average 9 miles per hour on your bike, how long will it take you to complete the trip?

3. If your bike tires have a diameter of 2 feet, calculate the circumference of the tires to the nearest thousandth.

4. To analyze the wear on your tires, calculate how many rotations of the tires are needed to complete your trip. (*Note*: 1 mile = 5280 feet.)

5. Participants in the Tour de France bike 3454 kilometers (km). How many hours would it take you to complete the race if you average 9 miles per hour? (*Note*: 1 mile = 1.609 km.)

6. Compare your time with the 2005 time by Lance Armstrong: 86.25 hours.

SUMMARY
ACTIVITY 6.3

1. A many sided closed plane figure may be viewed as a combination of basic plane figures: squares, rectangles, parallelograms, triangles, trapezoids, and circles.

2. To calculate the perimeter of a many sided plane figure,
   i. Determine the length of each part that contributes to the perimeter.
   ii. Add the lengths to obtain the figure's total perimeter.

EXERCISES
ACTIVITY 6.3

1. You plan to fly from New York City to Los Angeles via Atlanta and return from Los Angeles to New York City via Chicago.

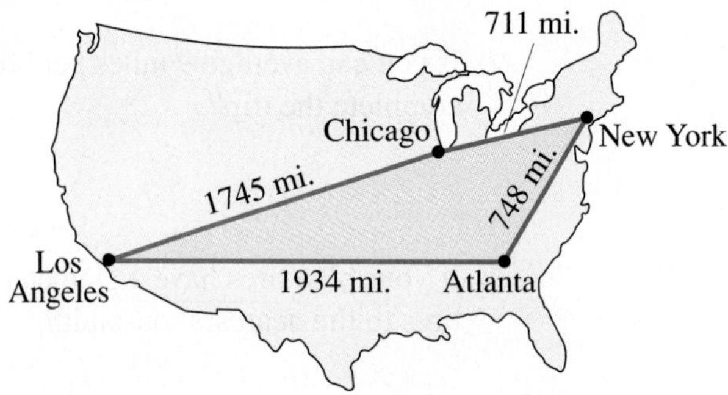

Use the data in the diagram to determine the total distance of your trip.

2. A Norman window is a rectangle with a semicircle on top. You decide to install a Norman window in your family room with the dimensions indicated in the diagram. What is its perimeter? Round to the nearest hundredth.

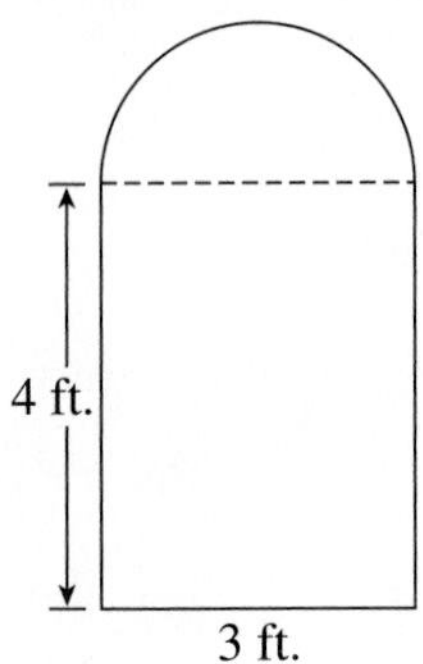

3. Stonehenge is an ancient site on the plains of southern England consisting of a collection of concentric circles (that is, circles with the same center) outlined with large sandstone blocks. Carbon dating has determined the age of the stones to be approximately 5000 years. Much curiosity and mystery has surrounded this site over the years. One theory about Stonehenge is that it was a ritualistic prayer site. However, even today, there is still controversy over what went on there. One thing everyone interested in Stonehenge can agree on is the mathematical description of the circles.

The diameters of the four circles are 288 feet for the largest circle, 177 feet for the next, 132 feet for the third, and 110 feet for the innermost circle.

a. For each time you walked around the outermost circle, how many times could you walk around the innermost circle? Round to the nearest hundredth.

b. How much longer is the trip around one of the two intermediate circles than around the other? Round to the nearset hundredth.

4. Calculate the perimeters of each of the following figures:

a.

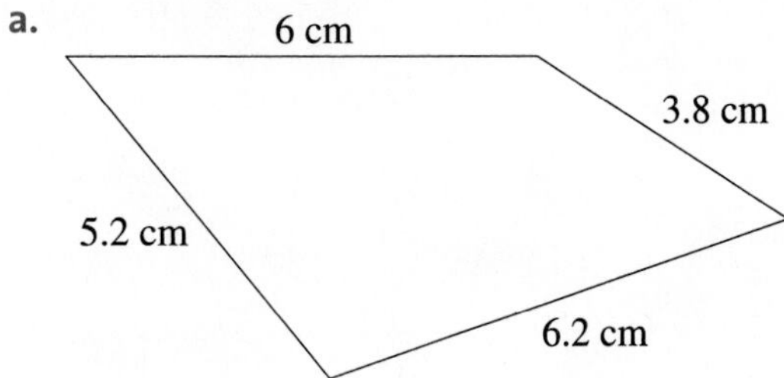

b.

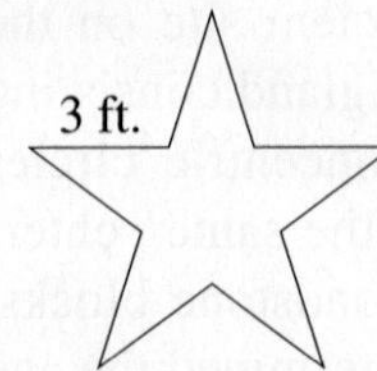

Note: All sides of the star are equal length.

c.

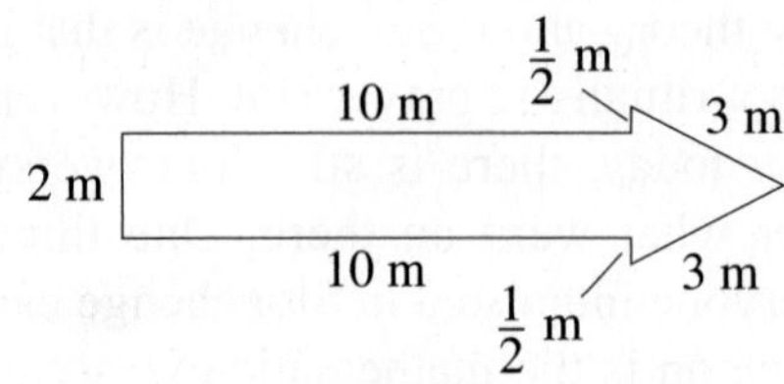

d.

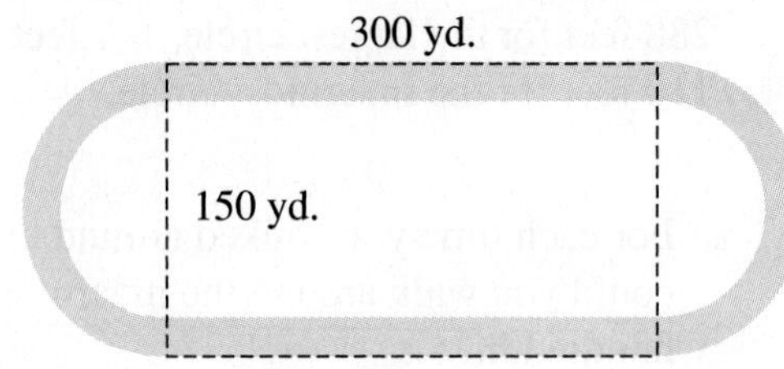

Note: The ends are semicircles.

ACTIVITY 6.4

Baseball Diamonds, Gardens, and Other Figures Revisited

OBJECTIVES

1. Write formulas for areas of squares, rectangles, parallelograms, triangles, trapezoids, and polygons.
2. Calculate areas of polygons using appropriate formulas.

In previous activities, you ran the bases of a baseball diamond, fenced in a rectangular garden, and explored the perimeters of triangular trusses, parallelograms, trapezoids, and circles. Now, you will look at those same geometric figures, but from a different perspective.

Squares

As groundskeeper for the local baseball team, you need to guarantee good-quality turf for the baseball diamond. This is the square area enclosed by the baselines. To estimate the amount of sod to plant in the baseball diamond, you need to know the size of this area.

Recall from Chapter 2 that the area of a square is measured by determining the number of unit squares (squares that measure 1 unit in length on each side) that are needed to completely fill the inside of the square.

For example, the square pictured below has sides that are 4 units in length. As illustrated, it takes four rows of four unit squares to fill the inside of the square. You say the area, A, of the square is 16 square units.

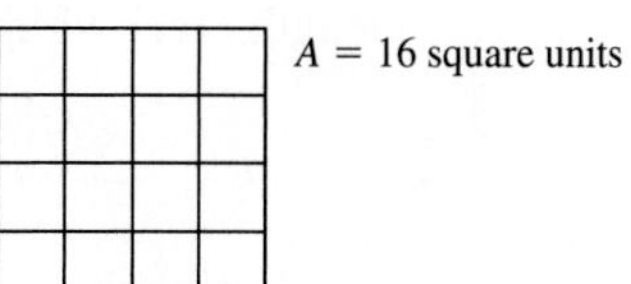

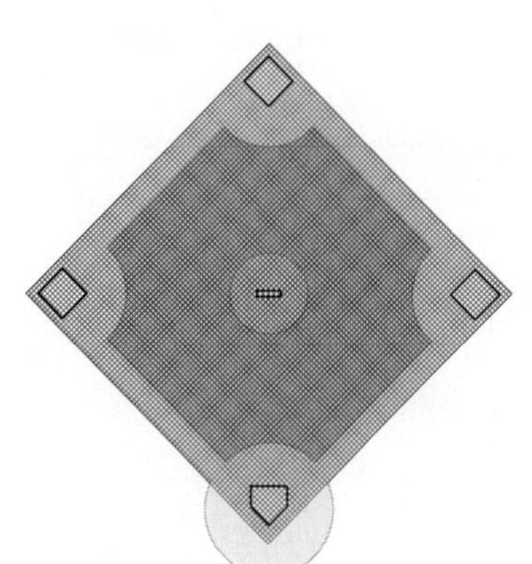

1. A regulation baseball diamond is a square with each side 90 feet long. How many square feet are needed to cover this baseball diamond? That is, how many 1-foot by 1-foot squares are inside the diamond?

PROCEDURE

Calculating the Area of a Square The formula for the area, A, of a square with sides of length s is

$$A = s \cdot s \text{ or } A = s^2$$

2. The Little League baseball diamond has sides measuring 60 feet. Use the formula to calculate the area of the Little League diamond.

DEFINITION

The **area** of a square, or any polygon, is the measure of the region enclosed by the sides of the polygon. Area is measured in square units.

Rectangles

Planting your rectangular 15-foot by 10-foot garden requires that you know how much space it contains.

3. How many square feet are required to cover your garden? Include a sketch to explain your answer.

PROCEDURE

Calculating the Area of a Rectangle The formula for the area of a rectangle with length l and width w is

$$A = lw.$$

4. Will doubling the length and width of your garden in Problem 3 double its area? Explain.

Parallelograms

Once you know the formula for the area of a rectangle, then you have the key for determining the formula for the area of a parallelogram. Recall that a parallelogram is formed by two intersecting pairs of parallel sides.

5. If you "cut off" a triangle and move it to the other side of the parallelogram forming the shaded rectangle, is the area different? Explain.

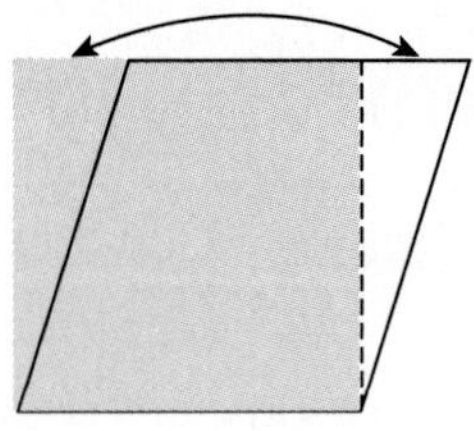

DEFINITION

The height of a parallelogram is the distance from one side (the base) to the opposite side, measured along a perpendicular line (see illustration).

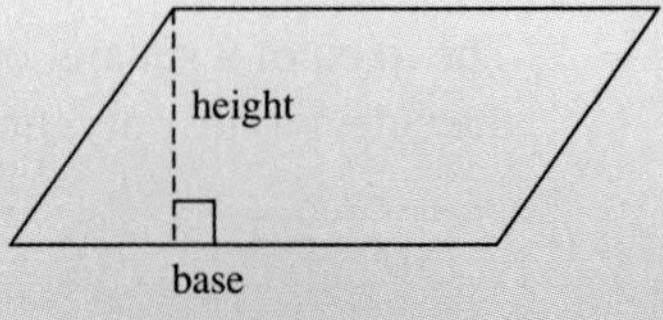

6. Use your observations from Problem 5 to calculate the area of the parallelogram with height 4 inches and base 7 inches.

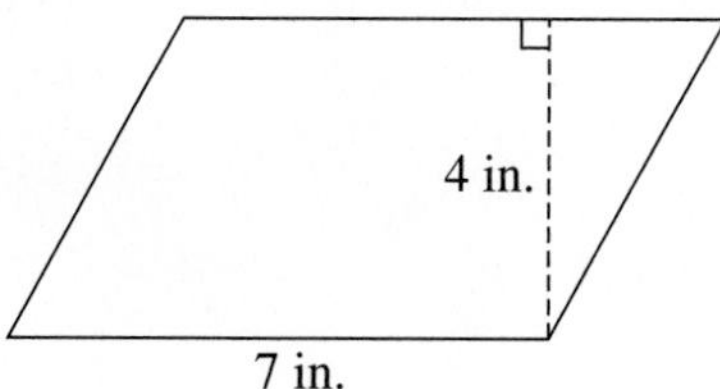

PROCEDURE

Calculating the Area of a Parallelogram The formula for the area of a parallelogram with base b (the length of one side) and height h (the perpendicular distance from the base b to its parallel side) is

$$A = bh.$$

Triangles

Every triangle can be pictured as one-half of a rectangle or parallelogram.

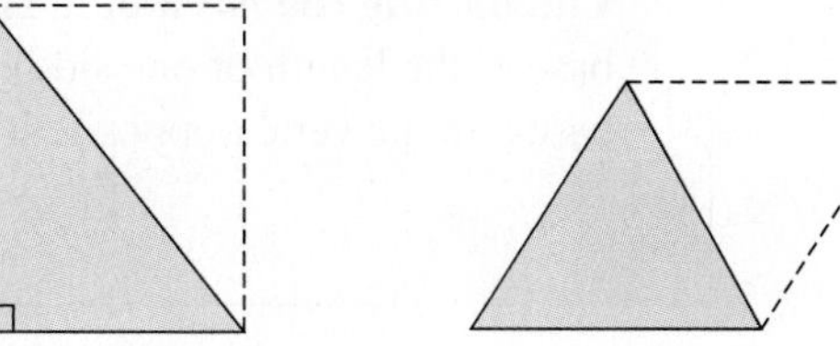

7. For each of the following triangles, draw a rectangle or parallelogram that encloses the triangle.

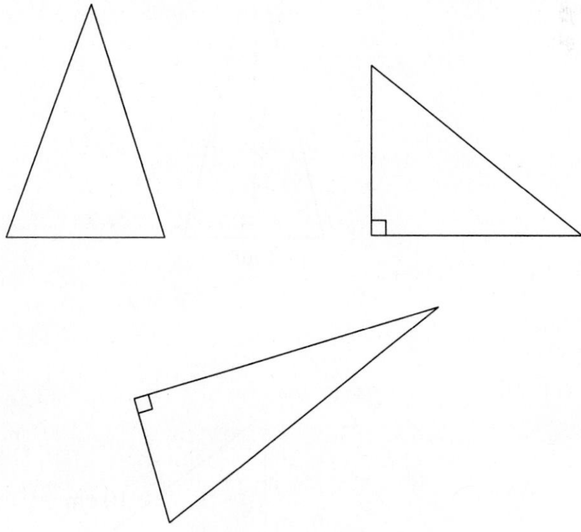

8. Use the idea that a triangle is one-half of a rectangle or parallelogram to write a formula for the area of a triangle in terms of its base b and height h. Explain how you obtained the formula.

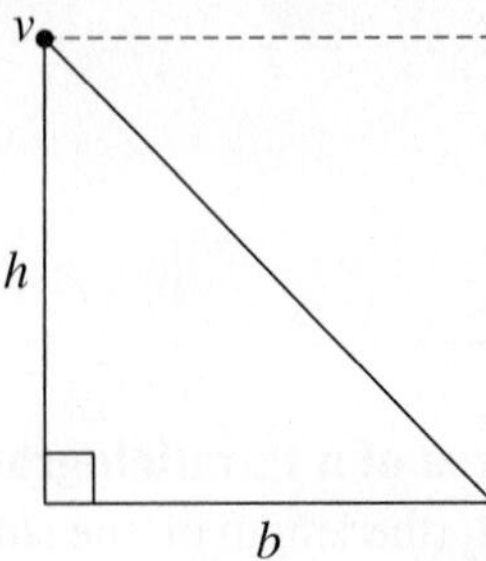

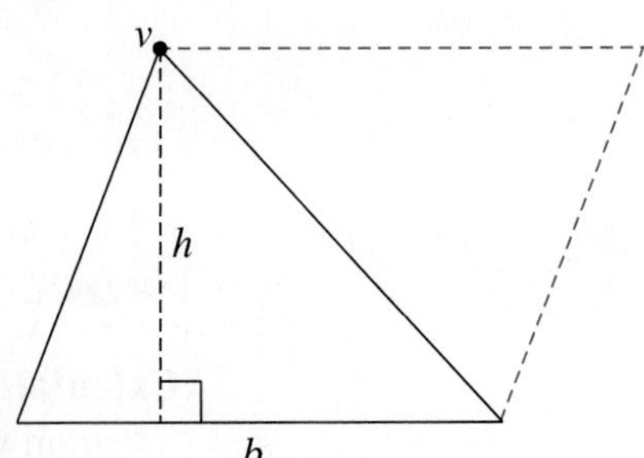

PROCEDURE

Calculating the Area of a Triangle The formula for the area of a triangle with base b (the length of one side) and height h (the perpendicular distance from the base b to the vertex opposite it) is

$$A = \frac{1}{2}bh.$$

9. Calculate the areas of the following triangles using a formula.

a.

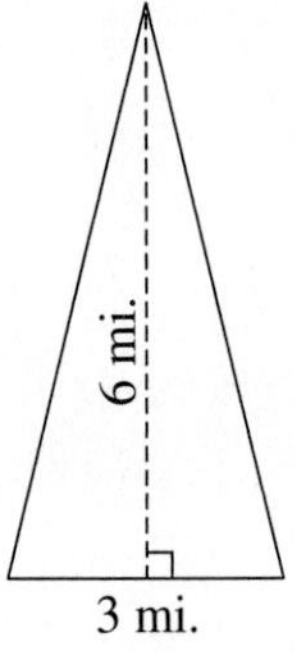

b.

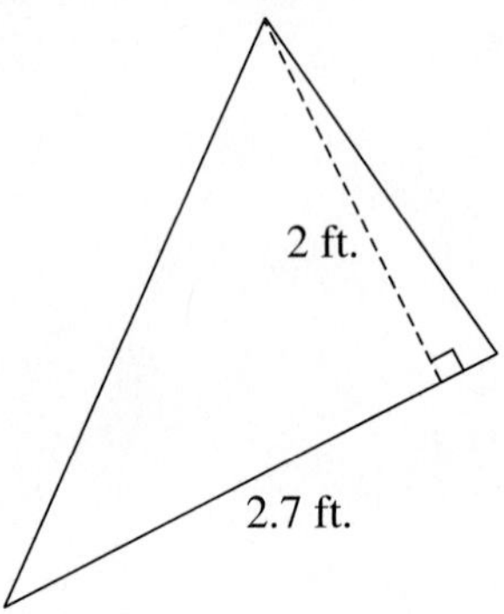

c.

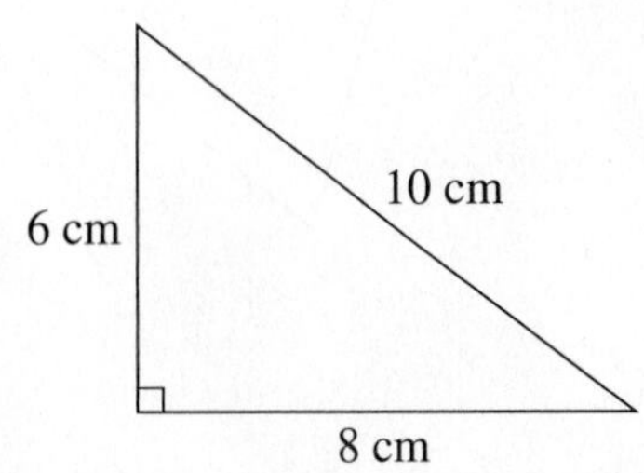

d.

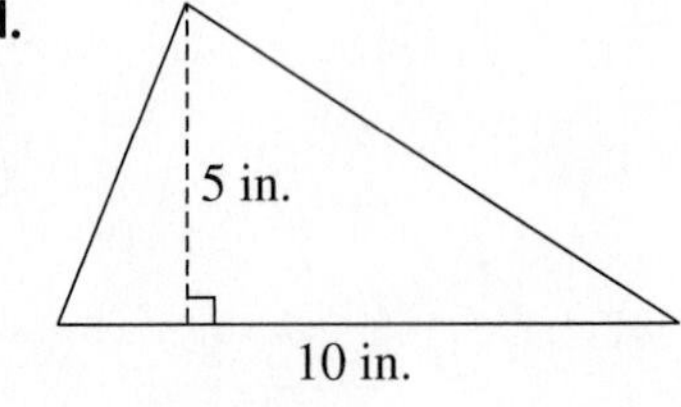

Trapezoids

Trapezoids may be viewed as one-half of the parallelogram that is formed by adjoining the trapezoid to itself turned upside down, as shown.

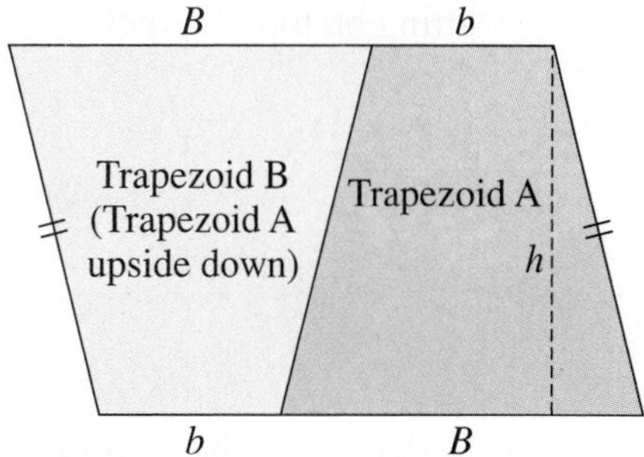

10. a. The parallelogram shown above has base $b + B$ and height h. Determine the area of this parallelogram.

b. Use the idea that a trapezoid is one-half of a parallelogram to write a formula for the area A of a trapezoid in terms of its bases b and B and its height h.

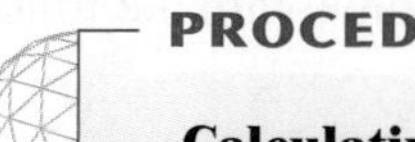

PROCEDURE

Calculating the Area of a Trapezoid The formula for the area of a trapezoid with bases b and B and height h is

$$A = \frac{1}{2}h(b + B).$$

11. Calculate the area of the trapezoid with height 5 feet and bases 6 feet and 11 feet.

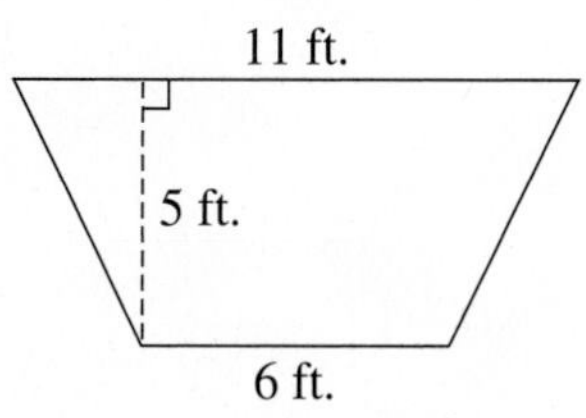

12. If the lower base of the trapezoid in Problem 11 is increased by 2 feet and the upper base is decreased by 2 feet, draw the new trapezoid and compute its new area.

Polygons

You can determine the area of a polygon by seeing that polygons can be broken up into other, more basic figures, such as rectangles and triangles. For example, the front of a garage shown here is a polygon that can also be viewed as a triangle sitting on top of a rectangle.

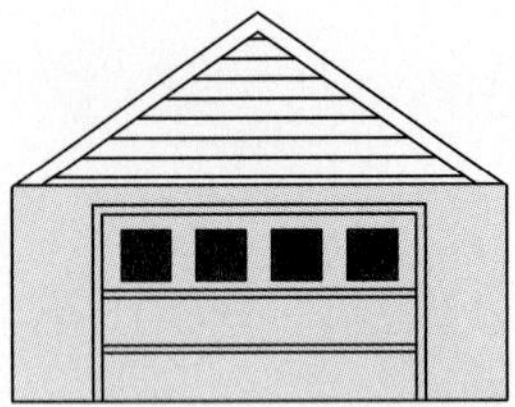

To determine the area of this polygon, use the appropriate formulas to determine each area, and then sum your answers to obtain the area of the polygon.

13. Calculate the area of the front of the garage:

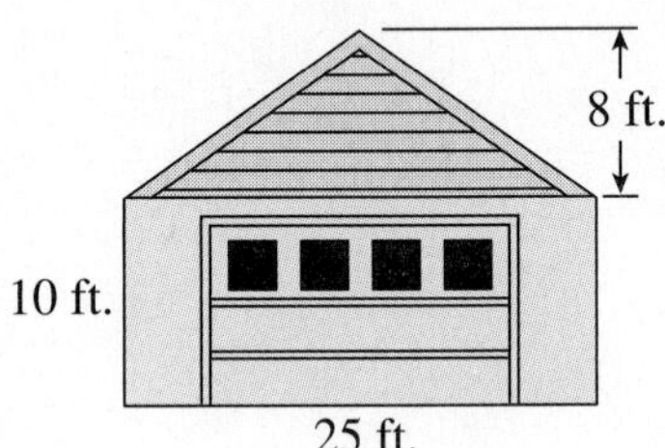

14. You are planning to build a new home with the following floor plan.

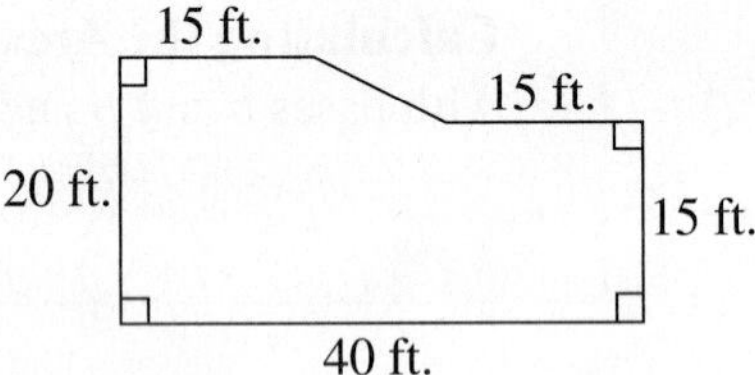

Calculate the total floor area.

15. Calculate the areas of each of the following polygons.

a.

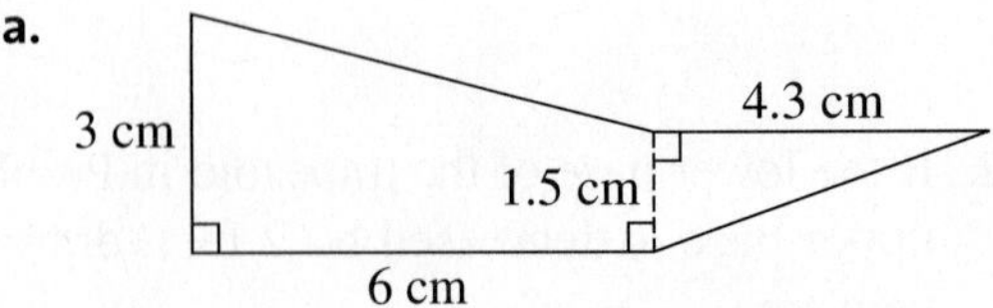

b.

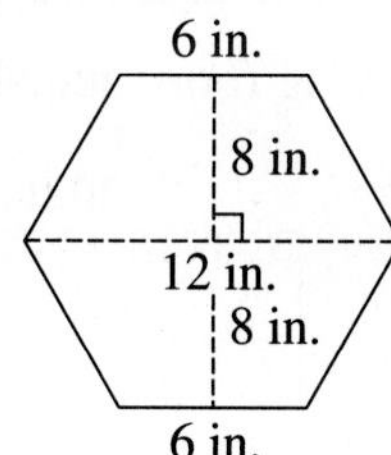

c.

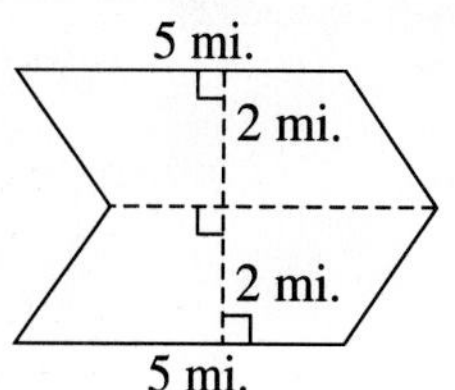

SUMMARY
ACTIVITY 6.4

Two-Dimensional Figure	Labeled Sketch	Area Formula
Square	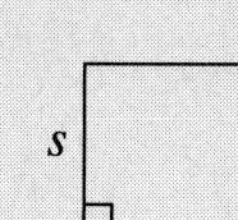	$A = s^2$
Rectangle	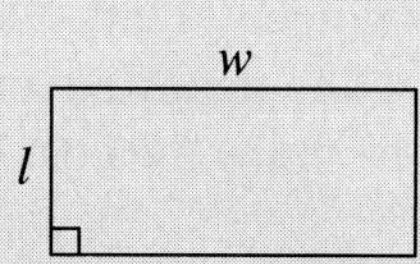	$A = lw$
Triangle	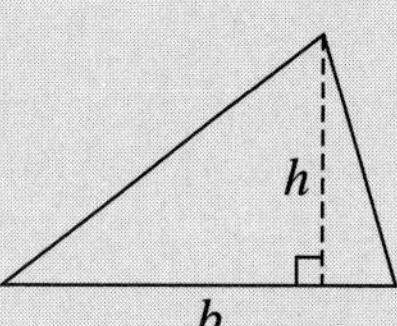	$A = \frac{1}{2}bh$
Parallelogram	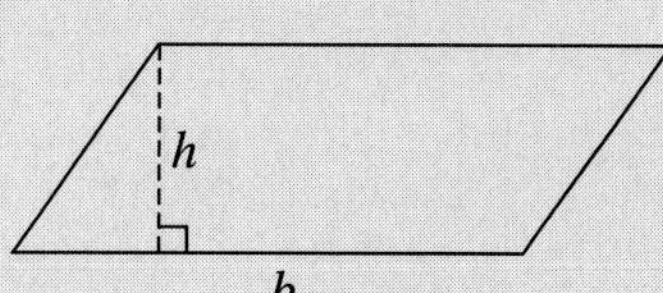	$A = bh$
Trapezoid	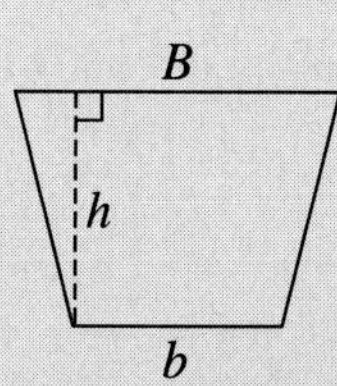	$A = \frac{1}{2}h(b + B)$
Polygon	A many-sided figure that is subdivided into familiar figures whose areas are given by formulas in this summary. 	A = sum of the areas of each figure $= A_1 + A_2 + A_3$ 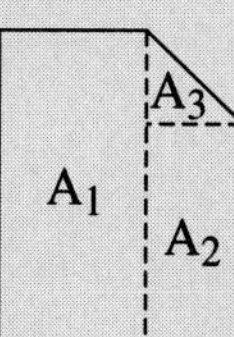

EXERCISES
ACTIVITY 6.4

1. You are carpeting your living room and sketch the following floor plan.

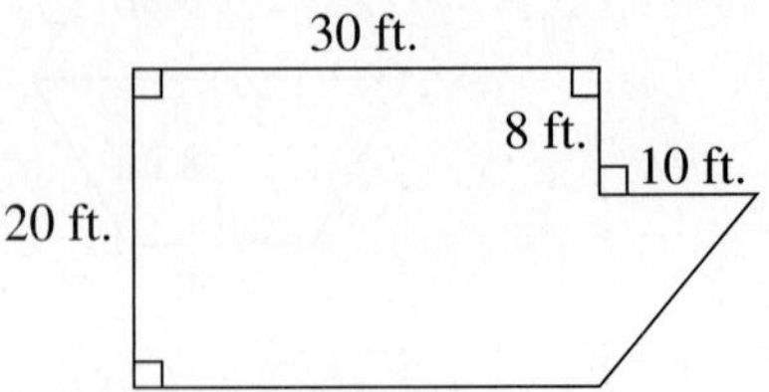

a. Calculate the area that you need to carpet.

b. Carpeting is sold in 10-foot-wide rolls. Calculate how much you need to buy. Explain.

2. You need to buy a solar cover for your 36-foot by 18-foot rectangular pool. A pool company advertises that solar covers are on sale for $1.77 per square foot. Determine the cost of the pool cover before sales tax.

3. Calculate the areas of each of the following figures.

a.

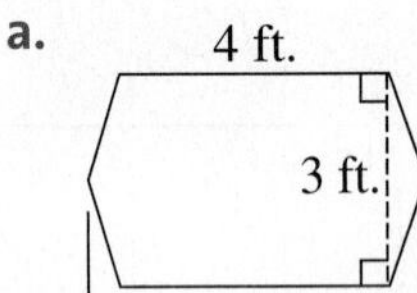

b.

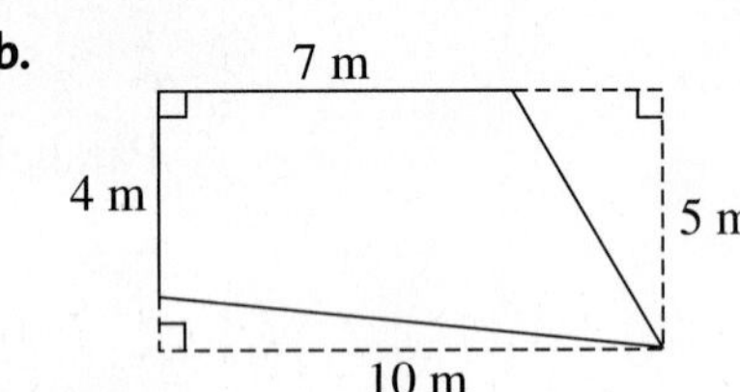

4. How would you break up the following star to determine what dimensions you need to know in order to calculate its area? Explain.

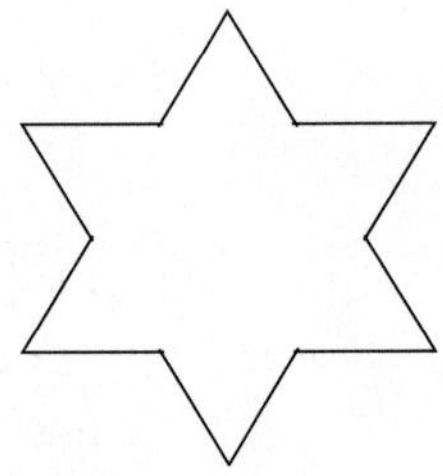

5. A standard basketball court is a rectangle with length 94 feet and width 50 feet. How many square feet of flooring would you need to purchase in order to replace the court?

6. You are planning to build a new garage on your home and need to measure the length and width of your cars to help you estimate the size of the double garage. Your car measurements are

   Car 1: 14 ft. 2 in. by 5 ft. 7 in.

   Car 2: 14 ft. 6 in. by 5 ft. 9 in.

   a. Based on these measurements, what would be a reasonable floor plan for your garage? Explain.

   b. What is the area of your floor plan?

ACTIVITY 6.5

How Big Is That Circle?

OBJECTIVES

1. Develop a formula for the area of a circle.
2. Use the formula to determine areas of circles.

Determining the area of a circle becomes a challenge because there are no straight sides. No matter how hard you try, you cannot neatly pack unit squares inside a circle to completely cover the area of a circle. The best you can do in this way is to get an approximation of the area. In this activity you will explore methods for estimating the area of a circle and develop a formula to calculate the exact area.

DEFINITION

The **area of a circle** is the measure of the region enclosed by the circumference of the circle.

1. a. To help understand the formula for a circle's area, start by folding a paper circle in half, then in quarters, and finally halving it one more time into eighths.

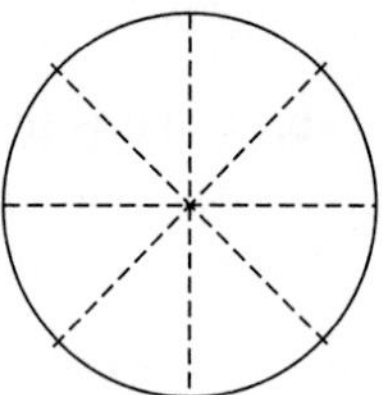

b. Cut the circle along the folds into eight equal pie-shaped pieces (called sectors) and rearrange these sectors into an approximate parallelogram (see accompanying figures).

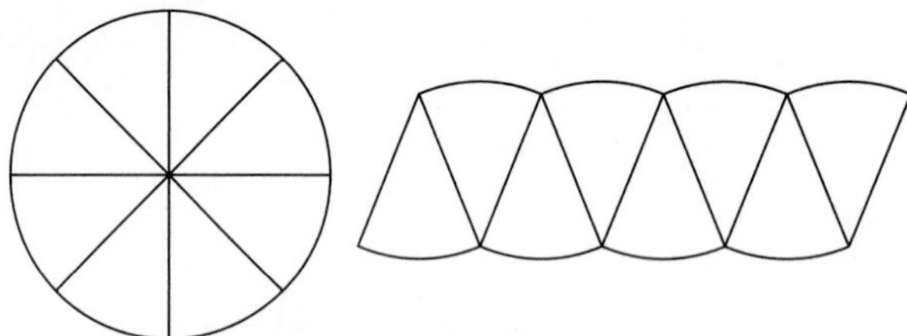

c. What measurement on the circle approximates the height of the parallelogram? Explain.

d. What measurement on the circle approximates the base of the parallelogram?

e. Recall that the area of a parallelogram is given by the product of its base and its height. Use the approximations from parts b and c to determine the approximate area of the parallelogram and the approximate area of the circle.

Imagine cutting a circle into more than eight equal sectors. Each sector would be thinner. When reassembled, as in Problem 1, the resulting figure will more closely approximate a parallelogram. Hence, the formula for the area of a circle, $A = \pi r^2$, is even more reasonable and accurate.

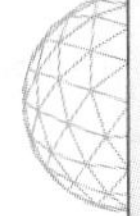

PROCEDURE

Calculating the Area of a Circle The formula for the area A of a circle with radius r is

$$A = \pi r^2.$$

2. You own a circular dartboard of radius 1.5 feet. To figure how much space you have as a target, calculate the area of the dartboard. Be sure to include the units of measurement in your answer.

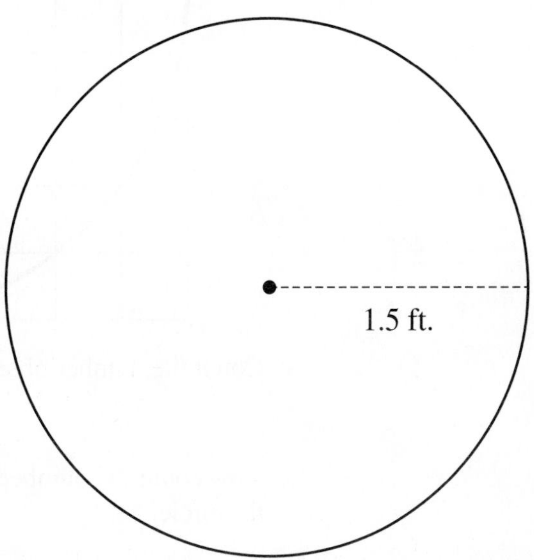

3. The diameter of a circle is twice its radius. Use this fact to rewrite the formula for the area of a circle using its diameter instead of the radius. Show the steps you took.

4. Calculate the areas of the following circles. Round to the nearest hundredth.

a.

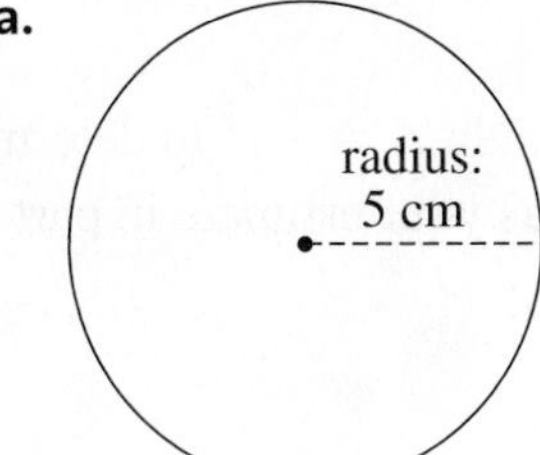

b.

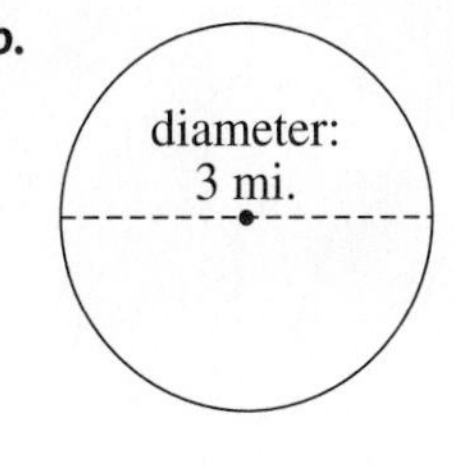

5. As discussed in the introduction to this activity, you can estimate the area of a circle by counting the number of squares that come close to filling the inside of a circle. In the illustration, use the distance between grid lines as the basic unit of measurement. The steps in this problem show you how.

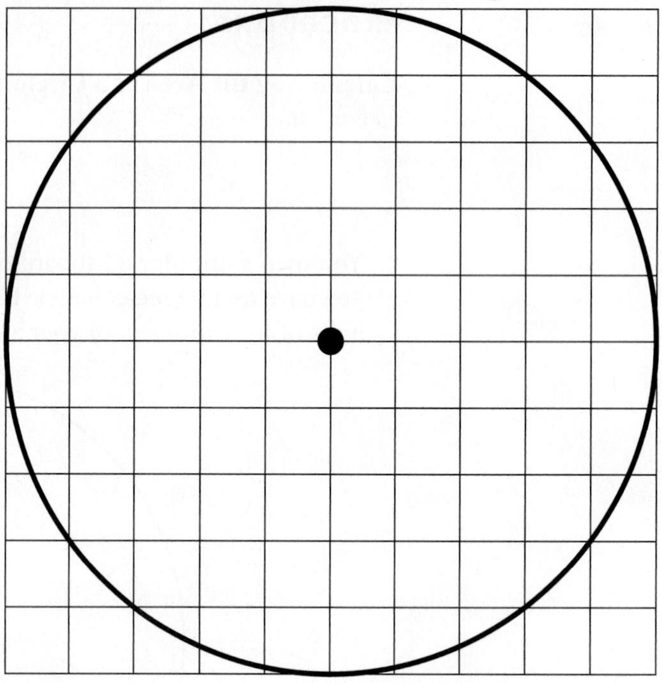

a. Count the number of small unit squares that are entirely inside the circle.

b. Now count the number of squares that are partly inside and partly outside the circle.

c. What is the total number of squares that are at least partly inside the square?

d. Your answer in part a is less than the actual area of the circle. Your answer in part c is more than the actual area of the circle. Find the average of these two numbers to get an estimate of the circle's area.

e. Determine the radius of the circle from the diagram.

f. Apply the formula $A = \pi r^2$ to determine the exact area of the circle. How far off was your estimate in part d?

g. A better estimate can be determined by using smaller squares. Note that in the following grid each square from the previous grid has been divided into four smaller squares. So when you count the squares, every four smaller squares will be the size of the unit square you used in parts a–f.

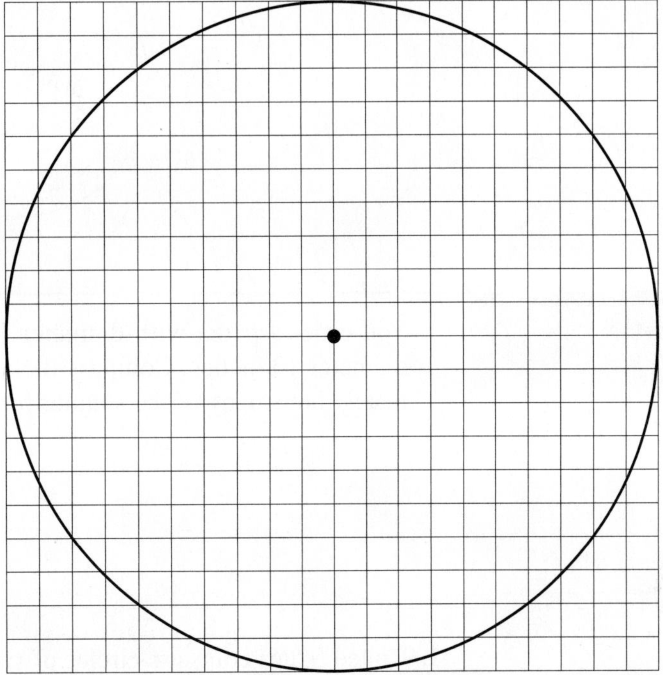

Follow the same procedure as before to estimate the area of the circle.

h. Divide the result in part g by 4 to obtain the area in terms of the larger squares.

i. How far is your new estimate from the actual area?

j. How much better is your new estimate than your estimate in part d?

SUMMARY
ACTIVITY 6.5

Two-Dimensional Figure	Labeled Sketch	Area Formula
Circle	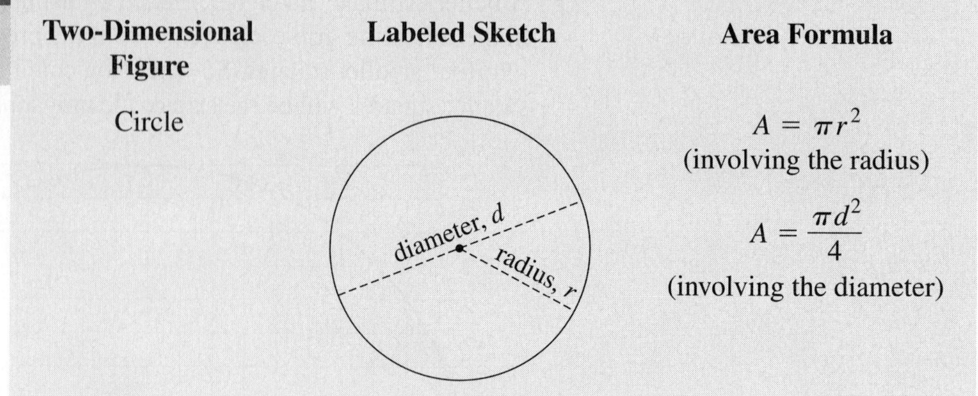	$A = \pi r^2$ (involving the radius) $A = \frac{\pi d^2}{4}$ (involving the diameter)

EXERCISES
ACTIVITY 6.5

1. You order a pizza with diameter 14 inches. Your friend orders a pizza with diameter 10 inches. Compare the areas of the two pizzas to estimate approximately how many of the smaller pizzas are equivalent to one larger pizza.

2. United States coins are circles of varying sizes. Choose a quarter, dime, nickel, and penny.

   a. Use a tape measure (or ruler) to estimate the diameter of each coin in centimeters.

   b. Calculate the area of each coin to the nearest hundredth.

3. Use an appropriate formula to calculate the area for each of the following figures. Round to the nearest hundredth.

   a.

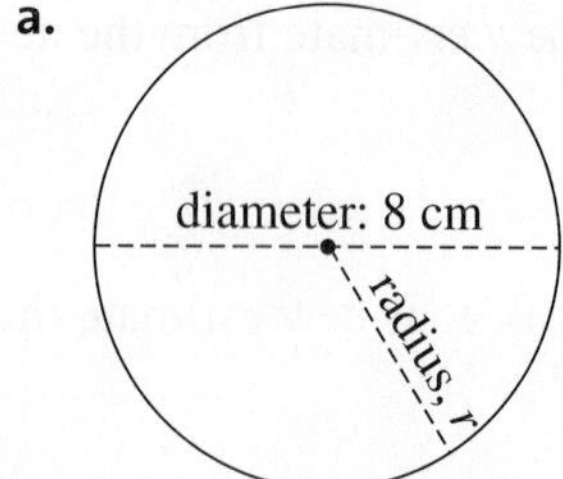

   b.

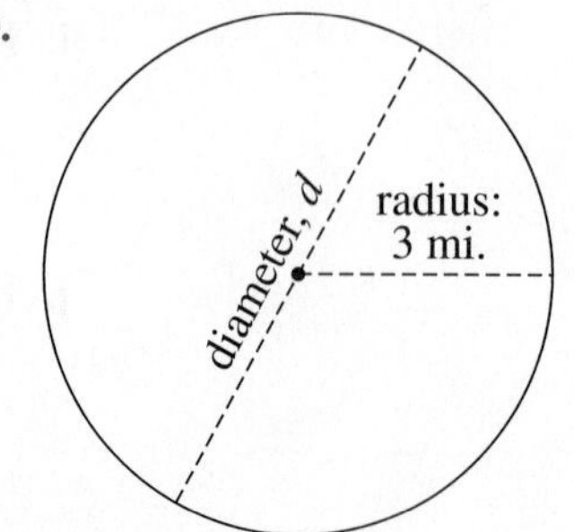

c.

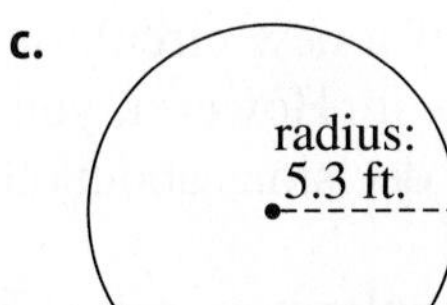

d.

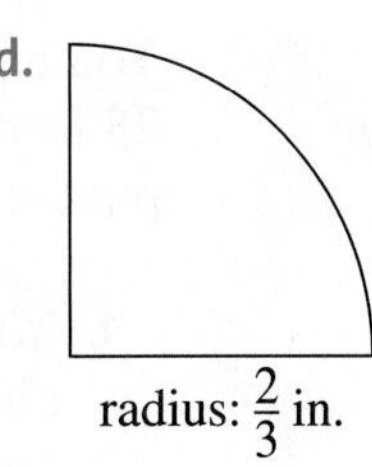

4. a. For each of the following three circles, estimate how many 1-cm by 1-cm unit squares and fractions of unit squares can fit in each circle by drawing these squares on the circles. Place these estimates in column 2 of the table in part c.

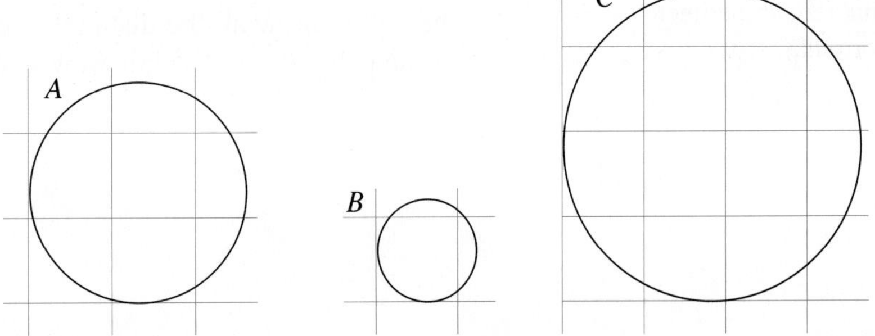

b. Measure the radius of each circle in centimeters, and record the results in column 3 of the table in part c.

c. Compute πr^2 for each circle and place your answers in column 4.

Measure for Measure

COLUMN 1 CIRCLE	COLUMN 2 ESTIMATED AREA	COLUMN 3 MEASURED RADIUS	COLUMN 4 CALCULATED πr^2
A			
B			
C			

d. Compare your answers in columns 2 and 4. What do you notice?

ACTIVITY 6.6

A New Pool and Other Home Improvements

OBJECTIVES

1. Solve problems in context using geometric formulas.
2. Distinguish between problems that require area formulas and those that require perimeter formulas.

You are the proud owner of a new circular swimming pool with a diameter of 25 feet and are eager to dive in. However, you quickly discover that having a new pool requires making many decisions about other purchases.

1. Your first concern is a solar pool cover. You do some research and find that circular pool covers come in the following sizes: 400, 500, and 600 square feet. Friends recommend that you buy a pool cover with very little overhang. Which size is best for your needs? Explain.

2. You decide to build a concrete patio around the circumference of your pool. It will be 6 feet wide all the way around. Provide a diagram of your pool with the patio. Show all the distances you know on the diagram. What is the area of the patio alone? Explain how you determined this area.

3. State law requires that all pools be enclosed by a fence to prevent accidents. You decide to completely enclose your pool and patio with a stockade fence. How many feet of fencing do you need? Explain.

4. Next, you decide to stain the new fence. The paint store recommends a stain that covers 500 square feet per gallon. If the stockade fence has a height of 5 feet, how many gallons of stain should you buy? Explain.

5. Lastly, you decide to plant a circular flower garden near the pool and patio but outside the fence. You determine that a circle of circumference 30 feet would fit. What is the length of the corresponding diameter of the flower garden?

Other Home Improvements

Now that you have a new pool, patio, and flower garden, you want to do some other home improvements in anticipation of enjoying your new pool with family and friends.

You have $1000 budgeted for this purpose and the list looks like this:

- Replace the kitchen floor.
- Add a wallpaper border to the third bedroom.
- Paint the walls in the family room.

To stay within your budget, you need to determine the cost of each of these projects. You expect to do this work yourself, so the only monetary cost will be for materials.

6. The kitchen floor is divided into two parts. The first section is rectangular and measures 12 by 14 feet. The second section is a semicircular breakfast area that extends off the 14-foot side. The cost of vinyl flooring is $21 per square yard plus 6% sales tax. The vinyl is sold in 12-foot widths.

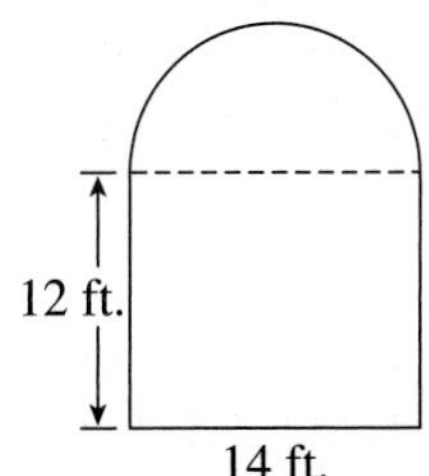

a. How long a piece of vinyl flooring will you need to purchase if you want only one seam, where the breakfast area meets the main kitchen, as shown? Remember that the vinyl is 12 feet wide.

b. How many square feet of flooring must you purchase? How many square yards is that (9 square feet = 1 square yard)?

c. How much will the vinyl flooring cost, including tax?

d. How many square feet of flooring will be left over after you're done? Explain.

7. The third bedroom is rectangular in shape and has dimensions of $8\frac{1}{2}$ by 13 feet. On each 13-foot side, there is a window that measures 3 feet 8 inches wide. The door is located on an $8\frac{1}{2}$-foot side and measures 3 feet wide from edge to edge. You are planning to put up a decorative horizontal wallpaper stripe around the room about halfway up the wall.

a. How many feet of wallpaper stripe will you need to purchase?

b. The border comes in rolls 5 yards in length. How many rolls will you need to purchase?

c. The wallpaper border costs $10.56 per roll plus 6% sales tax. Determine the cost of the border.

d. How many feet of wallpaper will be left over after you're done?

8. Your family room needs to be painted. It has a cathedral ceiling with front and back walls that measure the same, as shown in the diagram. The two side walls are 14 feet long and 12 feet high. Each of the walls will need two coats of paint. The ceiling will not be painted. For simplicity, ignore the fact that there are windows.

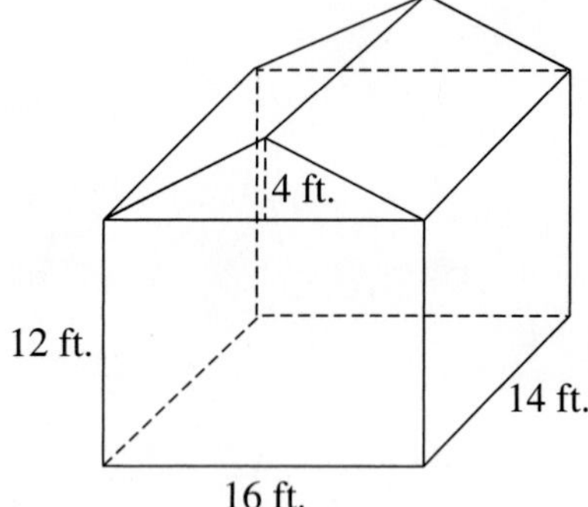

a. How many square feet of wall surface will you be painting? (Remember, all walls will need two coats.)

b. Each gallon of paint covers approximately 400 square feet. How many gallons of paint will you need to purchase?

c. The paint costs $19.81 per gallon plus 6% sales tax. What is the total cost of the paint you need for the family room?

9. What is the cost for all your home-improvement projects (not including the new pool, patio, and flower garden)?

10. Additional costs for items such as paint rollers and wallpaper paste amount to approximately $30. Can you afford to do all the projects? Explain.

SUMMARY
ACTIVITY 6.6

1. Area formulas are used when you are measuring the amount of space *inside* a figure.

2. Perimeter or circumference formulas are used when you are measuring the length *around* a figure.

EXERCISES
ACTIVITY 6.6

1. Your driveway is rectangular in shape and measures 15 feet wide and 25 feet long. Calculate the area of your driveway.

2. A flower bed in the corner of your yard is in the shape of a right triangle. The perpendicular sides of the bed measure 6 feet 8 inches and 8 feet 4 inches. Calculate the area of the flower bed. What are the units of this area?

3. The front wall of a storage shed is in the shape of a trapezoid. The bottom measures 12 feet, the top measures 10 feet, and the height is 6 feet. Calculate the area of the front wall of the shed.

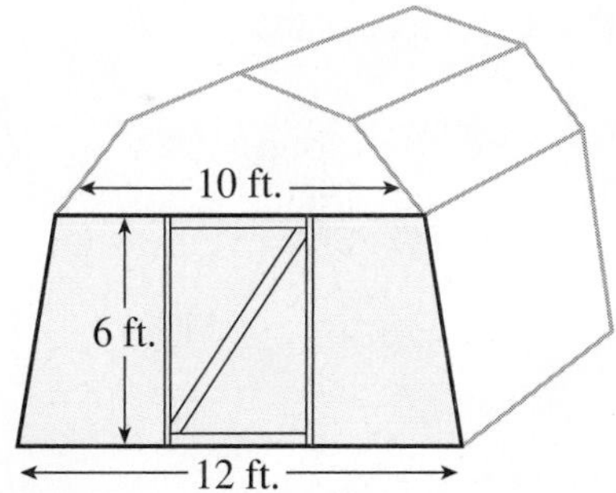

4. You live in a small, one-bedroom apartment. The bedroom is 10 by 12 feet, the living room is 12 by 14 feet, the kitchen is 8 by 6 feet, and the bathroom is 5 by 9 feet. Calculate the total floor space (area) of your apartment.

5. In your living room, there is a large rectangular window with dimensions of 10 feet by 6 feet. You love the sunlight but would like to redesign the window so that it admits the same amount of light but is only 8 feet wide. How tall should your redesigned window be?

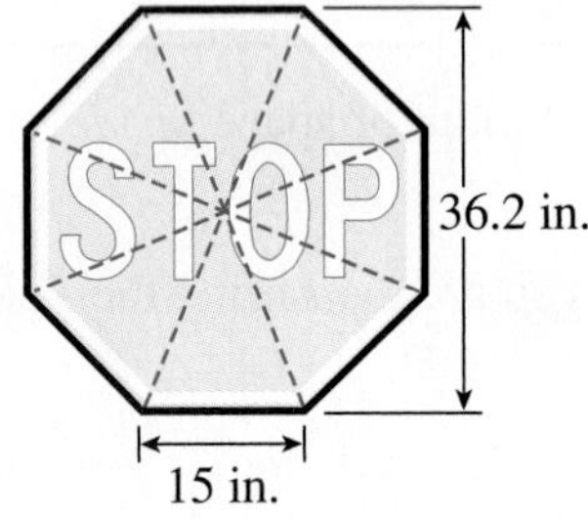

6. A stop sign is in the shape of a regular octagon (an eight-sided polygon with equal sides and angles). A regular octagon can be created using eight triangles of equal area. One triangle that makes up a stop sign has a base of 15 inches and a height of approximately 18.1 inches. Calculate the area of the stop sign.

7. How does the area of the stop sign in Exercise 6 compare with the area of a circle of radius 18.1 inches? Explain.

8. You are buying material to make drapes for your Norman window (in the shape of a rectangle with a semicircular top). So you need to calculate the area of the window. If the rectangular part of the window is 4 feet wide and 5 feet tall, what is the area of the entire window? Explain.

9. The diameter of Earth is 12,742 kilometers; the diameter of the Moon is 3476 kilometers.

   a. If you flew around Earth by following the equator at a height of 10 kilometers, how many trips around the Moon could you take in the same amount of time, at the same height from the Moon, and at the same speed? Explain.

   b. The circle whose circumference is the equator is sometimes called the "great circle" of Earth or of the Moon. Compare the areas of the great circles of Earth and the Moon.

LABORATORY ACTIVITY 6.7

How Big Is That Angle?

OBJECTIVES

1. Measure sizes of angles with a protractor.
2. Classify triangles as equiangular, equilateral, right, isosceles, or scalene.

EQUIPMENT

In this laboratory activity, you will need the following equipment:

1. A 12-inch ruler.
2. A protractor.

You may have wondered why a right angle measures 90 degrees. Why not 100 degrees? As with much of the mathematics that we use today, the measurement of angles has a rich history, going back to ancient times when navigation of the oceans and surveying the land were priorities. The ancient Babylonian culture used a base 60 number system, which at least in part led to the circle being divided into 360 equal sectors. The angle of each sector was simply defined to measure 1 degree and so today we say there are 360 degrees (abbreviated 360°) in a circle. In this activity you will be using degrees to measure angles. The illustration shows a sector having an angle of 20 degrees ($\frac{1}{18}$ of a circle).

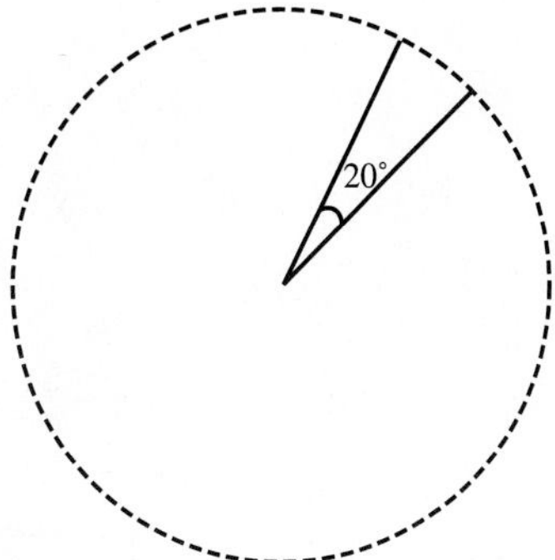

In the following diagram, the circle is divided into four equal sectors by two perpendicular lines. The angles of the four sectors must add up to 360 degrees. Since the angles are equal, dividing 360° by 4 results in each angle measuring 90°. Recall that 90° angles are called *right* angles.

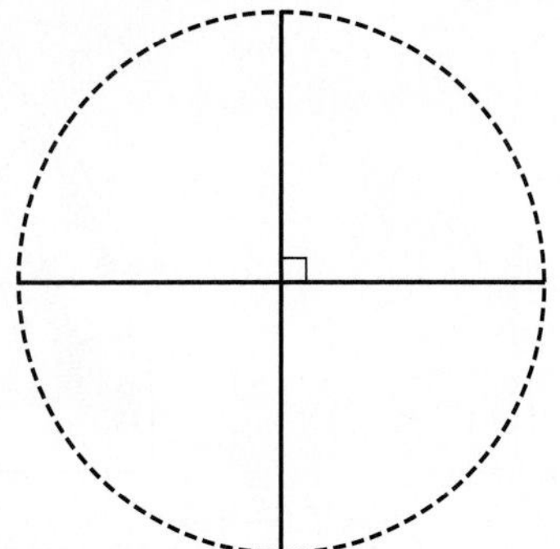

1. How many degrees are in the angle of a sector that is exactly half of a circle? Such an angle is called a *straight angle*.

Measuring Angles

PROCEDURE

Measuring Angles with a Protractor A protractor is a device for measuring the size of angles in degrees. Place the vertex of the angle at the center of the protractor (often a hole in the center of the baseline) and place one side of the angle along the baseline of the protractor. Where the other side of the angle meets the appropriate scale on the semicircle is the measure of the angle in degrees.

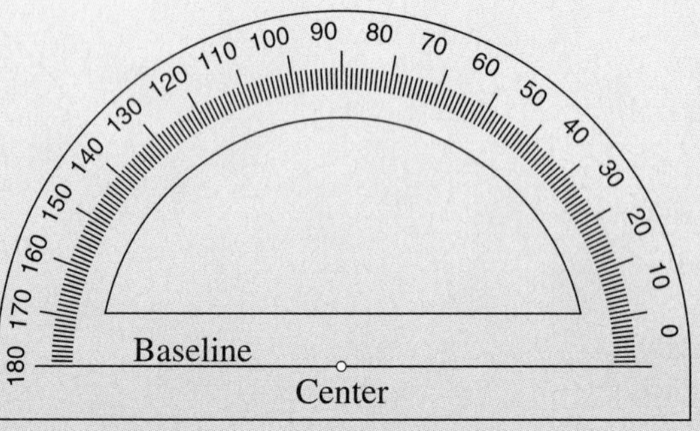

EXAMPLE 1 *The following angle measures* 75°.

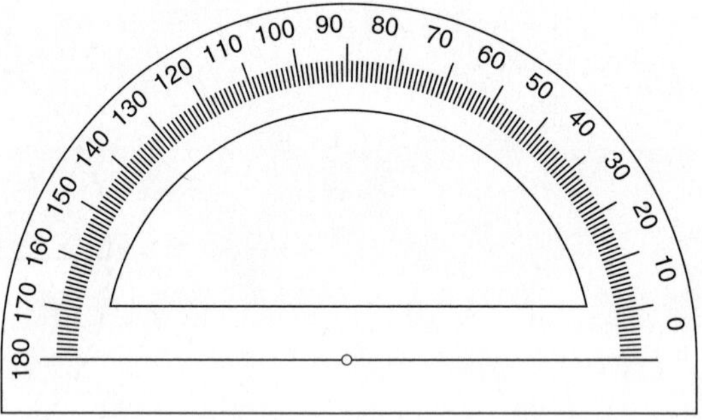

2. Use a protractor to measure the size of each of the following angles.

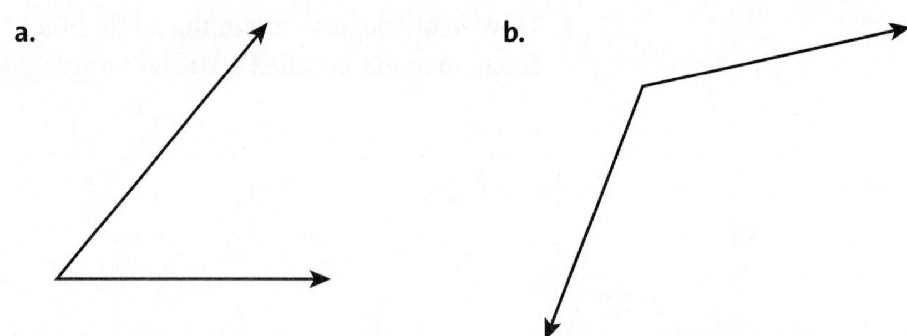

DEFINITIONS

An **acute angle** is any angle that is smaller than a right angle. Its degree measure is less than 90°.

An **obtuse angle** is any angle that is larger than a right angle. Its degree measure is greater than 90°.

3. Cut out a triangle from a piece of paper. Label the three angles X, Y, and Z.

a. Tear off two of the corners. Place the three vertices (plural for vertex) together at a single point so the three angles are next to each other. (See the illustration.)

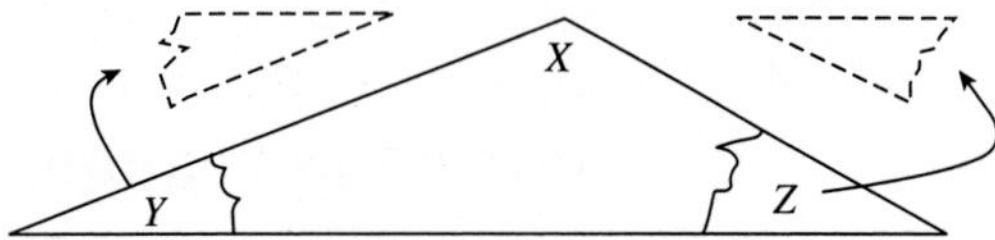

From what you can see, estimate the sum of the three angles of your triangle. Compare your answer to those of your classmates.

b. Use a protractor to measure each angle of your paper triangle, to the nearest degree. Record the sum of these three angles.

It is a well-known theorem in geometry that the sum of the measures of the angles of a triangle must equal 180°.

4. Verify this theorem by carefully measuring the angles of each of the following triangles, recording the results in the table on page 460. You will have to extend the sides of each triangle to use your protractor effectively.

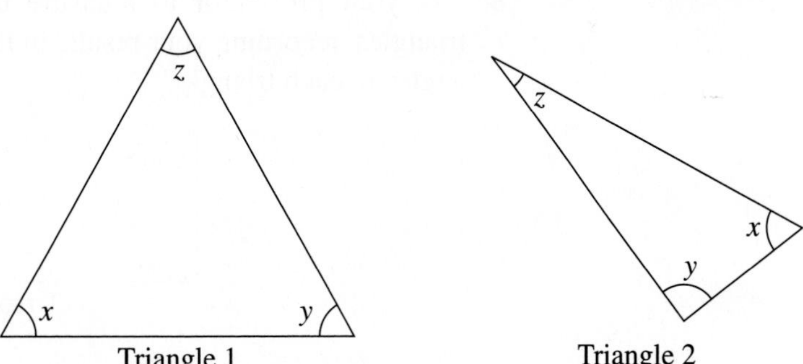

Triangle 1

Triangle 2

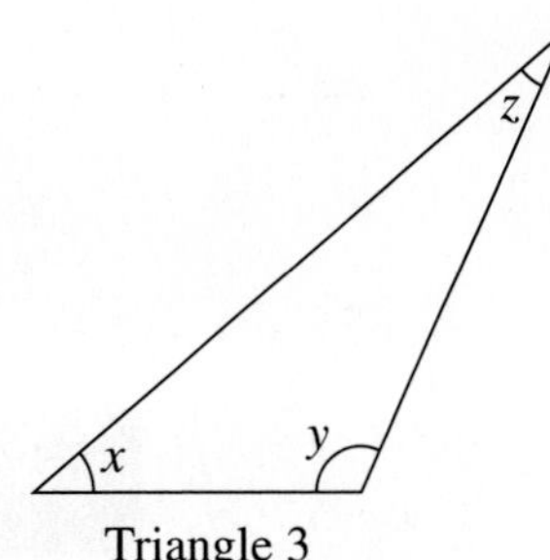

Triangle 3

	ANGLE *X*	ANGLE *Y*	ANGLE *Z*	SUM
TRIANGLE 1				
TRIANGLE 2				
TRIANGLE 3				

Classifying Triangles

Sometimes, it is useful to classify triangles in terms of special properties of their sides or angles. These classifications are summarized in the following table.

TRIANGLE CLASSIFICATIONS	DEFINITION
Equilateral	All three sides have the same length.
Isosceles	Exactly two sides have the same length. The two angles opposite the equal sides will also have the same measure.
Scalene	None of the sides have the same length.
Equiangular	All three angles have the same measure. An equiangular triangle is also an equilateral triangle.
Right	One angle measures 90°.
Acute	All angles measure less than 90°.
Obtuse	One angle measures greater than 90°.

5. Use your protractor to measure the acute angles in the following right triangles, recording your results in the table. What is the sum of the two acute angles in each triangle?

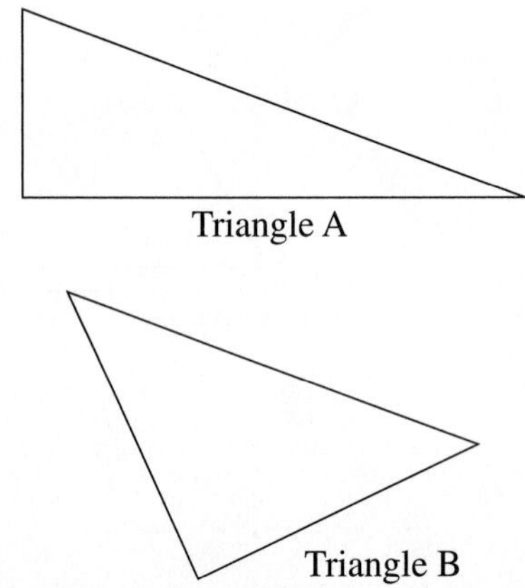

	ONE NONRIGHT ANGLE	OTHER NONRIGHT ANGLE	SUM
TRIANGLE A			
TRIANGLE B			

6. In a right triangle, what must be true about the two angles which are not right angles? Explain your answer.

7. a. In an equiangular triangle, what is the measure of each angle?

b. In an isosceles triangle, if one angle measures 110°, what is the measure of the other two equal angles?

c. Consider the following three triangles:

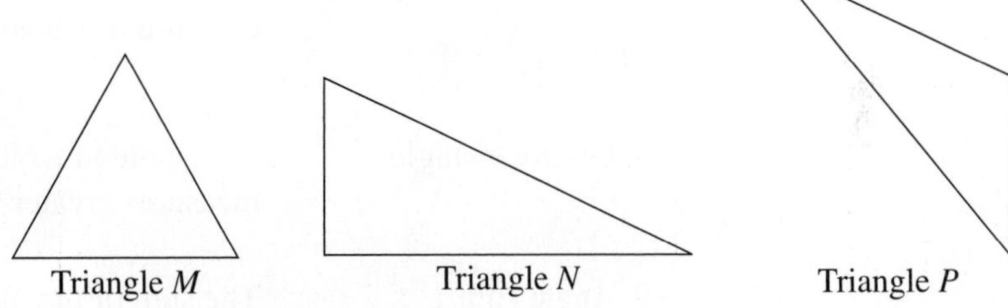

i. Choose the triangles that are scalene, and explain why. Then use a protractor to measure the angles.

ii. Choose the triangle(s) which is(are) obtuse and explain why.

iii. Choose the triangle(s) which is(are) acute and explain why.

SUMMARY
ACTIVITY 6.7

Concept/Skill	Description	Example
1. Using a protractor	Place the vertex of the angle to be measured at the center of the protractor and line up one side with the base line. Read the protractor scale to measure the angle.	See Example 1, page 458

2. Equilateral triangle	A triangle in which all three sides have the same length (also equiangular).
3. Isosceles triangle	A triangle in which two sides have the same length. The two angles opposite the equal sides will also have the same measure.
4. Scalene triangle	A triangle in which all the sides have different lengths.
5. Equiangular triangle	A triangle in which all three angles are the same measure (also equilateral).
6. Right triangle	A triangle in which one angle measures 90°.
7. Acute triangle	A triangle in which all angles measure less than 90°.
8. Obtuse triangle	A triangle in which one angle measures greater than 90°.
9. Angle sum of a triangle	The sum of the measures of the angles of a triangle equals 180°.

EXERCISES
ACTIVITY 6.7

1. Use a protractor to measure the angles in this triangle. Verify that the sum of the angles is 180°.

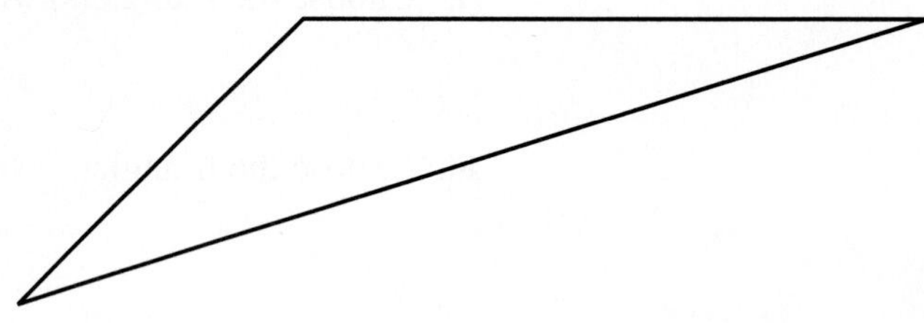

2. In a right triangle, one of the acute angles measures 47°. What is the size of the other acute angle?

3. In an isosceles triangle one of the angles measures 102°. What is the size of the other two angles?

Exercise numbers appearing in color are answered in the Selected Answers appendix.

4. What is true about the sizes of the three angles in a scalene triangle?

5. Give an example of the possible sizes of the three angles in a scalene triangle which is also an acute triangle.

6. What must be the sum of the four angles in a parallelogram? (*Hint:* Consider dividing the parallelogram into two triangles.)

7. a. Extend the idea of Exercise 6 to a pentagon (a five-sided polygon) to determine the sum of all 5 angles by dividing the pentagon into 3 triangles that all meet at one point on the pentagon.

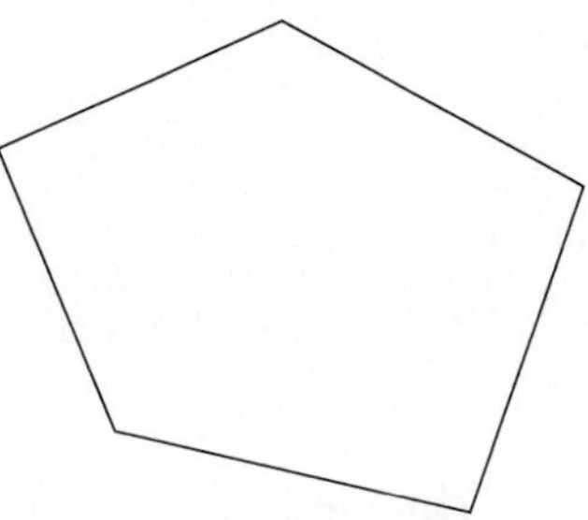

b. Extend the idea of part a to determine the sum of all eight angles in an octagon (an eight-sided polygon).

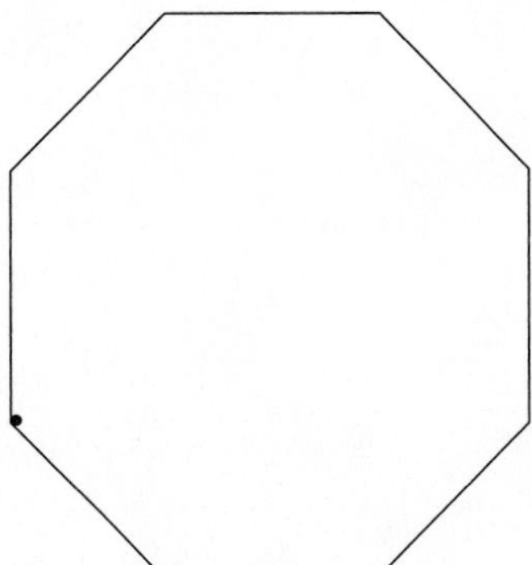

LABORATORY ACTIVITY 6.8

How About Pythagoras?

OBJECTIVES

1. Develop and use the Pythagorean Theorem for right triangles.
2. Calculate the square root of numbers other than perfect squares.
3. Apply the Pythagorean Theorem in context.

EQUIPMENT

In this laboratory activity, you will need the following equipment:

1. A 12-inch ruler.
2. A protractor.
3. A 12-inch piece of string.

In this activity you will experimentally verify a very important formula of geometry. The formula is used by surveyors, architects, and builders to check whether or not two lines are perpendicular, or if a corner truly forms a right angle.

A right triangle is simply a triangle that has a right angle. In other words, one of the angles formed by the triangle measures 90°. The two sides that are perpendicular and form the right angle are called the **legs** of the right triangle. The third side, opposite the right angle, is called the **hypotenuse**.

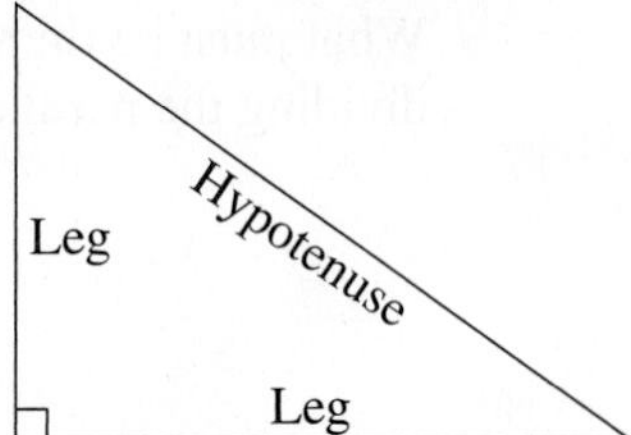

1. Use a protractor to construct three right triangles, one with legs of length 1 inch and 5 inches, a second with legs of length 3 inches and 4 inches, and a third with legs of length 2 inches each.

Triangle 1

Triangle 2

Triangle 3

Pythagorean Theorem

2. For each triangle in Problem 1, complete the following table. The first triangle was done for you as an example. Note that the lengths of the legs of the triangles are represented by a and b. The letter c represents the length of the hypotenuse.

a. Use a ruler to measure the length of each hypotenuse, in inches. Record the lengths in column c.

b. Square each length a, b, and c and record in the table.

	a	b	c	a^2	b^2	c^2
Triangle 1	1	5	5.1	1	25	26.0
Triangle 2	3	4				
Triangle 3	2	2				

3. There does not appear to be a relationship between a, b, and c, but investigate further. What is the relationship between a^2, b^2, and c^2?

The relationship demonstrated in Problem 2 has been known since antiquity. It was known by many cultures, but has been attributed to the Greek mathematician Pythagoras, who lived in the sixth century B.C.

The **Pythagorean theorem** states that, in a right triangle, the sum of the squares of the leg lengths is equal to the square of the hypotenuse length.

Symbolically, the Pythagorean theorem is written as

$$c^2 = a^2 + b^2 \quad \text{or} \quad c = \sqrt{a^2 + b^2},$$

where a and b are leg lengths and c is the hypotenuse length.

Note that this relationship is true for any right triangle. Also, if this relationship is true for a triangle, then the triangle is a right triangle.

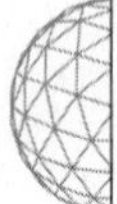

EXAMPLE 1 ***If a right triangle has legs $a = 4$ centimeters and $b = 7$ centimeters, you can calculate the length of the hypotenuse as follows:***

$$c = \sqrt{a^2 + b^2} = \sqrt{(4\text{ cm})^2 + (7\text{ cm})^2}$$
$$= \sqrt{16\text{ cm}^2 + 49\text{ cm}^2} = \sqrt{65\text{ cm}^2} \approx 8.06\text{ cm}$$

DEFINITION

The **square root** of a non-negative number, N, written $\sqrt{N}$, is the non-negative number M whose square is N, that is,

$$\sqrt{N} = M \text{ provided } M^2 = N.$$

Example: $\sqrt{25} = 5$ because $5^2 = 25$.

The number 25 is an example of a *perfect square* since the square root of 25 can be determined exactly.

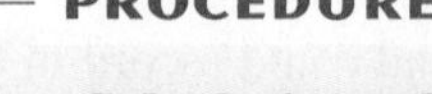

PROCEDURE

Calculating a Square Root If N is not a perfect square, then estimating the square root of N by hand involves "guess and check." You can determine a square root more efficiently by using a calculator with a square root key.

EXAMPLE 2 ***Estimate $\sqrt{10}$ by guess and check.***

Determining the square root of 10 involves finding a number whose square is 10. Note that 3 is too small, since $3^2 = 9$ and 4 is too large, since $4^2 = 16$. So, $3 < \sqrt{10} < 4$, and $\sqrt{10}$ is closer to 3. Try $3.1^2 = 9.61$ and $3.2^2 = 10.24$. Therefore, $3.1 < \sqrt{10} < 3.2$. Continuing in this manner, one can estimate that $\sqrt{10} \approx 3.16$, easily confirmed on your calculator.

4. Estimate the square root of 53, written $\sqrt{53}$. Then use your calculator to check the answer.

5. A right triangle has legs measuring 5 centimeters and 16 centimeters. Use the Pythagorean theorem to calculate the length of the hypotenuse.

6. A popular triangle with builders and carpenters has dimensions 3 units by 4 units by 5 units.

a. Use the Pythagorean theorem to show that this triangle is a right triangle.

b. Builders use this triangle by taking a 12-unit-long rope and marking it in lengths of 3 units, 4 units, and 5 units (see graphic).

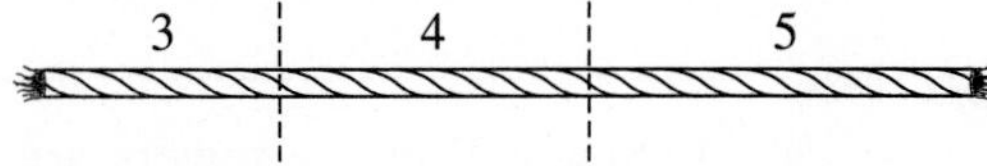

Then, by fitting this rope to a corner, they can quickly tell if the corner is a true right angle. Another special right triangle has legs of length 5 units and 12 units. Determine the perimeter of this right triangle. Explain how a carpenter can use this triangle to check for a right angle.

7. Suppose you wish to measure the distance across your pond but don't wish to get your feet wet! By being clever, and knowing the Pythagorean theorem, you can estimate the distance by taking two measurements on dry land, as long as your two distances lie along perpendicular lines. (See the illustration.)

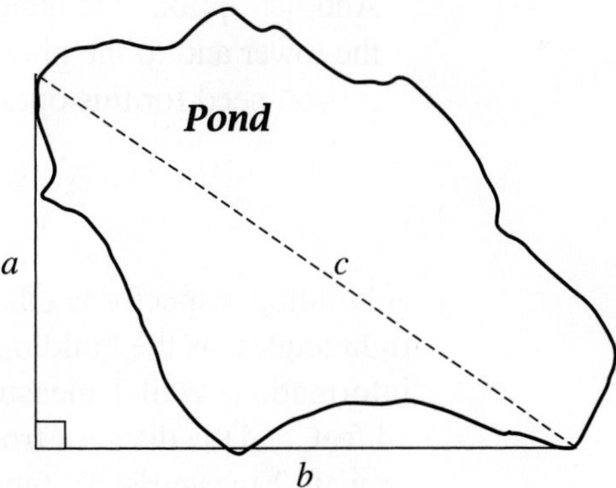

If your measurements for legs a and b are 260 feet and 310 feet, respectively, what is the distance, c, across the pond?

SUMMARY
ACTIVITY 6.8

Concept/Skill	Description	Example
1. Pythagorean theorem	In a right triangle, $c^2 = a^2 + b^2$.	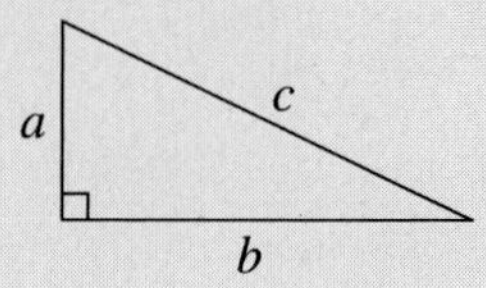
2. Determining a square root	If $N \geq 0$, then $\sqrt{N} = M$ provided $M^2 = N$; use a calculator to obtain an approximation.	

EXERCISES
ACTIVITY 6.8

1. New cell phone towers are being constructed on a daily basis throughout the country. Typically, they consist of a tall, thin tower supported by several guy wires. Assume the ground is level in the following. Round answers to the nearest foot.

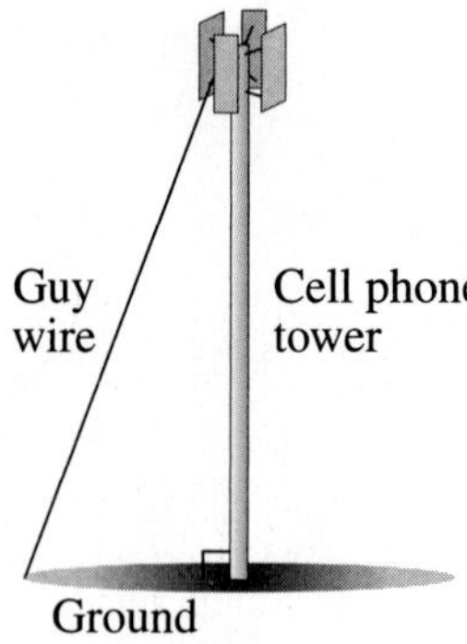

   a. The guy wire is attached on the ground at a distance of 100 feet from the base of the tower. The guy wire is also attached to the phone tower 300 feet above the ground. What is the length of the guy wire?

   b. You move the base of the guy wire so that it is attached 120 feet from the base of the tower. Now how much wire do you need for the one guy wire?

   c. Another option is to attach the base of the guy wire 100 feet from the base of the tower and to the phone tower 350 feet above the ground. How much wire do you need for this option?

2. A building inspector needs to determine if two walls in a new house are built at right angles, as the building code requires. He measures and finds the following information. Wall 1 measures 12 feet, wall 2 measures 14 feet, and the distance from the end of wall 1 to the end of wall 2 measures 18 feet. Do the walls meet at right angles? Explain.

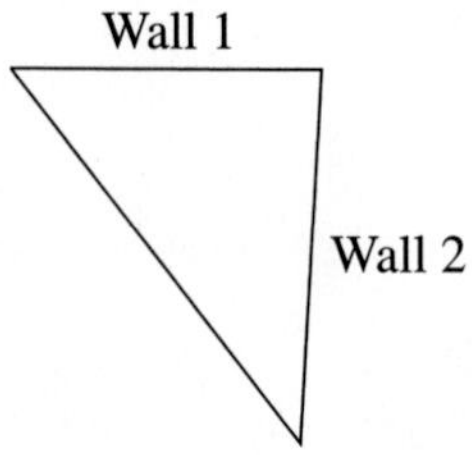

Exercise numbers appearing in color are answered in the Selected Answers appendix.

3. Trusses used to support the roofs of many structures can be thought of as two right triangles placed side by side.

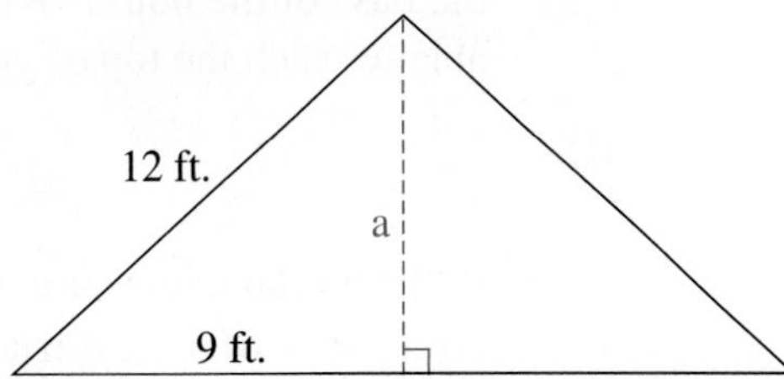

a. If the hypotenuse in one of the right triangles of a truss measures 12 feet and the horizontal leg in the same right triangle measures 9 feet, how high is the vertical leg of the truss?

b. To make a steeper roof, you may increase the vertical leg to 10 feet. Keeping the 9-foot horizontal leg, how long will the hypotenuse of the truss be now?

4. You can consider the truss in Exercise 3 as a single triangle. In this case, it is a good example of an isosceles triangle.

a. If the top angle of the truss is 120°, then what are the measures of the other two angles?

b. If you wish to have a steeper roof, with the base angles of the isosceles truss each measuring 42°, what is the measure of the top angle?

5. For the following right triangles, use the Pythagorean theorem to compute the length of the third side of the triangle:

a.

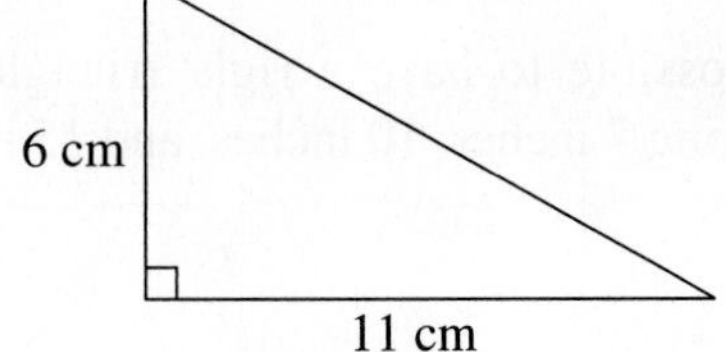

b.

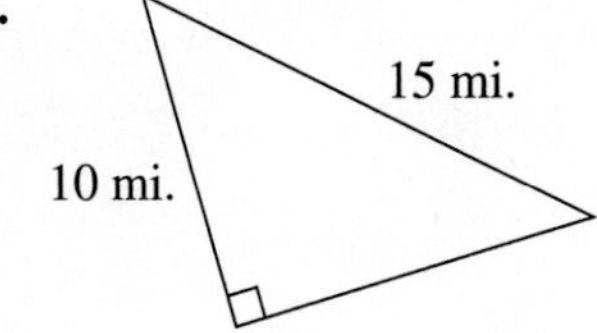

6. You are buying a ladder for your 30-foot-tall house. For safety, you would always like to ensure that the base of the ladder be placed at least 8 feet from the base of the house. What is the shortest ladder you can buy in order to be able to reach the top of your house?

7. Pythagorean triples are three positive integers that could be the lengths of three sides of a right triangle. For example, 3, 4, 5 is a Pythagorean triple since $5^2 = 3^2 + 4^2$.

a. Is 5, 12, 13 a Pythagorean triple? Why or why not?

b. Is 5, 10, 15 a Pythagorean triple? Why or why not?

c. Is 1, 1, 2 a Pythagorean triple? Why or why not?

d. Name another Pythagorean triple. Explain.

8. You own a summer home on the east side of Lake George in New York's Adirondack Mountains. To drive to your favorite restaurant on the west side of the lake, you must go directly south for 7 miles and then directly west for 3 miles. If you could go directly to the restaurant in your boat, how far is the boat trip?

9. You are decorating a large evergreen tree in your yard for the holidays. The tree stands 25 feet tall and 15 feet wide. You want to hang strings of lights from top to bottom draped on the outside of the tree. How long should the strings of lights be? Explain.

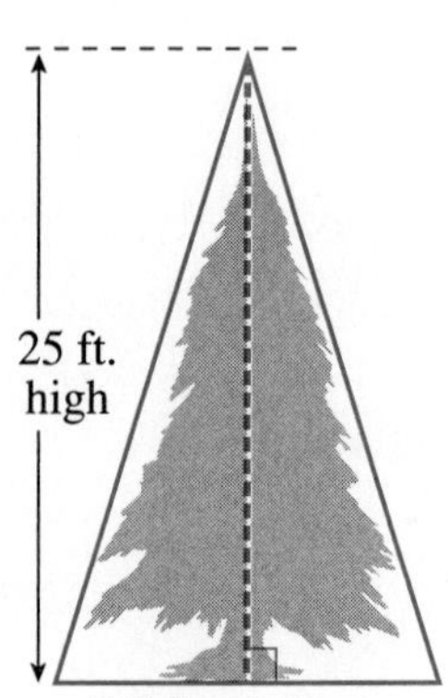

10. Is it possible to have a right triangle with sides measuring 7 inches, 10 inches, and 15 inches?

ACTIVITY 6.9

Moving Up with Math

OBJECTIVES

1. Recognize the geometric properties of similar triangles.
2. Use similar triangles in indirect measurement.

Imagine you live on the twelfth floor of an apartment building. Looking out your window, you wonder how high you are above the ground. Measuring directly would be difficult. But applying a basic formula from geometry will allow you to determine the height indirectly. By walking outside, to a point 100 feet away from your building, a right triangle is formed with the wall of your building and your line of sight to the twelth-floor window (see illustration).

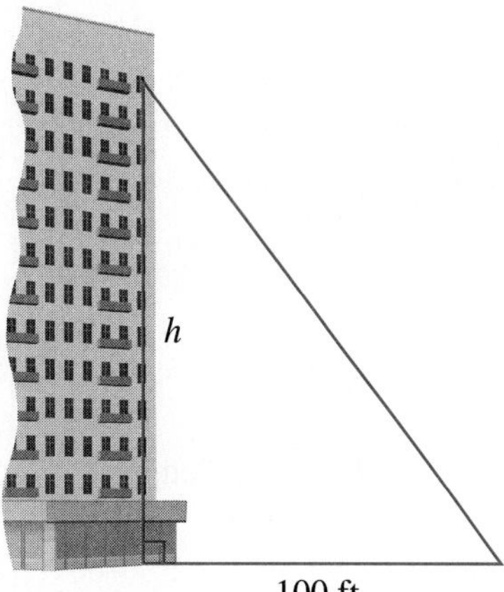

The Pythagorean theorem does not help since you only know the length of one side of the right triangle. The problem can be solved, however, by applying a result involving similar triangles.

DEFINITION

Two triangles are **similar** when they are exactly the same shape, meaning their angles are the same size. The pairs of matching angles are called **corresponding angles**. The two sides that are opposite corresponding angles are called **corresponding sides**.

The two triangles shown here are similar because the pairs of angles A and X, B and Y, and C and Z are the same size. Note that the pairs of corresponding sides are a and x, b and y, and c and z.

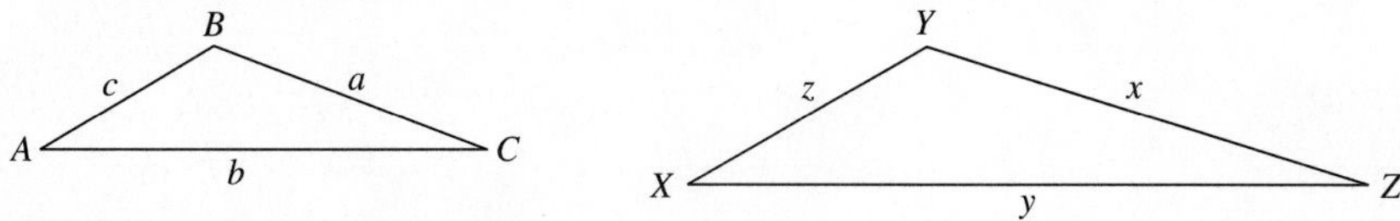

An important theorem that relates similar triangles involves the ratios of the corresponding sides of the triangles.

THEOREM

The ratio of the lengths of corresponding sides in similar triangles are equal. Using the similar triangles illustrated,

$$\frac{a}{x} = \frac{b}{y} = \frac{c}{z}.$$

The corresponding sides of similar triangles are said to be in **proportion**.

You can use this theorem and a simple drinking straw to determine the height of your twelfth-floor apartment.

After walking 100 feet away from your building, you use a drinking straw to get the line of sight to your twelfth-floor window. Your straw is 10 inches long. With your window in view through the straw, a horizontal distance of 6 inches can be measured from your eye to the end of the straw. The height of the end of the straw above your eye will measure 8 inches. The illustration shows this small right triangle.

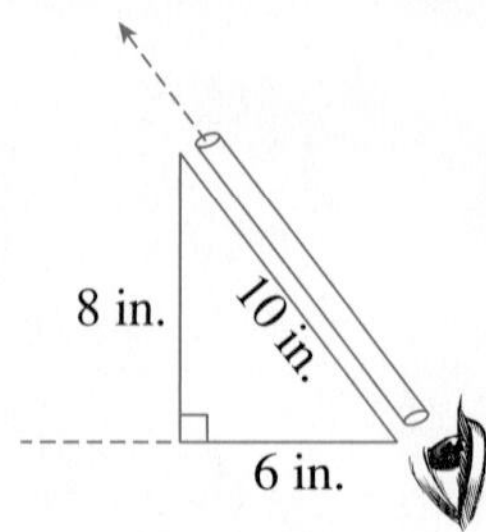

1. Do the lengths of the sides of this small triangle satisfy the Pythagorean theorem?

2. Explain why the small drinking straw triangle is similar to the large triangle involving your apartment building.

3. Use the information from Problems 1 and 2 to determine the height of your window.

   a. Set up the proportion involving the corresponding legs of the similar triangles (not the hypotenuse). Let h represent the unknown height of your window, in the large triangle.

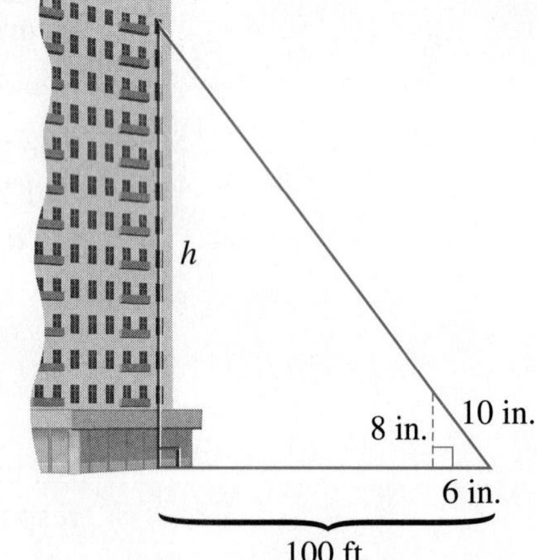

   b. Solve the proportion for h.

   c. Is your answer reasonable if each floor of your apartment building is approximately 10 to 12 feet high? Explain.

4. You have a friend who lives on the twentieth floor. Go through the same procedure as in Problem 3 to determine the height of your friend's window from the ground. The dimensions of the small "drinking straw triangle" are shown in the diagram.

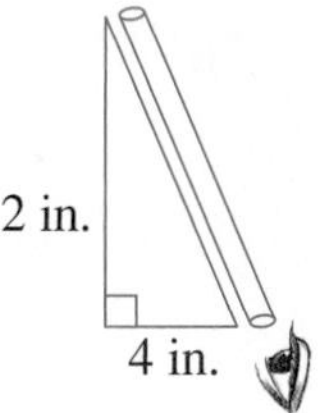

SUMMARY
ACTIVITY 6.9

1. Two triangles are **similar** provided they have equal corresponding angles.
2. For all similar triangles, the ratios of the lengths of corresponding sides are equal. In other words, corresponding sides of similar triangles are in **proportion**.

EXERCISES
ACTIVITY 6.9

1. You and your friends decide to set up an experiment to estimate the height of the math building on your campus. You wait until dark and then use a flashlight to project shadows onto the building. One of your friends sets the flashlight on the ground 50 feet from the building and shines it at you. You walk away from the flashlight and towards the building. Another friend tells you to stop when the height of your shadow reaches the top of the building. You have walked 12 feet. Because you know you are 6 feet tall, it is possible to determine the height of the building.

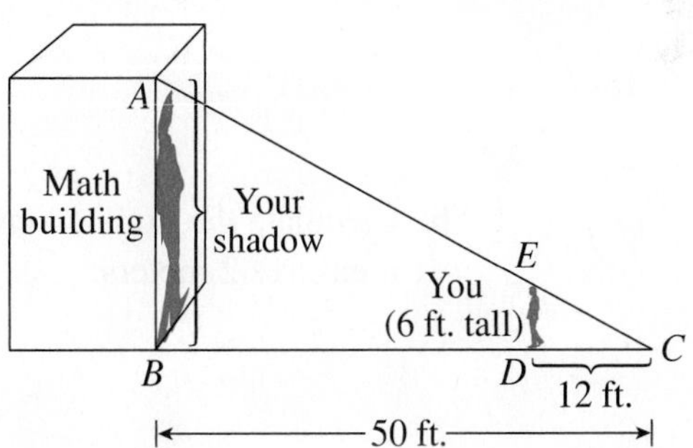

a. Where do you see two similar right triangles in the diagram? Explain.

b. Use the properties of similar right triangles to estimate the height of the math building.

2. You measure an isosceles triangle and label the lengths, as shown. In a similar isosceles triangle the longest side is 14 feet. Sketch the similar triangle and compute the lengths of its two equal sides.

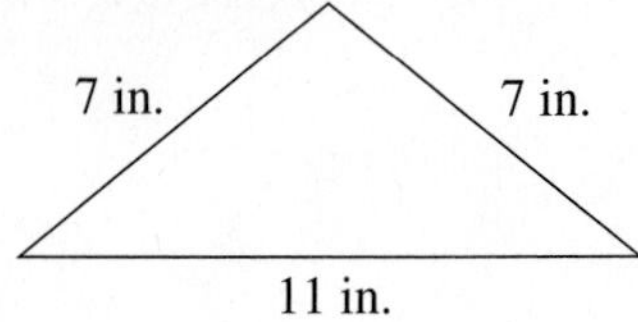

3. On campus, there is a very tall tree. Your math professor challenges the class to devise a way to use similar triangles to indirectly measure the height of the tree. Be specific in explaining your best strategy.

4. A triangle has sides that measure 3 feet, 5 feet, and 6 feet.

   a. Calculate the dimensions of a similar triangle whose longest side measures 15 inches.

   b. Calculate the dimensions of a similar triangle whose shortest side measures 2 meters.

d. Give an example of another right triangle that is similar to the given triangle.

5. A triangle has two angles measuring 42° and 73°.

a. Make a sketch of the triangle.

b. Calculate the third angle of the triangle.

c. Give the angle measurements of a triangle similar to this triangle.

6. You intend to put in a 4-foot-wide concrete walkway along two sides of your house, as shown.

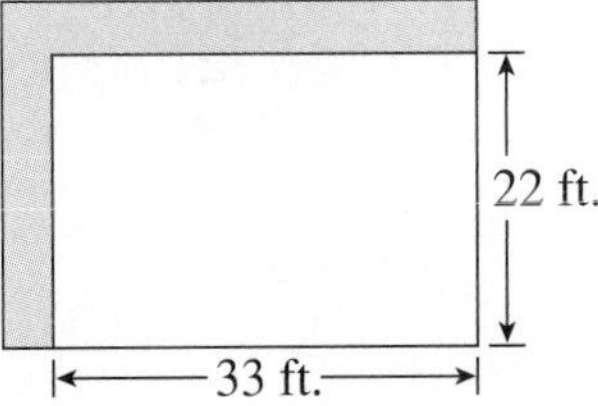

a. Determine the area covered by the walkway only.

b. If you decide to place a narrow flower bed along the outside of the walkway, how many feet of flowers should you plan for?

7. Triangle *A* has sides measuring 3 feet, 5 feet, and 7 feet; triangle *B* has sides measuring 4 feet, 4 feet, and 5 feet; triangle *C* has sides measuring 5 inches, 12 inches, and 13 inches.

a. Which of the three triangles is a right triangle? Explain.

b. A fourth triangle, *D*, is similar to triangle *B* but not identical to it. What are some possibilities for the lengths of the sides of triangle *D*? Explain.

c. If 2 feet is added to the lengths of each side of triangle *A*, will the resulting triangle be similar to triangle *A*? Explain.

d. If the length of each side of triangle *C* is tripled, will the resulting triangle be similar to triangle *C*?

e. Which of the three triangles, *A*, *B*, or *C*, are scalene? Explain.

8. Answer the following questions about right triangles.

a. Is it possible for a right triangle to be isosceles? Scalene? Equilateral? Explain.

b. Is it possible for a right triangle to be acute? Obtuse? Equiangular? Explain.

9. If you measure the circumference of a circle to be 20 inches, estimate the length of its radius.

10. You want to know how much space is available between a basketball and the rim of the basket. One way to find out is to measure the circumference of each and then use the circumference formula to determine the corresponding diameters. The distance you want to determine is the difference between the diameter of the rim and the diameter of the ball. Try it!

CLUSTER 2 The Geometry of Three-Dimensional Space Figures

ACTIVITY 6.10

Painting Your Way through Summer

OBJECTIVES

1. Recognize geometric properties of three-dimensional figures.
2. Write formulas for and calculate surface areas of boxes (rectangular prisms), cans (right circular cylinders), and balls (spheres).

You decide to paint houses for summer employment. To determine how much to charge, you do some experimenting to discover that you can paint approximately 100 square feet per hour. To pay college expenses, you need to make at least $900 per week during the summer to cover your profit and the cost of paint and brushes. You figure that it is reasonable to paint for approximately 40 hours per week.

1. Use the appropriate information from above to determine your hourly fee.

Your first job is to paint the exteriors of a three-building farm complex (a small barn, a storage shed, and a silo) with the following dimensions.

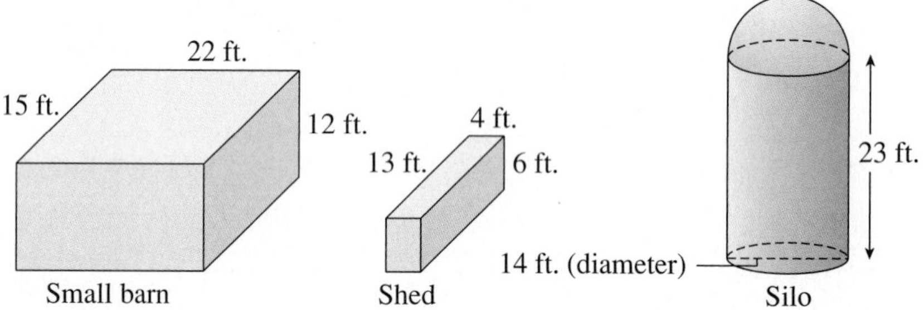

In order to determine your fee to the farmer, you need to estimate how many square feet of surface must be painted. Since there are few windows in the buildings, they may be ignored.

2. The small barn and shed are in the shape of rectangular boxes. Each surface is the shape of a rectangle. Such a three-dimensional figure is formally called a **rectangular prism**.

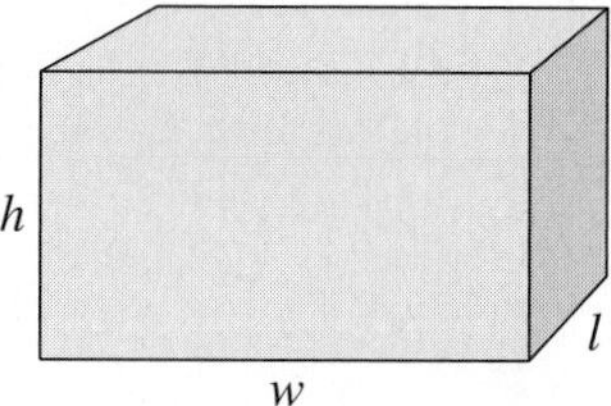

a. To determine the total surface area, S, of a rectangular prism, you need to add up the areas of the six rectangular surfaces. Using l for length, w for width, and h for height, express each area with a formula.

Area of front = ________. Area of back = ________.

Area of one side = ________. Area of other side = ________.

Area of top = ________. Area of bottom = ________.

b. Add the six areas from part a to obtain the formula for the surface area of a rectangular prism.

3. a. Write a formula for the surface area, S, of a rectangular prism if the bottom area is *not* included.

b. Assume that you will not paint the floors of the small barn and storage shed (but you *will* paint the special flat roofs). Compute the total surface area for the exteriors of these two buildings.

4. The silo consists of a can-shaped base, with half a ball shape for the roof. The three-dimensional can shape is formally called a **right circular cylinder**—circular because the base is a circle, right because the smooth rounded surface is perpendicular to the base.

a. To determine the total surface area of a right circular cylinder, you need to add up the areas of the circular top and bottom and the area of curved side. Using r for the radius and h for the height, express each area with a formula.

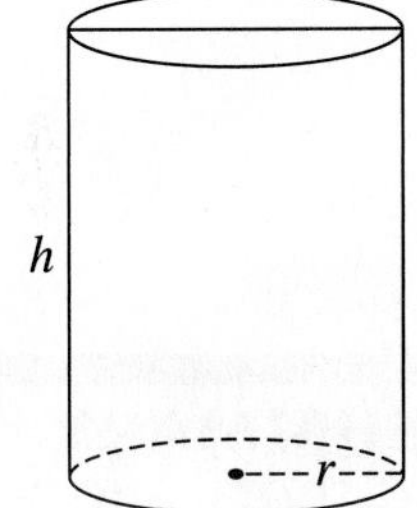

Area of circular top = ___

Area of circular bottom = ___

Area of side = _____

(*Hint*: To determine the area of the side, think about cutting off both circular ends and cutting the side of the can perpendicular to the bottom. Then, uncoil the side of the can into a big rectangle with height h and width the circumference of the circle.)

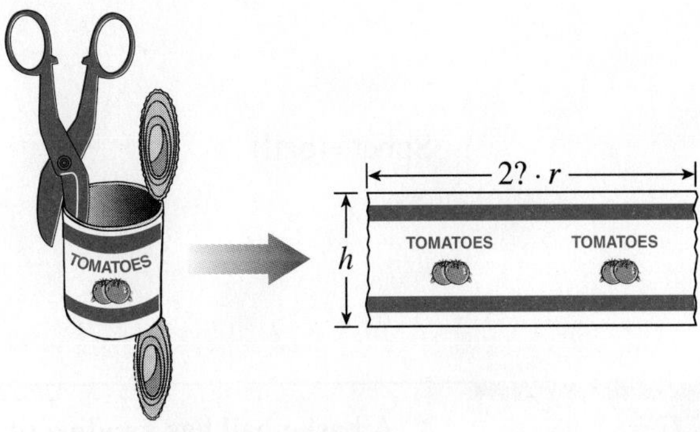

b. Add the three areas from part a to obtain the formula for the surface area of a right circular cylinder.

The roof of the silo is half of a sphere with radius r. The surface area, S, for a ball, or sphere, with radius r is a little more difficult to derive. You may study the formula in future courses. It is provided here.

$$S = 4\pi r^2$$

The formula states that the surface area of a sphere is equal to the sum of the areas of four circles of radius r.

5. Carefully use the surface area formulas for a sphere and a cylinder to compute the total exterior surface area for the silo. Again, assume you will not paint the floor, but will paint the semispherical roof.

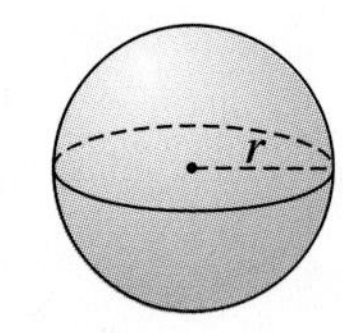

6. What is the total surface area for all three buildings that you must paint?

7. How long will it take you to paint all three buildings?

8. What will you charge for the entire job?

SUMMARY
ACTIVITY 6.10

Figure	Labeled Diagram	Surface Area Formula
Rectangular prism (box)		$S = 2wl + 2hl + 2wh$
Right circular cylinder (can)	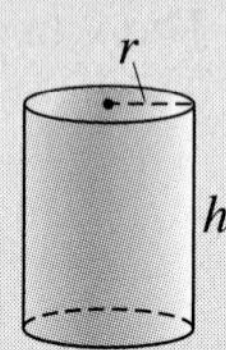	$S = 2\pi r^2 + 2\pi rh$
Sphere (ball)	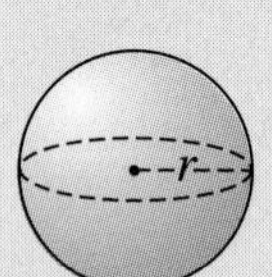	$S = 4\pi r^2$

EXERCISES
ACTIVITY 6.10

1. A basketball has a radius of approximately 4.75 inches.

   a. Compute the basketball's surface area.

   b. Why would someone want to know this surface area?

2. Hot air balloons require large amounts of both hot air and fabric material. A hot air balloon is spherical with a diameter of 25 feet,

   a. Does finding the surface area help you determine the amount of the hot air or the fabric material? Explain.

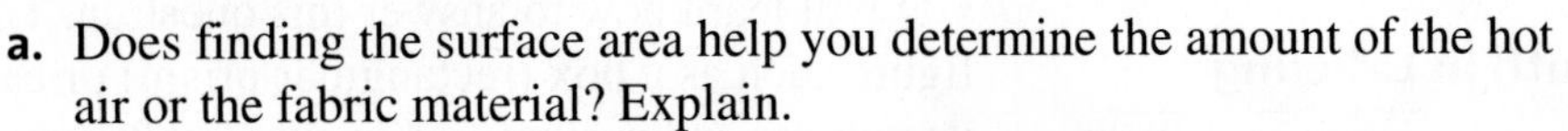

   b. Compute that surface area. What are the units?

3. Compute the surface areas of each of the following figures.

   a.

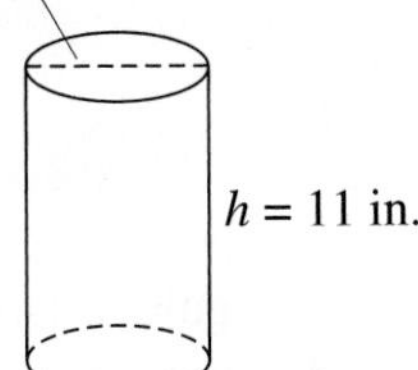

   b.

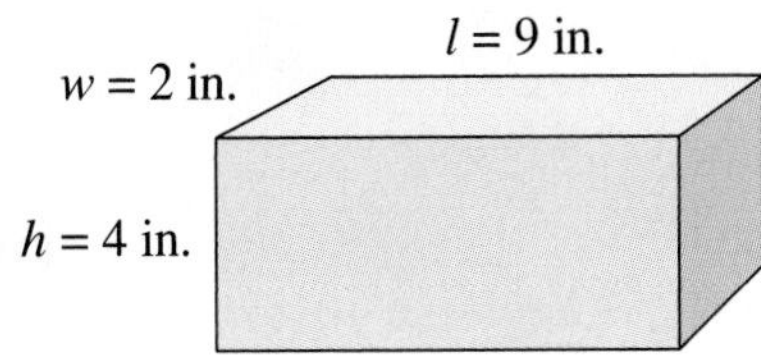

   c.

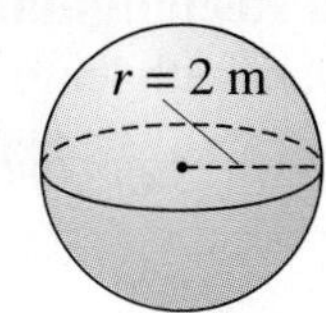

4. You need to wrap a rectangular box with dimensions 2 feet by 3.5 feet by 4.2 feet. What is the least amount of wrapping paper you must buy in order to complete the job?

5. A can of soup is 3 inches in diameter and 5 inches in height. How much paper is needed to make a label for the soup can?

6. Calculate the height of a right circular cylinder with a surface area of 300 square inches and a radius of 5 inches.

ACTIVITY 6.11

Truth in Labeling

OBJECTIVES

1. Write formulas for and calculate volumes of boxes and cans.
2. Recognize geometric properties of three-dimensional figures.

When buying a half-gallon of ice cream or a 12-ounce can of soda, have you ever wondered if the containers actually hold the amounts advertised? In this activity, you will learn how to answer this question. The space inside a three-dimensional figure such as a box (rectangular prism) or can (right circular cylinder) is called its **volume** and is measured in cubic units (or unit cubes).

1. A half-gallon of ice cream has estimated dimensions as shown in the figure. In order to determine the number of unit cubes (1 inch by 1 inch by 1 inch) in this box, explain why it is reasonable to multiply the area of the top or bottom by the height.

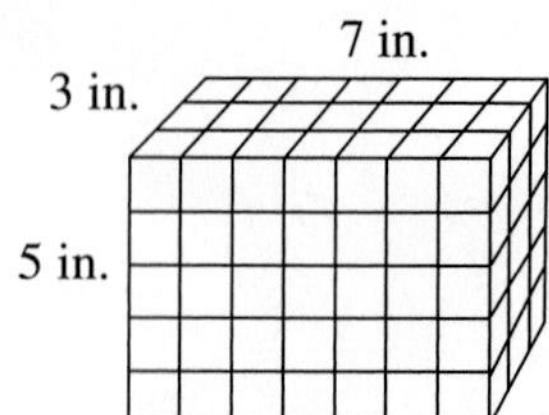

DEFINITION

The **volume** of a box (rectangular prism) is the measure, in cubic units, of the space enclosed by the sides, top, and bottom of a three-dimensional figure.

PROCEDURE

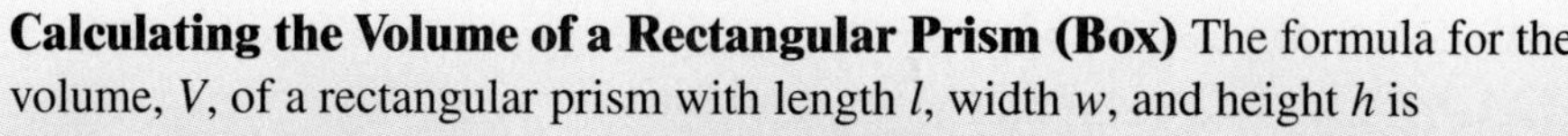

Calculating the Volume of a Rectangular Prism (Box) The formula for the volume, V, of a rectangular prism with length l, width w, and height h is

$$V = lwh.$$

2. Use the formula to calculate the volume of the half-gallon of ice cream.

3. One cubic inch contains 0.554 fluid ounces of ice cream. Estimate the amount of fluid ounces of ice cream in the carton.

4. There are 64 fluid ounces in a half-gallon. How close is your estimate to the half-gallon?

5. In a way similar to the volume of a box, it is reasonable to define the volume of a right circular cylinder (can) as the product of the area of its circular bottom and its height. If a can has height h and radius r, write a formula for its volume. Explain.

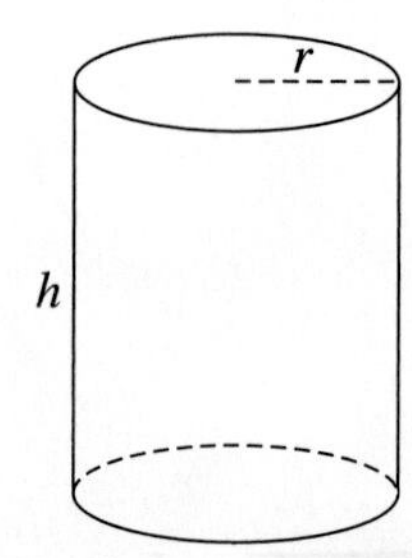

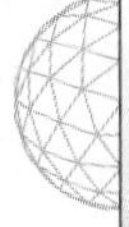

PROCEDURE

Calculating the Volume of a Right Circular Cylinder (Can) The formula for the volume, V, of a right circular cylinder with height h and radius r is

$$V = \pi r^2 h.$$

6. A can of soda is estimated to be $2\frac{1}{2}$ inches in diameter and $4\frac{3}{4}$ inches high. Use the volume formula to calculate the volume of a can of soda.

7. One cubic inch contains 0.554 fluid ounces of soda. Estimate the amount of soda in a can. How close to 12 ounces is your estimate?

8. Assume that you need a long slim container that can hold up to 100 cu. in.

a. You want the container to have a base 2 in. long and 4 in. wide. How high would the container be?

b. You need another box to hold 100 cu. in. but now with base 3 in. by 4 in. How high would it stand?

c. Design a cube (length = width = height) to store 100 cubic inches.

i. Let x represent the length = width = height. Write an equation for a volume of 100 cubic inches.

ii. To solve the equation in part i, you need to "uncube," that is, take the cube root of, both sides. You are looking for a number whose cube is 100. You note that 4 is too small, since $4^3 = 64$. Is 5 too large or too small?

iii. Estimate the cube root of 100.

iv. Most calculators have a cube root key or menu to calculate cube roots. The cube root of 100 is denoted as $\sqrt[3]{100}$. Use your calculator to compute the cube root of 100.

d. Design a cylindrical can that holds 100 cu. in. How tall will it be if the diameter of the base is 4 in.?

e. Design a "short" can (like a can of tuna fish) with a volume of 100 cu. in. that is 1 in. high.

SUMMARY
ACTIVITY 6.11

Three-Dimensional Figure	Labeled Sketch	Volume Formula
Rectangular prism		$V = lwh$
Right circular cylinder		$V = \pi r^2 h$

EXERCISES
ACTIVITY 6.11

1. Compute the volumes in each of the following figures.

   a. Diameter = 7 in.

   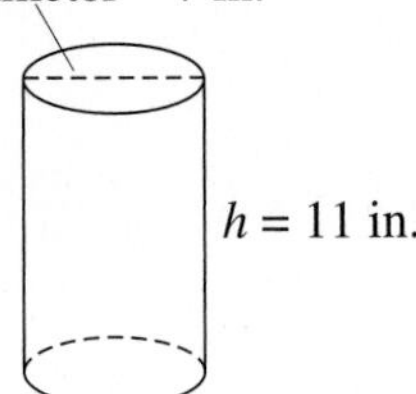

   b.

   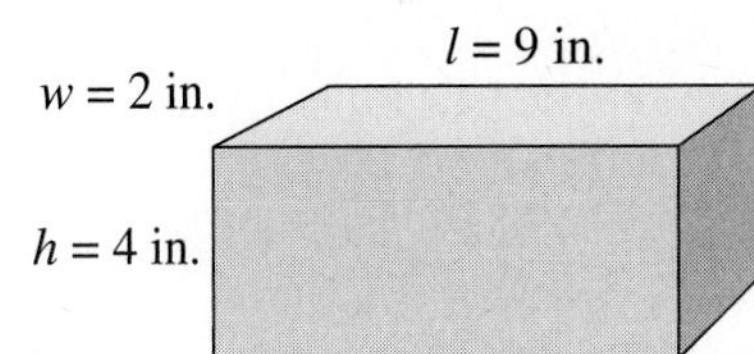

2. You want to buy a pickup truck and are interested in one with a large carrying capacity. One model features a rectangular prism-shaped cargo space, measuring 6 feet by 10 feet by 2 feet; another has a space with dimensions 5 feet by 11 feet by 3 feet. Which truck provides you with the most carrying capacity (that is, the most volume)? Explain.

3. A can of soup is 3 inches in diameter and 5 inches high. How much soup can fit into the can?

4. If the volume of a cube is given as 42 cubic feet, estimate its dimensions.

5. If the volume of a right circular cylinder is given as 50 cubic centimeters and if its radius measures 2 centimeters, calculate its height.

ACTIVITY 6.12

Analyzing an Ice Cream Cone

OBJECTIVE

1. Write formulas for and calculate volumes of balls (spheres) and cones.

A popular summertime treat is the ice cream cone. The geometry of this treat is interesting, since it involves a three-dimensional cone topped with spheres (of ice cream). Let's begin your analysis of the ice cream cone with the geometry of spheres.

Spheres

1. Visually, it seems clear that a golf ball is smaller than a tennis ball, which is smaller than a baseball, which is smaller than a basketball. One way to compare these balls is by their radii, r. Another way is to measure the volume, V, of each ball. Determining a formula for V in terms of r is an involved process, but it can be estimated visually.

   a. Think about one-half of a sphere that just fits inside a right circular cylinder.

   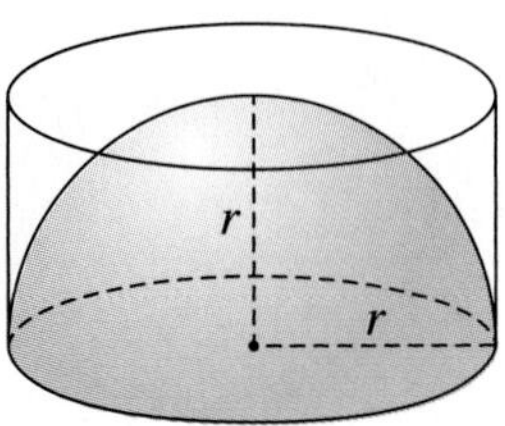

   The area of the circle passing through the center of the sphere at the base of the cylinder is given by $A = \pi r^2$. The height of the cylinder is r. What is a formula for the volume of the cylinder?

   b. Doubling the formula from part a produces a formula for the cylinder that *just* encloses the entire sphere. What is that formula?

   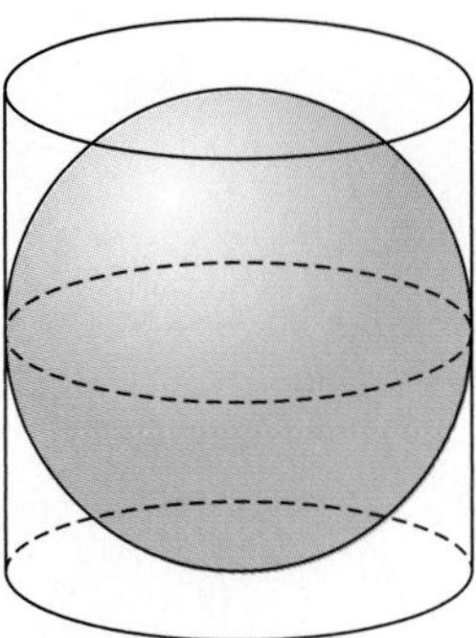

   c. It is visually clear that the formula from part b is an overestimate of the volume of the sphere. However, what is not so clear is that the exact volume for the sphere is given by the formula $V = \frac{4}{3}\pi r^3$. Explain why this formula for the volume of a sphere is possible.

PROCEDURE

Calculating the Volume of a Sphere The formula for the volume, V, of a sphere with radius r is

$$V = \frac{4}{3}\pi r^3.$$

2. The accompanying table contains the radii of golf balls, tennis balls, baseballs, and basketballs.

Round and Round

TYPE OF BALL	RADIUS, r	VOLUME, V
Golf	0.84 inch	
Tennis	1.28 inches	
Baseball	1.45 inches	
Basketball	4.75 inches	

a. Use the formula for volume of a sphere to compute the volumes of each type of ball and write your answers in the table. Round your answers to the nearest hundredth.

b. A basketball is how many times larger than a baseball?

3. You notice that the spherical scoop of ice cream on your cone has diameter 2 inches. How much ice cream does the scoop contain?

4. You read an ad for a beach ball that claims it has a volume of 50 cubic inches. Estimate its radius. Explain the procedure you used to determine your estimate.

5. A sphere has a circumference of 20 centimeters (as measured by the circumference of the "great circle" around the center of the sphere). Estimate its volume.

Cones

6. Determining the formula for the volume of a cone of radius r and height h is also involved, but can be estimated visually.

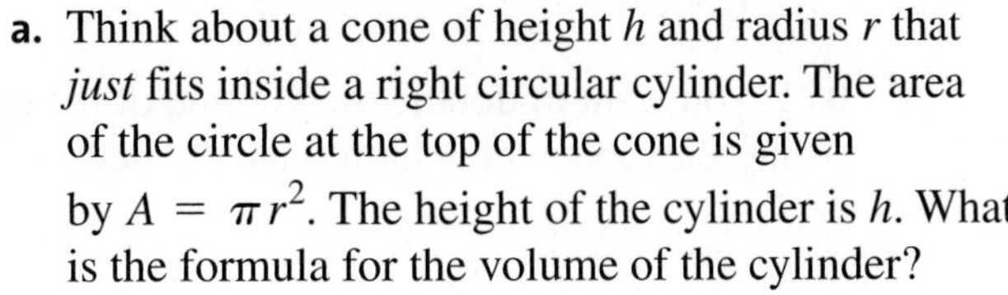

a. Think about a cone of height h and radius r that *just* fits inside a right circular cylinder. The area of the circle at the top of the cone is given by $A = \pi r^2$. The height of the cylinder is h. What is the formula for the volume of the cylinder?

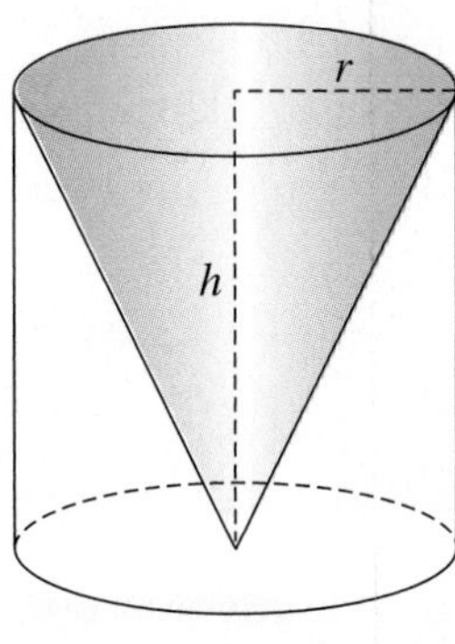

b. It is clear that $V = \pi r^2 h$ is an overestimate for the volume of the cone in the cylinder. Guess what fractional part of $\pi r^2 h$ is the volume of the cone and explain why your guess is reasonable.

c. The correct formula for the volume of a cone is $V = \frac{1}{3}\pi r^2 h$. Compare this with your answer to part b.

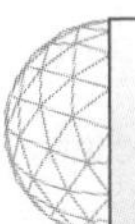

PROCEDURE

Calculating the Volume of a Cone The formula for the volume, V, of a cone with radius r and height h is

$$V = \frac{1}{3}\pi r^2 h.$$

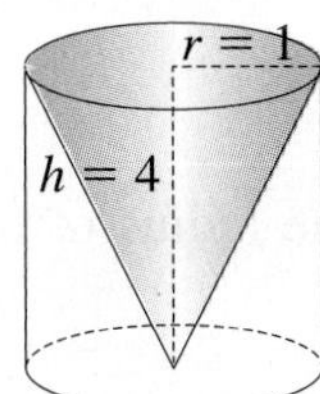

7. The cone for your ice cream has a radius of 1 inch and is 4 inches high.

a. Determine its volume.

b. If you filled the cone with soft ice cream so the ice cream is level with the top of the cone, would you have more ice cream than the spherical scoop in Problem 3?

8. Cones are also used to mark highway construction. To stabilize these cones, which measure 1 foot in diameter and 2.5 feet high, one option is to fill the cones with various materials. Compute the volume of the cone.

9. If you want to double the volume of the cone from Problem 7 without changing the radius, how high must the new cone be?

SUMMARY ACTIVITY 6.12

Three-Dimensional Figure	Labeled Sketch	Volume Formula
Sphere	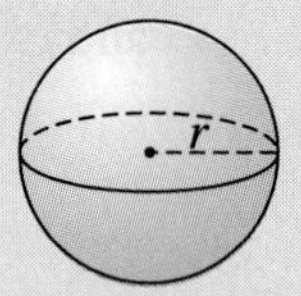	$V = \frac{4}{3}\pi r^3$
Cone	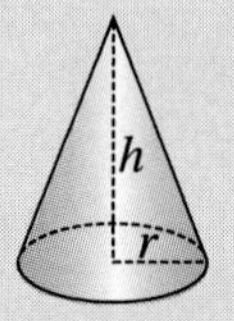	$V = \frac{1}{3}\pi r^2 h$

EXERCISES ACTIVITY 6.12

1. Hot air balloons require large amounts of hot air and fabric material. Suppose a hot air balloon is spherical with a diameter of 25 feet.

   a. Does finding the volume help you determine the amount of hot air or fabric material?

   b. Compute the volume.

2. Earth has a radius of approximately 6378 kilometers; the radius of Mars is approximately 3397 kilometers.

   a. Compute the volumes of Earth and Mars.

Exercise numbers appearing in color are answered in the Selected Answers appendix.

b. The volume of Earth is how many times larger than the volume of Mars?

3. Write a formula for the volume of a sphere in terms of its diameter.

4. Compute the volumes for four spheres of radii 1 foot, 2 feet, 3 feet, and 4 feet and compare the answers. How many times larger is the largest sphere than the smallest sphere?

5. A beachball has a radius of approximately 5.25 inches. How many cubic inches of air would you need to inflate the ball?

6. Write the formula for the volume of a cone in terms of its diameter and height.

7. You decide to start a soft ice cream cone business. You pack the cones with soft ice cream, and then top off each one with a "cone-shaped" amount of ice cream swiveled on top, about one-third the height of the cone.

If a regular-size cone has dimensions diameter 1.5 inches and height 3 inches, determine the volume of ice cream inside and on top of the cone.

8. Determine the volumes of the following cones and spheres. Round to the nearest tenth.

a.

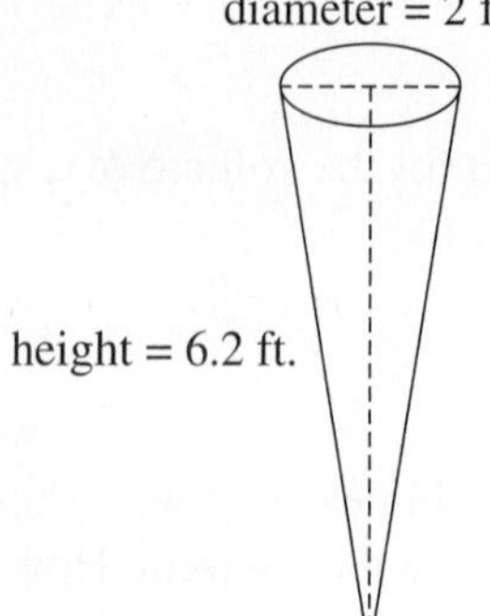

b.

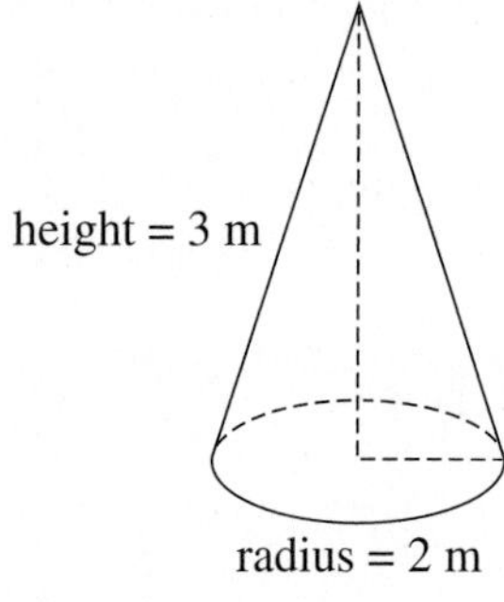

c.

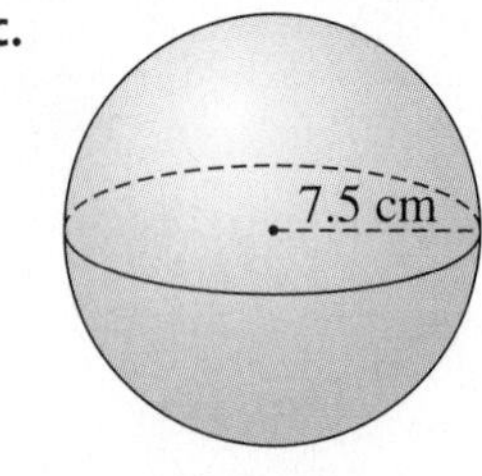

d.

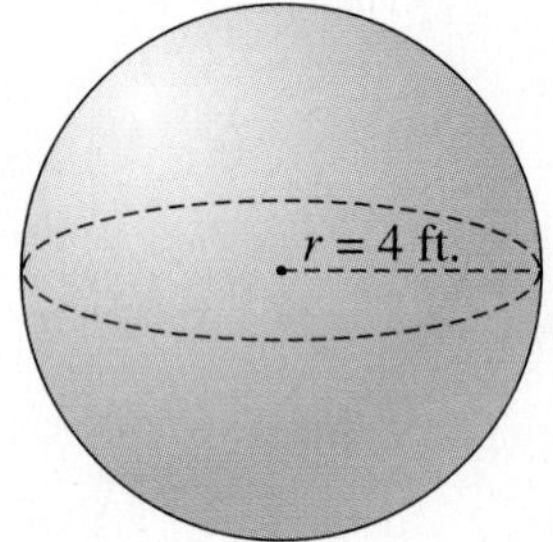

9. What other real-world examples of cones can you think of? Estimate their volumes.

LABORATORY ACTIVITY 6.13

Summertime

OBJECTIVES

1. Use geometry formulas to solve problems.
2. Use scale drawings in problem-solving.

EQUIPMENT

In this laboratory activity, you will need the following equipment:

1. A 12-inch ruler.

With summer approaching, your friend decides to invest in a new in-ground swimming pool with the dimensions given in the following scale drawing.

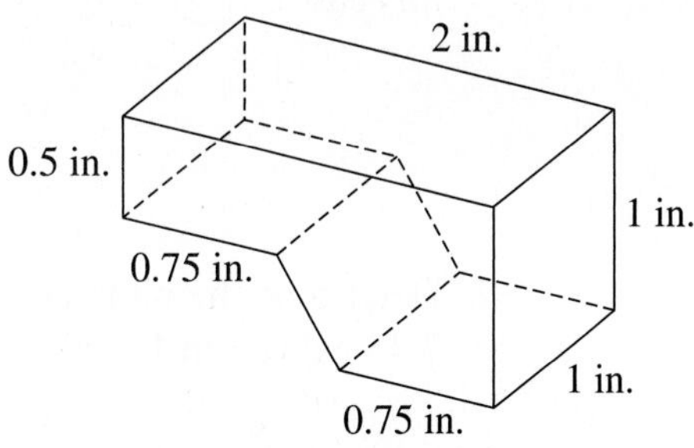

Scale: 0.5 in. = 6 ft.

1. Use the scale drawing measurements to determine the actual dimensions (in feet) of your friend's new pool. Lable the scale drawing with these actual dimensions.

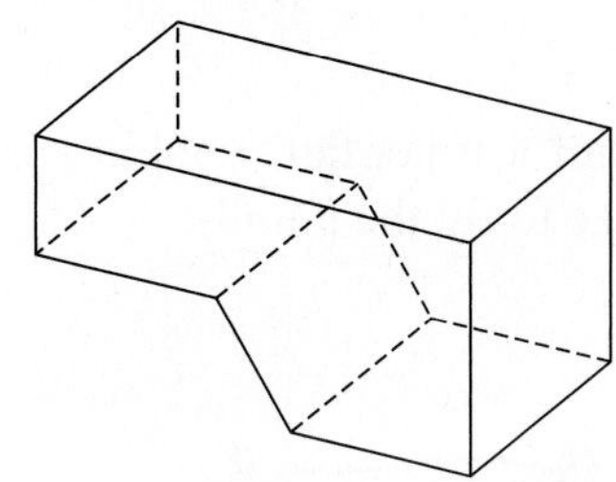

2. Calculate the perimeter of the top view of the pool.

3. Calculate the surface area of the top of the pool.

4. Calculate the area of the longer side of the pool.

5. Calculate the total volume of the pool. (*Hint:* To obtain the volume multiply the area of the longer side by the distance between the two longer side views.)

6. If the pool is to be filled with water to within 6 inches of the pool's top edge, calculate the amount of water needed to fill the pool. What are the units in this case?

7. Determine the number of gallons of water needed to fill the pool. (There are 7.48 gallons in 1 cubic foot of water.)

8. A garden hose can fill the pool at the rate of 4.5 gallons per minute. How many minutes will it take to fill the pool? How many hours? How many days?

9. A pool-filling company charges $0.03 per gallon for water delivered in a big tanker truck. How much will this company charge to fill the pool?

10. Because of the soil conditions in your area, your friend needs to know the weight of the water in the pool. Water weighs 62.4 pounds per cubic foot. Calculate the weight of the water in the pool to the nearest pound. Compare it with the weight of an average car.

11. The pool's liner is guaranteed for 5 years. A new pool liner sells for $6.19 per square yard. How much will a new liner for your friend's pool cost? (Note that the liner covers *all* inside surfaces—the sides and the bottom.)

12. To prevent accidents, state law requires that all pools be enclosed by a fence. How many feet of fencing are needed if a fence will be placed around the pool 4 feet from each side?

13. You decide to buy some beach balls of 10 cubic feet volume for playing in the pool. What is the diameter of these balls?

14. To decorate the area around your pool with flowers, you purchase conical urns of radius 1.5 feet and height 3 feet. How much soil must you buy to fill each urn?

CLUSTER 2 What Have I Learned?

1. What are the differences and similarities between the *surface area* and *volume* of a figure? Explain.

2. If the surface area of a figure is measured in square feet, then what are the units of the volume of that figure? Explain.

3. If the volume of a spherical figure is measured in cubic centimeters, then what are the units of the surface area of that sphere? Explain.

4. If different figures have the same surface areas, must their volumes be the same? Explain using an example.

5. If different figures have the same volume, must their surface areas be the same? Explain, using an example.

6. If you double the height of a can and keep all other dimensions the same, does the volume double? Explain, using an example.

7. If you double the diameter of a can and keep all other dimensions the same, does the volume double? Explain, using an example.

8. If you triple the radius of a sphere, what effect does that have on the volume? Explain, using an example.

9. If you take two identical cubes and insert the largest possible sphere into one of the cubes and the largest possible cone into the other, which figure has the larger volume? Explain, using an example.

CLUSTER 2 How Can I Practice?

1. Compute the volumes and surface areas for each of the following figures. Round your answers to the nearest hundredth.

a. 6 in.

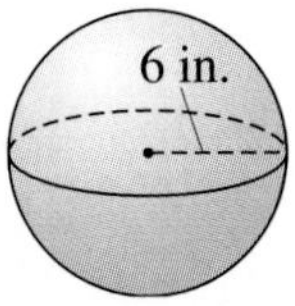

b. diameter = 4.5 ft.

height = 6.5 ft.

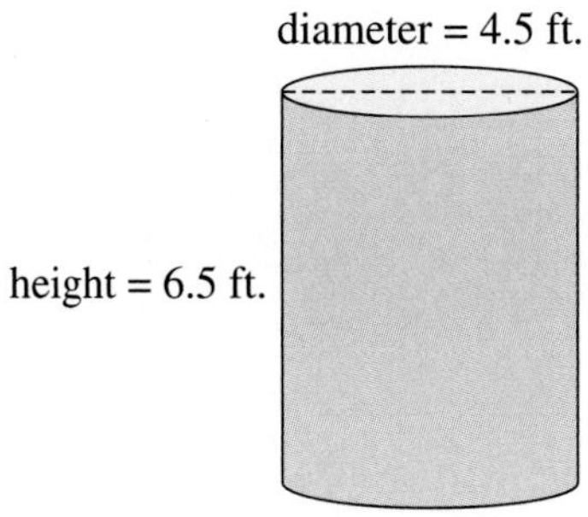

c. $w = 4.5$ ft. $l = 14.3$ ft.

$h = 5.3$ ft.

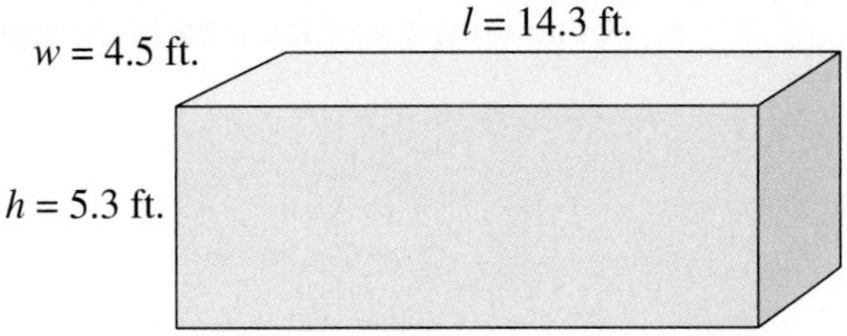

2. Compute the volume of a cone with the dimensions given.

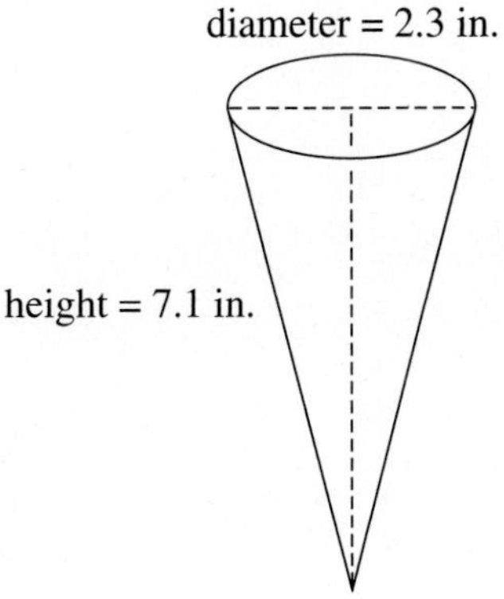

diameter = 2.3 in.

height = 7.1 in.

3. Sketch each of the following figures with the desired property or properties. In each case provide the dimensions for the radius, length, or height of the figure as appropriate. Round to nearest hundredth.

   a. A sphere with a volume of 20 cubic feet.

   b. A sphere with a surface area of 53 square centimeters.

   c. A can with a surface area of 20 square feet and radius of 1 foot.

   d. A can with a volume of 20 cubic feet and radius of 1 foot.

e. A cone with a volume of 20 cubic feet and radius of 1 foot.

f. A box with a surface area of 40 square feet with width 2 feet and length 4 feet.

g. A cubic box with a volume of 27 cubic feet.

CHAPTER 6 SUMMARY

The bracketed numbers following each concept indicate the activity in which the concept is discussed.

CONCEPT / SKILL	DESCRIPTION	EXAMPLE
Perimeter formulas [6.1]	Perimeter measures the length around the edge of the figure.	
Square [6.1]	$P = 4s$	$P = 4 \cdot 1 = 4$ ft.

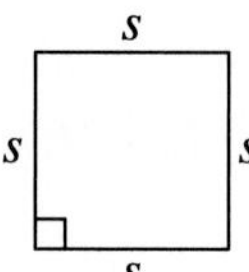

$s = 1$ ft.
$s = 1$ ft.
$s = 1$ ft.
$s = 1$ ft.

CONCEPT / SKILL	DESCRIPTION	EXAMPLE
Rectangle [6.1]	$P = 2l + 2w$	$P = 2 \cdot 2 + 2 \cdot 3 = 10$ in.

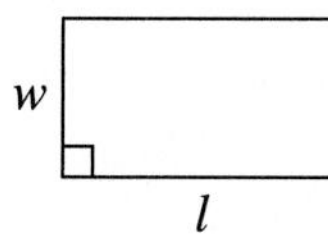

$w = 2$ in.
$l = 3$ in.

CONCEPT / SKILL	DESCRIPTION	EXAMPLE
Triangle [6.1]	$P = a + b + c$	$P = 3 + 4 + 6 = 13$ m

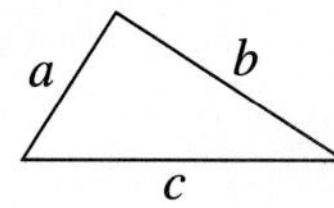

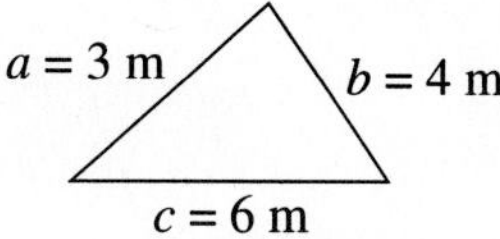

CONCEPT / SKILL	DESCRIPTION	EXAMPLE
Parallelogram [6.1]	$P = 2a + 2b$	$P = 2 \cdot 7 + 2 \cdot 9 = 32$ in.

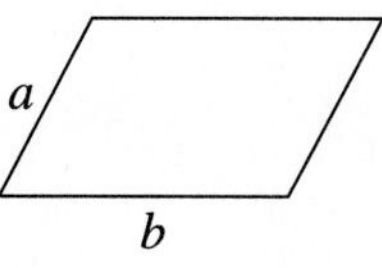

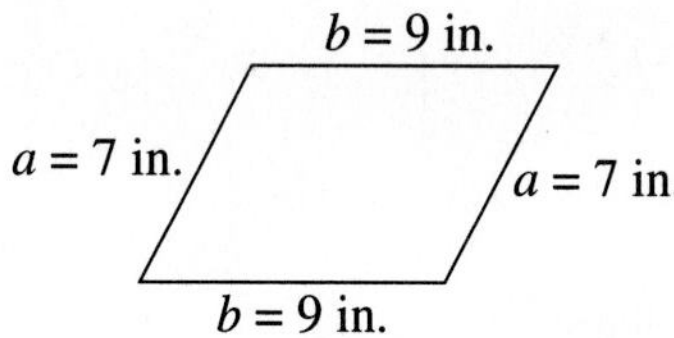

CONCEPT / SKILL	DESCRIPTION	EXAMPLE
Trapezoid [6.1]	$P = a + b + c + d$	$P = 2 + 3 + 1.5 + 1.5 = 8$ ft.

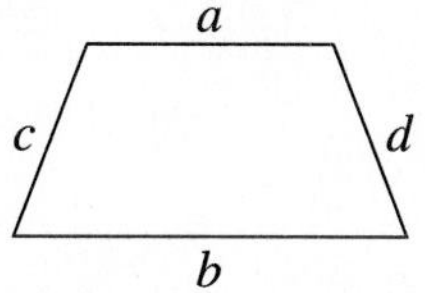

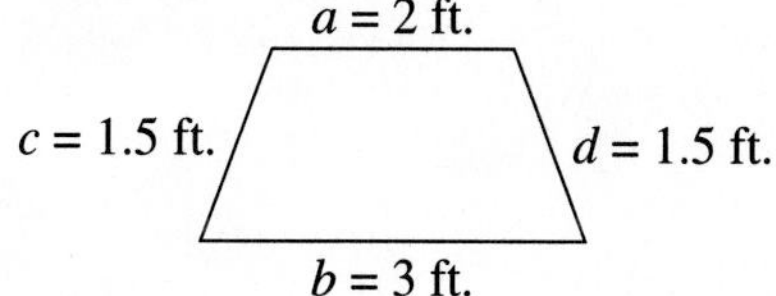

CONCEPT / SKILL	DESCRIPTION	EXAMPLE
Polygon [6.1]	$P =$ sum of the lengths of all the sides	$P = 1 + 4 + 5 + 7 + 2 = 19$ cm

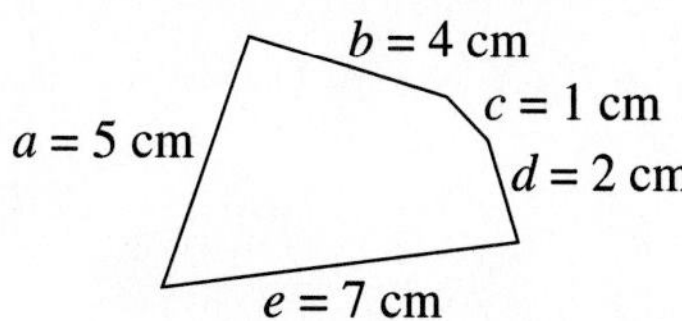

CONCEPT / SKILL	DESCRIPTION	EXAMPLE
Circle [6.2]	$C = 2\pi r$ 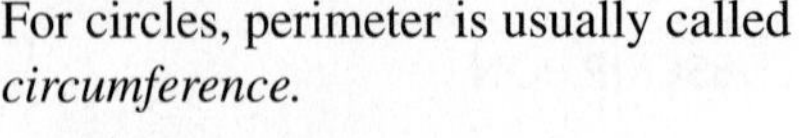 For circles, perimeter is usually called *circumference.*	$C = 2\pi \cdot 3 = 6\pi \approx 18.85$ ft. 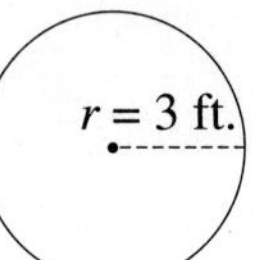
	$C = \pi d$ 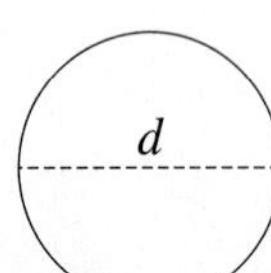	$C = \pi \cdot 5 \approx 15.71$ m 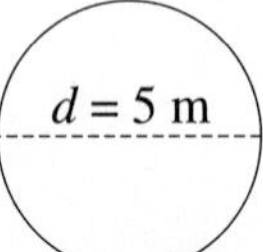
Area formulas [6.4]	Area is the measure of the space inside the figure.	
Square [6.4]	$A = s \cdot s$ or $A = s^2$ 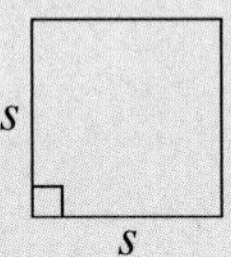	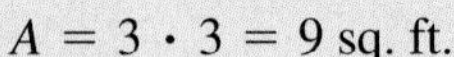 $A = 3 \cdot 3 = 9$ sq. ft. 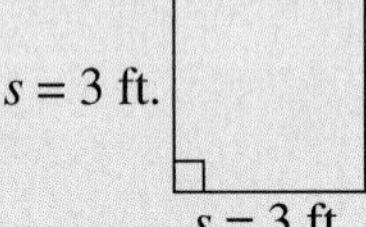
Rectangle [6.4]	$A = lw$ 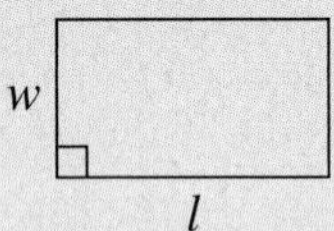	$A = 2 \cdot 3 = 6$ sq. in. 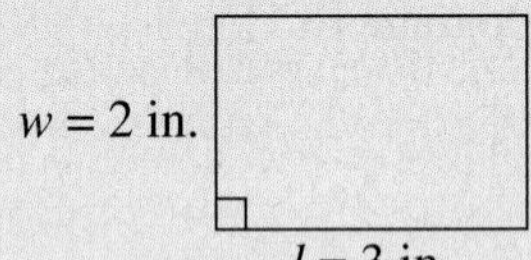
Triangle [6.4]	$A = \frac{1}{2}bh$ 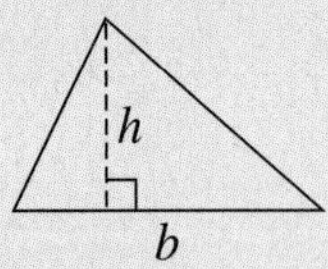	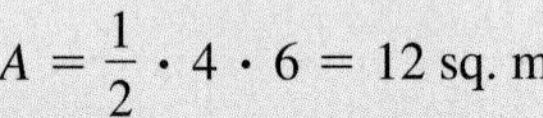 $A = \frac{1}{2} \cdot 4 \cdot 6 = 12$ sq. m 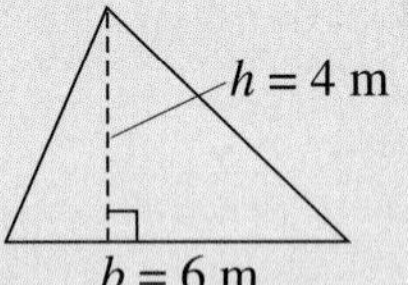
Parallelogram [6.4]	$A = bh$ 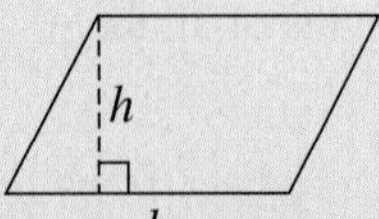	$A = 6 \cdot 3 = 18$ sq. in. 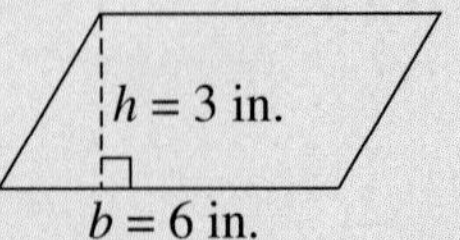

CONCEPT / SKILL	DESCRIPTION	EXAMPLE
Trapezoid [6.4]	$A = \frac{1}{2}h(b + B)$ 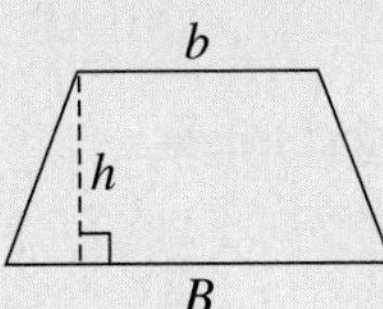	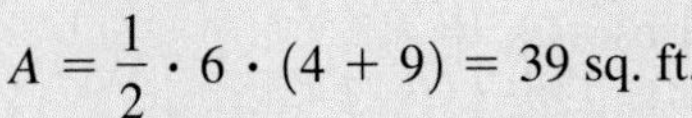 $A = \frac{1}{2} \cdot 6 \cdot (4 + 9) = 39$ sq. ft. 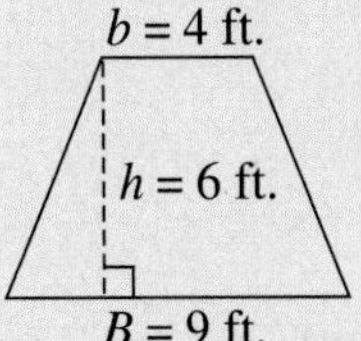
Polygon [6.4]	$A =$ sum of the areas of all the parts of the figure.	$A = 2 \cdot \frac{1}{2} \cdot 4 \cdot (6 + 12) = 72$ sq. m 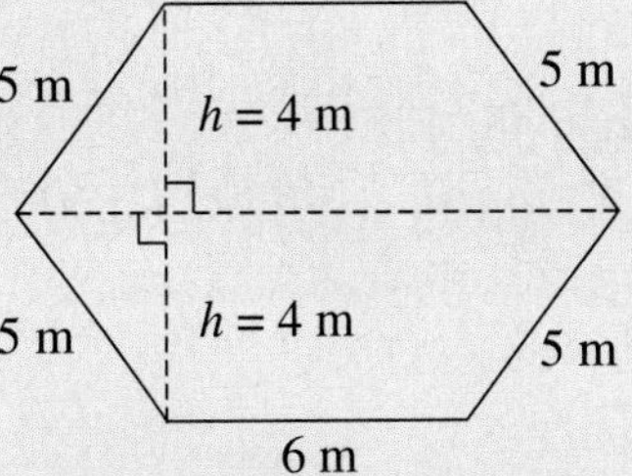
Circle [6.5]	$A = \pi r^2$ 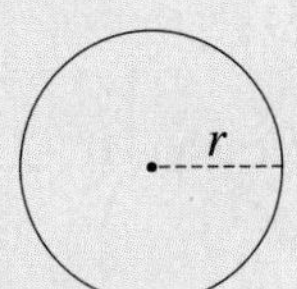	$A = \pi \cdot 3^2 = 9\pi \approx 28.27$ sq. ft. 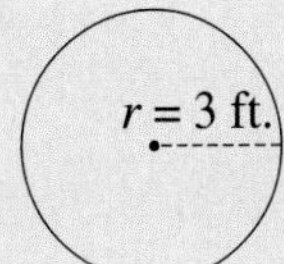
Circle [6.5]	$A = \pi\left(\frac{d}{2}\right)^2$ or $A = \frac{\pi d^2}{4}$ 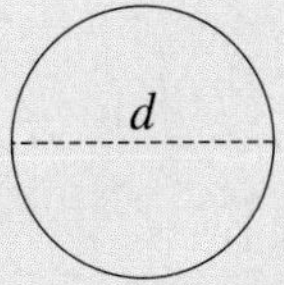	$A = \pi\left(\frac{5}{2}\right)^2 = \frac{\pi \cdot 5^2}{4} \approx 19.63$ sq. m 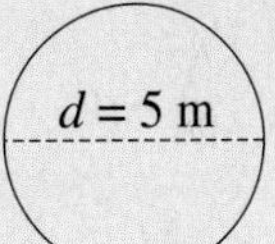
Using a protractor [6.7]	Place the vertex of the angle to be measured at the center of the protractor and one side of the angle along the baseline of the protractor. Where the other side of the angle meets the appropriate scale (smaller than 90° or larger than 90°) is the measure of the angle in degrees.	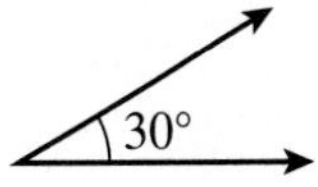
Classification of triangles [6.7]		
Equilateral [6.7]	All three sides have the same length.	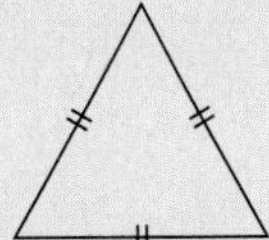

CONCEPT / SKILL	DESCRIPTION	EXAMPLE
Isosceles [6.7]	Two sides have the same length. (The two angles opposite the equal sides are also the same size.)	
Scalene [6.7]	Each side has a different length.	
Equiangular [6.7]	All three angles have the same measure.	
Right [6.7]	One angle measures 90°.	
Acute [6.7]	All angles measure less than 90°.	
Obtuse [6.7]	One angle measures greater than 90°.	
The sum of the angles of a triangle [6.7]	The sum of the angles of a triangle is 180°.	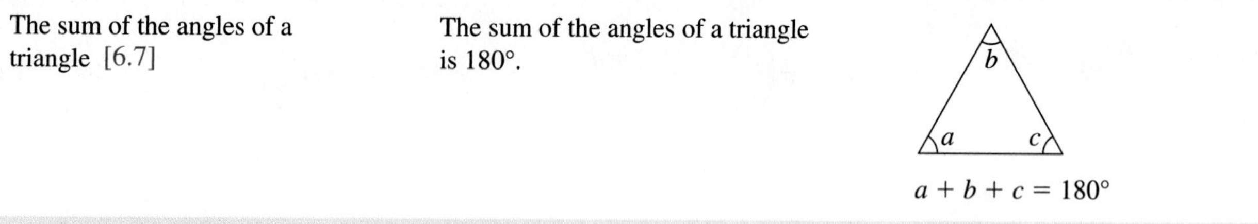 $a + b + c = 180°$
Pythagorean theorem [6.8]	$c^2 = a^2 + b^2$ This important formula for right triangles is useful for indirect measurement.	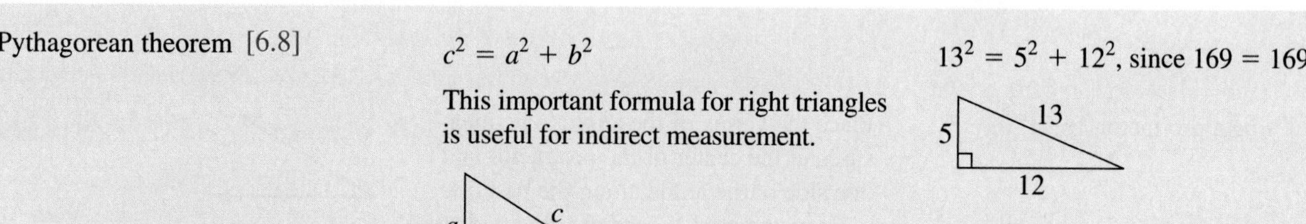 $13^2 = 5^2 + 12^2$, since $169 = 169$
Taking square roots [6.8]	If $N \geq 0$, then $\sqrt{N} = M$ providing $M^2 = N$. Calculators can be very useful in estimating square roots. Use $\sqrt{N}$.	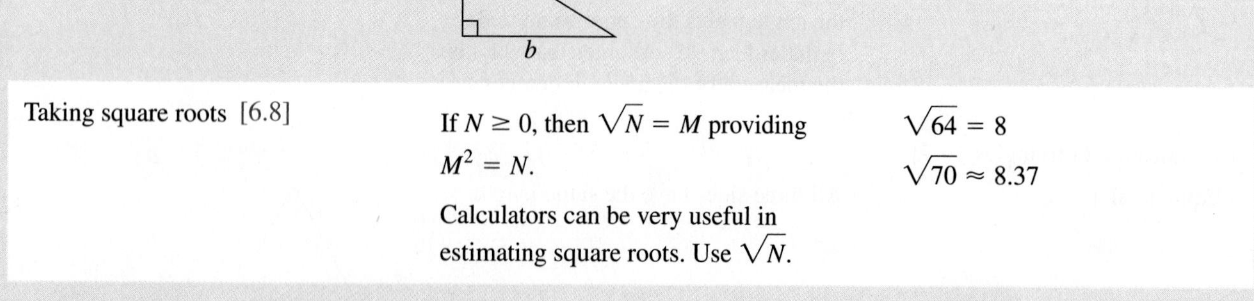 $\sqrt{64} = 8$ $\sqrt{70} \approx 8.37$

CONCEPT / SKILL	DESCRIPTION	EXAMPLE
Similar triangles [6.9]	Two triangles are similar provided their corresponding angles are equal; the lengths of their corresponding sides must be proportional (that is, their ratios must be equal).	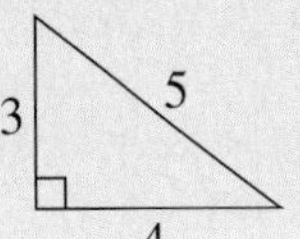 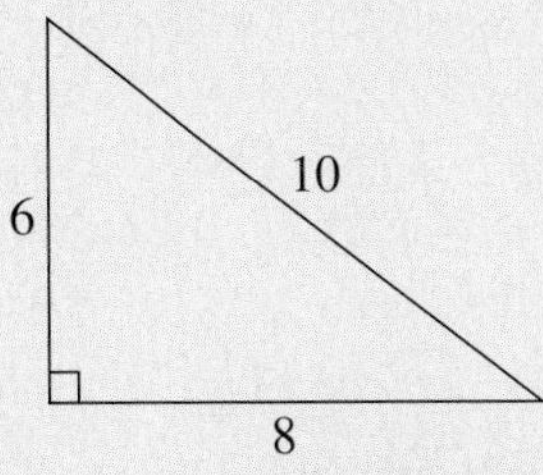 $\frac{3}{6} = \frac{4}{8}$ and $\frac{3}{6} = \frac{5}{10}$
Surface area formulas [6.10]	Surface area is the measure in square units of the exterior of the three-dimensional figure.	
Box (rectangular prism) [6.10]	$S = 2lw + 2lh + 2wh$	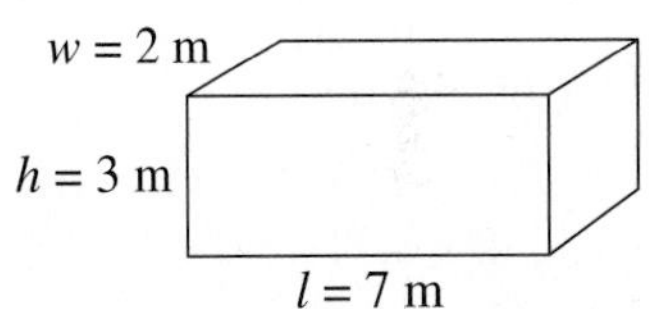 $S = 2 \cdot 7 \cdot 2 + 2 \cdot 7 \cdot 3 + 2 \cdot 2 \cdot 3$ $S = 28 + 42 + 12 = 82$ sq. m
Can (right circular cylinder) [6.10]	$S = 2\pi r^2 + 2\pi rh$	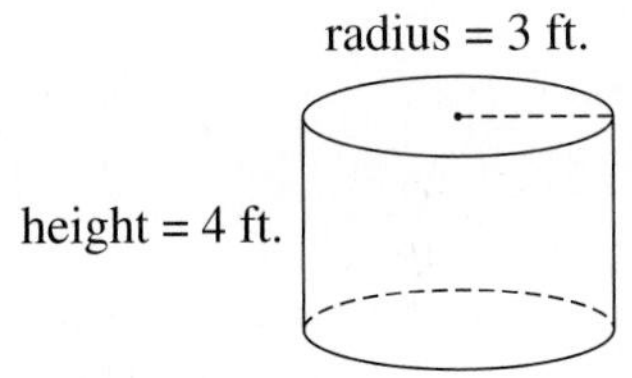 $S = 2\pi \cdot 3^2 + 2\pi \cdot 3 \cdot 4 = 42\pi$ $S \approx 131.95$ sq. ft.
Ball (sphere) [6.10]	$S = 4\pi r^2$ 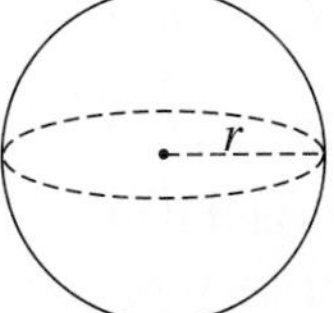	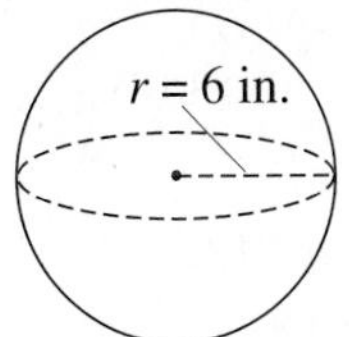 $S = 4\pi \cdot 6^2 = 144\pi \approx 452.39$ sq. in.

CONCEPT / SKILL	DESCRIPTION	EXAMPLE
Volume formulas [6.11], [6.12]	Volume is the measure in cubic units of the space inside a three-dimensional figure.	
Box (rectangular prism) [6.11]	$V = lwh$ 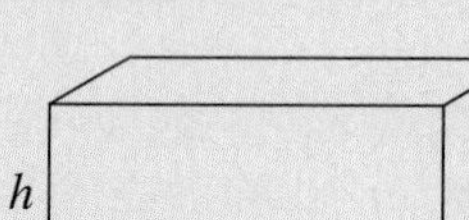	$V = 2 \cdot 4 \cdot 9 = 72$ cu. m 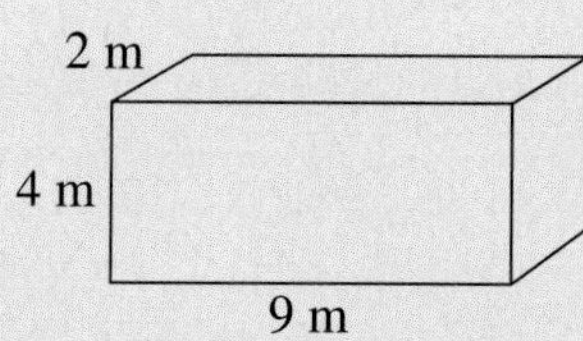
Can (right circular cylinder) [6.11]	$V = \pi r^2 h$ 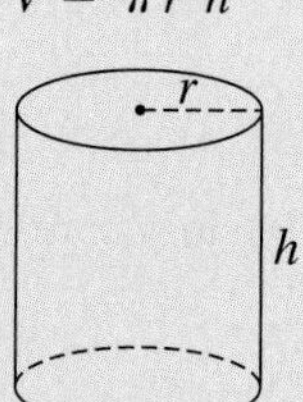	radius = 3 ft. 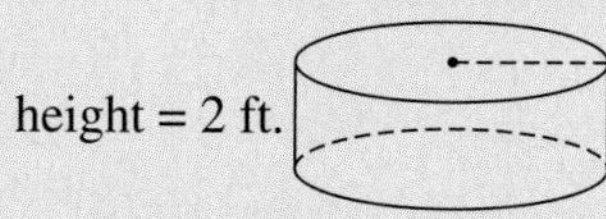 $V = \pi \cdot 3^2 \cdot 2 = 18\pi \approx 56.55$ cu. ft.
Ball (sphere) [6.12]	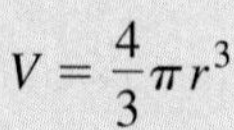 r	r = 7 cm; $V = \frac{4}{3} \cdot \pi \cdot 7^3 \approx 1436.76$ cu. cm
Cone [6.12]	$V = \frac{1}{3}\pi r^2 h$ 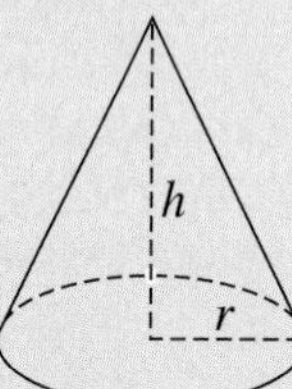	height = 8 ft.; radius = 6 ft.; $V = \frac{1}{3}\pi \cdot 6^2 \cdot 8 \approx 301.59$ cu. ft.
Taking cube roots [6.12]	$\sqrt[3]{N} = M$ providing $M^3 = N$. Calculators can be very useful in estimating cube roots. Use $\sqrt[3]{N}$.	$\sqrt[3]{64} = 4$ $\sqrt[3]{100} \approx 4.64$

CHAPTER 6 GATEWAY REVIEW

1. Consider the following two-dimensional figures.

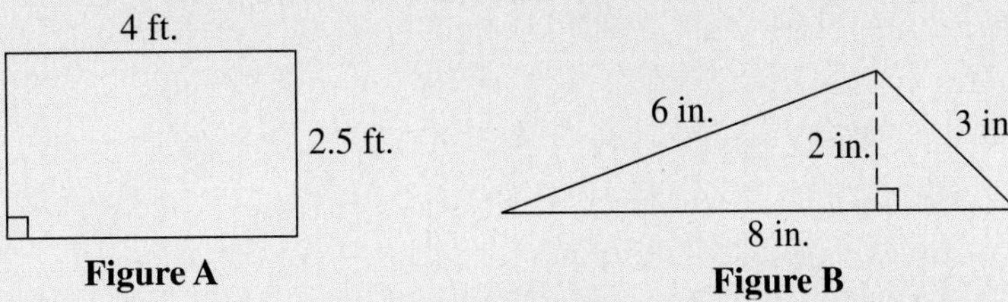

Figure A **Figure B**

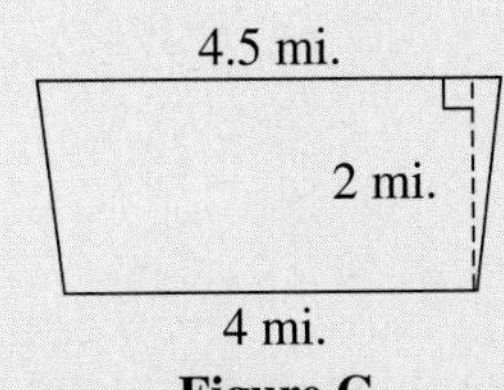

Figure C

   a. The area of Figure A = ______________

   b. The perimeter of Figure B = ______________

   c. The area of Figure B = ______________

   d. The area of Figure C = ______________

2. If a circle has area 23 square inches, what is the length of its radius?

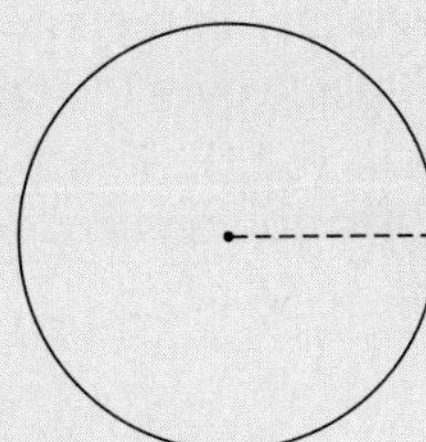

Answers to all Gateway exercises are included in the Selected Answers appendix.

3. You buy a candy dish with a 2-inch by 2-inch square center surrounded on each side by attached semicircles.

a. Draw the candy dish described above and label its dimensions.

b. Determine its perimeter and area.

c. If you place the dish on a 1-foot by 1-foot square table, how much space is left on the table for other items? Explain.

4. You buy a kite in the shape shown. When you open the box, you discover a tear in the kite's fabric. You decide to buy new fabric to place over the entire frame.

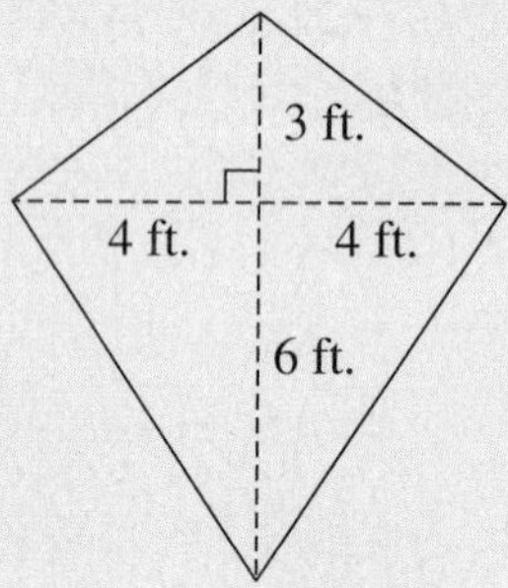

a. How much fabric do you need to buy?

b. You also decide to buy gold ribbon to line the perimeter of the kite. How much ribbon must you buy?

5. You are in the process of building a new home and the architect sends you the following floor plan for your approval.

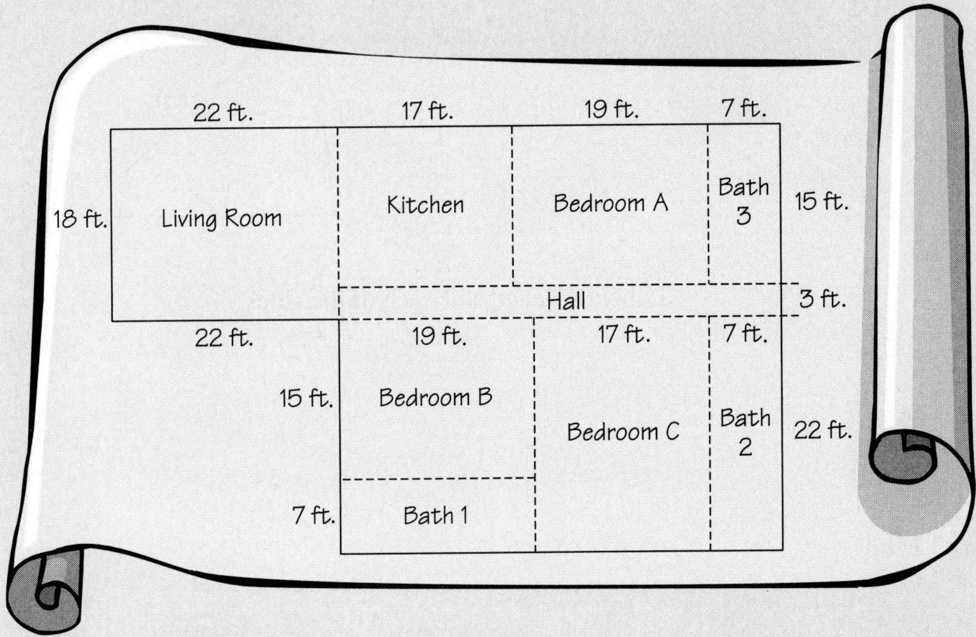

a. What is the perimeter of this floor plan?

b. What is the floor space, in square feet, of the floor plan?

c. Which bedroom has the largest area?

d. If you decide to double the area of the living room, what change in dimensions should you mark on the floor plan that you send back to the architect?

6. Your home is located in the center of a 300-foot by 200-foot rectangular plot of land. You are interested in measuring the diagonal of that plot. Use the Pythagorean theorem to make that indirect measurement.

7. Construct a triangle with its shortest side measuring 3 inches and similar to the following triangle.

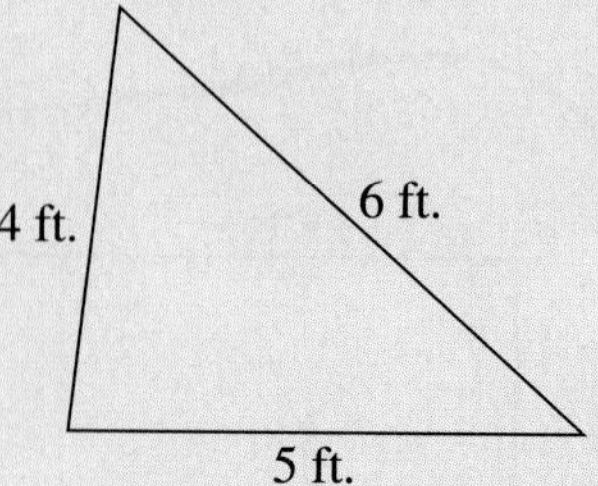

Label the lengths of each of the sides.

8. Calculate the volumes of the following three-dimensional figures.

a.

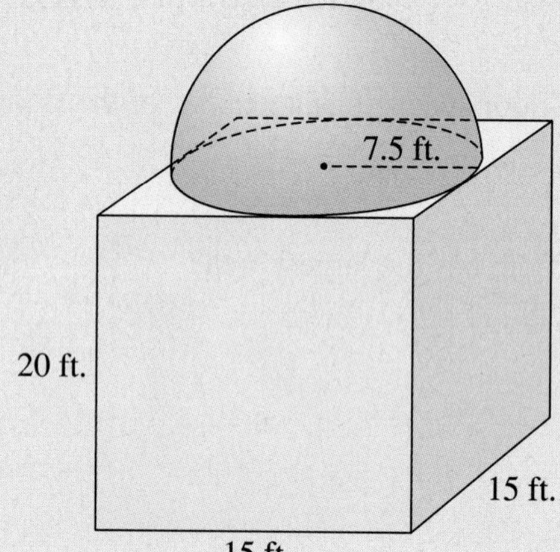

b.

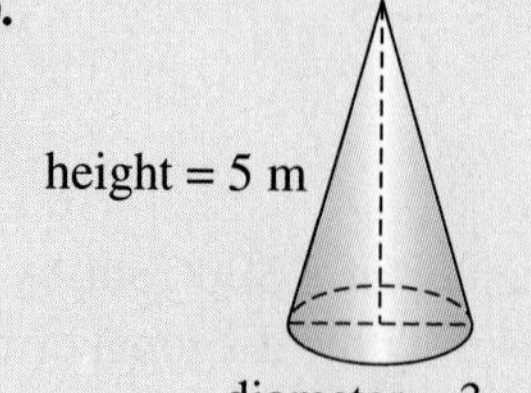

c.

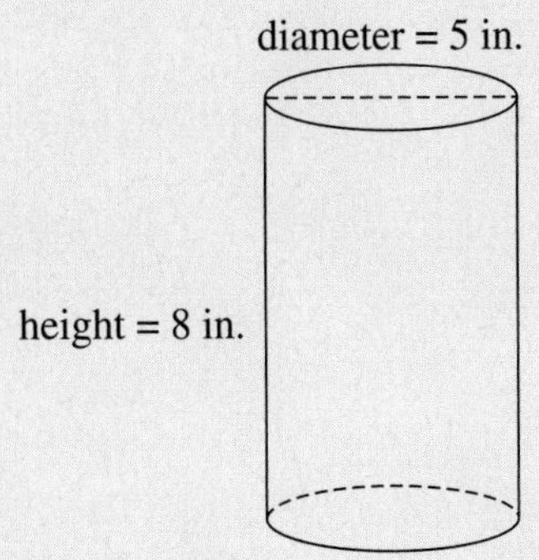

diameter = 5 in.

height = 8 in.

9. Biologists studying a lake determine that the lake's shape is approximately circular with a diameter of 1 mile and an average depth of 187 feet. Estimate how much water is in the lake. Note that for consistency in units, 1 mile = 5280 feet.

10. The Alaska pipeline is a cylindrical pipe with diameter 48 inches that carries oil for 800 miles through Alaska. It is an engineering marvel, with many safety and business concerns.

a. What is the total capacity of the pipeline for oil at any given time, assuming that it could be filled to capacity? Explain.

b. Estimate the amount of material needed to construct the pipeline. Explain.

MORE PROBLEM SOLVING WITH ALGEBRA AND MATHEMATICAL MODELS

Throughout this book you have solved problems from everyday life—science, business, sports, social issues, space design, and so on—that require mathematics. This approach emphasizes the fact that searching for solutions to some kinds of problems resulted in the development of the mathematics we use today. For example, the early Egyptians developed their geometry to recalculate yearly the boundaries of the lands owned by farmers because of the seasonal flooding of the Nile River.

When applied to solve problems or to describe a situation, the mathematics used became known as the **mathematical model** for the problem or situation. In this chapter you will develop your problem-solving skills further by using mathematical models in several different formats. You will recognize these formats from the previous chapters: symbolic equations, formulas, graphs, and tables.

ACTIVITY 7.1

Leasing a Copier

OBJECTIVES

1. Describe a mathematical situation as a set of verbal statements.
2. Translate verbal rules into symbolic equations.
3. Solve problems involving equations of the form $y = ax + b$.
4. Solve equations of the form $y = ax + b$ for the input x.
5. Evaluate expressions $ax + b$ in the equations of the form $y = ax + b$ to obtain an output y.

You are a college intern in a law office. The office manager asks you to get information about leasing a copy machine for the office. You will then use the information to estimate how much a machine will cost your office.

The sales representative from Eastern Supply Company recommends a 50-copy per minute copier for your office. The copier will cost $455 per month plus 1.5 cents a copy.

Do the following problems to figure out how much your office would pay for leasing a copier from the Eastern Supply Company.

1. a. The total monthly cost for leasing the copier depends upon the number of copies made. Identify the input variable and the output variable.

   b. Note that the monthly cost is expressed by the sum of two parts, a fixed cost of $455 and a cost of 1.5 cents per copy. The units of the two parts, dollars and cents, are different. One of the units has to be converted to the other unit. Convert 1.5 cents to dollars.

c. Write a verbal rule to determine the output (monthly cost) in terms of the input (number of copies).

d. Translate the verbal rule obtained in part c into an equation. Use n for the input variable and c for the output variable.

2. a. Use the equation obtained in Problem 1d to determine the monthly cost if 12,000 copies are made each month.

b. Calculate the monthly cost if 20,000 copies are made each month.

In Problem 1, the equation, $c = 0.015n + 455$ was developed to determine the leasing charges. In Problem 2, you determined the leasing charges by evaluating the expression $0.015n + 455$ for a known number of copies. The following example shows an algebraic approach to determine the number of copies that can be made for a known monthly cost.

EXAMPLE 1 ***Suppose the monthly budget for leasing the copier from Eastern Office Supply is \$800. How many copies can the law office make for that budgeted amount?***

SOLUTION

Step 1 Recognize that \$800 is the output (monthly cost) value and that the unknown value is the input (number of copies) corresponding to the \$800.

Step 2 Replace c in the equation $c = 0.015n + 455$ with 800 to obtain the equation

$$800 = 0.015n + 455.$$

Step 3 Solve the equation to determine the input value.

a.
$$\begin{array}{rcl} 800 &=& 0.015n + 455 \\ -455 & & \quad\;\; - 455 \\ \hline 345 &=& 0.015n \end{array}$$

Subtract 455 from each side of the equation to obtain $0.015n$ as a single term on the right side of the equation.

b.
$$\frac{345}{0.015} = \frac{\cancel{0.015}n}{\cancel{0.015}}$$

Divide each term in the equation by 0.015.

$$23000 = n$$

23,000 copies can be made.

Step 4 Check the solution:

$c = 0.015(23000) + 455$

$c = 345 + 455$

$c = 800$

Step 5 Interpret the solution of the equation by stating: The law office can make 23,000 copies per month on a budget of \$800.

3. How many copies can the office make if the copier budget is \$950?

4. The equation $120 = 3x + 90$ is solved for x as follows.

$$\begin{aligned} & 120 = 3x + 90 \\ \textbf{Step 1:}\quad & \underline{-90 \qquad\quad -90} \\ & 30 = 3x \\ \textbf{Step 2:}\quad & \frac{30}{3} = \frac{3x}{3} \\ & 10 = x \quad \text{or} \quad x = 10 \end{aligned}$$

a. In step 1, what operation is used to remove 90 from each side of the equation and why?

b. In step 2, what operation is used to remove 3 from the term $3x$ and why?

5. Solve the equation $12 = 5x - 8$ for x. At each step, state the operation you used and state why you used it.

6. In each of the following, an equation is given along with a specific output value. Replace the output variable y in the given equation with the specified value and solve the resulting equation for the unknown input x. Check each answer.

a. $y = 3x - 5$ and $y = 10$

b. $y = 30 - 2x$ and $y = 24$

c. $y = .75x - 21$ and $y = -9$

d. $y = -2x + 15$ and $y = -3$

Verbal rules and equations can also be used to make lists of paired input/output values. For example, if a firm has to pay sales tax on office supplies, the firm's purchasing agent may use an equation to determine the final cost after taxes. Also, for quick reference, a table listing the most common costs before and after taxes can be made (see Problem 7).

7. Suppose your law office has to pay a 4% sales tax on office supplies it purchases. That means that, as in Activity 5.4, the final cost of the item is determined by multiplying the price of the item by the growth factor, 104% = 1.04, that is associated with the 4% sales tax.

a. Use y to represent the final cost of a purchase and x to represent the cost of the items before tax. Write an equation to calculate the final cost of a purchase.

b. Use the equation from part a to complete the following table.

COST BEFORE TAX, x (in dollars)	FINAL COST AFTER TAX, y (in dollars)
100	
	260
375	
	624

8. In each of the following problems, use the equation to complete the table.

a. $y = 4x - 11$

x	y
6	
	53

b. $y = -5x - 80$

x	y
12	
	35

SUMMARY
ACTIVITY 7.1

1. To determine the output y for a specified input x in an equation of the form $y = ax + b$:

Step 1: Replace the input variable x in the expression $ax + b$ with the specified numerical value.

Step 2: Evaluate the numerical expression to obtain the numerical value for y.

Example: Solve $y = 2x + 1$ for y, given $x = 3$.

$$y = 2 \cdot 3 + 1$$

$$y = 7$$

2. To solve an equation of the form $y = ax + b$ for the unknown value of the input x, given a specified numerical value of the output y:

Use inverse operations to isolate the unknown x. Apply the inverse operations in the following order.

Step 1: Add the opposite of b to each side of the equation.

Step 2: Divide each term in the resulting equation by a to isolate x and then read the value for x.

Example: Solve $5 = 2x - 1$ for x.

$$\begin{array}{r} 5 = 2x - 1 \\ +1 \qquad +1 \\ \hline 6 = 2x \end{array}$$

$$\frac{6}{2} = \frac{2x}{2}$$

$$x = 3$$

EXERCISES
ACTIVITY 7.1

1. You are planning on taking some courses at your local community college on a part-time basis for the upcoming semester. For a student who carries fewer than 12 credits (the full-time minimum), the tuition is \$155 for each credit hour taken. All part-time and full-time students must pay a \$20 parking fee for the semester. This fixed fee is added directly to your tuition bill.

   a. Write an equation to determine the total tuition bill, t, for a student carrying fewer than 12 credit hours. Use n for the number of credit hours.

   b. Use the equation you wrote in part a to complete the following table.

Paying for Credit

NUMBER OF CREDIT HOURS, n	1	2	3	4	5	6
TOTAL TUITION, t						

   c. Determine the tuition bill if you take 9 credit hours.

   d. You have \$1000 to spend on tuition. How many credit hours can you carry for the semester?

2. Use an algebraic approach to solve each of the following equations for x. Check your answers by hand or with a calculator.

   a. $10 = 2x + 12$

   b. $-27 = -5x - 7$

   c. $3x - 26 = -14$

   d. $24 - 2x = 38$

   e. $5x - 15 = 15$

   f. $-4x + 8 = 8$

Exercise numbers appearing in color are answered in the Selected Answers appendix.

g. $12 + \frac{1}{5}x = 9$

h. $\frac{2}{3}x - 12 = 0$

i. $0.25x - 14.5 = 10$

j. $5 = 2.5x - 20$

3. A long-distance telephone rate plan costs \$4.95 a month plus 10 cents per minute, or part thereof, for any long-distance call made during the month.

   a. Write a verbal rule to determine the total monthly cost for your long-distance calls.

   b. Translate the verbal rule in part a into an equation using c to represent the total monthly cost and n to represent the total number of long-distance minutes for the month.

   c. Determine the monthly cost if 250 minutes of long-distance calls are made.

   d. If you can spend up to \$50 for long-distance calls in one month, what will your total number of long-distance minutes be for the month? Check your answer.

4. The social psychology class is planning a fund-raising project to benefit local charities. As a member of the fund-raiser budget committee, you suggest a \$10 per person admission donation for food, nonalcoholic beverages, and entertainment. The committee determines that the total fixed costs for the event (food, drinks, posters, and tickets) will total \$2100. The college is donating the use of the gymnasium for the evening.

   a. The total money collected from admissions (revenue) depends on the number, n, of students who attend. Write an expression in terms of n that represents the total revenue if n students attend.

   b. Profit is the money remaining after expenses are deducted from the revenue. Write an equation expressing the profit, p, in terms of the number, n, of students who attend.

c. If the gymnasium holds a maximum of 700 people, what is the maximum amount of money that can be donated to charity?

d. Suppose that the members of the fund-raiser budget committee want to be able to donate at least $1500 to local charities. How many students must attend in order to have a profit of $1500?

5. The value of many things we own, such as a car, computer, or appliance, depreciates (goes down) over time. When an asset's value decreases by a fixed amount each year, the depreciation is called straight-line depreciation. Suppose a car has an initial value of $12,400 and depreciates $820 per year.

a. Let v represent the value of a car after t years. Write a symbolic rule that expresses v in terms of t.

b. What is the value of the car after 4 years?

c. How long will it take for the value of the car to decrease below $2000?

6. The cost, c, in dollars, of mailing a priority overnight package weighing 1 pound or more is given by the formula $c = 2.085x + 15.08$, where x represents the weight of the package in pounds.

a. Determine the cost of mailing a package that weighs 10 pounds.

b. Determine the weight of the package if it cost $56.78 to mail.

7. Archaeologists and forensic scientists use the length of human bones to estimate the height of individuals. A person's height, h, in centimeters can be determined from the length of the femur, f (the bone from the knee to the hip socket), in centimeters using the following formulas:

Male: $h = 69.089 + 2.238f$ Female: $h = 61.412 + 2.317f$

a. A partial skeleton of a male is found. The femur measures 50 centimeters. How tall was the man?

b. What is the length of the femur for a female who is 150 centimeters tall?

8. Let p represent the perimeter of an isosceles triangle that has two equal sides of length a and a third side of length b. Determine the length of the equal sides of an isosceles triangle having a perimeter of $\frac{3}{4}$ meter and third side measuring $\frac{1}{3}$ meter.

9. The recommended weight for an adult male is given by the formula $w = \frac{11}{2}h - 220$, where w represents the recommended weight in pounds and h represents the height of the person in inches. Determine the height of an adult whose recommended weight is 165 pounds.

10. Complete the following tables using algebraic methods.

a. $y = 2x - 10$

x	y
4	
	14

b. $y = 20 + 0.5x$

x	y
3.5	
	−10

c. $y = -3x + 15$

x	y
$\frac{2}{3}$	
	−3

d. $y = 12 - \frac{3}{4}x$

x	y
−8	
	−6

ACTIVITY 7.2

Windchill

OBJECTIVES

1. Evaluate expressions to determine the output for a formula.
2. Solve formulas for a specified variable.

Have you ever wondered why people blow on hot things to cool them down? Blowing creates the effect of wind. The faster the wind blows, the faster things lose their heat. This cooling effect is called **windchill**. For example, on two different days, the temperature may be 25°F, but depending on the wind, you may feel colder one of the days. If there is no wind, a temperature of 25°F may not feel very cold. However, if the wind is blowing hard, the same 25°F may feel like −10°F.

Being curious about the mathematical relationship between temperature and wind speed, you search the Internet and find a windchill chart at www.nws.noaa.gov/om/windchill/, developed by the National Weather Service. A portion of the chart, or table, is reproduced here.

WINDCHILL CHART

Wind Speed (mph)

Air Temperature (°F)	5	10	25	30
35	31	27	23	22
30	25	21		15
25	19		9	8
20	13	9		1
15	7	3	−4	−5
10	1	−4		−12
5	−5		−17	−19
0	−11	−16	−24	−26
−5	−16	−22	−31	−33
−10	−22	−28	−37	−39
−15	−28	−35		−46
−20	−34	−41	−51	−53

The table indicates, for example, that when the air temperature is 15°F and the wind speed is 10 mph, the windchill temperature is 3°F. This means that it will feel like 3°F when the wind is blowing at 10 mph.

1. **a.** What is the windchill temperature when the air temperature is 5°F and there is a 25 mph wind?

b. What was the wind speed on the day the air temperature was 25°F and the windchill was 9°F?

c. Examine the columns of the windchill chart and list some of the patterns you discover. According to the patterns you observe, estimate the missing values in the chart.

2. The windchill temperature, w, produced by a 30-mph wind can be approximated by the formula

$$w = 1.36t - 25.86,$$

where t represents the air temperature in degrees Fahrenheit.

a. Complete the following table using the given formula. Round to the nearest whole number.

It's a Breeze

AIR TEMPERATURE, t	−15	5	30
EQUIVALENT WINDCHILL TEMPERATURE, w (30-mph wind)			

b. How do these windchill temperatures compare to the values given in the chart for a 30-mph wind?

3. Use the formula $w = 1.36t - 25.86$ in the following problems.

a. Determine the windchill temperature if the air temperature is 7°F when there is a 30-mph wind.

b. On a cold day in New York City, the wind is blowing at 30 mph. If the windchill temperature was reported to be −18°F, then what was the air temperature on that day?

Note that in the formula, $w = 1.36t - 25.86$, the variable w is on one side of the equation, isolated from the expression $1.36t - 25.86$. This allows for determining values of w directly. Simply substitute the desired value for t in the expression and evaluate. For example, in Problem 3a, 7 was substituted for t in $1.36t - 25.86$ to obtain a windchill of approximately −16°F.

$$w = 1.36(7) - 25.86 \approx -16°\text{F}.$$

However, in Problem 3b, to determine t for a given value of w the equation $w = 1.36t - 25.86$ had to be solved for t, after substituting -18 for w into the equation.

If you had to determine the corresponding t value for several different w values, you would have to solve an equation each time. It is more efficient to solve the original formula $w = 1.36t - 25.86$ for t.

EXAMPLE 1 ***Solve the formula $w = 1.36t - 25.86$ for t. The procedure is similar to solving the equation $-18 = 1.36t - 25.86$ for t.***

$-18 = 1.36t - 25.86$	$w = 1.36t - 25.86$	
$+25.86 \qquad +25.86$	$+25.86 \qquad +25.86$	**Add 25.86 to each side**
$7.86 = 1.36t$	$w + 25.86 = 1.36t$	
$\dfrac{7.86}{1.36} = \dfrac{1.36t}{1.36}$	$\dfrac{w + 25.86}{1.36} = \dfrac{1.36t}{1.36}$	**Divide each side by 1.36**
$5.8 = t$	$\dfrac{w + 25.86}{1.36} = t$	

The new formula is $t = \dfrac{w + 25.86}{1.36}$, which is approximately equivalent to $t = 0.735w + 19.015$.

To solve the equation $w = 1.36t - 25.86$ for t means to isolate the variable t, with coefficient 1, on one side of the equation, with all other expressions on the opposite side of the = sign.

4. Rework Problem 3b using the new formula that is solved for t. How does your answer here compare to your answer in Problem 3b?

5. a. If the wind speed is 15 mph, the windchill can be approximated by the formula $w = 1.28t - 19.38$, where t is the air temperature in degrees Fahrenheit. Solve the formula for t.

b. Use the new formula from part a to determine the air temperature, t, if the windchill temperature is -10°F.

Weather Balloon

6. A weather balloon carrying instruments that measure temperature is launched at sea level. After the balloon is launched, the data collected shows that the temperature dropped 0.15°F for each meter that the balloon rose.

a. If the temperature at sea level is 50°F, determine the temperature at a distance of 60 meters above sea level.

b. Write a verbal rule to determine the temperature at a given distance above sea level on a 50°F day.

c. If t represents the temperature (°F) a distance of m meters above sea level, translate the verbal rule in part b into a formula.

d. Use the formula from part c to complete the following table.

Weather or Not

METERS ABOVE SEA LEVEL, m	50	75	100
TEMPERATURE, t (°F)			

7. a. Solve the formula $t = 50 - 0.15m$ for m.

b. Water freezes at 32°F. Determine the distance above sea level that water will freeze. Use the new formula from part a.

Crickets and Temperature

A familiar late-evening sound during the summer is the rhythmic chirping of a male cricket. Of particular interest is the snowy tree cricket, sometimes called the temperature cricket. It is very sensitive to temperature, speeding up or slowing down its chirping as the temperature rises or falls.

Data show that the number, n, of chirps per minute of the snowy tree cricket is related to the temperature t (°F) by the formula

$$t = \frac{1}{4}n + 40.$$

8. a. If a cricket chirps 60 times in 1 minute, what is the temperature?

b. Solve the equation $t = \frac{1}{4}n + 40$ for n.

c. If the temperature is 80°F, use the formula from part b to determine the expected number of chirps made by the cricket in 1 minute.

Additional Practice

9. Solve each of the following formulas for the indicated variable.

a. $A = \frac{1}{2}bh$ for h

b. $p = c + m$ for m

c. $P = 2l + 2w$ for l

d. $R = 165 - 0.75a$ for a

SUMMARY
ACTIVITY 7.2

To solve an equation for a specific variable means to isolate that variable, with coefficient 1, on one side of the equation.

EXERCISES
ACTIVITY 7.2

1. The speed, S, of an ant (in centimeters per second) is related to the temperature, t (in degrees Celsius), by the formula

$$s = 0.167t - 0.67.$$

a. If an ant is moving at 4 centimeters per second, what is the temperature?

b. Solve the equation $s = 0.167t - 0.67$ for t.

c. Use the new formula from part b to answer part a.

Exercise numbers appearing in color are answered in the Selected Answers appendix.

2. The National Weather Service reports the daily temperature in degrees Fahrenheit. The scientific community, as well as Canada and most of Europe, reports temperature in degrees Celsius. The Celsius, C, and Fahrenheit, F, temperature readings are related by the formula

$$F = 1.8C + 32.$$

a. Determine the Fahrenheit reading corresponding to the temperature at which water boils, 100°C.

b. Solve the formula $F = 1.8C + 32$ for C.

c. Use the formula from part b to determine the Celsius temperature when the outdoor temperature is 80°F.

3. The number of women enrolled in college has been steadily increasing. The following table gives the enrollment of women, in millions, in a given year.

Enrolling Along

YEAR	1970	1975	1980	1985	1990	1995	2000
WOMEN ENROLLED (in millions)	3.5	5.0	6.0	6.6	7.4	8.0	8.9

Let t represent the number of years since 1970. The number of women, n (in millions), enrolled in college can be approximated by the formula

$$n = 0.17t + 3.9.$$

a. Use the formula to estimate the year that the number of women enrolled in college will reach 10 million.

b. Solve the equation $n = 0.17t + 3.9$ for t.

c. Use the formula in part b to estimate the year that the number of women enrolled in college will reach 10 million.

4. The pressure, p, of sea water (in pounds per square foot) at a depth of d feet below the surface is given by the formula

$$p = 15 + \frac{15}{33}d.$$

a. On November 14, 1993, Francisco Ferreras achieved a record depth for breath-held diving. During the dive, he experienced a pressure of 201 pounds per square foot. What record depth did he reach?

b. Solve the equation $p = 15 + \frac{15}{33}d$ for d.

c. Use the new formula from part b to answer part a.

Solve each of the following formulas for the given variable.

5. $E = IR$ for I

6. $C = 2\pi r$ for r

7. $P = 2a + b$ for b

8. $P = R - C$ for R

9. $P = 2l + 2w$ for w

10. $R = 143 - 0.65a$ for a

11. $A = P + Prt$ for r

12. $y = mx + b$ for m

ACTIVITY 7.3

Comparing Energy Costs

OBJECTIVES

1. Write symbolic equations from information organized in a table.
2. Produce tables and graphs to compare outputs from two different mathematical models.
3. Solve equations of the form $ax + b = cx + d$.

You hired an architect to design a house. She gave you the following information regarding the installation and operating costs for two types of heating systems: solar and electric. You will use the information to compare the costs for each heating system over a period of years. The following questions will guide you in making the comparisons.

Some Like It Hot

TYPE OF SYSTEM	OPERATING COST PER YEAR	INSTALLATION COST
Solar	\$200	\$19,000
Electric	\$1400	\$7,000

1. a. Use the information in the table to write an equation to represent the total cost of using solar heat in terms of the number of years it is in use. Let x represent the number of years of use and let S represent the total cost.

   b. Write an equation to represent the total cost of using electric heat in terms of the number of years it is in use. Let x represent the number of years of use and let E represent the total cost.

You can use the equations you wrote in Problem 1 to compare costs for the two systems in several ways. One way is to produce a table of costs for specific periods of use.

2. a. Use the equations from Problem 1 to complete the following table.

NUMBER OF YEARS IN USE, x	0	5	10	15
TOTAL COST FOR SOLAR HEAT, S (\$)				
TOTAL COST FOR ELECTRIC HEAT, E (\$)				

   b. Use the information in the table to compare the total costs of each system after each 5-year period.

   c. From comparing the total costs in the table, which system do you think is the better one for the house you are building?

You can also do comparisons by graphing the information from the table.

3. a. On the following grid, plot the data points for the solar heating system that you determined in Problem 2. Then connect the points by drawing a line through them.

   b. On the same grid, also plot the data points for the electric heating system and connect the points by drawing a line through them.

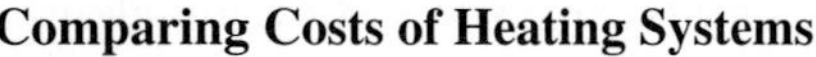

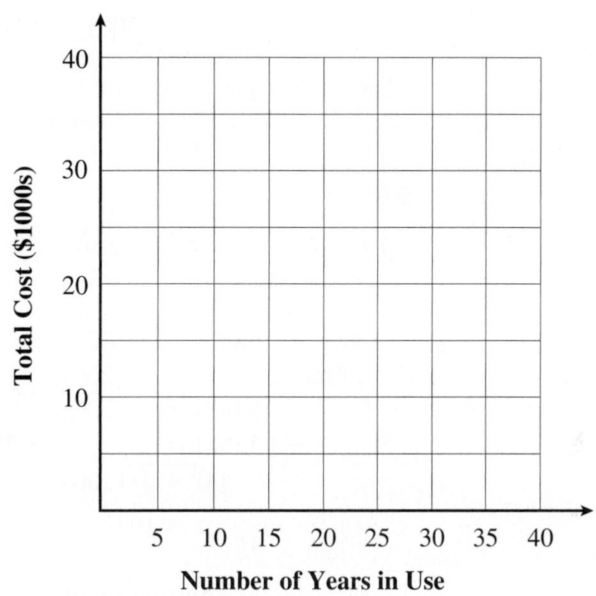

   c. From the graph, when are the total costs for the heating systems the same? How much is the cost?

   d. Which heating system costs less at 15 years? at 20? Explain.

   e. From comparing the total costs on the graph, which system do you think is the better one for the house you are building?

In Problems 2 and 3 you observed that the table and the graph led to the same result: the costs are the same at 10 years of use and that the cost of the solar system was less after more than 10 years. The point at which both systems cost the same is useful in making decisions about which heating method is better in terms of cost. When used to compare costs, this point is often referred to as the **break-even point**. You can calculate the break-even point using algebra, as the following example shows.

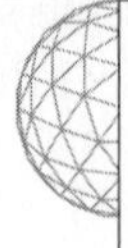

EXAMPLE 1 ***The comparison model for the two heating systems in the preceding problems includes two equations.***

$$S = 200x + 19{,}000$$
$$E = 1400x + 7000$$

The break-even point occurs when the outputs from the solar and electric heating systems are the same at the same time (for the same input value). This means $S = E$, or equivalently,

$$200x + 19{,}000 = 1400x + 7000.$$

This single equation can be solved for x to determine the number of years of use when the costs will be the same for both systems.

SOLUTION

First, add and/or subtract terms appropriately so that all terms involving the variable are on one side of the equal sign and all other terms are on the other side.

$$\begin{aligned} 200x + 19{,}000 &= 1400x + 7000 \\ -200x \quad\quad &= -200x \\ \hline 19{,}000 &= 1200x + 7000 \\ -7{,}000 &= \quad\quad -7000 \\ \hline 12{,}000 &= 1200x \\ x &= 10 \text{ yr.} \end{aligned}$$

Subtract $200x$ from both sides and combine like terms.

Subtract 7000 from each side and combine like terms.

Divide each term by 1200, the coefficient of x.

In checking the solution to the equation, you will also determine the total cost at 10 years.

$$200(10) + 19{,}000 = 21{,}000$$
$$1400(10) + 7000 = 21{,}000$$

The check shows that 10 is the correct solution and that the total cost after 10 years is \$21,000. So the break-even point occurs at 10 years.

4. You are interested in purchasing a new car. You have narrowed the choice to a Honda Accord LX (4 cylinder) and a Passat GLS (4 cylinder). You are concerned about the depreciation of the cars' values over time and want to make some comparisons. You search the Internet and obtain the following information:

Driven Down. . .

MODEL OF CAR	MSRP (Manufacturer's Suggested Retail Price)	STRAIGHT LINE DEPRECIATION EACH YEAR
Accord LX	\$22,600	\$2,240
Passat GLS	\$25,000	\$2,640

a. Let A represent the value, in dollars, of the Accord LX after x years of ownership. Use the information in the table to write an equation to determine A in terms of x.

b. Write an equation to determine the value, P, in dollars, of the Passat GLS after x years of ownership.

c. Use the equations from parts a and b to write a single equation to determine when the value of the Accord LX will equal the value of the Passat GLS.

d. Solve the equation in part c to determine the year in which the cars will have the same value.

e. Complete the following table to compare the values of the cars over several years.

NUMBER OF YEARS YOU OWN CAR	VALUE OF ACCORD LX ($)	VALUE OF PASSAT GLS ($)
1		
4		
8		

5. Solve each of the following equations for x. Check your answers, by hand or with a calculator.

a. $2x + 9 = 5x - 12$

b. $21 - x = -3 - 5x$

c. $2x - 6 = -8$

d. $2x - 8 + 6 = 4x - 7$

SUMMARY
ACTIVITY 7.3

General Strategy for Solving Equations for an Unknown, *x*

1. Rewrite the equation to obtain a single term containing x on one side of the equation. Do this by adding and/or subtracting terms to move the x terms to one side of the equation and all the other terms to the other side. Then combine like terms that appear on the same side of the equation.

2. Solve for the unknown x by dividing each side of the equation by the coefficient of x.

3. Check the result to be sure that the value of the unknown produces a true statement.

EXERCISES
ACTIVITY 7.3

1. Finals are over and you are moving back home for the summer. You need to rent a truck to move your possessions from the college residence hall. You contact two local rental companies and get the following information for the one-day cost of renting a truck.

   Company 1: $39.95 per day plus $0.19 per mile

   Company 2: $19.95 per day plus $0.49 per mile

   Let x represent the number of miles driven in one day.

   a. Write an equation that represents the total cost in dollars of renting a truck for one day from Company 1.

   b. Write an equation that represents the total cost in dollars of renting a truck for one day from Company 2.

   c. Using the equations in parts a and b, write a single equation to determine the mileage for which the cost would be the same from both companies.

   d. Solve the equation in part c.

   e. You actually live 90 miles from the campus. Which rental company would be the better deal?

2. Two companies sell software products. In 2006, Company 1 had total sales of \$17.2 million. Its marketing department projects that sales will increase by \$1.5 million per year for the next several years. Company 2 had total sales of \$9.6 million for software products in 2006 and predicts that its sales will increase on the average \$2.3 million each year. Let x represent the number of years since 2006.

a. Write an equation that represents the total sales, in millions of dollars, of Company 1 since 2006.

b. Write an equation that represents the total sales, in millions of dollars, of Company 2 since 2006.

c. Write a single equation to determine when the total sales of the two companies will be the same.

d. Solve the equation in part c.

3. You are considering installing a security system in your new house. You get the following information from two local home security dealers for similar security systems.

Dealer 1: \$3,560 to install and \$15 per month monitoring fee.

Dealer 2: \$2,850 to install and \$28 per month for monitoring.

Note that the initial cost of the security system from Dealer 1 is much higher than from Dealer 2, but the monitoring fee is lower.

Let x represent the number of months that you have the security system.

a. Write an equation that represents the total cost of the system with Dealer 1.

b. Write an equation that represents the total cost of the system with Dealer 2.

c. Write a single equation to determine when the total cost of the systems will be equal.

d. Solve the equation in part c.

e. If you plan to live in the house and use the system for 10 years, which system would be less expensive?

4. The life expectancies for men and women in the United States can be approximated by the following formulas

$$\text{Women:}\quad E = 0.126t + 76.74$$
$$\text{Men:}\quad E = 0.169t + 69.11,$$

where E represents the length of life in years and t represents the year of birth, measured as the number of years since 1975.

a. Write a single equation that can be used to determine in what year of birth the life expectancy of men and women would be the same.

b. Solve the equation in part a.

c. What is the life expectancy for the year of birth determined in part b?

In Exercises 5–10, solve each of the given equations for x. Check your answers by hand or with a calculator.

5. $5x - 4 = 3x - 6$

6. $3x - 14 = 6x + 4$

7. $0.5x + 9 = 4.5x + 17$

8. $4x - 10 = -2x + 8$

9. $0.3x - 5.5 = 0.2x + 2.6$

10. $4 - 0.025x = 0.1 - 0.05x$

ACTIVITY 7.4

Volume of a Storage Tank

OBJECTIVES

1. Use the property of exponents to multiply powers having the same base.
2. Use the property of exponents to raise a power to a power.
3. Use the distributive property and properties of exponents to write an expression as an equivalent expression in expanded form.

The volume V (in cubic feet) of a partially cylindrical gasoline storage tank is represented by the formula

$$V = r^2(4.2r + 37.7),$$

where r is the radius (in feet) of the cylindrical part of the tank.

1. Determine the volume of the tank if its radius is 3 feet.

First Property of Exponents

Suppose you were asked to write the expression $r^2(4.2r + 37.7)$ as an equivalent expression without parentheses. Using the distributive property, you would multiply each term within the parentheses by r^2.

$$r^2(4.2r + 37.7)$$

The first product is $r^2(4.2r)$ and the second product is $r^2(37.7)$. In the first product you need to multiply r^2 and r. Recall that in the expression r^2, the exponent 2 indicates that the base r is used as a factor two times. In the expression $r = r^1$, the exponent 1 indicates that the base r is used as a factor once.

$$r^2 \cdot r = \underbrace{r \cdot r \cdot r}_{\text{Base } r \text{ is used as a factor 3 times.}} = r^3$$

So the volume of the tank can also be expressed as $4.2r^3 + 37.7r^2$.

2. a. Complete the following table:

INPUT r	OUTPUT FOR $r^2 \cdot r$	OUTPUT FOR r^3
2		
4		
5		

b. How does the table demonstrate that $r^2 \cdot r$ is equivalent to r^3?

3. Simplify each expression by rewriting the base with a single exponent.

a. $x \cdot x^4$

b. $w^2 \cdot w^5$

c. $a^2 \cdot a^3 \cdot a^4$

d. $x \cdot x^2 \cdot x^3$

4. What pattern do you observe in Problem 3?

The results of Problems 2, 3, and 4 lead to the first property of exponents.

First Property of Exponents

If m and n represent positive integers, then

$$b^m \cdot b^n = b^{m+n}.$$

5. Does $x^3t^4 = xt^7$? Explain.

6. a. Multiply: $(2x^3)(3x^4)$

b. What is the coefficient of the product in part a? Explain how you obtained this coefficient.

PROCEDURE

Multiplying a Series of Factors

1. Multiply the numerical coefficients.
2. Simplify the product of the variable factors with the same base by applying the first property of exponents. That is, if m and n represent positive integers, then $b^m \cdot b^n = b^{m+n}$.

7. Multiply the following.

a. $(-3x^2)(4x^3)$

b. $(5a^3)(3a^5)$

c. $(a^3b^2)(ab^3)(b)$

d. $(3.5x)(-0.1x^4)$

e. $(r^2)(4.2r)$

8. Use the distributive property and the first property of exponents to write each of the following expressions in expanded form.

a. $x^3(x^2 + 3x - 2)$

b. $-2x(x^2 - 3x + 4)$

c. $2a^3(a^3 + 2a^2 - a + 4)$

d. $w^2(3.5w + 2.1)$

Second Property of Exponents

Recall that the volume of a solid (three-dimensional figure) is a measure of the amount of space it encloses. Many common solids have formulas that are used to determine their volume (see Chapter 6). For example, the volume V of a cube is the product of its width, length, and height, all of which have the same length, say a.

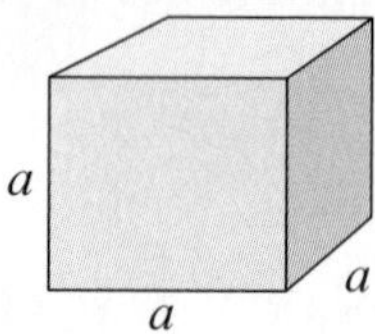

9. a. Write a formula for the volume of a cube, where a represents the length of one of its edges. (Remember that edges of a cube are the same length.)

b. Determine the volume of a cube with edge 2 centimeters.

Suppose that the length, a, of each edge of a cube is squared. The volume of the new cube can be written as $V = (a^2)^3$.

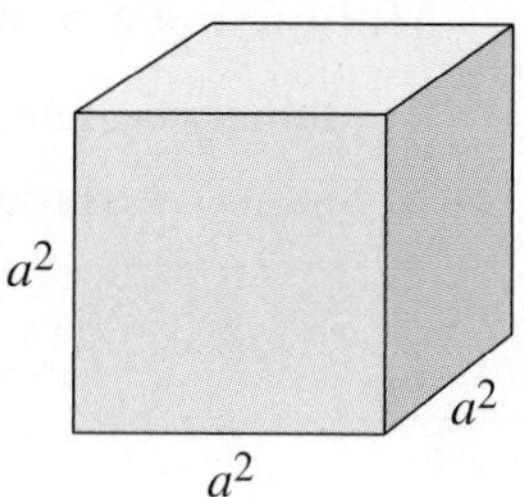

You can rewrite the expression $(a^2)^3$ more simply. First, note that the expression $(a^2)^3$ indicates that the base a^2 is used as a factor three times. Therefore,

$$(a^2)^3 = \underbrace{a^2a^2a^2}_{\substack{\text{Base } a^2 \text{ used} \\ \text{as a factor} \\ \text{3 times.}}} = \underbrace{a^{2+2+2}}_{\substack{\text{Property 1} \\ \text{of exponents}}} = a^6$$

The procedure just completed indicates how $(a^2)^3$ can be simplified *without* expanding. Do you see how? Problem 10 provides additional examples that you can use to confirm your observation or help you to discover the property.

10. Perform the given operation.

a. $(t^3)^5$ **b.** $(y^2)^4$ **c.** $(x^6)^3$

The pattern demonstrated by Problem 10 leads to the second property of exponents.

Second Property of Exponents

If m and n represent positive integers, then $(b^m)^n = b^{mn}$.

11. Use the properties of exponents to simplify each of the following expressions.

a. $(3^2)^4$

b. $(y^{11})^5$

c. $2(a^5)^3$

d. $x(x^2)^3$

e. $-3(t^2)^4$

f. $(5xy^2)(3x^4y^5)$

12. Expand each of the following expressions by using the distributive property to remove the parentheses and the properties of exponents to multiply the terms.

a. $2x(x^2 + 5)$

b. $y^2(3y - 2y^3)$

c. $ab^2(3a - 2ab + 3b)$

d. $4m^5n^3(3m^2n - 5m^3n^6)$

SUMMARY
ACTIVITY 7.4

1. First property of exponents

If m and n represent positive integers, then

$$b^m \cdot b^n = b^{m+n}.$$

2. Second property of exponents

If m and n represent positive integers, then

$$(b^m)^n = b^{mn}.$$

3. To multiply a series of factors such as $3x^4(x^2)^3 2x^3$,

i. Remove parentheses by applying the second property of exponents.

ii. Multiply the numerical coefficients.

iii. Apply the first property of exponents to variable factors that have the same base.

Therefore, $3x^4(x^2)^3 \cdot 2x^3 = 3x^4x^6 \cdot 2x^3 = 6x^{13}$.

EXERCISES
ACTIVITY 7.4

Use the properties of exponents to simplify the expressions in Exercises 1–13.

1. $a \cdot a^3$

2. $3x \cdot x^4$

3. $y^2 \cdot y^3 \cdot y^4$

4. $3t^4 \cdot 5t^2$

5. $-3w^2 \cdot 4w^5$

6. $3.4b^5 \cdot 1.05b^3$

7. $(a^5)^3$

8. $4(x^2)^4$

9. $-(x^{10})^5$

10. $(-3x^2)(-4x^7)(2x)$

11. $(-5x^3)(0.5x^6)(2.1y^2)$

12. $(a^2bc^3)(a^3b^2)$

13. $(-2s^2t)(t^2)^3(s^4t)$

Use the distributive property and the properties of exponents to expand the algebraic expressions in Exercises 14–21.

14. $2x(x + 3)$

15. $y(3y - 1)$

16. $x^2(2x^2 + 3x - 1)$

17. $2a(a^3 + 4a - 5)$

18. $5x^3(2x - 10)$

19. $r^4(3.5r - 1.6)$

20. $3t^2(6t^4 - 2t^2 - 1.5)$

21. $1.3x^7(-2x^3 - 6x + 1)$

22. a. You are drawing up plans to enlarge a square patio. You want to triple the lengths of one pair of opposite sides and double the lengths of the other opposite sides. If x represents a side of the square patio, write an equation for the new area, A, in terms of x.

b. You discover from the plan that to allow for bushes, you must take off 3 feet from each side that was doubled. Write an expression in terms of x to represent the length of those sides.

c. Use the result from part b to write an equation without parentheses to represent the new area, A, of the patio. Remember that the lengths of the other sides of the original square patio were tripled.

23. A rectangular bin has the following dimensions:

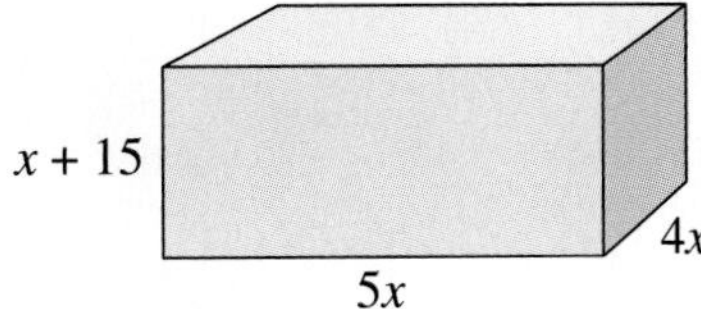

a. Write an expression that represents the area of the base of the bin.

b. Use the result from part a to write an equation in expanded form that represents the volume, V, of the bin.

24. A cube measures b^4 units on a side. Write an equation in terms of b that represents the volume of the cube.

25. A square measures $3xy^2$ units on each side. Write an equation that represents the area, A, of the square.

26. A car travels 4 hours at an average speed of $2a - 4$ miles per hour. Let d represent the distance traveled. Write an equation that expresses d in terms of the time traveled and the average speed. Leave your final equation in expanded form.

ACTIVITY 7.5

Math Magic

OBJECTIVES

1. Recognize an algebraic expression as a code of instructions to obtain an output.
2. Simplify algebraic expressions.

Algebraic expressions arise every time a sequence of arithmetic operations is applied to an input variable. The sequence of operations transforms a given value for the input variable into a single corresponding output value, as you have seen previously in Chapter 2.

Perhaps you have watched magicians on TV who astonish their audiences by correctly guessing the number that a volunteer has secretly picked. In the following example of math magic, assume that the input values in the table are the numbers secretly chosen by three different volunteers.

INPUT x (number selected)	OUTPUT y (result of sequence of operations)
3	
4	
6	

Each volunteer is asked to perform the following sequence of calculations, starting with his or her secret number.

Step 1: Add 6 to your chosen number.

Step 2: Double the result.

Step 3: Subtract 2.

Step 4: Subtract your chosen number from the result in step 3.

Step 5: This last result is the output value, which is then told to the magician.

1. **a.** Imagine you are a volunteer. Perform the sequence of arithmetic operations for each input value and record the end result (the output value) in the preceding table.

   b. Do you see a pattern between the input values and their corresponding output values?

   c. What operation does the magician need to perform in her head to correctly guess each volunteer's secret number?

If you use algebra to analyze the magician's trick in Problem 1, you will see why the output value is always 10 more than the input. The following example shows you how.

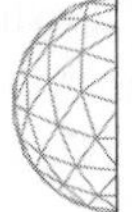

EXAMPLE 1 ***To use algebra to analyze the magician's trick in Problem 1, let x represent any chosen number. Then, using x, carry out the instructions as follows.***

Step 1	add 6 to your chosen number:	$x + 6$
Step 2	double the result:	$2(x + 6)$ $2x + 12$
Step 3	subtract 2:	$(2x + 12) - 2$ $2x + 10$
Step 4	subtract your chosen number from the result in step 3:	$(2x + 10) - x$ $x + 10$

Step 4 shows that the output from this sequence of arithmetic operations is determined by $x + 10$. That is, the output is always 10 more than the secret number x. Using y to represent the output, $y = x + 10$.

To see why the magician subtracted 10 from the volunteer's output to guess the secret number correctly, solve the equation $y = x + 10$ for x to obtain the equation

$$x = y - 10.$$

This equation shows that the secret number, x, is obtained by subtracting 10 from the output that resulted from the sequence of arithmetic operations.

In Example 1, the expressions were simplified at each step. It is also possible to wait and simplify after the last step. Example 2 shows how.

EXAMPLE 2

Step 1	add 6 to x:	$x + 6$
Step 2	double the result:	$2(x + 6)$
Step 3	subtract 2:	$2(x + 6) - 2$
Step 4	subtract your chosen number:	$2(x + 6) - 2 - x$
Step 5	simplify the result:	$2x + 12 - 2 - x$ $x + 10$

The equation relating the output to the input is $y = x + 10$.

2. a. Here is another trick you can try on a friend. Ask your friend to pick a number, keep it secret, and use it in the following sequence of instructions to obtain an output.

Step 1: Add 2 to the secret number.

Step 2: Double the result.

Step 3: Subtract 4 to obtain the output value.

b. To analyze the trick, carry out the steps in part a using x to represent the secret number. Simplify the expressions, either as you go along or after step 3.

c. Write an equation relating the output variable y to the input variable x.

d. Solve the equation in part c to show what you have to do to guess correctly your friend's secret number.

3. a. Perform the following sequence of arithmetic operations on an input variable x to determine the corresponding output variable y. Simplify the expressions, either as you go along or after the last step.

	Going Along	After Last Step
Step 1: Subtract 3 from the input, x.		
Step 2: Multiply the result by 5.		
Step 3: Add 10.		
Step 4: Divide the result by 5.		

b. Write a simple equation relating the output variable y to the input variable x.

c. Solve the equation in part b for the input variable.

4. a. Determine the algebraic expression that is equivalent to the following sequence of instructions.

   Step 1: Add 7 to the input.

   Step 2: Multiply the result by 3.

   Step 3: Subtract the sum of twice the input and 12.

   b. Simply the expression you obtained in step 3.

   c. Write an equation for the output variable y in terms of the input variable x.

   d. Write an equation that shows how to obtain the input variable from the output variable.

Many problems based on arithmetic can be solved by determining an expression that provides an algebraic model. The expression can become part of a formula or equation that can be solved. In the math magic problems, you determined an algebraic expression that modeled the trick.

In Example 2, $2x + 12 - 2 - x$ was the expression that modeled the arithmetic performed on each secret number, represented by x. When simplified to $x + 10$, it provided a formula for determining the output, represented by y. The resulting equation, $y = x + 10$, can then be solved for x to determine the secret number for any output value y.

The following problem can be solved in a similar manner.

5. In tennis, the length of a rectangular singles court is 3 feet less than 3 times its width.

   a. Let w represent the width of a singles court. Write an expression using w that represents the length of the court.

   b. Recall that the perimeter of a rectangle is the sum of twice the length and twice the width. Write a formula that represents the perimeter, P, of a singles court, using w to represent the width and the expression in part a to represent the length.

   c. Simplify the expression in the formula you wrote in part b.

d. If the perimeter of a singles court is actually 210 feet, what equation would you solve to find the actual width and length of the court?

e. Solve the equation from part d to find the width and length.

SUMMARY
ACTIVITY 7.5

To simplify an algebraic expression,

a. Remove parentheses, if necessary, by applying the distributive property.

b. Combine like terms.

EXERCISES
ACTIVITY 7.5

1. a. Determine the algebraic expression that is equivalent to the following sequence of instructions.

Step 1: Subtract 5 from the input.

Step 2: Multiply the result by 4.

Step 3: Subtract 3 times the input.

Step 4: Add 8.

b. Simply the expression you obtained in step 4.

c. Write an equation for the output variable y in terms of the input variable x.

d. Write an equation that shows how to obtain the input variable from the output variable.

2. Design your own magic trick, writing the step-by-step sequence of arithmetic operations, the corresponding algebraic expression, and the equation that can be solved to determine the original input value (secret number). Try your trick on some friends.

Exercise numbers appearing in color are answered in the Selected Answers appendix.

3. **a.** Select an integer and perform the following sequence of calculations.

Step 1: Multiply by 3.

Step 2: Add 8.

Step 3: Subtract the original number.

Step 4: Divide by 2.

Step 5: Subtract 4.

b. In part a, what is the relationship between the integer you selected and the final result of the five calculations?

c. Let x represent the integer you selected. Translate each step in part a into an algebraic expression. You can simplify after each step or wait until you have written a simple polynomial using all steps.

d. If you have not done so, simplify the expression in part c. What does your result tell you about the relationship between the integer you select (represented by x) and the result of the sequence of calculations?

In Exercises 4–10, simplify the algebraic expression.

4. $3x + 2(5x - 4)$

5. $3.1(a + b) + 8.7a$

6. $6x + 2(x - y) - 5y$

7. $(3 - x) + (2x - 1)$

8. $4(x + 3) + 5(x - 1)$

9. $3(2 - x) - 4(2x + 1)$

10. $10(0.3x + 1) - (0.2x + 3)$

In Exercises 11–16, evaluate the expression for the given values.

11. $3x^2 + 2x - 1$ for $x = 4$

12. $2(l + w)$ for $l = 3.5$ and $w = 2.8$

13. Prt for $P = 2100$, $r = 8\%$, $t = 3$

14. $0.3d^2 + 4d$ for $d = 2.1$

15. $4x^3 + 15$ for $x = 10$

16. $3(f - 20)$ for $f = -5$

17. Opposite sides of a square are each increased by 5 units and the other opposite sides are each decreased by 3 units. If x represents a side of the original square, write an equation that represents the perimeter P of the newly formed rectangle.

18. You want to boast to a friend about the stock that you own without telling him how much money you originally invested in the stock. You watch the stock market once a month for 4 months and record the following.

MONTH	1	2	3	4
STOCK VALUE	Increased $50	Doubled	Decreased $100	Tripled

a. Letting x represent the value of your original investment, write an algebraic expression to represent the value of your stock after the first month.

b. Use the result from part a to determine the value at the end of the second month, simplifying when possible. Continue until you determine the value of your stock at the end of the fourth month.

c. Do you have good news to tell your friend? Explain to him what has happened to the value of your stock over 4 months.

d. Instead of simplifying after each step, write a single algebraic expression that represents the value of the stock at the end of 4 months.

e. Simplify the expression in part d. How does the simplified algebraic expression compare with the result in part b?

ACTIVITY 7.6

Mathematical Modeling

OBJECTIVES

1. Develop an equation to model and solve a problem.
2. Solve problems using formulas as models.
3. Recognize patterns and trends between two variables using a table as a model.
4. Recognize patterns and trends between two variables using a graph as a model.

A model of an object is usually an alternative version of the actual object, produced in another medium. You are familiar with model airplanes and model cars. Architects and engineers build models of buildings and bridges. Computer models are used to simulate complicated phenomena like the weather and global economies. To be useful, a model must accurately describe an object or situation, in order to better understand the actual object.

In this course, you have been using **mathematical models.** A mathematical model uses equations, formulas, tables, or graphs to describe the important features of an object or situation. Such models can then be used to solve problems, make predictions, and draw conclusions about the given situation.

The process of developing a mathematical model is called **mathematical modeling**. The development of mathematical models to help solve simple to extremely complicated problems is a very important use of mathematics.

Equations as Mathematical Models

In Activity 7.1, you developed the equation $c = 0.015n + 455$, where c represents the monthly cost in dollars of leasing a copier and n represents the number of copies. The equation represents the mathematical model used by the copier company to determine charges for leasing the machine. The equation is an efficient format for solving problems regarding the leasing of copy machines by the company.

1. As part of a community service project at your college, you are organizing a fund-raiser at the neighborhood roller rink. Money raised will benefit a summer camp for children with special needs. The admission charge is \$4.50 per person, \$2.00 of which is used to pay the rink's rental fee. The remainder is donated to the summer camp fund.

   a. The first step in developing a mathematical model is to identify the problem and develop a well-defined question about what you want to know. State a question that you want answered in this situation.

   b. The next step is to identify the variables involved. What two variables are involved in this problem?

   c. Which variable can best be designated as the input variable? As the output variable?

   d. Next, look for relationships and connections between the variables involved in the situation. State in words the relationship between the input and output variables.

e. Now, translate the features and relationships you have identified into an equation. Use appropriate letters to represent the variables, stating what each represents.

f. If 91 tickets are sold, use the equation model developed in part e to determine the amount donated to the summer camp fund.

g. If the maximum capacity of the rink is 200 people, what is the maximum amount that can be donated?

Formulas as Mathematical Models

In previous chapters, you used formulas that model situations in geometry, business, and science. In almost every field of study, you are likely to encounter formulas that relate two or more variables represented by letters. Problems 2 and 3 feature formulas used in the health field.

2. In order for exercise to be beneficial, medical researchers have determined that the desirable heart rate, R, in beats per minute, can be approximated by the formulas

$$R = 143 - 0.65a \text{ for women}$$
$$R = 165 - 0.75a \text{ for men,}$$

where a represents the person's age in years.

a. If the desirable heart rate for a woman is 130 beats per minute, how old is she?

b. If the desirable heart rate for a man is 135 beats per minute, how old is he?

3. The basal energy rate is the daily amount of energy (measured in calories) needed by the body at rest to maintain the basic life functions. The basal energy rate differs for individuals, depending on their gender, age, height, and weight. The formula for the basal energy rate for males is

$$B = 655.096 + 9.563W + 1.85H - 4.676A,$$

where: B is the basal energy rate (in calories)
W is the weight (in kilograms)
H is the height (in centimeters)
A is the age (in years)

 a. A male patient is 70 years old, weighs 55 kilograms, and is 172 centimeters tall. He is prescribed a total daily calorie intake of 1000 calories. Determine if the patient is being properly fed.

 b. A male is 178 centimeters tall and weighs 84 kilograms. If his basal energy rate is 1500 calories, how old is the male?

Tables as Mathematical Models

You have encountered tables of ordered pairs of values throughout this course. Those tables can be viewed as a type of mathematical model that represents relationships between variables. For example, the windchill table studied in Activity 7.2 allows one to predict the windchill when temperature and wind speed are known.

4. The following table gives the windchill temperature (how cold your skin feels) for various air temperatures when there is a 20 mph wind.

Blowin' in the Wind

AIR TEMPERATURE °F	40	30	20	10	0	−10	−20
WINDCHILL TEMPERATURE (20-mph wind)	30	17	4	−9	−22	−35	−48

 a. What relationship do you observe between the windchill temperature and the air temperature?

b. Estimate the windchill temperature for an air temperature of −30°F.

c. If the wind speed is 30 mph, how would you expect the windchill temperatures in the table to change?

The following Windchill Chart, published by the National Weather Service, displays windchill temperature for many different wind speeds and air temperatures.

WINDCHILL CHART

Wind (mph) \ Air Temperature (°F)	40	35	30	25	20	15	10	5	0	−5	−10	−15	−20	−25	−30	−35	−40	−45
5	36	31	25	19	13	7	1	−5	−11	−16	−22	−28	−34	−40	−46	−52	−57	−63
10	34	27	21	15	9	3	−4	−10	−16	−22	−28	−35	−41	−47	−53	−59	−66	−72
15	32	25	19	13	6	0	−7	−13	−19	−26	−32	−39	−45	−51	−58	−64	−71	−77
20	30	24	17	11	4	−2	−9	−15	−22	−29	−35	−42	−48	−55	−61	−68	−74	−81
25	29	23	16	9	3	−4	−11	−17	−24	−31	−37	−44	−51	−58	−64	−71	−78	−84
30	28	22	15	8	1	−5	−12	−19	−26	−33	−39	−46	−53	−60	−67	−73	−80	−87
35	28	21	14	7	0	−7	−14	−21	−27	−34	−41	−48	−55	−62	−69	−76	−82	−89
40	27	20	13	6	−1	−8	−15	−22	−29	−36	−43	−50	−57	−64	−71	−78	−84	−91
45	26	29	12	5	−2	−9	−16	−23	−30	−37	−44	−51	−58	−65	−72	−79	−86	−93
50	26	19	12	4	−3	−10	−17	−24	−31	−38	−45	−52	−60	−67	−74	−81	−88	−95
55	25	18	11	4	−3	−11	−18	−25	−32	−39	−46	−54	−61	−68	−75	−82	−89	−97
60	25	17	10	3	−4	−11	−19	−26	−33	−40	−48	−55	−62	−69	−76	−84	−91	−98

Frostbite Times: 30 minutes, 10 minutes, 5 minutes

5. a. In the Windchill Chart, locate the row that lists windchill temperatures for a 30 mph wind and use it to complete the following table.

AIR TEMPERATURE °F	40	30	20	10	0	−10	−20
WINDCHILL TEMPERATURE (30-mph wind)							

b. If the wind speed is 40 mph, what is the windchill temperature if the air temperature is −20°F?

c. Approximately how long could you be exposed to a 20-mph wind when the air temperature is 0°F? (That would be the frostbite time in the chart.)

Graphs as Mathematical Models

Graphs are very effective mathematical models that can be used to visualize patterns and trends between two variables.

6. Medicare is a government program that helps senior citizens pay for medical expenses. As the U.S. population ages, the expense and quality of health service becomes an increasing concern. The following graph presents Medicare expenditures from 1967 through 2004. Use the graph to answer the following questions.

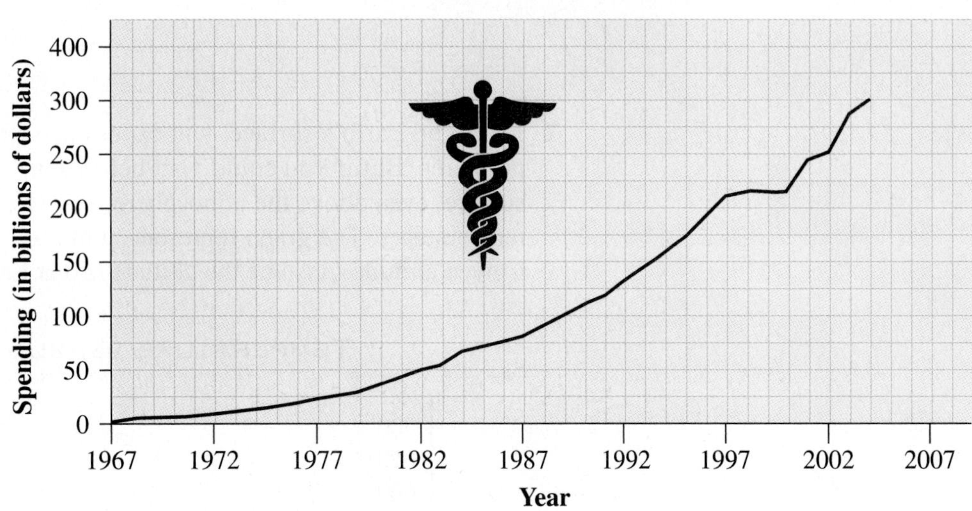

Source: Centers for Medicare and Medicaid Services

a. Use the graph to estimate the Medicare expenditures for the years listed in the following table.

YEAR	MEDICARE EXPENDITURE (billions of dollars)
1972	
1977	
1982	
1987	
1992	
1997	
2001	
2004	

b. Estimate the year in which expenditures reached \$25 billion.

c. Estimate the year in which expenditures reached \$100 billion.

d. Approximately how much did Medicare expenditures increase between 1987 and 1997?

e. During which 10-year period did Medicare expenditures change the least?

f. During what period was there essentially no change in Medicare expenditures?

7. Living on Earth's surface, you experience a relatively narrow range of temperatures. But if you could visit below Earth's surface or high up above the surface, even above the atmosphere, you would experience a wider range of temperatures. The graph represents a model that predicts the temperature for a given altitude. Assume the altitude is 0 at Earth's surface.

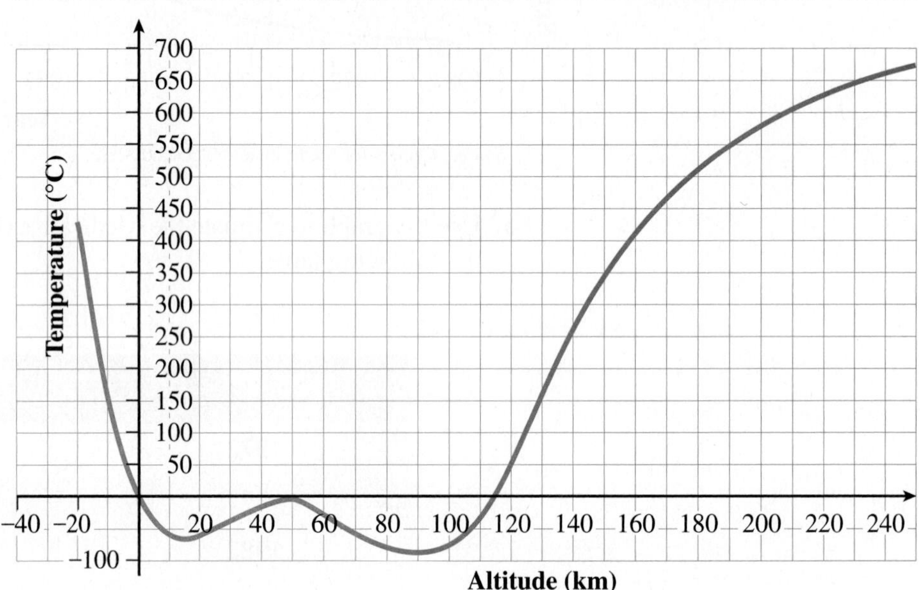

a. What is the temperature of Earth 10 kilometers below the surface?

b. The ozone layer is approximately 50 kilometers above Earth's surface. What is the approximate temperature in the ozone layer?

c. Is it warmer or cooler above and below the ozone layer?

d. Describe how the temperature changes as one moves up through the atmosphere.

e. How deep under Earth's surface does one need to go to reach a temperature of 400°C?

f. How high above Earth's surface does one need to go to reach a temperature of 400°C?

SUMMARY
ACTIVITY 7.6

1. A **mathematical model** uses equations, formulas, tables, or graphs to describe the important features of an object or situation. Such models can then be used to solve problems, make predictions, and draw conclusions about the given situation.
2. Mathematical modeling is the process of developing a mathematical model for a given situation.

EXERCISES
ACTIVITY 7.6

1. The value of almost everything you own, such as a car, computer, or appliance, depreciates (goes down) over time. When the value decreases by a fixed amount each year, the depreciation is called straight-line depreciation.

   Suppose your car has an initial value of \$16,750 and depreciates \$1030 per year.

   a. State a question that you might want answered in this situation.

   b. What two variables are involved in this problem?

   c. Which variable do you think should be designated as the input variable?

   d. Complete the following table.

YEARS CAR IS OWNED	1	2	3	4	5
VALUE OF CAR (in dollars)					

   e. State in words the relationship between the value of the car and the number of years the car is owned.

f. Use appropriate letters to represent the variables involved and translate the written statement in part e to an equation.

g. If you plan to keep the car for 7 years, determine the value of the car at the end of this period. Explain the process you used.

2. In 1966, the U.S. Surgeon General's warnings began appearing on cigarette packages. A model that predicts the percentage of the adult U.S. population that was still smoking in the years since 1966 can be represented by the formula $p = -0.62t + 42.47$, where t is the number of years after 1965 and p is the percentage of the adult population that smoked.

a. Determine the percentage of the population that smoked in 1981 ($t = 16$).

b. Using the formula, in what year would the percentage of smokers be 15% ($p = 15$)?

3. In 1965, 51.9% of all males (18 or older) smoked. The percentage, p, of males who smoke in t years after 1965 is modeled by the formula $p = -0.89t + 51.1$.

a. Determine the percentage of males smoking in the year 2000.

b. In what year would the percentage of male smokers 18 or older be 20%?

4. The average annual out-of-pocket expenses for health care for an individual can be modeled by the formula

$$c = 37.7a - 170,$$

where c is the average amount of money spent, in dollars, and a is the person's age.

a. Determine the average annual out-of-pocket health care expenses for a 30-year-old.

b. Using the formula, how old is an individual with out-of-pocket health care expenses of \$2000?

5. The following formula is used by the National Football League (NFL) to calculate quarterback ratings.

$$R = \frac{6.25A + 250C + 12.5Y + 1000T - 1250I}{3A}$$

where: R = quarterback rating
A = passes attempted
C = passes completed
Y = passing yardage
T = touchdown passes
I = number of interceptions

In the 2005–2006 regular season, quarterbacks Tom Brady, of the New England Patriots, and Brett Favre, of the Green Bay Packers, had the following player statistics.

Take a Pass

PLAYER	PASSES ATTEMPTED	PASSES COMPLETED	PASSING YARDAGE	NUMBER OF TOUCHDOWN PASSES	NUMBER OF INTERCEPTIONS
Tom Brady	530	334	4110	26	14
Brett Favre	607	372	3881	20	29

a. Determine the quarterback rating for Tom Brady for the 2005–2006 NFL football season.

b. Determine the quarterback rating for Brett Favre.

c. Visit www.nfl.com and select stats to obtain the rating of your favorite quarterback.

6. a. You want to invest so as to receive the best return on your money. You have two options.

Option 1: Invest at 6% simple annual interest for 10 years.

Option 2: Invest at 5% interest compounded annually.

The following table models the growth of $2000 over the 10-year period for the two options.

Interesting Choices

NUMBER OF YEARS	1	2	3	4	5	6	7	8	9	10
6% SIMPLE ANNUAL INTEREST	2120	2240	2360	2480	2600	2720	2840	2960	3080	3200
5% COMPOUNDED ANNUALLY	2100	2205	2315.30	2431	2552.60	2680.20	2814.20	2954.90	3102.70	3257.80

Describe any trends or patterns that you observe in the data.

b. The value of your investment in Option 1 can be determined by the following formula.

$$A = P + Prt$$

where: A = value of the investment
P = principal or amount invested
r = annual percentage rate (expressed as a decimal)
t = number of years invested

Use the formula to determine the value of your $2000 investment in Option 1 after 20 years. What is the total amount of interest earned?

c. The value of your investment in Option 2 can be determined by

$$A = P(1 + r)^t$$

where: A = value of the investment
P = principal
r = annual percentage rate (expressed as a decimal)
t = number of years invested

Use the formula to determine the value of your $2000 investment in Option 2 after 20 years.

d. Which option would you choose if you were planning to invest the principal for 20 years? Explain.

7. The following table gives the number of people infected by the flu over a given number of months.

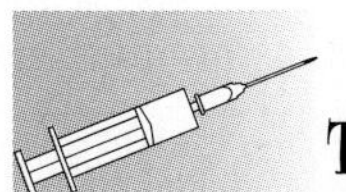

The Spread of the Flu

NUMBER OF MONTHS	0	1	2	3	4	5
NUMBER OF PEOPLE INFECTED	1	5	13	33	78	180

a. Describe any trends or patterns that you observe in the table.

b. Graph the ordered pairs in the table on the following grid. Connect successive points with line segments.

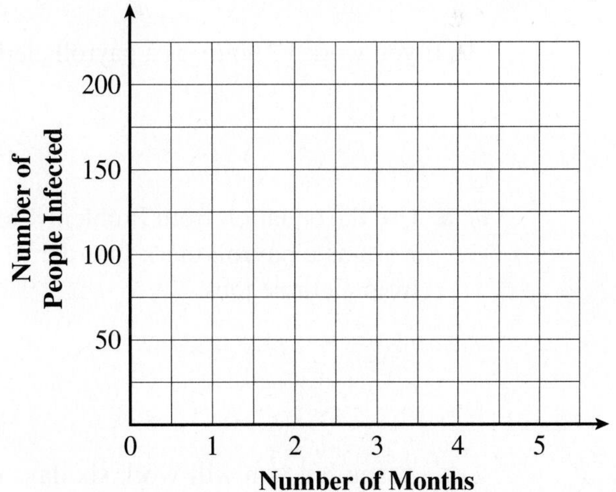

c. Describe any trends or patterns that you observe on the graph. Compare with those you made based on the table.

What Have I Learned?

You are able to get three summer jobs to help pay for college expenses. You work 20 hours per week in your job as a cashier and earn \$6.50 per hour. The second and third jobs are both at a local hospital. You earn \$8.50 per hour as a payroll clerk and \$6.00 per hour as an aide. You always work 3 hours more per week as an aide than you do as a payroll clerk. Your weekly salary is determined by the number of hours that you work at each job.

Use what you learned in this chapter to answer the following questions about your summer job.

1. Explain how you would calculate the total amount earned each week.

2. Let x represent the number of hours that you work as a payroll clerk. Represent the number of hours that you work as an aide in terms of x.

3. Write an equation in terms of x that you can use to calculate the total amount E that you earn in a week from all three jobs.

4. Simplify the equation you obtained in Problem 3.

5. If you work 12 hours as a payroll clerk, how much will you earn that week?

6. **a.** Use the equation from Problem 4 to determine how many hours you must work as a payroll clerk in order to have a total salary of \$500 in 1 week from all three jobs.

   b. Assuming you will work six days each week, approximately how many hours must you work each day to earn \$500 a week?

The part-time jobs at the hospital have been eliminated and you need to find another job. You read about a full-time summer position in sales at the Furniture Barn. You would earn \$260 *per week plus* 20% *commission on sales over* \$1000. *You decide to quit the cashier's job and take the sales position at the Furniture Barn.*

7. Explain how you would calculate the total amount earned each week.

8. Let F represent the total amount earned each week and let x represent the amount of sales for the week. Write an equation in terms of x to determine F.

9. Use the equation you wrote in Problem 8 to determine how much your sales must be to have a gross salary of \$500 for the week.

How Can I Practice?

In Exercises 1–14, solve each equation for x. Check your results by hand or with a calculator.

1. $4x - 7 = 9$

2. $9 - 2x = 23$

3. $\frac{3}{4}x + 2 = 5$

4. $15 = 5 - 2.5x$

5. $2x - 5 = 4x + 7$

6. $3(2x + 1) = 9$

7. $2(x + 1) = 5x - 3$

8. $14 - 6x = -2x + 3$

9. $2.1x + 15 = 3.5 - 1.9x$

10. $4(2x + 1) = 2(7x + 7)$

11. $3(x - 2) + 10 = 5x$

12. $0.25(x - 2) = 0.2(x + 10)$

13. $\frac{1}{2}x - 6 = 3x + 4$

14. $\frac{1}{3}(x - 1) = 3(2x - 2)$

Answers to all How Can I Practice? exercises are included in the Selected Answers appendix.

In Exercises 15–18, use the distributive property to write each of the following products in expanded form.

15. $-5(4x - 3)$

16. $3x(2a + 4b + c)$

17. $4.5(3x - 0.2)$

18. $-4(5x - 4y + 10)$

In Exercises 19–23, solve the equation for the specified letter.

19. $P = 2a + b$ for a

20. $P = rt$ for t

21. $f = v + at$ for t

22. $3x - 4y = 8$ for x

23. $A = P(1 + rt)$ for r

24. A worker's weekly earnings are given by the formula $E = S + \frac{3}{2}rn$, where E represents the weekly earnings, S the weekly salary, r the hourly rate, and n the number of hours worked overtime. Solve the formula for r.

25. Since 1960, the median age of men at their first marriage has steadily increased. If a represents the median age of men at their first marriage, then a can be approximated by the formula

$$a = 0.11t + 22.5,$$

where t represents the number of years since 1960.

a. What is the median age of men who are married for the first time in 2005?

b. According to the formula, in what year will the median age of men be 30 at first marriage?

c. Solve the formula $a = 0.11t + 22.5$ for t.

d. Answer part b using the new formula in part c.

26. The cost of printing a brochure to advertise your lawn-care business is a flat fee of \$10 plus \$0.03 per copy. Let c represent the total cost of printing and x represent the number of copies.

a. Write an equation that will relate c and x.

b. Use the equation from part a to complete the following table. Begin with 1000 copies, increase by increments of 1000, and end with 5000 copies.

Duplicated Effort

NUMBER OF COPIES, x	TOTAL COST, c
1000	

c. What is the total cost of printing 8000 copies?

d. You have \$300 to spend on printing. How many copies can you have printed for that amount?

In Exercises 27–34, use the properties of exponents to simplify the product.

27. $-x^3 \cdot x^5 \cdot x^2$

28. $(-2x^3)(5x^4)$

29. $(2.75x^3)(-0.2x^4)$

30. $(5s^2t^3)(-3st^2)$

31. $x^3(x^2 + 2x - 1)$

32. $(y^3)^5$

33. $(t^4)^2$

34. $3(x^2)^3(x^4)(2x)$

In Exercises 35–38, use the distributive property to perform the multiplication and combine like terms if possible.

35. $2x + 3x(x - 4) + 5x^2$

36. $3.5(2x + 4) - (2.7x + 10.6)$

37. $2x^3(4x^2 - 3x + 2)$

38. $5x(2 - 4x) - 3(x^2 - 4x - 1) + 5x^2$

39. A rectangular box has dimensions given by the following expressions.

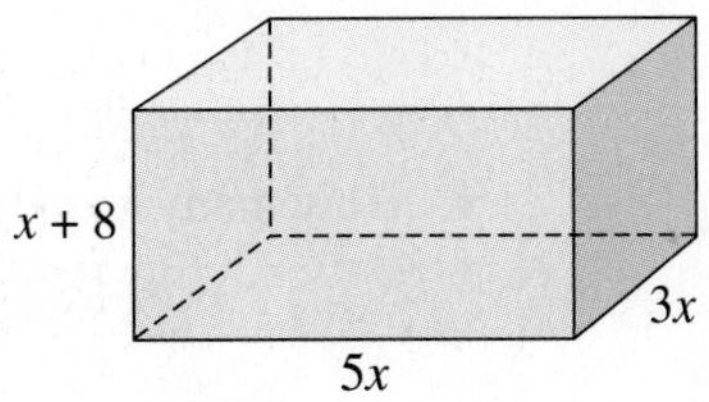

a. Write an equation that represents the area A of the box's base.

b. Write an equation that represents the volume V of the box.

In Exercises 40–43, evaluate the expression for the given value(s).

40. $3t^2$ for $t = 4$

41. $2x^2 - 3y$ for $x = 3$ and $y = 2.5$

42. $P(1 + rt)$, where $P = 5$, $r = 7$, and $t = 2$

43. $180 - t - r$, where $t = 15$, and $r = -25$

In Exercises 44–47, solve the equation for x. Check your answers by hand or with a calculator.

44. $21 + 3(x - 4) = 24$

45. $2(4x - 3) = 3(2x + 6)$

46. $2 - 5(x + 5) = 3(x - 2) - 1$

47. $0.16x + 0.24(10 - x) = 1.8$

48. You contact the local print shop to produce a commemorative booklet for your college theater group's twenty-fifth anniversary. The print shop has quoted you a price of $750 to typeset the booklet and 25 cents for each copy produced.

a. Write an equation that gives the total cost C in terms of the number, x, of booklets produced.

b. Determine the total cost of producing 500 booklets.

c. How many booklets can be produced for $1000?

d. Suppose the booklets are sold for 75 cents each. Write an equation for the total revenue R from the sale of x booklets.

e. How many booklets must be sold to break even? That is, for what value of x is the total cost of production, C, equal to the total amount of revenue, R?

f. Recall that the profit is revenue minus the cost $(R - C)$. How many booklets must be sold to make a \$500 profit?

49. You live 7.5 miles from work, where you have free parking. Some days, you must drive to work so that you can call on clients. On other days, you can take the bus. It costs you 20 cents per mile to drive the car and \$1.50 round-trip to take the bus. Assume that there are 22 working days in a month and that x represents the number of days you take the bus to work.

a. Write an expression in terms of x for the cost of taking the bus each month.

b. Write an expression in terms of x that represents the number of days that you drive.

c. Write an expression in terms of x for the cost of driving each month.

d. Write an equation in terms of x that determines the total cost of transportation, C.

e. How many days can you drive if you budget \$42 a month for transportation?

f. How much should you budget for the month if you would like to take the bus only half of the time?

CHAPTER 7 SUMMARY

The bracketed numbers following each concept indicate the activity in which the concept is discussed.

CONCEPT / SKILL	DESCRIPTION	EXAMPLE
Solve $ax + b = c$ for x [7.1]	• Undo the addition of b by adding the opposite of b to each side of the equation. • Undo the multiplication of the variable by the nonzero coefficient a by dividing each side by a.	Solve for x: $2x - 5 = 11$ $2x - 5 = 11$ $+5 \quad +5$ $2x = 16$ $\frac{2x}{2} = \frac{16}{2}$; $x = 8$
Evaluate expressions [7.2], [7.6]	To evaluate expressions for a given numerical value, substitute the value for the variable and perform the arithmetic using the order of operations.	Evaluate $2x + 10$ for $x = -2$. Substitute -2 for x: $2(-2) + 10$ $= -4 + 10$ $= 6$
Solve a formula for a given variable [7.2]	To solve the formula $y = at + b$ for t means to isolate the variable t, with coefficient 1, on one side of the equation, with all other terms on the opposite side.	Solve $y = at + b$ for t. $y = at + b$ $-b \quad -b$ $y - b = at$ $\frac{y-b}{a} = \frac{at}{a}$ $\frac{y-b}{a} = t$
Distributive property [7.3], [7.4]	$a(b + c) = ab + ac$ Extension of distributive property: $a(b + c + d + \ldots) =$ $ab + ac + ad + \ldots$	$3(2x^2 + 4x - 1)$ $= 6x^2 + 12x - 3$
General strategy for solving an equation for an unknown x [7.3]	1. Remove parentheses, if necessary. 2. Combine like terms on the same side of the equation. 3. Isolate the variable term. 4. Divide out the nonzero coefficient of the variable. 5. Check.	$2(x + 4) + 3x = 2x - 1$ $2x + 8 + 3x = 2x - 1$ $5x + 8 = 2x - 1$ $-2x - 8 \quad -2x - 8$ $3x = -9$ $\frac{3x}{3} = \frac{-9}{3}$ $x = -3$ *Check:* $2(-3 + 4) + 3(-3) = 2(-3) - 1$ $2 - 9 = -6 - 1$ $-7 = -7$

CONCEPT / SKILL	DESCRIPTION	EXAMPLE
First Property of Exponents [7.4]	If m and n represent positive integers, then $$b^m \cdot b^n = b^{m+n}.$$	$(x^3)(x^4) = x^{3+4} = x^7$
Multiply two or more factors [7.4]	1. Multiply the numerical coefficients. 2. Apply first property of exponents where powers have the same base.	$(3x^2y^4)(-2x^5y)$ $= 3(-2)x^2x^5y^4y$ $= -6x^7y^5$
Second Property of Exponents [7.4]	If m and n represent positive integers, then $$(b^m)^n = b^{mn}.$$	$(x^4)^3 = x^{4 \cdot 3} = x^{12}$
Simplify an algebraic expression [7.5]	To simplify: 1. Apply distributive property to remove parentheses, if necessary. 2. Combine like terms.	$2x(5x^2 + 4x - 6) + 7x^3 - 4x$ $= 10x^3 + 8x^2 - 12x + 7x^3 - 4x$ $= 17x^3 + 8x^2 - 16x$
Mathematical Model [7.6]	Description of mathematical features of an object or situation that uses equations, formulas, tables, and/or graphs.	See Activity 7.6
Mathematical Modeling [7.6]	The process of developing a mathematical model for a given situation.	See Activity 7.6

21. You have an opportunity to be the manager of a day camp for the summer. You know that your fixed costs for operating the camp are \$600 per week, even if there are no campers. Each camper costs the management \$10 per week. The camp charges each camper \$40 per week.
Let x represent the number of campers.

a. Write an equation in terms of x that represents the total cost, C, of running the camp per week.

b. Write an equation in terms of x that represents the total income (revenue), R, from the campers per week.

c. Write an equation in terms of x that represents the total profit, P, from the campers per week.

d. How many campers must attend the camp to break even with revenue and costs?

e. The camp would like to make a profit of \$600. How many campers need to enroll to make that profit?

f. How much money would the camp lose if only 10 campers attend?

APPENDIX

LEARNING MATH OPENS DOORS: TWELVE KEYS TO SUCCESS

1. Are you in the right math course for your skill level?

Students are sometimes placed in the wrong math course for a variety of reasons. Your answers to the following questions will help you and your instructor determine if you are in the correct math course for your skill level. This information will also help your instructor understand your background more quickly, thus providing you with a better learning experience.

a. Did you register for this math course because you took a placement exam and, as a result of your exam score, you selected or were placed in this course?

b. If you were not placed in this class because of a placement exam, please list reasons you are taking this course.

c. Name ________________________________

d. Phone and/or e-mail address ________________________________

e. List the mathematics courses you have taken at the *college level.*

Course Title	Grade
1. ____________________________	________
2. ____________________________	________

f. List the mathematics courses you have taken at the *high school level.*

Course Title	Grade
1. ____________________________	________
2. ____________________________	________
3. ____________________________	________

g. Have you taken this course before? Yes ________ No ________

h. When did you take your last mathematics course? ________________

i. Are you a full-time or part-time student? ________________

j. Do you have a job? ________ If yes, how many hours per week do you work? ________

k. Do you take care of children or relatives at home? ________

You should now share your information with your instructor to make sure you are in the correct class. It is a waste of a semester of time and money if the course is too easy for you or too difficult for you. Take control of and responsibility for your learning!

2. What is your attitude about mathematics?

a. Write one word that describes how you feel about learning mathematics.

____

b. Was the word that you wrote a positive word, a negative word, or a neutral word? ________________

If your word was a positive word, you are on your way to success in this course. If your word was negative, then before progressing any further, you may want to determine why you have negative feelings toward mathematics.

c. Write about a positive or negative experience that you have had related to mathematics.

____

____

____

____

If you have not had success in the past, try something different. Here are a few suggestions:

d. Make a list of all the materials (pencils, notebook, calculator) that you might need for the course. Make sure you have all the materials required for the course.

____

____

e. Find a new location to study mathematics. List two or three good places you can study.

1. ________________ 2. ________________ 3. ________________

f. Study with a classmate. Help each other organize the material. Ask each other questions about assignments. Write down names, phone numbers/e-mail addresses of two or three other students in your class to study with.

1. ______________________________

2. ______________________________

3. ______________________________

g. How did you study mathematics in the past? Write a few sentences about what you did outside of class to learn mathematics.

____

____

____

h. What will you change from your past to help you become more successful? Write down your strategy for success.

____

____

____

3. Do you attend all classes on time and are you organized?

Class work is vital to your success. Make it a priority to attend every class. Arrive in sufficient time and be ready to start when class begins. Start this exercise by recording the months and dates for the entire semester in each box. Then write in each box when and where your class meets corresponding to your schedule. Keep track of exam and quiz dates, deadlines, study sessions, homework, and anything else that can help you succeed in the course.

Schedule routine medical and other appointments so you don't miss class. Allow time for traffic jams, finding a parking space, bus delays, and other emergencies that may occur. Arriving late interferes with your learning and the learning of your fellow students. Make sure assignments are completed and handed in on time. Attending class and being on time and organized is in your control.

✔ Place a check in each day that you are in class. Circle the check to indicate that you were on time. Place an *a* in each day you miss class.

Sunday	Monday	Tuesday	Wednesday	Thursday	Friday	Saturday

4. When is that assignment due?

In mathematics, assignments are usually given each class time. Assignments are meant to reinforce what you learned in class. If you have trouble completing your assignment, take charge and get help immediately. The following table will help you organize your assignments and the dates they are due. Keeping a record of your assignments in one location will help you know at a glance what and when your assignment is due.

Date Assigned	Assignment	Specific Instructions	Dates	
			Due	Completed

Date Assigned	Assignment	Specific Instructions	Dates	
			Due	Completed

5. Do you keep track of your progress?

Keeping track of your own progress is a great way to monitor the steps you are taking toward success in this course. Different instructors may use different methods to determine your grade.

a. Explain how your instructor will determine your final grade for this course.

____

____

____

____

b. Use the following table to keep track of your grades in this course. (You may want to change the headings to reflect your instructor's grading system; an Excel spreadsheet may also be helpful for this exercise.)

Date	Type of Assessment	Topic(s) and/ or Chapters	Number of Points Correct	Total Number of Points	Comments

c. Determine your final average using your instructor's grading system.

6. How well do you know your textbook?

Knowing the structure of your textbook helps you use the book more effectively to reach your learning goals. The Preface and "To the Student" sections at the beginning of the book provide guidance and list other valuable resources. These include CDs/DVDs, Web sites, and computer software with tutorials and/or skill practice for added help in the course. Use the following guidelines to learn the structure and goals of the textbook.

a. According to your syllabus, which chapters will you be studying this semester?

b. Find and underline the titles of the chapters in the table of contents.

c. Read the Preface and summarize the key points.

d. In the Preface, you should have found the student supplements for the course. List them.

e. What are the key points the authors make to you in the "To the Student" section?

f. Each chapter is divided into smaller parts called clusters. Name clusters in Chapter 4.

g. Each cluster is made up of smaller sections called activities. What is the title of the activity in Chapter 3, Cluster 1, Activity 1?

h. Within an activity you will see information that has a box around it. Why is this information important? Look at Chapter 2, Cluster 1, Activity 1 before you respond.

i. Look at the end of each activity. There you will find the Exercises section. Doing these exercises should help you better understand the concepts and skills in the activity. Why are some of the numbers of the exercises in color?

j. Look at the end of Cluster 1 in Chapter 4. You will see a section entitled "What Have I Learned?" This section will help you review the key concepts in the cluster. What are the concepts taught in this cluster?

k. Look at the end of Cluster 1 in Chapter 4. What is the section called right after "What Have I Learned?" This section will help you practice all of the skills in the cluster. What are the skills taught in this cluster?

l. At the end of each chapter you will see a Gateway and a Summary. Look at the end of Chapter 2 and write how each of these sections will help you in this course.

m. Locate and briefly review the Glossary. Select a mathematical term from the Glossary and write the term and its definition here.

7. Where does your time go?

Managing your time is often difficult. A good rule to follow when studying is to allow 2 hours of study time for each hour of class time. It is best not to have marathon study sessions, but, rather, to break up your study time into small intervals. Studying a little every day rather than "cramming" once a week helps to put the information into your long-term memory. Ask yourself this question: Is going to school a top priority? If you answered yes, then your study time must be a priority if you are going to be successful.

a. In the past, approximately how many hours per week did you study mathematics? Reflect on whether it was enough time to be successful.

b. The commitments in your life can be categorized. Estimate the number of hours that you spend each week on each category. (There are 168 hours in a week.)

Class Hours (Class) _________ Chore Hours (C) _________

Study Hours (ST) _________ Sleep Hours (Sl) _________

Work Hours (W) _________ Personal Hours (P) _________

Commute Hours (T) _________ Leisure Hours (L) _________

Other (O) _________

c. Use the information from part b to fill in the grid with a workable schedule using the appropriate letter to represent each category.

Time	Mon.	Tues.	Wed.	Thurs.	Fri.	Sat.	Sun.
12 midnight							
1:00 A.M.							
2:00							
3:00							
4:00							
5:00							
6:00							
7:00							
8:00							
9:00							
10:00							
11:00							
12:00 noon							
1:00 P.M.							
2:00							
3:00							
4:00							
5:00							
6:00							
7:00							
8:00							
9:00							
10:00							
11:00							

d. Is your schedule realistic? Does it represent a typical week in *your* life? If not, make adjustments to it now.

e. Circle the hours in your schedule that you will devote to studying.

f. Shade or highlight the hours that you will use for studying mathematics.

g. Do you have some times scheduled each day to study mathematics? How much time is devoted to mathematics each day?

h. Studying math soon after class time will help you review what you learned in class. Did you place any hours to study math close to your actual class time?

i. Scheduling time before class can help you prepare for the class. Did you schedule any study time just before class time?

j. Review your schedule once more and make any changes.

k. Periodically review your schedule to see if it is working and whether you are following it. How are you doing?

8. Where will you use mathematics?

Students often ask math instructors, "Where will I ever use this?" Here is an opportunity to show where you have used mathematics, including what you have learned from this course.

a. Describe one way in which you use math in your everyday life.

____

b. Find an article in a newspaper that contains a graph or some quantitative information.

c. Describe ways in which you use math in your job.

____

____

d. Describe a problem in a course you are taking or took (not a mathematics course) that involved a math concept or skill. Include a description of the math concept or skill that is involved.

____

____

After you have been in this course for a few weeks, answer parts e, f, and g.

e. Write a problem from one of your courses (not a mathematics course) or from your work experience where you used concepts or skills that you have learned in *this* mathematics class.

____

____

f. Show how you solved the problem.

____

____

g. Write the math skill(s) or concept(s) that you learned in this math course that helped you solve this problem.

____

____

9. Do you need extra help?

Receiving extra help when you need it may mean the difference between success and failure in your math course. If you are struggling, don't wait, seek help immediately. When you go for extra help, identify specific areas in which you need help. Don't just say "I'm lost."

a. Your instructor usually will note office hours right on or near the office door or they may be listed in the syllabus. Take a few extra minutes and locate the office. List your instructor's office hours below. Write the office room number next to the hours. Circle the office hours that work with your schedule.

b. Does your college have a center where you can go for tutoring or extra help? If there is one, what is the room number? Take a minute and locate the center. When is the center open? Determine what hours would be good for you if you should need help.

c. Students in your class may also help. Write the phone numbers or e-mail addresses of two or three students in your class whom you could call for help. These students may form a study group with you. When would it be a good time for all of you to meet? Set up the first meeting.

d. You can also help yourself by helping others learn the information. By working a problem with someone else, it will help you to reinforce the concepts and skills. Set aside some time to help someone else. What hours are you available? Tell someone else you are available to help. Write the person's name here.

Extra help is also there for you during class time. Remember to ask questions in class when you do not understand. Don't be afraid to raise your hand. There are probably other students who have the same question. If you work in groups, your group can also help. Use the following chart to record the times you have needed help.

Date	Topic you needed help with	Where did you find help?	Who helped you?	Were your questions answered?	Do you still have questions?

10. Have you tried creating study cards to prepare for exams?

a. Find out from your instructor what topics will be on your exam. Ask also about the format of the exam. Will it be multiple choice, true/false, problem solving, or some of each?

b. Organize your quizzes, projects, homework, and book notes from the sections to be tested.

c. Read these quizzes, projects, homework, and book notes carefully and pick out those concepts and skills that seem to be the most important. Write this information on your study card. Try to summarize in your own words and include an example of each important concept.

d. Make sure you include:

Vocabulary Words, with definitions:

Key Concepts, explained *in your own words:*

Skills, illustrated with an example or two:

e. Reviewing this study card with your instructor may be a good exam preparation activity.

11. How can you develop effective test-taking strategies?

a. A little worry before a test is good for you. Your study card will help you feel confident about what you know. Wear comfortable clothing and shoes to your exam and make yourself relax in your chair. A few deep breaths can help! Don't take stimulants such as caffeine; they only increase your anxiety!

b. As soon as you receive your test, write down on your test paper any formulas, concepts, or other information that you might forget during the exam. This is information that you would recall from your study card.

c. You may want to skim the test first to see what kind of test it is and which parts are worth the most points. This will also help you to allocate your time.

d. Read all directions and questions carefully.

e. You may want to do some easy questions first to boost your confidence.

f. Try to reason through tough problems. You may want to use a diagram or graph. Look for clues in the question. Try to estimate the answer before doing the problem. If you begin to spend too much time on one problem, you may want to mark it to come back to later.

g. Write about your test-taking strategies.

	What test-taking strategies will you use?	*What test-taking strategies did you use?*
Test 1		
Test 2		
Test 3		
Test 4		

12. How can you learn from your exams?

a. Errors on exams can be divided into several categories. Review your exam, identify your mistakes and determine the category for each of your errors.

Type of Error	Meaning of Error	Question Number	Points Deducted
A. Concept	You don't understand the properties or principles required to answer the question.		
B. Application	You know the concept, but cannot apply it to the question.		
C. Skill	You know the concept and can apply it, but your skill process is incorrect.		
D. Test-taking	These errors apply to the specific way you take tests. **1.** Did you change correct answers to incorrect answers? **2.** Did you miscopy an answer from scrap paper? **3.** Did you leave an answer blank? **4.** Did you miss more questions at the beginning or in the middle or at the end of your test? **5.** Did you misread the directions?		
E. Careless	Mistakes you made that you could have corrected had you reviewed your answers to the test.		
	TOTAL:		

b. You should now correct your exam and keep it for future reference.

i. What questions do you still have after making your corrections?

ii. Where will you go for help to get answers to these questions?

iii. Write some strategies that you will use the next time you take a test that will help you reduce your errors.

SELECTED ANSWERS

Chapter 1

Activity 1.1 Exercises: 1. \$34,285. **3.** 1,306,313,812. **4.** \$67,000; \$67,100. **8. a.** odd; 22,225 ends with an odd digit. **9. c.** 51 is composite since 51 has factors other than 1 and 51, namely 3 and 17. **11. a.** \$31,000; \$29,000

Activity 1.2 Exercises: 1. a. 5; **c.** 50; **4. a.** 2; **b.** 5. **6. a.** 9; **g.** $80 + 290 + 220 + 110 + 170 + 50 + 60 + 70 + 160 = 1210$ waste sites total.

Activity 1.3 Exercises: 1. a. $\begin{array}{r} 11 \\ 80 \\ +200 \\ \hline 291 \end{array}$ **b.** $\begin{array}{r} 16 \\ 50 \\ +700 \\ \hline 766 \end{array}$ **c.** $\begin{array}{r} 10 \\ 200 \\ +300 \\ \hline 510 \end{array}$

4. b. $49 + 71 = 120$. **5. a.** $200 + 100 + 200 = 500$. **7. a.** 32; **d.** 752; **f.** 312. **10. b.** $113 + 770 + 564 = 1447$. **11. a.** $370 - 360 = 10$; **b.** $400 - 400 = 0$; **c.** $370 - 360 = 10$. **12. f.** $28 - 13$; **g.** $85 + 8$

Activity 1.4 Exercises: 1. b. 4232; **h.** 206535. **3. a.** $12(30 + 6) = 12 \cdot 30 + 12 \cdot 6 = 360 + 72 = 432$. **6. a.** 1656; **b.** 1656; **c.** yes; **d.** The commutative property is demonstrated. **9. a.** $20 \cdot 20 = 400$ hot dog rolls;

b. $\begin{array}{r} 24 \\ \times 16 \\ \hline 144 \\ 24 \\ \hline 384 \end{array}$ The actual number of rolls is 384.

c. The estimate is slightly higher.

Activity 1.5 Exercises: 1. a. quotient 8, remainder 0; **f.** quotient 25, remainder 12; **h.** undefined. **3. b.** Each student will get $1500 \div 4 = 375$ index cards. **4. a.** 3; **b.** no; **c.** no. **6. a.** $4000 \div 50 = 80$;

b. $\begin{array}{r} 74 \\ 52\overline{)3850} \\ 364 \\ \hline 210 \\ 208 \\ \hline 2 \end{array}$ **c.** My estimate is higher.

Activity 1.6 Exercises: 1. 10,000,000,000,000,000. **3.** 53, 59. **5. d.** $22 \cdot 1 = 11 \cdot 2 = 22$. **6. a.** $30 \cdot 1 = 15 \cdot 2 = 10 \cdot 3 = 6 \cdot 5 = 2 \cdot 3 \cdot 5 = 30$; $105 \cdot 1 = 35 \cdot 3 = 21 \cdot 5 = 15 \cdot 7 = 3 \cdot 5 \cdot 7 = 105$. **8. d.** $3 \cdot 2^5$. **10. d.** 32; **e.** 144. **11. a.** 5^{11}. **12. e.** 15. **13. d.** yes, 16 ft. by 16 ft.

Activity 1.7 Exercises: 1. b. 145; **d.** 55. **2. d.** 60. **4. a.** $884 - 34 = 850$. **5. c.** $12 + 45 - 4 = 53$. **7. e.** $100/10 = 10$; **f.** $8 \cdot 20 - 4 = 160 - 4 = 156$. **8. b.** $80 - 27 = 53$. **9. e.** $56 - 18 + 5 = 43$; **j.** $5^2 = 25$.

What Have I Learned? 4. a. addition and multiplication: $3 + 5 = 8 = 5 + 3$; $3 \cdot 5 = 15 = 5 \cdot 3$; **b.** subtraction and division: $5 - 1 = 4$ and $1 - 5 \neq 4$; $8 \div 2 = 4$ and $2 \div 8 \neq 4$. **5. a.** yes, $90 + 17 = 107$ and $69 + 38 = 107$; **c.** no, $48 - 17 = 31$ and $69 - 4 = 65$.

How Can I Practice?
3. $132{,}000{,}000{,}000 \div 690{,}000{,}000 \approx 191$ times. **5. a.** 92,956,000. **6. c.** 4660. **8. c.** $98 - 15 = 83$; **i.** $95 - 25 = 70$. **9. a.** 225. **13.** 31, 37, 41, 43, 47. **15. b.** $3 \cdot 3 \cdot 7 = 3^2 \cdot 7$. **18. d.** 9^6. **19. b.** approximately 5; **c.** 15. **20. d.** $(36 - 18)/6 = 18/6 = 3$; **e.** $45 - 5 \cdot 8 + 5 = 45 - 40 + 5 = 10$; **h.** $49 + 49 = 98$.

Gateway Review 1. one hundred forty-three dollars. **2.** twenty-two million, five hundred twenty-eight thousand, seven hundred thirty-seven. **3.** 108,091; 108,901; 108,910; 109,801; 180,901. **4.** 0 is thousands and 9 is tens. **5. a.** odd; ends with odd digit; **b.** even; ends with even digit; **c.** odd; ends with odd digit. **6. a.** composite, since $145 = 5 \cdot 29$; **b.** prime, since $61 = 1 \cdot 61$, the only factors; **c.** prime, the first one; **d.** composite, since $121 = 11 \cdot 11$. **7. a.** 1,253,000; **b.** 900. **8. a.** Emily; Daniel; **b.** Joshua; **c.** Isabella; **9. a.** 1973; 53 in. **b.** 1954; 14 in. **c.** between 20 and 30 inches: 26 years; between 30 and 40 inches: 23 years. So, the number of years of 20 to 30 inches of rain per year is more than from 30 to 40 inches. **10. a.** 2; **b.** See grid; **c.** 5;

d. See grid. **e.**

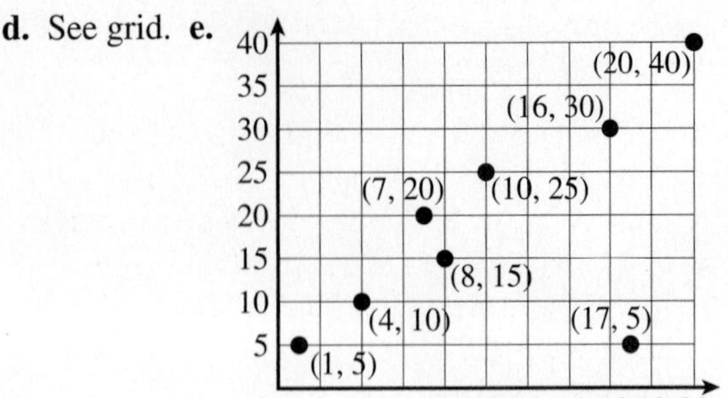

11. a. 8951; **b.** 896; **c.** 177; **d.** 662; **e.** 475. **12.** 113 = 72 + 41. **13. a.** associative property; **b.** commutative property; **c.** The two expressions are equal because adding 0 does not change the value of the sum. **14.** They are not equal; subtraction is not commutative. **15. a.** 510 + 90 + 120 + 350 = 1070 (estimate); **b.** 1066; **c.** Estimate was a little higher than actual since 1070 is larger than 1066. **16. a.** 30 + 22; **b.** 67 − 15; **c.** 125 − 44; **d.** 250 − 175; **e.** 25 · 36; **f.** 55 ÷ 11; **g.** 13^2; **h.** 2^5; **i.** $\sqrt{49}$; **j.** 27(50 + 17). **17. a.** 61 + 61 + 61 + 61 + 61 = 305; **b.** It is the same: 305. **c.** 5(61) = 5(60 + 1) = 300 + 5 = 305. **18. a.** 4(29) = 4(30 − 1) = 120 − 4 = 116; **b.** It is faster to mentally multiply 4 · 30 − 4 · 1 to obtain 116. **c.** 6 · 98 = 6(100 − 2) = 600 − 12 = 588. **19. a.** associative property; **b.** commutative property; **c.** Because multiplying any number by 1 does not change the value of the number. **20. a.** 72 − 12 = 60; 60 − 12 = 48; 48 − 12 = 36; 36 − 12 = 24; 24 − 12 = 12; 12 − 12 = 0 So, the quotient is 6 since there were six subtractions of 12 with a remainder of 0. **b.** 86 − 16 = 70; 70 − 16 = 54; 54 − 16 = 38; 38 − 16 = 22; 22 − 16 = 6 So, the quotient is 5 since there were five subtractions of 16 with a remainder of 6. **21.** 12 ÷ 6 = 2 but 6 ÷ 12 is not a whole number. So, since the answers are different, division is not commutative. **22. a.** 182,352; **b.** Quotient is 15 and remainder is 6. **c.** 861; **d.** 861; **e.** 1; **f.** 0; **g.** undefined; **h.** Quotient is 273 and remainder is 20. **23. a.** 300 · 80 = 24,000 (estimate); **b.** 24,675; **c.** Estimate is lower. **24. a.** 2000 ÷ 40 = 50 (estimate); **b.** Quotient is 44 and remainder is 2. **c.** Estimate is higher. **25.** No, I made a mistake since 3 · 20 = 60. The correct answer is 21. **26.** 1 ÷ 0 is undefined, whereas 0 ÷ 1 = 0. So, the answers are different. **27. a.** The answer is that whole number; for example, 15 ÷ 1 = 15. **b.** The result is 1. It applies to any number except 0, since you cannot divide by 0. **28. a.** 1, 2, 5, 7, 10, 14, 25, 35, 50, 70, 175, 350; **b.** 1, 3, 9, 27, 81; **c.** 1, 2, 3, 4, 6, 9, 12, 18, 36. **29. a.** $2 \cdot 5^2 \cdot 7$; **b.** 3^4; **c.** $2^2 \cdot 3^2$. **30. a.** 1; **b.** 49; **c.** 32; **d.** 1. **31. a.** 9^2; **b.** 3^4; **c.** 25^2; **d.** 5^4. **32. a.** 3^{10}; **b.** 5^{45}; **c.** 21^{19}. **33. a.** a perfect square since $5^2 = 25$; **b.** not a perfect square since 125 = 5 · 5 · 5; **c.** a perfect square since $11^2 = 121$; **d.** a perfect square since $10^2 = 100$; **e.** not a perfect square, since 200 = 2 · 10 · 10. **34. a.** 4; **b.** 6; **c.** 17. **35. a.** not; **b.** yes: $20^2 = 400$; **c.** not; **d.** yes: $100^2 = 10{,}000$; **e.** not.

36. a. 48 − 3(4) + 9 = 48 − 12 + 9 = 45; **b.** 16 + 4 · 4 = 16 + 16 = 32; **c.** 243/(35 − 8) = 243/27 = 9; **d.** (160 − 5)/10 = 110/10 = 11; **e.** 7 · 8 − 9 · 2 + 5 = 56 − 18 + 5 = 43; **f.** 8 · 9 = 72; **g.** 36 + 64 = 100; **h.** $(9 - 8)^2 = 1^2 = 1$. **37.** They do not, since 6 + 10 ÷ 2 = 6 + 5 = 11 and (6 + 10) ÷ 2 = 16 ÷ 2 = 8.

Chapter 2

Activity 2.1 Exercises:

1. a. $P = 4s$
$P = 4(5)$
$P = 20$ cm

b. $A = lw$
$A = 23(35)$
$A = 805$ sq. in.

c. $P = 2l + 2w$
$P = 2(6) + 2(8)$
$P = 12 + 16$
$P = 28$ cm

d. $A = s^2$
$A = 15^2$
$A = 225$ sq. in.

3. a. $P = a + b + c$

b. i. $P = a + b + c$
$P = 4 + 7 + 10$
$P = 21$ ft.

ii. $P = a + b + c$
$P = 24 + 31 + 47$
$P = 102$ cm

iii.

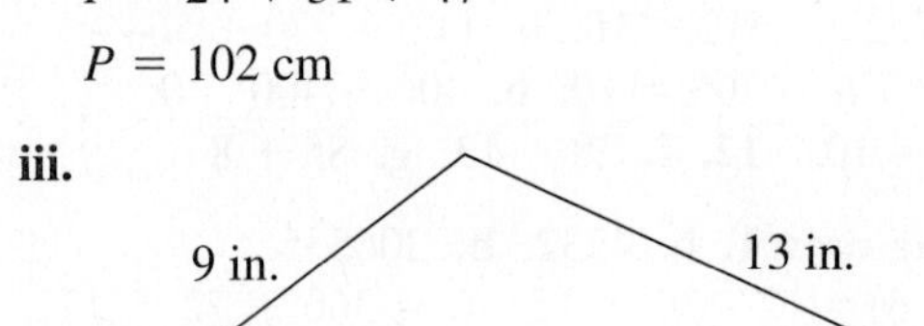

$P = a + b + c$
$P = 9 + 19 + 13$
$P = 41$ in.

5. a.

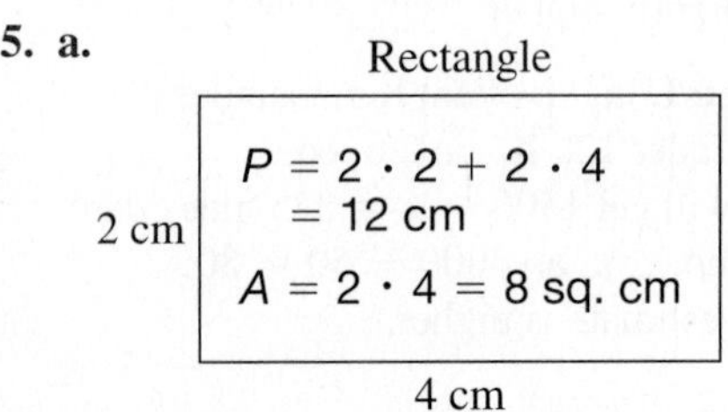

b.

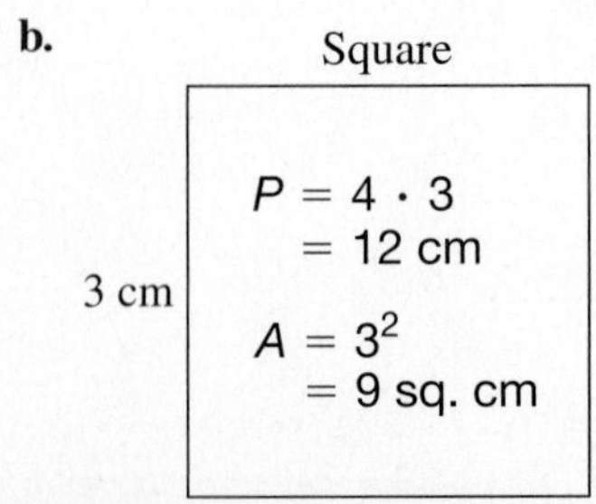

c.

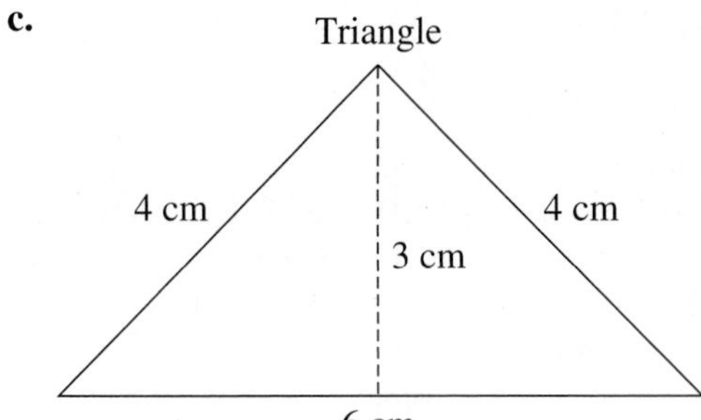

$P = 4 + 4 + 6 = 14$ cm

$A = 6 \cdot 3 \div 2 = 9$ sq. cm

d. Rectangle

3 cm

$P = 2 \cdot 3 + 2 \cdot 8 = 22$ cm

$A = 3 \cdot 8 = 24$ sq. cm

8 cm

7. P = profit $\quad P = R - C$
R = revenue $\quad P = 400{,}000 - 156{,}800$
C = cost $\quad P = \$243{,}200$ is the profit.

9. D = annual depreciation $\quad D = (c - v) \div y$
c = original cost $\quad D = (25{,}000 - 2000) \div 10$
v = remaining value $\quad D = \$2300$ is the annual
y = estimated life in years $\quad$ depreciation of the car.

11. a. $C = 5(86 - 32) \div 9$
$C = 5(54) \div 9$
$C = 270 \div 9$
$C = 30°$

b. $C = 5(41 - 32) \div 9$
$C = 5(9) \div 9$
$C = 45 \div 9$
$C = 5°$

Activity 2.2 Exercises: 2. a. 1, 5, 9, 13, 17, 21
b. The x variable is the input.
c.

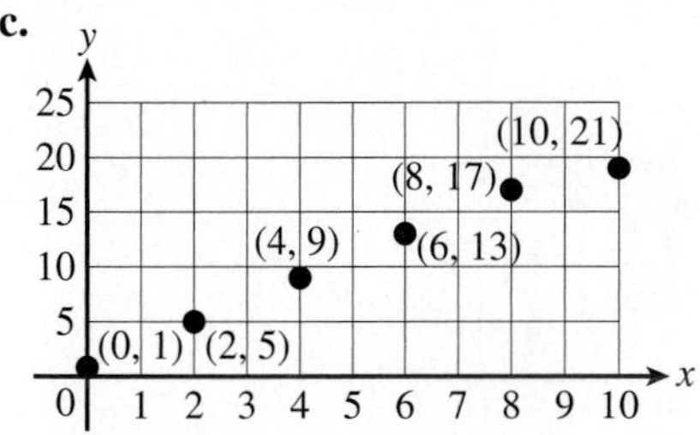

5. a. 0, 64, 96, 96, 64, 0
b.

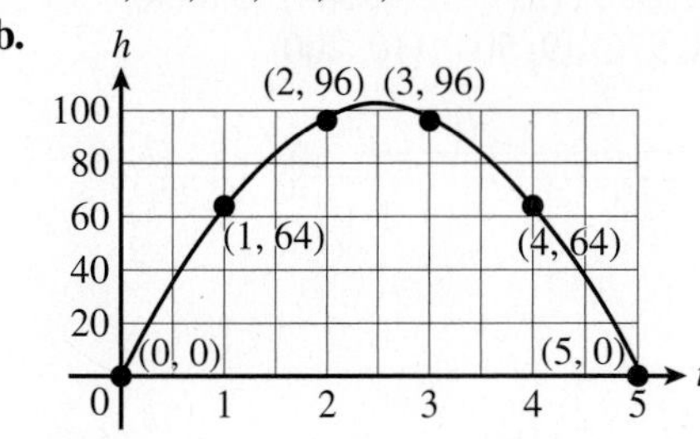

c. I think it will reach 100 feet. **d.** At 2.5 seconds the ball will be 100 feet above the ground. **e.** The ball will hit the ground at 5 seconds, since that is when the height is again 0.
7. a. x, T **b.** x, F **c.** F, C **d.** s, A

Activity 2.3 Exercises: 1. a. My bid is equal to my opponent's bid plus \$20. **b.** The input variable is my opponent's bid, which I will call x. The output variable is my bid, which I will call y. **c.** $y = x + 20$
d. $y = 575 + 20 = 595$. I should bid \$595.
e.
$$1245 = x + 20$$
$$1245 - 20 = x + 20 - 20$$
$$1225 = x$$
My opponent bid \$1225.
3. a. $x = 148$ **c.** $w = 139$ **e.** $t = 0$ **g.** $x = 289$
i. $W = 5845$ **k.** $x = 1950$ **m.** $y = 0$ **o.** $x = 648$
5. a. Add \$1500 to the manager's old salary to obtain the new salary. **b.** $y = x + 1500$, x represents the old salary, and y represents the new salary after the increase.
c. The input variable is x, the old salary. The output variable is y, the new salary. **d.** 3700, 3900, 4100, 4300, 4500 **e.** $x = 2600$
f.

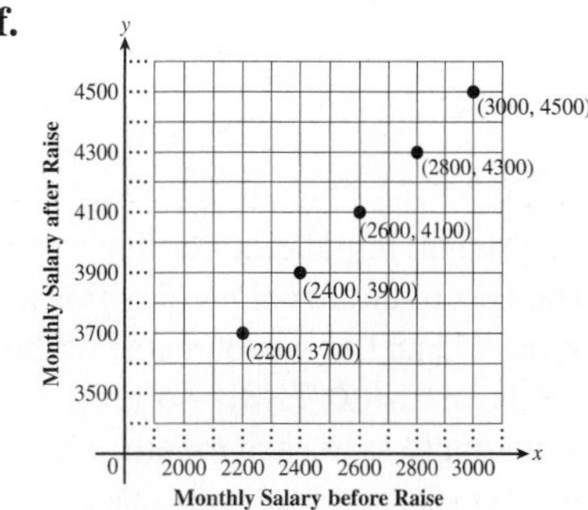

g. The points fall on a straight line.

Activity 2.4 Exercises: 1. a. $x = 9$ **c.** $80 = y$ **e.** $13 = z$
g. $w = 2$ **i.** $t = 0$ **k.** $y = 42$
3. a.
$$400g = A$$
$$400(5) = 2000 \text{ sq. ft.}$$
b.
$$400g = A$$
$$400g = 2400$$
$$g = 6 \text{ gal.}$$

c. *Wall Areas:* 448, 736, 528, 448, 416; Total: 3200
$$400g = 3200$$
$$400g \div 400 = 3200 \div 400$$
$$g = 8 \text{ gal.}$$
4. a. m = number of miles $\quad 5280m = f$
f = number of feet $\quad 5280(4) = f$
$f = 21{,}120$ ft.

c. m = number of meters
c = number of centimeters
$$100m = c$$
$$100(49) = c$$
$$c = 4900 \text{ cm}$$
5. a. $x = 132$ **c.** $w = 128$ **e.** $s = 25$ **g.** $g = 2$
k. $53 = y$ **m.** $w = 12$ **o.** $6413 = t$

Activity 2.5 Exercises: 1. a. Three terms are in the expression. **b.** The coefficients are 5, 3, and 2. **c.** The like terms are $5t$ and $2t$. **2. a.** $6x + 12y$ **c.** $4x + 5y$
e. $39p + 42n$

3. a. $18x = 720$ **c.** $320 = 8s$ **e.** $18x = 270$
$x = 40$ $40 = s$ $x = 15$

g. $2y = 178$ **i.** $2368 = 4t$ **k.** $36 = x + 8$
$y = 89$ $592 = t$ $28 = x$

5. a. $12p + 15p + 21p = 18{,}960$
b. $12p + 15p + 21p = 18{,}960$
$48p = 18{,}960$
$48p \div 48 = 18{,}960 \div 48$
$p = 395$

Activity 2.6 Exercises:

1. $x + 425 = 981$
$-425 \quad -425$
$x = 556$

3. $x - 541 = 198$
$-541 \quad +541$
$x = 739$

5. $x \div 9 = 63$
$x = 63 \cdot 9$
$x = 567$

7. $13 + x = 51$
$-13 \quad -13$
$x = 38$

9. $642 = 6x$
$642 \div 6 = 6x \div 6$
$107 = x$

11. x = number of days driving; rental cost per day = \$75; total budget = \$600. The cost of the rental per day times the number of days driving is equal to the total cost of the rental. $75x = 600$, $x = 8$ days; Check: $75(8) = 600$; $600 = 600$. **13.** x = amount to save each month; $t = 5$ months; total cost of books and fees = \$1200. The amount that I will save each month times the number of months that I will save is equal to the total cost of books. $5x = 1200$, $x = \$240$, Check: $5(240) = 1200$; $1200 = 1200$.

15. better set: \$35; cheaper set: \$20; total receipts: \$525. x represents the number of better sets sold; $2x$ represents the number of cheaper sets sold. The number of sets sold times the cost per set equals the total receipts for that set. $35x + 20(2x) = 525$, $35x + 40x = 525$, $75x = 525$, $x = 7$; Check: $35(7) + 20(2)(7) = 525$;
$245 + 280 = 525$;
$525 = 525$.

17. $P = 2l + 2w$ $540 = 10w + 2w$
$P = 2(5w) + 2w$ $540 = 12w$
$P = 540$ $45 = w$

The dimensions of the field are $45 \cdot 5(45)$ or 45 ft · 225 feet.
Check: $540 = 2(5)(45) + 2(45)$
$540 = 450 + 90$
$540 = 540$

How Can I Practice?

1. $P = 2l + 2w$ $A = lw$
$P = 2(25) + 2(13)$ $A = 25(13)$
$P = 50 + 26$ $A = 325$ sq. cm
$P = 76$ cm

3. $A = (b \cdot h) \div 2$
$A = 8(4) \div 2$
$A = 32 \div 2$
$A = 16$ sq. in.

5. a. $t = 12h$ **c.** $D = 5280m$

6. a. $t = 12h$ **c.** $t = d \div r$
$t = 12(35)$ $t = 480 \div 40$
$t = 420$ total earnings $t = 12$ hr.

7. Substitute 40° for C in the formula $F = (9C \div 5) + 32$
$F = (9(40) \div 5) + 32$ Multiply 9 times 40, divide by 5 and add 32. $F = 104°$

9. a. 1, 8, 19, 34, 53, 76 **b.** y is the output variable.
c.

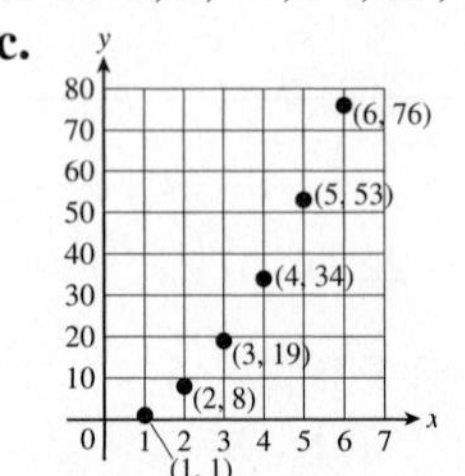

10. a. $42x = 462$
$42x \div 42 = 462 \div 42$
$x = 11$

c. $518 = 14s$
$518 \div 14 = 14s \div 14$
$37 = s$

e. $t - 14 = 90$
$+14 \quad +14$
$t = 104$

g. $15x = 765$
$15x \div 15 = 765 \div 15$
$x = 51$

Gateway Review

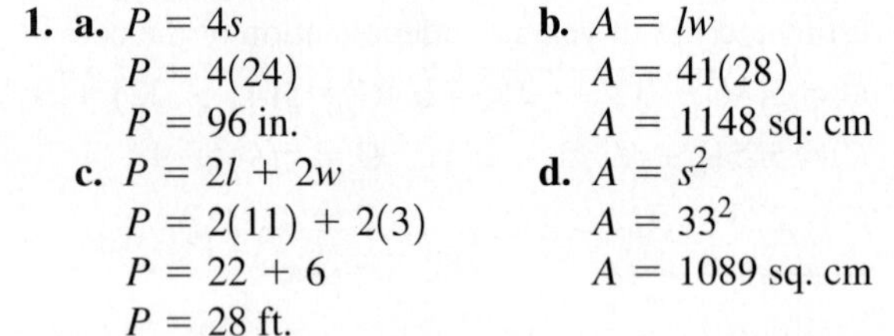

1. a. $P = 4s$ **b.** $A = lw$
$P = 4(24)$ $A = 41(28)$
$P = 96$ in. $A = 1148$ sq. cm
c. $P = 2l + 2w$ **d.** $A = s^2$
$P = 2(11) + 2(3)$ $A = 33^2$
$P = 22 + 6$ $A = 1089$ sq. cm
$P = 28$ ft.

2. a. C is the input variable. **b.** (0, 32), (10, 50), (20, 68), (30, 86), (40, 104), (50, 122), (60, 140), (70, 158), (80, 176), (90, 194), (100, 212).
c.

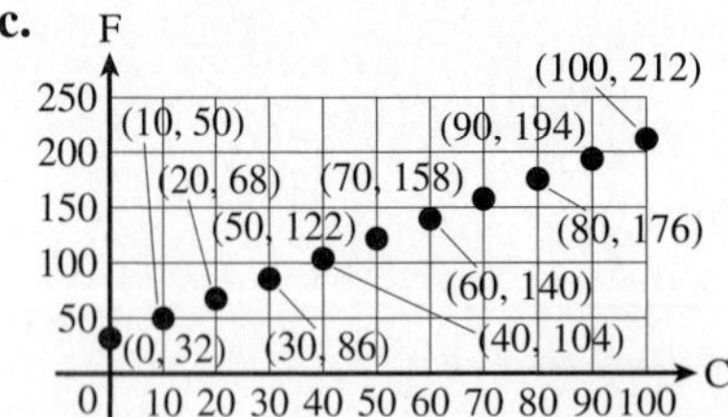

3. (0, 0), (1, 184), (2, 336), (3, 456), (4, 544), (5, 600), (6, 624), (7, 616), (8, 576), (9, 504), (10, 400).

b.

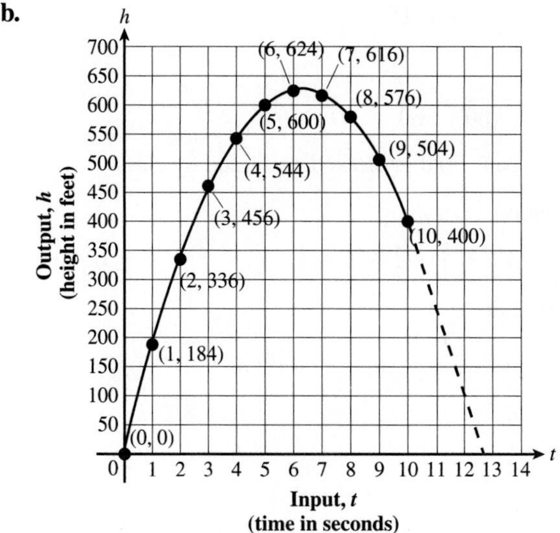

c. The ball will be 400 feet above the ground at about 2.5 seconds and again at 10 seconds. **d.** The ball will go up to approximately 625 feet in the air. **e.** The ball will hit the ground at approximately 12.5 seconds.

4. a. $x = 86$ **b.** $y = 55$
c. $12 = t$ **d.** $x = 54$
e. $x = 5$ **f.** $168 = w$
g. $13 = y$ **h.** $w = 47$

5. x = amount you need to save

$$240 + 420 + 370 + x = 2500$$
$$1030 + x = 2500$$
$$-1030 \qquad -1030$$
$$x = \$1470 \text{ (amount left to be saved)}$$

6. $r \cdot t = A$

r = rate (20 pages of text per hour)
A = amount (260 pages)

$$20t = 260$$
$$20t \div 20 = 260 \div 20$$
$$t = 13 \text{ hr.}$$

7. x = cost of each pizza

$$3x + 6x + 5x = 126$$
$$(3 + 6 + 5)x = 126$$
$$14x = 126$$
$$14x \div 14 = 126 \div 14$$
$$x = \$9 \text{ (cost of each pizza)}$$

8. a. $351 + x = 718$, $x = 367$ **b.** $624 = 48x$, $x = 13$

c. $27x = 405$, $x = 15$ **d.** $x - 38 = 205$, $x = 243$

e. $x + 2x = 273$
$3x = 273$
$x = 91$

9. 160 words per minute is a rate. 800 words on each page and 50 pages of text means 800 times 50 pages is the total number of words that will be read (40,000).

I am asked to find the time.
The formula rate · time = amount will apply.
$r \cdot t = A$
$160t = 40{,}000$ Divide both sides of the equation by 160.
$t = 250$ min.
Check: $160(250) = 40{,}000$ It will take 250 min. to read 50 pages.

Chapter 3

Activity 3.1 Exercises: 2. a. -120 points, **b.** -145 ft., **c.** \$50, **d.** -15 yd., **e.** $-\$75$; **3. b.** $<$, **d.** $>$, **e.** $<$; **4. c.** 32, **d.** 7; **5. a.** 6 is the opposite of -6; **6. b.** -100 feet is further below sea level.

Activity 3.2 Exercises: 1. -2; **4.** -5;
5. $-21 + 18 = -3$; **8.** $-54 + 72 = -126$;
11. $-4 + (-5) + (-3) = -12$; **15.** $-4 + 5 + (-6) = -5$;
17. $-7 - 5 = -12$. The nighttime temperature is -12°F.
19. $6 + 5 = 11$. The noon temperature was 11°F.
21. $-600 - 500 = -1100$. My new elevation is -1100 feet; **24.** Total $= 20 + 35 + 10 - 20 - 40 + 5 - 10 + 30 = 100 - 70 = 30$. The profit is 30 million dollars.

Activity 3.3 Exercises: 3. a. $139 - I = -7$,
b. $139 - I + I = -7 + I$
$139 + 7 = I$
$146 = I$
I weighed 146 pounds at the beginning of the month.
5. a. $S = P - D$, **b.** $35 = P - 8$
$35 + 8 = P$
$\$43 = P$, the regular price;
7. b. $S + D = P$
$D = P - S$;
8. b. $x - 9 = 16$, $x = 25$, **d.** $10 - x = 6$
$10 - x + x = 6 + x$
$10 = 6 + x$
$10 - 6 = x$
$4 = x$;

Activity 3.4 Exercises: 1. b. $T = M + 2$,
c. $T = M + 2$
$T = 36 + 2$
$T = 38$ The total cost is \$38.
3. a. $x - y = 5$.

What Have I Learned? 4. b. negative, **e.** negative;

How Can I Practice? 1. a. -3; **2. a.** 13, **b.** 15;
3. a. 51, **c.** -17, **e.** -84, **h.** -90, **i.** 9;
4. c. 14, **d.** -12, **g.** -53, **i.** -7;
5. b. -181, **d.** 39, **f.** -14;
6. a. $-18 + 7 = -11$, **d.** $-17 + 5 = -12$;
7. c. $23 - (-62) = 23 + 62 = 85$;
8. a. $x + 21 - 21 = -63 - 21$ $x = -84$
Check: $-84 + 21 = -63$,

e. $x - 17 + 17 = 19 + 17$
$x = 36$
Check: $36 - 17 = 19$,

f. $x - 35 + 35 = 42 + 35$
$x = 77$
Check: $77 - 35 = 42$;

9. c. $x - 8 = 6$
$x = 14$ The number is 14.

d. $x - 3 = 12$
$x = 15$ The number is 15.

14. $7 - (-6) = 7 + 6 = 13$. The change in temperature was 13°F.

17. $1000 - 1200 + 800 = -200 + 800 = 600$. Her score became \$600.

Activity 3.5 Exercises: 1. $\frac{2}{3} = \frac{2 \cdot 4}{3 \cdot 4} = \frac{8}{12}$? = 8;
4. $\frac{8}{28} = \frac{8 \div 4}{28 \div 4} = \frac{2}{7}$; **6.** $\frac{3}{7} = \frac{15}{35}$ and $\frac{2}{5} = \frac{14}{35}$ so $\frac{3}{7} > \frac{2}{5}$;
7. $1\frac{2}{3}$; **10.** $5\frac{3}{4} = \frac{5 \cdot 4 + 6}{4} = \frac{23}{4}$; **12.** $\frac{2}{12} = \frac{1}{6}$ and $\frac{3}{18} = \frac{1}{6}$ so you painted the same amount.

14. $\frac{5}{100} = \frac{1}{20}$ of the applications are accepted.

Activity 3.6 Exercises: 1. $\frac{6}{8} = \frac{3}{4}$; **2.** $\frac{3}{5}$; **8.** $\frac{16}{12} = \frac{4}{3} = 1\frac{1}{3}$;
9. $7\frac{3}{5}$; **12.** $7 + \frac{7}{5} - 5 - \frac{4}{5} = 2\frac{3}{5}$;
13. $-7 - \frac{4}{9} + 5 + \frac{2}{9} = -2 - \frac{2}{9} = -2\frac{2}{9}$;
17. $-20 - \frac{14}{13} = -20\frac{14}{13} = -21\frac{1}{13}$;
20. $12\frac{1}{8} + 10\frac{3}{8} + 12\frac{1}{8} + 10\frac{3}{8} = 44\frac{8}{8} = 45$ ft.
I will need 45 feet of wallpaper border.

Activity 3.7 Exercises: 1. $\frac{1}{6} + \frac{3}{6} = \frac{4}{6} = \frac{2}{3}$;
3. $-\frac{4}{10} + \frac{9}{10} = \frac{5}{10} = \frac{1}{2}$;
6. $12\frac{15}{20} + 6\frac{8}{20} = 18\frac{23}{20} = 18 + 1\frac{3}{20} = 19\frac{3}{20}$;
9. $14\frac{9}{12} - 6\frac{5}{12} = 8\frac{4}{12} = 8\frac{1}{3}$; **10.** $10\frac{7}{7} - 6\frac{3}{7} = 4\frac{4}{7}$;
14. $-7\frac{15}{24} + 2\frac{4}{24} = -7 + 2 - \frac{15}{24} + \frac{4}{24} = -5\frac{11}{24}$;
16. $2\frac{2}{7} + 3\frac{3}{8} = 2\frac{16}{56} + 3\frac{21}{56} = 5\frac{37}{56}$;
17. $-5\frac{18}{45} + \left(-6\frac{20}{45}\right) = -11\frac{38}{45}$;
19. $2\frac{2}{3} + 1 + \frac{1}{2} + \frac{5}{8} = 2 + 1 + \frac{16}{24} + \frac{12}{24} + \frac{15}{24}$
$= 3 + \frac{43}{24}$
$= 4\frac{19}{24}$ cups;

21. a. $x + \frac{1}{5} + \frac{1}{3} + \frac{1}{4} = 1$,
b. $x + \frac{1}{5} + \frac{1}{3} + \frac{1}{4} = 1$
$x + \frac{1 \cdot 12}{5 \cdot 12} + \frac{1 \cdot 20}{3 \cdot 20} + \frac{1 \cdot 15}{4 \cdot 15} = 1$
$x + \frac{12 + 20 + 15}{60} = 1$
$x = 1 - \frac{47}{60} = \frac{60}{60} - \frac{47}{60} = \frac{13}{60}$
So, $\frac{13}{60}$ of your final grade is determined by class participation.

23. $1\frac{1}{2} + 1\frac{1}{2} + \frac{3}{4} + \frac{3}{4} + \frac{3}{4} + \frac{3}{4} + \frac{3}{4} = 3 + \frac{15}{4} = 6\frac{3}{4}$ in.
Yes, it will fit.

25. $\frac{3}{10} + \frac{2}{10} = c$
$\frac{5}{10} = c$
$\frac{1}{2} = c$;

28. $b = \frac{2}{3} - \frac{1}{2}$
$b = \frac{4}{6} - \frac{3}{6}$
$b = \frac{1}{6}$;

30. $x = -7\frac{1}{3} - 5\frac{1}{4}$
$x = -7\frac{4}{12} - 5\frac{3}{12}$
$x = -12\frac{7}{12}$;

What Have I Learned? 3. a. The statement is true by definition of improper fractions. **c.** The statement is false because the negative sign of the mixed number applies to both the integer part and the fractional part.

How Can I Practice? 1. a. $\frac{3 \cdot 3}{11 \cdot 3} = \frac{9}{33}$, **d.** $\frac{7 \cdot 3}{12 \cdot 3} = \frac{21}{36}$;
2. a. $\frac{3 \cdot 9 + 7}{9} = \frac{34}{9}$, **d.** $\frac{10 \cdot 13 + 8}{13} = \frac{138}{13}$,
e. $\frac{1 \cdot 31 + 27}{31} = \frac{58}{31}$; **3. b.** $2\frac{2}{13}$, **c.** $7\frac{5}{12}$;
4. c. $\frac{36 \div 3}{39 \div 3} = \frac{12}{13}$, **e.** $\frac{54 \div 2}{82 \div 2} = \frac{27}{41}$;
6. a. JFK $\frac{1}{16}$; LaGuardia $\frac{1}{8} = \frac{2}{16}$; Newark $\frac{1}{2} = \frac{8}{16}$. Newark Airport had the best visibility;

7. c. $1\frac{5}{12} + 3\frac{9}{12} = 4\frac{14}{12}$
$= 4 + 1\frac{2}{12} = 5\frac{2}{12} = 5\frac{1}{6}$,

d. $8\frac{54}{63} + 4\frac{35}{63} = 12\frac{89}{63}$
$= 12 + 1\frac{26}{63} = 13\frac{26}{63}$,

g. $-8\frac{16}{24} + \left(-25\frac{7}{24}\right)$
$= -8 - 25 - \frac{16}{24} - \frac{7}{24}$
$= -33 - \frac{23}{24} = -33\frac{23}{24}$,

j. $-9\frac{15}{70} + \left(-11\frac{12}{70}\right)$
$= -9 - 11 - \frac{15}{70} - \frac{12}{70}$
$= -20 - \frac{27}{70} = -20\frac{27}{70}$;

8. b. $11\frac{17}{54}$, **d.** $11\frac{32}{72} - 7\frac{15}{72} = 4\frac{17}{72}$, **g.** $13\frac{3}{8}$,
k. $13\frac{2}{15} - 8\frac{12}{15} = 12\frac{17}{15} - 8\frac{12}{15} = 4\frac{5}{15} = 4\frac{1}{3}$,

m. $29\frac{21}{36} - 21\frac{30}{36}$
$= 29 - 21 + \frac{21}{36} - \frac{30}{36}$
$= 8 - \frac{9}{36} = 7 + \frac{36}{36} - \frac{9}{36}$
$= 7\frac{27}{36} = 7\frac{3}{4}$,

o. $17\frac{27}{42} - 9\frac{16}{42}$
$= 17 - 9 + \frac{27}{42} - \frac{16}{42}$
$= 8 + \frac{11}{42} = 8\frac{11}{42}$,

q. $-6\frac{4}{5} + 7\frac{11}{15}$
$= -6 + 7 - \frac{12}{15} + \frac{11}{15}$

$= 1 - \frac{1}{15}$
$= \frac{15}{15} - \frac{1}{15} = \frac{14}{15}$;

11. c. $x = 17\frac{4}{7} + 13\frac{5}{6}$
$x = 17\frac{24}{42} + 13\frac{35}{42}$
$x = 30\frac{59}{42}$
$x = 31\frac{17}{42}$,

d. $x = 19\frac{2}{3} - 8\frac{4}{9}$
$x = 19\frac{6}{9} - 8\frac{4}{9}$
$x = 11\frac{2}{9}$;

Activity 3.8 Exercises: 3. a. \$0.889597 ≈ \$0.89, **b.** \$0.889597 ≈ \$0.9; **5. a.** fifty-two thousandths, **d.** 0.0064, **f.** 2041.0673; **7. a.** false, since $2 > 0$ in the hundred-thousandths place;

Activity 3.9 Exercises: 1. 360 cm.; **4.** 4.705 m.; **7.** 4255 cg.; **10.** 0.742 ℓ; **12.** 124 milligrams is smaller, since 0.15 gram = 150 milligrams.

Activity 3.10 Exercises: 1. b. 17.319, **e.** −0.0516, **f.** −3.212; **3.** \$2.99 + \$2.99 + \$2.69 + \$4.49 + \$3.29 + \$2.09 + \$2.49 = \$21.03. \$20 is not enough for everything on the list. Any item could be eliminated;

5. b.

Australia	108.847
China	110.028
France	110.159
Romania	114.283
Russia	113.235
Spain	111.572
Ukraine	112.309
USA	113.584

Gold: Romania; Silver: USA; Bronze: Russia

Activity 3.11 Exercises: 2. a. 31.79, **e.** 1.227, **g.** −20.404, **i.** 15.216, **k.** −0.00428, **l.** −0.2019;
5. Let *y* represent the amount of detergent you have used.
$1.75 - y = 0.5$
$1.75 - y + y = 0.5 + y$
$1.75 = 0.5 + y$
$1.75 - 0.5 = 0.5 - 0.5 + y$
$1.25 = y$
1.25 pints of detergent were already used.
7. $x = 91.4 - 11$
$x = 80.4$
The difference between the two kinds is 80.4 meters.
9. b. $0.01003 - x + x = 0.0091 + x$
$0.01003 - 0.0091 = 0.0091 - 0.0091 + x$
$0.00093 = x$,
d. $1.626 - 1.626 + b = 14.503 - 1.626$
$b = 12.877$,
f. $0.0114 + z = 0.0151$
$0.0114 - 0.0114 + z = 0.0151 - 0.0114$
$z = 0.0037$;

How Can I Practice? 1. c. Thirteen thousand, fifty-three and three thousand eight hundred ninety-one ten thousandths; **2. b.** 611,712.00068, **e.** 9,000,000,005.28; **3. b.** $0.0983 < 0.0987 < 0.384 < 0.392 < 0.561$; **4. c.** 182.1000, **d.** 60.0; **5. b.** 968.366, **d.** 245.212, **f.** −138.709, **i.** 378.2877; **6. c.** 245.1309, **f.** −80.488, **g.** 55.708, **h.** −30.763; **7. b.** $-6.13 + (-5.017) = -11.147$, **d.** $-4.802 - (-19.99) = -4.812$;
8. b. $x - 13.14 + 13.14 = 69.17 + 13.14$
$x = 82.31$,
d. $35.17 - 35.17 + x = 12.19 - 35.17$
$x = -22.98$;
19. \$25.37 + \$39.41 + \$52.04 = \$116.82. I spent \$116.82; **20.** 2105.96 − (311.93 + 64.72 + 161.11) = 1568.20. My take-home pay is \$1568.20.

Gateway Review 1. a. 15, **b.** 23, **c.** 0, **d.** 42, **e.** 67; **2. a.** −9, **b.** −6, **c.** 16, **d.** −14, **e.** 12, **f.** −8, **g.** −53, **h.** −6; **3. a.** $3 - (-3) = 6$, **b.** $-5 + -2 = -7$, **c.** $-(-3) - (-7) = 10$, **d.** $-2 + (-5) = -7$; **4. a.** $x = -2$, **b.** $x = -19$, **c.** $x = -15$, **d.** $x = 43$, **e.** $x = -38$;
5. a. $x + 18 = -7$
$x = -25$,
b. $x + 11 = 29$
$x = 18$,
c. $15 + x = -28$
$x = -43$,
d. $x - 20 = 39$
$x = 59$,
e. $8 - x = 12$
$-4 = x$,
f. $x - 17 = 42$
$x = 59$,
g. $x - 13 = -34$
$x = -21$;
6. a. $12\frac{3}{5}$, **b.** 9, **c.** $5\frac{2}{15}$, **d.** $5\frac{12}{13}$;
7. a. $\frac{23}{5}$, **b.** $\frac{19}{7}$, **c.** $\frac{57}{11}$, **d.** $\frac{83}{8}$;
8. a. >, **b.** >, **c.** <, **d.** >, **e.** >, **f.** <;
9. a. $\frac{11}{7}$ or $1\frac{4}{7}$, **b.** $-\frac{6}{15} = -\frac{2}{5}$, **c.** $\frac{7}{11}$, **d.** $\frac{83}{56}$ or $1\frac{27}{56}$, **e.** $\frac{58}{39}$ or $1\frac{19}{39}$, **f.** $-\frac{49}{36}$ or $-1\frac{13}{36}$, **g.** $\frac{1}{15}$, **h.** $-\frac{2}{33}$; **10. a.** 25, **b.** $19\frac{2}{3}$, **c.** $13\frac{4}{5}$, **d.** $24\frac{17}{24}$, **e.** $19\frac{17}{18}$, **f.** $16\frac{3}{4}$, **g.** $11\frac{3}{8}$, **h.** $5\frac{1}{2}$, **i.** $11\frac{2}{13}$, **j.** $1\frac{3}{4}$, **k.** $5\frac{35}{36}$, **l.** $2\frac{1}{6}$, **m.** $5\frac{9}{10}$, **n.** $7\frac{1}{28}$, **o.** $4\frac{1}{3}$, **p.** $3\frac{3}{7}$, **q.** $4\frac{1}{2}$, **r.** $-14\frac{7}{18}$, **s.** $-1\frac{19}{24}$, **t.** $15\frac{7}{45}$, **u.** $-13\frac{5}{36}$, **v.** $-12\frac{5}{18}$, **w.** $-7\frac{4}{5}$; **11. a.** $x = 9\frac{19}{24}$, **b.** $x = -7\frac{22}{63}$, **c.** $x = -2\frac{7}{12}$, **d.** $x = -33\frac{1}{2}$, **e.** $x = -24\frac{1}{14}$, **f.** $x = 8\frac{31}{36}$;
12. a. eight hundred forty-nine million, eighty-three thousand, six hundred fifty-nine and seven hundred twenty-five ten-thousandths, **b.** thirty-two billion, four million, three hundred eighty-nine thousand, four hundred twelve and twenty-three thousand, four hundred eighteen hundred-thousandths, **c.** two hundred thirty-five million, eight hundred sixty-four and five hundred eighty-seven thousand two hundred thirty-four millionths, **d.** seven hundred eighty-four million, six hundred thirty-two thousand, five hundred forty-one and eight hundred nineteen hundred thousandths;

13. a. 65,073,412.0682, **b.** 89,000,549,613.048, **c.** 7,612,011.05601; **14. a.** $\frac{85}{10{,}000} = \frac{17}{2000}$, **b.** $\frac{3834}{1000} = \frac{1917}{500}$, **c.** $\frac{425}{100} = \frac{17}{4}$, **d.** $\frac{150{,}125}{10{,}000} = \frac{1201}{80}$, **e.** $\frac{712}{100} = \frac{178}{25}$; **15. a.** 692,895.10, **b.** 692,895.098, **c.** 692,895.1, **d.** 692,895, **e.** 692,900, **f.** 690,000; **16. a.** $4.00078 < 4.0078 < 4.07008 < 4.0708 < 4.078 < 4.78$, **b.** $3.00085 < 3.00805 < 3.0085 < 3.0805 < 3.085 < 3.85$, **c.** $8.00046 < 8.00406 < 8.0046 < 8.0406 < 8.046 < 8.46$; **17. a.** -17.62, **b.** -29.07, **c.** -32.12, **d.** -23.37, **e.** -40.97; **18. a.** $12.68 + 7.05 = 19.73$, **b.** $98.99 - 14.85 = 84.14$, **c.** $-17.32 + (-4.099) = -21.419$, **d.** $3.98 - 0.125 = 3.855$; **19. a.** $x = 25.55$, **b.** $x = -46.05$, **c.** $x = 48.5$, **d.** $x = -18.97$, **e.** $x = 7.47$; **20. a.** $-9 + 3 = -6$. The noon temperature was -6°F. **b.** $3 - 5 = -2$. The evening temperature was -2°F. **c.** $-3 - (-8) = 5$. The change was 5°F, **d.** $-14 - (-9) = -5$°F. The change was -5°F

21. $12\frac{5}{12} + 8\frac{1}{4} + 12\frac{5}{12} + 8\frac{1}{4} = 41\frac{1}{3}$ You would use $41\frac{1}{3}$ feet of border. **22.** $\frac{1}{3} + 1\frac{3}{5} = \frac{5}{15} + 1\frac{9}{15} = 1\frac{14}{15}$ You ran $1\frac{14}{15}$ miles each way. **23.** $6\frac{1}{3} + 2\frac{1}{8} + 2\frac{1}{8} = 10\frac{7}{12}$ feet taken up by the couch and end tables; $14 - 10\frac{7}{12} = 3\frac{5}{12}$ feet remaining. The bookcase will fit.

24. $179.99 + 14.85 = 194.84$. The price of the TV was \$194.84. **25.** $20 - 18.35 = 1.65$. You would receive \$1.65 in change. **26.** $67.95 - 10.19 = 57.76$. The sale price of the dress is \$57.76.

27. $1504.75 - 157.32 - 115.11 - 45.12 = 1187.20$. The cook's take-home pay is \$1187.20. **28.** $62.05 - 25.12 - 13.59 + 40 - 117.5 - 85.38 + 359.13 = 219.59$ You will have \$219.59 in your bank account.

29.

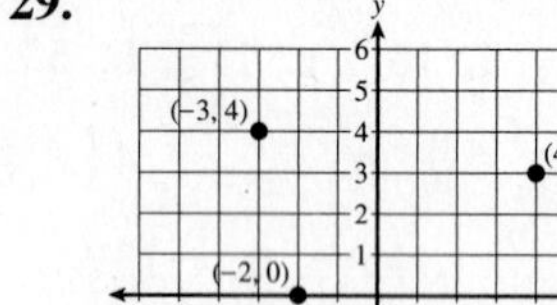
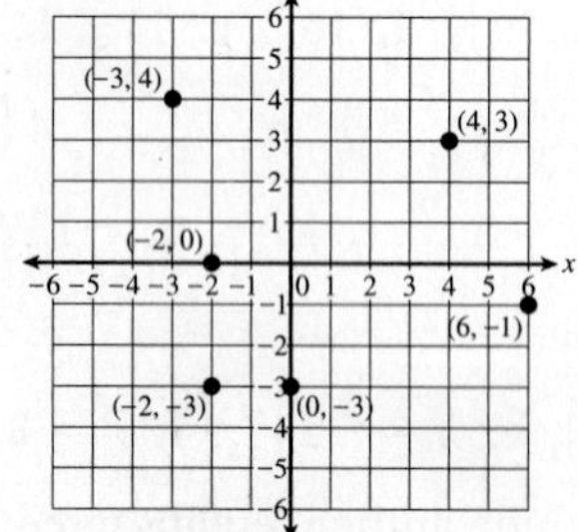

30. 0.795 g; **31.** 2750 m; **32.** 0.025 ℓ; **33.** 1050 g; **34.** 2350 m = 2,350,000 mm; **35.** 85 mℓ; **36.** 455 + 806 + 423 + 795 = 2479 m = 2.479 km; **37.** 1250 g − 285 g = 965 g or 0.965 kg; **38.** $95 + 150 + 180 + x = 1920$ $x = 1920 - 425 = 1495$ There are 1495 mℓ left, or 1.495 ℓ.

Chapter 4

Activity 4.1 Exercises: 1. a. -42 **c.** -1600 **e.** 0 **g.** 240 **i.** 5 **l.** 0. **2. b.** 192 **d.** -55 **e.** -36 **f.** 36.

4. Approximate depth of the water supply $= 5(-25) = -125$ ft. The daily drilling rate is -25 ft per day.

8. $22(5) + 7(-2) + 6(0) = 96$.

11. $4(-230) + 2(350) = -\$220$, a debit.

Activity 4.2 Exercises:

1. $72 + 6(-1) + 10(0) + 2(+1) = 68$. Tiger's score was 68. **3. b.** $3 + 8 = 11$

d. $4 \cdot 9 - 50 = 36 - 50 = -14$ **j.** $(6 - 20) \div (-2) = -14 \div (-2) = 7$

l. $-5 + 9 \div (-3) = -5 + -3 = -8$ **o.** $9 \cdot 25 \div (-1) = 225 \div (-1) = -225$

4. a. $5 \cdot (-3) + 5 \cdot 4 = -15 + 20 = 5$ **c.** $4 \cdot (-6) - 4 \cdot 2 = -24 - 8 = -32$

5. a. The value is $3 \cdot (-5)^2 + 4(-3) = 3 \cdot 25 + -12 = 75 - 12 = 63$.

c. The value is $-5(3)(14 \cdot 3 \cdot -2 - 3(-2)^2) = -15(-84 - 12) = -15 \cdot -96 = 1440$

d. The value is $\frac{3(-2)(-8) - 4(4)}{4(-2)(4)} = \frac{32}{-32} = -1$.

7. b. $(-30)x = -150$; $x = 5$. **d.** $2x = -28$; $x = -14$. **8. b.** $s = -9$ **e.** $x = 3$ **10.** $19d = -57$; $d = -3$, and the average drop in temperature is 3°F.

How Can I Practice? 1. c. 160 **f.** 9 **h.** -6 **k.** -30. **2. d.** $40 + -2 \cdot 2 = 36$ **f.** $-9 - 36 = -45$ **g.** $18 - 6 + 6 = 18$ **i.** $49 - 14 + 5 = 40$. **3. b.** The answer is $(-6)^2 - (-6)(4) = 36 + 24 = 60$. **c.** The answer is $2(-2 + 3) - 5(-4) = 2 \cdot 1 + 20 = 22$. **7. a.** $3x = -15$, $x = \frac{-15}{3} = -5$ **d.** $90 = (-3)(-6)x$, $90 = 18x$, $x = \frac{90}{18} = 5$

Activity 4.3 Exercises: 2. a. $\frac{6}{35}$ **d.** $-3\frac{3}{4}$ **e.** $\frac{-4}{9}$ **h.** $\frac{16}{21}$. **3. a.** $\frac{10}{5} = 2$ **d.** $\frac{-12 \cdot 15}{26 \cdot 8} = -\frac{45}{52}$ **e.** $\frac{-3 \cdot 9}{10 \cdot -5} = \frac{27}{50}$ **f.** $\frac{23}{19 \cdot -46} = -\frac{1}{38}$. **4. a.** $\frac{9}{10}$

b. He can expect $2500 \cdot \frac{9}{10} = 2250$ plants to grow.

7. $\$24{,}350 \cdot \frac{1}{10} = \$2435$ for tuition.

10. b. $\frac{2}{3}x = -4$; solving, $\frac{3}{2} \cdot \frac{2}{3}x = -4 \cdot \frac{3}{2}$, so $x = -4 \cdot \frac{3}{2} = -6$. **c.** $\frac{x}{7} = 9$; solving, $\frac{x}{7} \cdot 7 = 9 \cdot 7$, so $x = 63$.

11. b. $\frac{-10}{1} \cdot \frac{-x}{10} = \frac{4}{15} \cdot \frac{-10}{1}$; $x = \frac{4}{15} \cdot \frac{-10}{1} = \frac{-8}{3}$ or $-2\frac{2}{3}$. **d.** $\frac{-24}{5} \cdot \frac{-5}{24}y = \frac{15}{32} \cdot \frac{-24}{5}$; $y = \frac{15}{32} \cdot \frac{-24}{5} = \frac{-9}{4}$ or $-2\frac{1}{4}$. **f.** $\frac{27}{13} \cdot \frac{13}{27}x = \frac{20}{45} \cdot \frac{27}{13}$; $x = \frac{20}{45} \cdot \frac{27}{13} = \frac{12}{13}$.

Activity 4.4 Exercises: 1. a. $\frac{33}{5} = 6\frac{3}{5}$ **b.** $-\frac{48}{5} = -9\frac{3}{5}$ **d.** 129. **2. b.** $\frac{-15}{8} \cdot \frac{4}{3} = -\frac{5}{2} = -2\frac{1}{2}$ **d.** $\frac{38}{3} \cdot \frac{6}{19} = 4$. **3. c.** $\frac{16}{49}$ **d.** $\frac{32}{243}$ **h.** $\sqrt{25} = 5$.

4. b. $\frac{2}{3}\left(\frac{2}{3} - 2\right) + 3 \cdot \frac{2}{3}\left(\frac{2}{3} + 4\right) =$
$\frac{2}{3} \cdot \frac{-4}{3} + 2 \cdot \frac{14}{3} = \frac{76}{9} = 8\frac{4}{9}$

d. $6\left(\frac{21}{16}\right)\left(\frac{1}{8}\right) - \left(\frac{1}{8}\right)^2 = \frac{63}{64} - \frac{1}{64} = \frac{62}{64} = \frac{31}{32}$.

5. $A = \left(\frac{23}{4}\right)^2 = \frac{529}{16} = 33\frac{1}{16}$ sq. ft.

7. b. $\frac{49}{8}w = -\frac{49}{2} \Rightarrow w = -\frac{49}{2} \cdot \frac{8}{49} = -4$ **c.** $-\frac{9}{5}t = -\frac{99}{10} \Rightarrow t = -\frac{99}{10} \cdot \frac{-5}{9} = \frac{11}{2} = 5\frac{1}{2}$.

9. a. $d = 16\left(\frac{1}{2}\right)^2 = 4$ ft.

How Can I Practice? 1. b. $\frac{30}{143}$ **d.** $\frac{12}{35}$ **e.** $\frac{6}{5}$ or $1\frac{1}{5}$ **h.** $\frac{14}{-16} = -\frac{7}{8}$. **2. a.** $\frac{-3}{7} \cdot \frac{7}{9} = \frac{-1}{3}$ **d.** $\frac{24}{45} \cdot \frac{18}{15} = \frac{16}{25}$ **f.** $\frac{-35}{10} \cdot \frac{1}{-3} = \frac{7}{6}$ or $1\frac{1}{6}$. **3. b.** $\frac{-17}{6} \cdot \frac{7}{3} = \frac{-119}{18}$
$= -6\frac{11}{18}$;
d. $\frac{36}{5} \cdot \frac{5}{12} = 3$ **f.** $\frac{-33}{4} \cdot \frac{-2}{3} = \frac{11}{2} = 5\frac{1}{2}$. **4. b.** $\frac{9}{11}$ **c.** $-\frac{27}{125}$. **5. b.** $\left(\frac{1}{4}\right)^2 - \left(\frac{1}{4}\right)\left(\frac{-4}{3}\right) = \frac{1}{16} + \frac{1}{3} = \frac{19}{48}$.

6. b. $x = \frac{-42}{5}$ or $-8\frac{2}{5}$ **e.** $x = -\frac{1}{20}$ **f.** $x = \frac{-9}{16} \cdot \frac{-4}{3} = \frac{3}{4}$. **9.** You will need $\frac{1}{3} \cdot \frac{129}{2} = \frac{129}{6} = 21\frac{3}{6} = 21\frac{1}{2}$ ounces of split peas.

13. a. $V = 15 \cdot 30 \cdot \frac{11}{2} = 2475$ cu. ft.

b. Weight is $2475 \cdot \frac{312}{5} = 154{,}440$ lb.

Activity 4.5 Exercises: 2. a. 102.746 **c.** -10.965 **f.** 70,250 **g.** 5000. **4.** $\frac{68700}{1000} \cdot 7.48 = \$513.88$ in taxes.
6. a. \$120 million ÷ 60 shows = 2 million per show
b. \$121.2 million ÷ 60 shows = 2.02 million per show.

Activity 4.6 Exercises: 1. $= 38.16 + 4.66 = 42.82$.
3. $10.31 + 8.05 \cdot 0.4 - 0.0064$
$= 10.31 + 3.22 - 0.0064$
$= 13.5236$

6. $(3.14)(0.25)^2 = 3.14(0.0625) = 0.19625$ sq. in.

10. a. $\frac{x}{5.3} = -6.7$; solving for x, $x = (5.3)(-6.7) = -35.51$. **c.** $42.75 = x \cdot (-7.5)$; solving for x, $x = \frac{42.75}{-7.5} = -5.7$. **e.** $(-9.4)x = 47$; solving for x, $x = \frac{47}{-9.4} = -5$. **11. a.** $x = \frac{15.3}{3} = 5.1$ **c.** $a = \frac{-44.2}{-5.2} = 8.5$ **f.** $x = \frac{-8}{4.2} \approx -1.90$.

How Can I Practice? Exercises: 1. -160. **4.** 500. **6.** 1.1345. **9.** $7 \cdot 0.5 = 3.5$.
12. $\frac{1}{2}(0.7)(3.45 + 5.009) = \frac{1}{2}(0.7)(8.45) = 2.9575$.
14. $9.9 \div 0.33 - 2.7 \cdot 4 = 30 - 10.8 = 19.2$.
16. $x \cdot (-9.76) = 678.32$; solving, $x = \frac{678.32}{-9.76} = -69.5$.
17. $\frac{x}{-9.5} = 78.3$; solving, $x = (78.3)(-9.5) = -743.85$.
19. $x = \frac{58.65}{-2.3} = -25.5$.
22. a. $204/540 = 0.378$; in 1921, Babe Ruth's batting average was .378. **b.** $156/476 = 0.328$; Barry Bonds's batting average was .328 in 2001. **c.** Babe Ruth's average is higher by .050. **24.** 5.01, 12.00, 6.99, 14.68, 6.66, 16, 45.34, 2.83.

Gateway Review Exercises: 1. a. -152
b. $\frac{2 \cdot 2}{3 \cdot 3} \cdot \frac{-5 \cdot 3}{7 \cdot 2} = \frac{-10}{21}$ **c.** 7 **d.** 15,400
e. $\frac{14}{5} \cdot \frac{10}{21} = \frac{7 \cdot 2 \cdot 5 \cdot 2}{5 \cdot 7 \cdot 3} = \frac{4}{3} = 1\frac{1}{3}$ **f.** 0.003912.
2. a. $\frac{64}{169}$ **b.** -16 **c.** $\frac{11}{12}$. **3.** $(-1)^{13} = -1$.
4. $(-5)^2 = (-5)(-5) = 25$; $-5^2 = -(5)(5) = -25$.
5. a. $(-0.432) + (-0.9) = -1.332$
b. $\frac{10 - 24 - 9}{6 - 4} = \frac{-23}{2}$.
6. a. $2\left(\frac{-3}{2}\right) - 6\left(\frac{-3}{2} - 3\right) = -3 + 9 + 18 = 24$
b. $(-3 \cdot -2 \cdot -6) - (-2)^2 = -36 - 4 = -40$
c. $\pi(2.4)^2 \cdot (0.9) = 5.184\pi \approx 16.27776$ using $\pi \approx 3.14$.
7. a. $x = \frac{-72}{-12} = 6$ **b.** $x = 14 \cdot -6 = -84$

c. $x = \frac{-4}{9 \cdot -18} = \frac{2 \cdot 2}{9 \cdot 9 \cdot 2} = \frac{2}{81}$ **d.** $n = \frac{73.84}{8} = 9.23$

e. $s = \frac{-5 \cdot 33}{9 \cdot 20} = \frac{-5 \cdot 3 \cdot 11}{3 \cdot 3 \cdot 5 \cdot 4}$ **f.** $x = \frac{12.9}{0.0387} \approx 333.33$

$= \frac{-11}{12}$.

8. a. $\frac{x}{-15} = -7; x = (-7)(-15) = 105.$

b. $\left(\frac{-11}{12}\right)x = 2\frac{1}{16}$;

$x = \frac{33}{16} \cdot \frac{12}{-11} = \frac{3 \cdot 11 \cdot 4 \cdot 3}{4 \cdot 4 \cdot -11} = -\frac{9}{4} = -2\frac{1}{4}$

c. $1.08 = x(0.2); x = \frac{1.08}{0.2} = 5.4.$

9. Using dimensional analysis,

$\frac{19}{2}\text{ gal.} \cdot \frac{16 \text{ cups} \cdot 1 \text{ serving}}{1 \text{ gal.} \cdot \frac{3}{4} \text{ cup}}$

$= \frac{19 \cdot 16 \cdot 1 \cdot 4}{2 \cdot 1 \cdot 3} \text{ servings} = \frac{608}{3} \approx 202;$

202 servings, plus a little left over.

10. $\frac{638{,}800 - 680{,}845}{1990 - 1930} = \frac{-42{,}045}{60} = -700.75$; so the average population decrease per year was approximately 701 persons per year.

11. $\frac{47{,}224 \text{ sq.mi.}}{18{,}196{,}601 \text{ persons}} \approx 0.00259521 \approx 0.003$ sq. mi. per person. **12.** $\frac{7}{10} \cdot 2.9$ million $= 2.03$ million people had farm-related occupations in 1820.

13. $\$13.79 + 270 \cdot \$0.0059714 + 270 \cdot \$0.0049600 = \$16.74$. **14.** Estimating, $\frac{7000}{10} = 700$ mph. If $d = 7318$ miles and $t = 11\frac{3}{4}$ hours, then the formula $d = r \cdot t$ yields $7318 = r \cdot 11\frac{3}{4}$. So, solving for r,

$r = \frac{7318}{\frac{47}{4}} = \frac{7318 \cdot 4}{47} \approx 623$ mph, close to the estimate.

15. a. revenue $= \$15.85b + \$9.95h$ **b.** revenue $= (15.85)(26) + (9.95)(13) = \$541.45.$

16. a. \$81.00, \$121.50, \$162.00, \$202.50 **b.** I multiplied the number of hours by \$6.75 per hour. **c.** $150 = 6.75x$; solving, $x = \frac{150}{6.75} \approx 22.22$ hours. I would need to work 23 hours. **17. a.** $x + 2$ **b.** $x + 2 + 2x$

c. $x + 2 + 2x + x - 4 + \frac{x}{4} = 4\frac{1}{4}x - 2$

d. $d = (52)\left(4\frac{1}{4}x - 2\right)$

e. $d = (52)\left(\frac{17}{4} \cdot 7 - 2\right) = (52)\left(\frac{119 - 8}{4}\right) = 1443$ mi.

Chapter 5

Activity 5.1 Exercises: 2. a. See answers to part b for the matching.

b. $\frac{12}{27} = \frac{20}{45} \approx 0.444 = 44.4\%$, $\frac{28}{36} = \frac{21}{27} \approx 0.778 = 77.8\%$, $\frac{45}{75} = \frac{42}{70} = 0.6 = 60\%$, $\frac{64}{80} = \frac{60}{75} = 0.80 = 80\%$, $\frac{35}{56} = \frac{25}{40} = 0.625 = 62.5\%$; **4. a.** 0.296; **5. a.** $\frac{1720}{3200}$, **b.** $\frac{43}{80}$, **c.** ≈ 0.538, **d.** $\approx 53.8\%$;

7.

	NUMBER OF GAMES	RATIO OF WINS TO GAMES PLAYED
Regular season:	99 + 63 = 162	$\frac{99}{162} \approx 0.611 = 61.1\%$
Playoff season:	11 + 1 = 12	$\frac{11}{12} \approx 0.917 = 91.7\%$

The White Sox played better in the playoffs relatively speaking.

9. Brand A: $\frac{2940}{13{,}350} \approx 0.22 = 22\%$

Brand B: $\frac{730}{1860} \approx 0.392 = 39.2\%$

Brand A dishwasher has a better repair record.

Activity 5.2 Exercises: 1. a. 8, **b.** 27, **c.** 15, **e.** 18; **l.** 60,000; **3.** $25.8 \div 64\% = 25.8 \div 0.64 = 40.3125$ million. Therefore, about 40.3 million people worldwide are infected with AIDS. **5.** 8% of $22{,}500 = 0.08 \cdot 22{,}500 = 1800$ dollars is the state sales tax. **7.** $\frac{22{,}000}{0.45} \approx 48{,}889$; Approximately 49,000 registered voters; **10.** Total number of calls $= \frac{50}{0.05} = 1000$; I need to make about 1000 phone calls; **12.** Bookstore pays 20% of $90 = 0.2 \cdot 90 = 18$ dollars. Bookstore nets $65 - 18 = 47$ dollars;

Activity 5.3 Exercises: 1. a. $3560 - 3200 = 360$ students, **b.** $\frac{360}{3200} = \frac{9}{80} = 11.25\%$; The full-time enrollment increased by 11.25% last year. **3.** Actual increase: $6.50 - 6.25 = \$0.25$; percent increase: $\frac{0.25}{6.25} = 0.04 = 4\%$;

5. Amount of decrease: $2400 - 1800 = 600$ calories; percent decrease: $\frac{600}{2400} = 0.25 = 25\%$;

7. b. $25 \cdot 2 = 50$; actual increase: $50 - 25 = 25$; percent increase: $\frac{25}{25} = 1 = 100\%$; **8.** $10 \cdot 3 = 30$; actual increase: $30 - 10 = 20$; percent increase: $\frac{20}{10} = 2 = 200\%$. For a quantity that triples in size, the percent increase is 200%.

Activity 5.4 Exercises: 2. a. Growth factor: 1.08375, **b.** Total cost: $17{,}944 \cdot 1.08375 = \$19{,}446.81$; **5.** 1990 population: $15{,}982{,}378 \div 1.235 \approx 12{,}941{,}197$; **7.** She must earn $50{,}000 \cdot 1.35 = \$67{,}500$; **8.** The investment will be worth $3000 \cdot 1.0505 = \$3151.50$. **10. a.** $24 \cdot 1.59 = 38.16$ million to $28 \cdot 1.59 = 44.52$ million. Therefore, the range estimated for 2000 was between 38 million and 45 million transistors on a chip. **b.** $42 \cdot 1.59 = 66.78$ million; so approximately 67 million transistors per chip were predicted for 2000–2001.

Activity 5.5 Exercises: 2. The length of the article must be reduced by 2 pages: $\frac{2}{8} = \frac{1}{4} = 0.25 = 25\%$; **3.** Decay factor: $531{,}285 \div 551{,}226 \approx 0.964$; percent decrease: $1 - 0.964 = 0.036 = 3.6\%$; **6.** Decay factor: $100\% - 30\% = 70\% = 0.70$; sale price: $129.95 \cdot 0.70 = \$90.97$; **8. a.** $5000 \div 0.95 \approx 5263$ tigers is the lower estimate for the tiger population, **b.** $7000 \div 0.95 \approx 7368$ tigers is the upper estimate for the tiger population; **9.** $100\% - 5\% = 95\% = 0.95$ is the decay factor. $558.60 \cdot 0.95 = \$530.67$ is the discounted online fare for the fully refundable ticket. $273.60 \cdot 0.95 = \$259.92$ is the discounted online fare for the restricted ticket.

Activity 5.6 Exercises: 1. a. 30% off means 0.70 is the decay factor. 20% off means 0.80 is the decay factor, **b.** Sale price of the suit is $\$300 \cdot 0.70 \cdot 0.80 = \$168$; **3.** Associated factors are $100\% + 25\% = 125\% = 1.25$ $100\% - 75\% = 25\% = 0.25$. The remaining inventory is $2000 \cdot 1.25 \cdot 0.25 = 500$ toys. **5.** The current budget is $600{,}000 \cdot 0.95 \cdot 0.95 \cdot 0.95 = 600{,}000 \cdot 0.95^3 = \$514{,}425$.

Activity 5.7 Exercises: 1. 100 yd. · 3 ft./yd. = 300 ft.

3. $80\ \cancel{\text{parts}} \cdot \frac{3\text{ min.}}{16\ \cancel{\text{parts}}} = 15$ min. It will take 15 minutes.

5. $\frac{\$11.50}{1\ \cancel{\text{hr.}}} \cdot \frac{40\ \cancel{\text{hr.}}}{1\ \cancel{\text{wk.}}} \cdot \frac{52\ \cancel{\text{wk.}}}{1\ \cancel{\text{yr.}}}(5\ \cancel{\text{yr.}}) = \$119{,}600$ sum of the total gross salaries for the next 5 years.

7. $29{,}035\ \cancel{\text{ft.}} \cdot \frac{1\text{ mi.}}{5280\ \cancel{\text{ft.}}} \approx 5.5$ mi.

$29{,}035\ \cancel{\text{ft.}} \cdot \frac{1\ \cancel{\text{mi.}}}{5280\ \cancel{\text{ft.}}} \cdot \frac{1.609\text{ km}}{1\ \cancel{\text{mi.}}} \approx 8.85$ km

$29{,}035\ \cancel{\text{ft.}} \cdot \frac{1\ \cancel{\text{mi.}}}{5280\ \cancel{\text{ft.}}} \cdot \frac{1.609\ \cancel{\text{km}}}{1\ \cancel{\text{mi.}}} \cdot \frac{1000\text{ m}}{1\ \cancel{\text{km}}} \approx 8848$ m;

9. $4.5\ \ell \cdot \frac{1.06\text{ qt.}}{1\ \ell} = 4.77$ qt.

$4.77\ \cancel{\text{qt.}} \cdot \frac{2\text{ pt.}}{1\ \cancel{\text{qt.}}} = 9.54$ pt.;

13. $24\text{ carat} \cdot \frac{1\text{ g}}{5\text{ carat}} = 4.8$ g

$4.8\text{ g} \cdot \frac{0.035\text{ oz.}}{1\text{ g}} = 0.168$ oz.

Activity 5.8 Exercises: 2. a. $9x = 2 \cdot 108$; $x = \frac{2 \cdot 108}{9} = 24$;

4. $\frac{\$6.99}{8\text{ qt.}} = \frac{x}{12\text{ qt.}}$; $x = \frac{12 \cdot 6.99}{8}$; $x = \$10.49$. Twelve quarts of skim milk prepared this way costs \$10.49;

7. $\frac{x}{1859} = \frac{9}{10}$; $x = \frac{9 \cdot 1859}{10}$; $x \approx 1673$. 1673 consumers in the sample reported problems with transactions online.

What Have I Learned? Exercises: 1. I answered more questions correct on the practice exam (32) than on the actual exam (16). I scored $\frac{32}{40} = \frac{4}{5} = 0.8 = 80\%$ on the practice exam. I scored $\frac{16}{20} = \frac{4}{5} = 0.8 = 80\%$ on the actual exam. No, the relative scores were the same. So, I did the same on the actual exams as I did on the practice exam; **3.** Let $x =$ the number of Florida residents over 65 years old: $x = \frac{183}{1000} \cdot 15{,}982{,}378 \approx 2{,}924{,}775$. Approximately 2,924,775 Florida residents were 65 years or older in 2000; **6. a.** Decay factor: $100\% - 10\% = 90\% = 0.90$ $199 \cdot 0.90 = 179.1$ lb. My relative will weigh 179.1 pounds, **b.** $179.1 \cdot 0.90 \approx 161.2$ lb., $161.2 \cdot 0.90 \approx 145$ lb. He must lose 10% of his body weight 3 times to reach 145 pounds.

How Can I Practice? Exercises: 1. a. 0.25, **e.** 2.50, **f.** 0.003; **5.** Current staff: $1500 \div 0.40 = 3750$ employees; **7. a.** Final cost: $400 \cdot 0.70 \cdot 0.80 = \$224$, **b.** Decay factor: $0.70 \cdot 0.80 = 0.56$. Equivalent percent discount: $1 - 0.56 = 0.44 = 44\%$; **9.** Former rent: $900 \div 1.20 = \$750$. **11.** Decay factor: $100\% - 14\% = 86\% = 0.86$. Number of 2000 Explorers: $28{,}000 \div 0.86 \approx 32{,}558$ SUVs.

13. $\frac{\$22.50}{1\ \cancel{\text{hr.}}} \cdot \frac{40\ \cancel{\text{hr.}}}{1\ \cancel{\text{wk.}}} \cdot \frac{52\ \cancel{\text{wk.}}}{1\ \cancel{\text{yr.}}} \cdot 2\ \cancel{\text{yr.}} = \$93{,}600$; I would earn \$93,600 in 2 years.

Gateway Review Exercises: 1. a. $\frac{4 \cdot 5 \cdot 7}{4} = 35$, **b.** $\frac{4}{5} \cdot \frac{1}{2} = \frac{2}{5}$, **c.** $0.27 \cdot 44 = 11.88$, **d.** 6500, **e.** $\approx 38{,}083.33$, **f.** ≈ 6.22;

2. a. $4 \cdot 45 = 9x$, $\frac{4 \cdot 5 \cdot 9}{9} = x$, $20 = x$ **b.** $4x = 4 \cdot 5$, $x = 5$ **c.** $3 = 2x$, $\frac{3}{2} = x$

d. $2.3 \cdot 4 = 1.7x$, $\frac{2.3 \cdot 4}{1.7} = x$, $5.41 \approx x$ **e.** $\frac{1}{2} \cdot 6 = 7x$, $\frac{3}{7} = x$ **f.** $x = 12$;

3. Females: 70% of 1400 $= 0.7 \cdot 1400 = 980$
Males: 30% of 1000 $= 0.3 \cdot 1000 = 300$
Total $= 980 + 300 = 1280$; Percent of student body $= \dfrac{1280}{1400 + 1000} = \dfrac{1280}{2400} \approx 0.533 = 53.3\%$; **4.** Total mailing list $= 600 \div \dfrac{2}{3} = 600 \cdot \dfrac{3}{2} = 900$ envelopes;
5. Number of layoffs: $110{,}000 - 77{,}000 = 33{,}000$
Percent of workforce: $\dfrac{33{,}000}{110{,}000} = 0.3 = 30\%$. **6.** Growth factor: $611{,}666 \div 470{,}816 \approx 1.299$; Percent increase: 29.9%; **7.** Decay factor: $100\% - 75\% = 25\% = 0.25$. Amount remaining: $650 \cdot 0.25 = 162.5$ mg.; **8.** Sales revenue: $400{,}000 \cdot 1.20 \cdot 1.30 = \$624{,}000$; **9.** Growth factor: $100\% + 13.5\% = 113.5\% = 1.135$. Number of 2004 Silverados: $53{,}800 \div 1.135 \approx 47{,}401$;
10. a. $4900 \cdot 1.12 = \$5488.00$, **b.** $3590 \cdot 1.25 = \$4487.50$, **c.** $4487.50 \cdot 1.13 = \$5070.88$;
11. $10 \text{ m} \cdot \dfrac{1 \text{ ft.}}{0.3048 \text{ m}} \approx 32.81$ ft.;
12. 1. O'Neal: $229 \div 408 \approx 0.561 = 56.1\%$, 2. Garnet: $87 \div 167 \approx 0.521 = 52.1\%$, 3. Bryant: $170 \div 342 \approx 0.497 = 49.7\%$;
13. $\dfrac{56 \text{ mi.}}{1 \text{ gal.}} \cdot 10.6 \text{ gal.} = 593.6$ mi. I can travel almost 594 miles. **14.** $\dfrac{6 \cancel{\text{mi.}}}{1 \cancel{\text{hr.}}} \cdot \dfrac{5280 \text{ ft.}}{1 \cancel{\text{mi.}}} \cdot \dfrac{1 \cancel{\text{hr.}}}{60 \cancel{\text{min.}}} \cdot \dfrac{1 \cancel{\text{min.}}}{60 \text{ sec.}} = 8.8$ ft. per sec. I run at 8.8 feet per second. **15. a.** $x = 2$, **b.** $x = 90$, **c.** $x = 12$; **16.** $\dfrac{\$5.28}{8 \text{ lb.}} = \dfrac{x}{100 \text{ lb.}}$

$x = \dfrac{\$5.28 \cdot 100 \cancel{\text{lb}}}{8 \cancel{\text{lb}}} = \$66.00$.

Chapter 6

Activity 6.1 Exercises: 1. a. $P = 2 \cdot 25 + 2 \cdot 15 = 80$ ft.
b. $P = 15 + 25 + 10 + 25 + 10 + 10 + 25 = 120$ ft.
c. $P = 25 + 45 + 25 + 10 + 10 + 25 + 10 + 10 = 160$ ft. **3. a.** $P = 1038 + 965 + 1042 = 3045$ mi.
b. $3045 \text{ mi.} \cdot \dfrac{1 \text{ hr.}}{600 \text{ mi.}} = 5.075$ hr.
5. b. $P = 2 \cdot 2.5 + 2 \cdot 6 = 17$ in.
d. $P = 3 + 1 + 1 + 1 + 3 + 1 + 1 + 1 = 12$ mi.
7. Since $P = 2l + 2w$, then $75 = 2(10) + 2w$.
Solving this equation,
$75 = 20 + 2w$
$55 = 2w$
$\dfrac{55}{2} = w$ So, $w = \dfrac{55}{2} = 27.5$ meters.

Activity 6.2 Exercises: 2. If the radius measures 1 cm, then $C = 2 \cdot \pi \cdot 1 \approx 6.28$ cm. If the radius measures 3 cm, then $C = 2 \cdot \pi \cdot 3 = 6 \cdot \pi \approx 18.85$ cm. If the radius measures 6 cm, then $C = 2 \cdot \pi \cdot 6 = 12 \cdot \pi \approx 37.70$ cm. If the radius measures 10 cm, then $C = 2 \cdot \pi \cdot 10 = 20 \cdot \pi \approx 62.83$ cm.

4. a. $C = \pi \cdot 3 \approx 9.4$ cm **d.** $\dfrac{1}{4}C = \dfrac{1}{4} \cdot 2 \cdot \pi \cdot 2 = \pi \approx 3.1$ in. **5.** Since $C = 2\pi r$, then $63 = 2\pi r$. Solving for r, $r = \dfrac{63}{2\pi} \approx 10.03$ inches.

Activity 6.3 Exercises:
2. $P = 4 + 3 + 4 + \dfrac{1}{2} \cdot \pi \cdot 3 = 11 + 1.5 \cdot \pi \approx 15.71$ ft.
4. a. $P = 6 + 3.8 + 6.2 + 5.2 = 21.2$ cm
c. $P = 2 + 10 + \dfrac{1}{2} + 3 + 3 + \dfrac{1}{2} + 10 = 29$ m.

Activity 6.4 Exercises:
1. a. $A = 20 \cdot 30 + \dfrac{1}{2} \cdot 10 \cdot 12 = 600 + 60 = 660$ sq. ft.
b. $A = 30 + 30 + 6 = 66$ feet of 10-foot widths; run the carpet horizontally to cover the main part of the room. Then cut the remaining 6-foot length diagonally to fit the triangular part.
3. a. $A = 4 \cdot 3 + 2 \cdot \dfrac{1}{2} \cdot 3 \cdot \dfrac{1}{2} = 12 + 1.5 = 13.5$ sq. ft.
5. $A = 94 \cdot 50 = 4700$ sq. ft.

Activity 6.5 Exercises: 1. The larger pizza has area $\pi \cdot \left(\dfrac{14}{2}\right)^2 \approx 153.94$ square inches; the smaller pizza has area $\pi \cdot \left(\dfrac{10}{2}\right)^2 \approx 78.54$ square inches. So you can fit approximately two smaller pizzas into the larger one.
2. a. The diameters for a quarter, nickel, penny, and dime are: 2.4 cm, 2.1 cm, 1.9 cm, and 1.8 cm.
b. The area of a quarter is $\pi \cdot 1.2^2 \approx 4.52$ sq. cm.
The area of a nickel is $\pi \cdot 1.05^2 \approx 3.46$ sq. cm.
The area of a penny is $\pi \cdot .95^2 \approx 2.84$ sq. cm.
The area of a dime is $\pi \cdot .9^2 \approx 2.54$ sq. cm.
3. b. $A = \pi \cdot 3^2 \approx 28.27$ sq. mi.
d. $A = \dfrac{1}{4} \cdot \pi \cdot \left(\dfrac{2}{3}\right)^2 \approx 0.35$ sq. in.

Activity 6.6 Exercises: 1. Area $= 15 \cdot 25 = 375$ sq. ft.
3. $A = \dfrac{1}{2} \cdot 6(10 + 12) = 66$ sq. ft.
6. $A \approx 8\left(\dfrac{1}{2} \cdot 15 \cdot 18.1\right) = 1086$ sq. in.

Activity 6.7 Exercises: 2. Solving $x + 47 = 90$, the other acute angle must be $x = 90° - 47° = 43°$. **4.** They all must add up to 180° and all be different sizes, since the triangle is equilateral or isosceles. **6.** The sum of the four angles must be the same as the sum of six angles in two triangles, which will be $180° + 180° = 360°$.

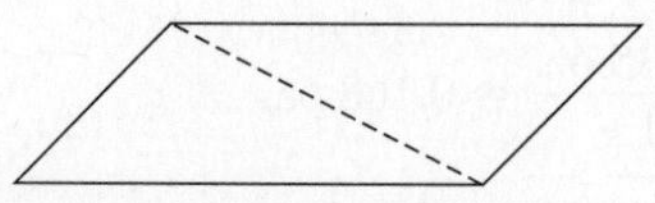

Activity 6.8 Exercises:

2. If they did meet at right angles, then the Pythagorean theorem would be true, and $18^2 = 12^2 + 14^2$. But $18^2 = 324$ and $12^2 + 14^2 = 144 + 196 = 340$. And $324 \neq 340$. So, the walls do not meet at right angles.
3. a. $12^2 = a^2 + 9^2$, so $a^2 = 12^2 - 9^2 = 144 - 81 = 63$. Therefore, $a = \sqrt{63} \approx 7.94$ feet.
b. $c^2 = 10^2 + 9^2 = 181$, so $c = \sqrt{181} \approx 13.45$ feet.
5. a. $c^2 = 6^2 + 11^2 = 157$, so $c = \sqrt{157} \approx 12.53$ centimeters. **7. a.** Yes, since $13^2 = 5^2 + 12^2$ because $169 = 169$. **b.** No, since $15^2 \neq 5^2 + 10^2$ because $225 \neq 125$. **8.** $c = \sqrt{3^2 + 7^2} = \sqrt{58} \approx 7.62$ miles for the boat trip. **10.** No, since $15^2 \neq 7^2 + 10^2$ (that is, $225 \neq 49 + 100$).

Activity 6.9 Exercises: 2. Setting up the proportion, $\frac{d}{7} = \frac{14}{11}$. Solving, $11d = 14 \cdot 7$, so $d = \frac{98}{11} \approx 8.91$ feet.

4. a. Let x represent the length of the smallest side and y represent the length of the medium side.

$$\frac{x}{3} = \frac{15}{6} \qquad \frac{y}{5} = \frac{15}{6}$$
$$6x = 45 \qquad 6y = 75$$
$$x = \frac{45}{6} \text{ in.} \qquad y = \frac{75}{6} \text{ in.}$$

So the dimensions are $7\frac{1}{2}$ inches by $12\frac{1}{2}$ inches by 15 inches.

What Have I Learned? Exercises: 5. No, consider the following example:

2 cm × 8 cm rectangle: Has perimeter 20 cm and area 16 square cm.

1 cm × 9 cm rectangle: Has perimeter 20 cm and area 9 square cm.

So two rectangles can have the same perimeter and different areas.

How Can I Practice? Exercises:

1. a. Perimeter $= 2 \cdot 5 + 2 \cdot 10 = 30$ ft.
Area $= 4 \cdot 10 = 40$ sq. ft.

2. a. Perimeter $= \frac{1}{2} \cdot \pi \cdot 2 + 2 \cdot 4 + 2 \approx 13.14$ ft.
Area $= \frac{1}{2} \cdot \pi \cdot 1^2 + 4 \cdot 2 = \frac{\pi}{2} + 8 \approx 9.57$ sq. ft.

3. a. Area $= \frac{1}{2} \cdot 6 \cdot 4 = 12$ sq. m

7. a. Triangle C is a right triangle because it is the only one that satisfies $a^2 + b^2 = c^2$: $13^2 = 169 = 12^2 + 5^2$.

b.

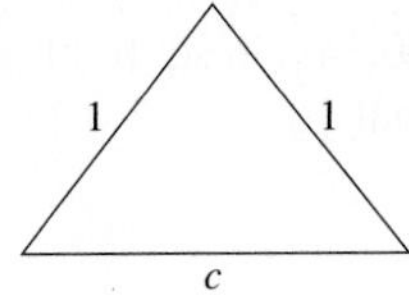

(Answers may vary.) Setting up the proportion, $\frac{c}{5} = \frac{1}{4}$. So $c = 1.25$. So the sides of triangle D are 1 foot, 1 foot, and 1.25 feet.

c. No, since the new triangle will have sides measured by 5 feet, 7 feet, and 9 feet. However, $\frac{5}{3} \neq \frac{7}{5}$ since $1.67 \neq 1.4$.

Activity 6.10 Exercises:

1. a. $S = 4\pi \cdot 4.75^2 \approx 284$ sq. in.

3. a. $S = 2\pi \cdot \left(\frac{7}{2}\right)^2 + 2\pi \cdot \left(\frac{7}{2}\right) \cdot 11 \approx 319$ sq. in.

4. $S = 2 \cdot 2 \cdot 3.5 + 2 \cdot 3.5 \cdot 4.2 + 2 \cdot 2 \cdot 4.2$
$= 14 + 29.4 + 16.8 = 60.2$ sq. ft.

6. The surface area formula is $S = 2\pi \cdot r^2 + 2\pi \cdot rh$. So,

$$300 = 2\pi \cdot 5^2 + 2\pi \cdot 5h$$
$$300 \approx 157.08 + 31.42h$$
$$142.92 \approx 31.42h$$
$$\frac{142.92}{31.42} \approx h$$
$$h \approx 4.55 \text{ in.}$$

Activity 6.11 Exercises:

1. a. $V = \pi \cdot \left(\frac{7}{2}\right)^2 \cdot 11 \approx 423$ cu. in.

b. $V = 9 \cdot 2 \cdot 4 = 72$ cu. in.

3. $V = \pi \cdot \left(\frac{3}{2}\right)^2 \cdot 5 = \pi \cdot \frac{9}{4} \cdot 5 \approx 35$ cu. in.

4. Let x represent the length = width = height. Then,
$x^3 = 42$
$x = \sqrt[3]{42} \approx 3.48$
So the dimensions are 3.48 feet by 3.48 feet by 3.48 feet.

Activity 6.12 Exercises:

2. a. $V_E = \frac{4}{3}\pi \cdot 6378^3 \approx 1.087 \cdot 10^{12}$ cu. km

$V_M = \frac{4}{3}\pi \cdot 3397^3 \approx 1.642 \cdot 10^{11}$ cu. km

3. $V = \frac{4}{3}\pi \cdot \left(\frac{d}{2}\right)^3 = \frac{\pi \cdot d^3}{6}$

5. $V = \frac{4}{3}\pi \cdot 4.75^3 \approx 448.92$ cu. in. of air

7. a. $V = \frac{1}{3}\pi \cdot 1^2 \cdot 6.2 \approx 6.5$ cu. ft.

d. $V = \frac{4}{3}\pi \cdot 4^3 \approx 268.1$ cu. ft.

What Have I Learned? 6. Yes; for example, consider a can of height 1 foot and radius 1 foot.

$V = \pi \cdot 1^2 \cdot 1 = \pi$ cu. ft. Next, if you double the height to 2 feet, then $V = \pi \cdot 1^2 \cdot 2 = 2\pi$ cu. ft., which is double the original volume.

Because of the placement of h as a factor of $\pi \cdot r^2$ in the formula for volume of a cylinder, replacing the height with its double will always just double the volume.

7. $V = \pi \cdot r^2 \cdot h = \pi \cdot \left(\frac{d}{2}\right)^2 \cdot h = \frac{\pi \cdot d^2 \cdot h}{4}$, the formula for volume of a cylinder in terms of its diameter and height. So doubling d means replacing d by $2d$ in the formula. The new formula becomes

$V = \frac{\pi \cdot (2d)^2 \cdot h}{4} = \pi \cdot d^2 \cdot h$. So, comparing volumes, the new volume is four times larger than the initial volume. Therefore, the volume does not double; it quadruples.

How Can I Practice?

1. a. $V = \frac{4}{3}\pi \cdot 6^3 \approx 904.78$ cu. in.
   $S = 4\pi \cdot 6^2 \approx 452.39$ sq. in.

   b. $V = \pi \cdot \left(\frac{4.5}{2}\right)^2 \cdot 6.5 \approx 103.38$ cu. ft.
   $S = 2\pi \cdot \left(\frac{4.5}{2}\right)^2 + 2\pi \cdot \left(\frac{4.5}{2}\right) \cdot 6.5 \approx 31.81 + 91.89$
   ≈ 123.70 sq. ft.

   c. $V = 14.3 \cdot 4.5 \cdot 5.3 \approx 341.06$ cu. ft.
   $S \approx 2 \cdot 14.3 \cdot 4.5 + 2 \cdot 14.3 \cdot 5.3 + 2 \cdot 4.5 \cdot 5.3$
   $= 128.7 + 149.99 + 47.7 = 326.39$ sq. ft.

3. b. 2.05 ft. $53 = 4\pi \cdot r^2$. Solving, $r^2 \approx 4.22$. So, $r \approx \sqrt{4.22} \approx 2.05$ centimeters.

Gateway Review

1. a. 10 sq. ft. b. 17 in. c. $B = \frac{1}{2} \cdot 8 \cdot 2 = 8$ sq. in.
   d. $C = \frac{1}{2} \cdot 2(4 + 4.5) = 8.5$ sq. mi.

2. $23 = \pi \cdot r^2$. Solving, $r^2 \approx 7.32$. Therefore, $r \approx \sqrt{7.32} \approx 2.71$ inches.

3.

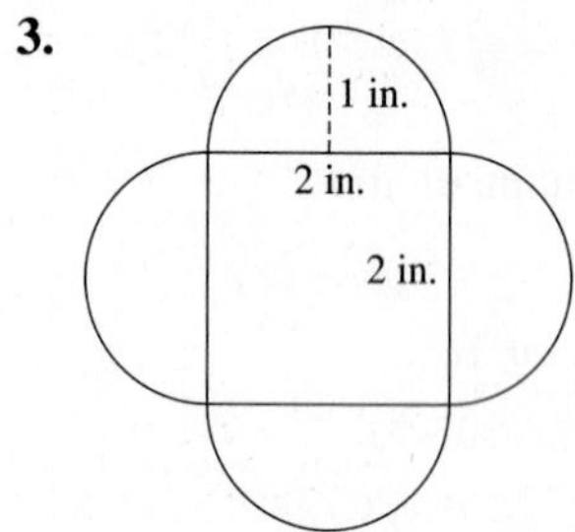

b. $P = 4 \cdot \frac{1}{2} \cdot 2 \cdot \pi \cdot 1 = 4\pi \approx 12.57$ in.
$A = 2 \cdot 2 + 4 \cdot \frac{1}{2} \cdot \pi \cdot 1^2 \approx 10.28$ sq. in.

c. $A_{\text{table}} = 12 \cdot 12 = 144$ sq. in. So, there is still space for $144 - 10.28 = 133.72$ square inches.

4. a. $A = 2 \cdot \frac{1}{2} \cdot 4 \cdot 3 + 2 \cdot \frac{1}{2} \cdot 4 \cdot 6 = 36$ sq. ft.
   So, you need to buy at least 36 square feet of material.

   b. You need to determine the hypotenuses of the four right triangles:
   $c_1 = \sqrt{3^2 + 4^2} = 5$ ft.
   $c_2 = \sqrt{6^2 + 4^2} = \sqrt{52} \approx 7.21$ ft.
   So, $P \approx 2 \cdot 5 + 2 \cdot 7.21 = 24.42$ ft.
   Buy at least 24.42 feet of ribbon.

5. a. $P = 22 + 17 + 19 + 7 + 15 + 3 + 22 + 7 + 17 + 19 + 7 + 15 + 22 + 18 = 210$ ft.
   b. $A = 43 \cdot 40 + 18 \cdot 22 = 2116$ sq. ft.
   c. The area for bedroom A is $15 \cdot 19 = 285$ square feet. The area for bedroom B is $15 \cdot 19 = 285$ square feet. The area for bedroom C is $17 \cdot 22 = 374$ square feet. So, bedroom C is the largest.
   d. (Answers will vary.) The area of the living room is $18 \cdot 22 = 396$ square feet. Doubling this area would require 792 square feet. So, perhaps the easiest way to do this is to increase the length of the living room to 36 feet and keep the width the same, 22 feet. This produces an area of 792 square feet.

6. $c = \sqrt{300^2 + 200^2} = \sqrt{130{,}000} \approx 360.56$ feet, the measurement of the hypotenuse of the right triangle.

7. Setting up the proportions for similar triangles.

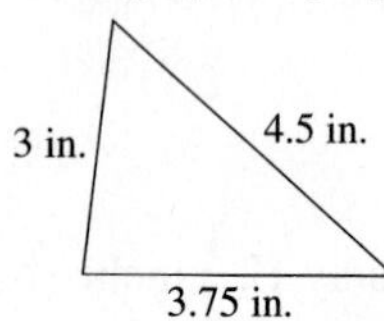

$\frac{a}{6} = \frac{3}{4}$

Solving, $a = \frac{6 \cdot 3}{4} = 4.5$ inches.

$\frac{b}{5} = \frac{3}{4}$

Solving, $b = \frac{5 \cdot 3}{4} = 3.75$ inches.

8. a. $V = 20 \cdot 15 \cdot 15 + \frac{1}{2} \cdot \frac{4}{3} \cdot \pi \cdot 7.5^3 \approx 5383.57$ cu. ft.
   b. $V = \frac{1}{3} \cdot \pi \cdot \left(\frac{3}{2}\right)^2 \cdot 5 \approx 11.78$ cu. m
   c. $V = \pi \cdot \left(\frac{5}{2}\right)^2 \cdot 8 \approx 157.08$ cubic in.

9. $V = \pi \cdot 2640^2 \cdot 187 \approx 4{,}094{,}485{,}458$ cu. ft. of water in the lake

10. a. For consistency in units, the diameter of 48 inches was changed to 4 feet, so the radius of the pipe is 2 feet. The 800 miles was changed to
    $800 \text{ miles} \cdot \frac{5280 \text{ feet}}{1 \text{ mile}} = 4{,}224{,}000.$
    So, $V = \pi \cdot 2^2 \cdot 4{,}224{,}000 \approx 53{,}080{,}349.48$ cubic feet of oil.
    b. $S = 2\pi \cdot 2 \cdot 4{,}224{,}000 \approx 53{,}080{,}349.48$ sq. ft. of material

Chapter 7

Activity 7.1 Exercises:

2. a. $-2 = 2x$, $x = -1$, **c.** $3x = 12$, $x = 4$, **e.** $5x = 30$, $x = 6$,

g. $\frac{1}{5}x = -3$, $x = -15$, **i.** $0.25x = 24.5$, $x = 98$;

3. a. Multiply the number of minutes by 0.10 and then add 4.95 to get the answer. **b.** $c = 0.10n + 4.95$,
c. $c = 0.10 \cdot 250 + 4.95 = \$29.95$,
d.
$$\begin{aligned} 50 &= 0.10n + 4.95 \\ 45.05 &= 0.10n \\ n &= 450.5 \end{aligned}$$
I will have 450 minutes for calls.

4. a. $10n$, **b.** $p = 10n - 2100$,
c. $p = 10 \cdot 700 - 2100 = \$4900$,
d.
$$\begin{aligned} 1500 &= 10n - 2100 \\ 3600 &= 10n \\ n &= 360; \end{aligned}$$

6. a. $c = 2.085 \cdot 10 + 15.08 = \$35.93$,
b.
$$\begin{aligned} 56.78 &= 2.085x + 15.08 \\ 41.70 &= 2.085x \\ x &= 20 \text{ lb.}; \end{aligned}$$

8.
$$\begin{aligned} \frac{3}{4} &= 2a + \frac{1}{3} \\ 3 &= 8a + \frac{4}{3} \\ 9 &= 24a + 4 \\ \frac{5}{24} &= a. \end{aligned}$$
Check: $\frac{5}{24} + \frac{5}{24} + \frac{1}{3} = \frac{18}{24} = \frac{3}{4}$ The equal sides measure $\frac{5}{24}$ meter.

10. a.

X	Y
4	−2
12	14

$$\begin{aligned} y &= 2 \cdot 4 - 10 \\ y &= 8 - 10 \\ y &= -2 \end{aligned}$$
$$\begin{aligned} 14 &= 2x - 10 \\ 24 &= 2x \\ x &= 12, \end{aligned}$$

c.

X	Y
$\frac{2}{3}$	13
6	−3

$$\begin{aligned} y &= -3 \cdot \frac{2}{3} + 15 \\ y &= 13 \end{aligned}$$
$$\begin{aligned} -3 &= -3x + 15 \\ -18 &= -3x \\ x &= 6; \end{aligned}$$

Activity 7.2 Exercises:

1. a.
$$\begin{aligned} 4 &= 0.167t - 0.67 \\ 4.67 &= 0.167t \\ t &\approx 27.96 \\ &\approx 28°\text{C}, \end{aligned}$$
b. $\frac{s + 0.67}{0.167} = t$, **c.** $t = \frac{4 + 0.67}{0.167} = \frac{4.67}{0.167} \approx 28°\text{C}$;

3. a.
$$\begin{aligned} 10 &= 0.17t + 3.9 \\ 6.1 &= 0.17t \\ t &= 35.88 \approx 36 \text{ yr.} \end{aligned}$$
$1970 + 36 = 2006$ (the year),

b. $\frac{n - 3.9}{0.17} = t$, **c.** $t = \frac{10 - 3.9}{0.17} = \frac{6.1}{0.17} \approx 36$ yr.; $1970 + 36 = 2006$ (the year); **5.** $\frac{E}{R} = I$; **7.** $P - 2a = b$; **9.** $\frac{P - 2l}{2} = w$; **11.** $\frac{A - P}{Pt} = r$

Activity 7.3 Exercises:

1. a. cost $= 0.19x + 39.95$, **b.** cost $= 0.49x + 19.95$,
c. $0.19x + 39.95 = 0.49x + 19.95$,
d.
$$\begin{array}{rcl} 0.19x + 39.95 &=& 0.49x + 19.95 \\ -0.19x & & -0.19x \\ \hline 39.95 &=& 0.30x + 19.95 \\ -\ 19.95 & & -\ 19.95 \\ \hline 20.00 &=& 0.30x \end{array}$$
$x = \frac{20}{.30} = 66.66 \approx 67$ mi.,
e. Company 1 is better if you drive more than 67 miles.

3. a. cost $= 3560 + 15x$, **b.** cost $= 2850 + 28x$,
c. $3560 + 15x = 2850 + 28x$,
d.
$$\begin{array}{rcl} 3560 + 15x &=& 2850 + 28x \\ -\ 15x & & -\ 15x \\ \hline 3560 &=& 2850 + 13x \\ -\ 2850 & & -\ 2850 \\ \hline 710 &=& 13x \end{array}$$
$\frac{710}{13} = \frac{13x}{13}$ $x \approx 55$ months,
e. 10 years is equivalent to 120 months. Dealer 1 is better for over 55 months of use.

6.
$$\begin{array}{rcl} 3x - 14 &=& 6x + 4 \\ -\ 3x & & -\ 3x \\ \hline -14 &=& 3x + 4 \\ -4 & & -4 \\ \hline -18 &=& 3x \\ x &=& -6; \end{array}$$

8.
$$\begin{array}{rcl} 4x - 10 &=& -2x + 8 \\ +\ 2x & & +2x \\ \hline 6x - 10 &=& 8 \\ +\ 10 & & +10 \\ \hline 6x &=& 18 \\ x &=& 3; \end{array}$$

10.
$$\begin{array}{rcl} 4 - 0.05x &=& 0.1 - 0.05x \\ +\ 0.05x & & +\ 0.05x \\ \hline 4 + 0.025x &=& 0.1 \\ -\ 4 & & -\ 4 \\ \hline 0.025x &=& -3.9 \\ x &=& -156; \end{array}$$

Activity 7.4 Exercises: 1. a^4; **3.** y^9; **5.** $-12w^7$; **7.** a^{15}; **9.** $-x^{50}$; **11.** $-5.25x^9y^2$; **13.** $-2s^6t^8$; **15.** $3y^2 - y$; **17.** $2a^4 + 8a^2 - 10a$; **19.** $3.5r^5 - 1.6r^4$; **21.** $-2.6x^{10} - 7.8x^8 + 1.3x^7$; **23. a.** $5x(4x) = 20x^2$, **b.** $V = 20x^2(x + 15) = 20x^3 + 300x^2$; **25.** $A = (3xy^2)^2 = 9x^2y^4$;

Activity 7.5 Exercises:

4. $3x + 10x - 8$
$13x - 8$;

5. $3.1a + 3.1b + 8.7a$
$11.8a + 3.1b$;

7. $3 - x + 2x - 1$
$x + 2$;

9. $6 - 3x - 8x - 4$
$-11x + 2$;

11. $3(4)^2 + 2(4) - 1$
$48 + 8 - 1$
55;

13. $2100(0.08)(3) = 504$;

15. $4(10)^3 + 15$
$4000 + 15 = 4015$;

17. $2(x + 5) + 2(x - 3)$
$2x + 10 + 2x - 6$
$4x + 4$;

Activity 7.6 Exercises:

2. a. $p = -0.620(16) + 42.47 = 32.55\%$,
b. $15 = -0.62t + 42.47$
$-27.47 = -0.62t$
$t \approx 44$ yr.
$1965 + 44 = 2009$;
4. a. $C = 37.7(30) - 170 = \$961$,
b. $2000 = 37.7a - 170$
$2170 = 37.7a$
$a \approx 58$ years old;

How Can I Practice?

1. $4x = 16$
$x = 4$;
2. $-2x = 14$
$x = -7$;
3. $\frac{3}{4}x = 3$
$x = 4$;
4. $10 = -2.5x$
$x = -4$;
5. $-12 = 2x$
$x = -6$;
6. $6x + 3 = 9$
$6x = 6$
$x = 1$;
7. $2x + 2 = 5x - 3$
$5 = 3x$
$x = \frac{5}{3} = 1\frac{2}{3}$;
8. $14 - 3 = 4x$
$11 = 4x$
$x = \frac{11}{4} = 2\frac{3}{4}$;
9. $4x = -11.5$
$x = \frac{-11.5}{4} = -2.875$;
10. $8x + 4 = 14x + 14$
$-10 = 6x$
$x = \frac{-10}{6} = -1\frac{2}{3}$;
11. $3x - 6 + 10 = 5x$
$3x + 4 = 5x$
$4 = 2x$
$x = 2$;
12. $0.25x - 0.5 = 0.2x + 2$
$0.05x = 2.5$
$x = 50$;
13. $-10 = 2.5x$
$-4 = x$
14. $\frac{1}{3}x - \frac{1}{3} = 6x - 6$
$5\frac{2}{3} = 5\frac{2}{3}$
$1 = x$;
15. $-20x + 15$;
16. $6ax + 12bx + 3cx$; **17.** $13.5x - 0.9$;
18. $-20x + 16y - 40$; **19.** $\frac{P - b}{2} = a$ **20.** $\frac{P}{r} = t$;
21. $\frac{f - v}{a} = t$; **22.** $x = \frac{4y + 8}{3}$;
23. $A = P + Prt$
$A - P = Prt$
$\frac{A - P}{Pt} = r$;
24. $E - S = \frac{3}{2}rn$
$\frac{2(E - S)}{3n} = r$;
25. a. $t = 2005 - 1960 = 45$
$a = 0.11(45) + 22.5 = 27.45$ years,
b. $30 = 0.11t + 22.5$
$7.5 = 0.11t$
$t = \frac{7.5}{0.11} \approx 68$
$1960 + 68 = 2028$,
c. $t = \frac{a - 22.5}{0.11}$,
d. $t = \frac{30 - 22.5}{0.11} \approx 68$; $1960 + 68 = 2028$,
26. a. $c = 10 + 0.03x$,
b.

NUMBER OF COPIES, x	TOTAL COST, c
1000	40
2000	70
3000	100
4000	130
5000	160

c. $c = 10 + 8000(0.03) = 250$ dollars,
d. $300 = 10 + 0.03x$; $290 = 0.03x$; $x = \frac{290}{0.03} = 9666.\overline{6}$.
I can have almost 9700 copies printed for \$300.
27. $-x^{10}$; **28.** $-10x^7$; **29.** $-0.55x^7$; **30.** $-15s^3t^5$;
31. $x^5 + 2x^4 - x^3$; **32.** y^{15}; **33.** t^8;
34. $6x^6x^4x = 6x^{11}$;
35. $2x + 3x^2 - 12x + 5x^2$
$8x^2 - 10x$;
36. $7x + 14 - 2.7x - 10.6$
$4.3x + 3.4$;
37. $8x^5 - 6x^4 + 4x^3$;
38. $10x - 20x^2 - 3x^2 + 12x + 3 + 5x^2$
$-18x^2 + 22x + 3$;
39. a. $A = (5x)(3x) = 15x^2$ **b.** $V = 15x^2(x + 8)$;
40. $3(4)^2 = 3(16) = 48$;
41. $2(3)^2 - 3(2.5) = 18 - 7.5 = 10.5$;
42. $5(1 + 7(2)) = 5(1 + 14) = 5(15) = 75$;
43. $180 - 15 - (-25) = 180 - 15 + 25 = 190$;
44. $21 + 3x - 12 = 24$
$9 + 3x = 24$
$3x = 15$
$x = 5$;
45. $8x - 6 = 6x + 18$
$2x = 24$
$x = 12$;
46. $2 - 5x - 25 = 3x - 6 - 1$
$-5x - 23 = 3x - 7$
$-16 = 8x$
$x = -2$;
47. $0.16x + 2.4 - 0.24x = 1.8$
$-0.08x + 2.4 = 1.8$
$-0.08x = -0.6$
$x = \frac{-0.6}{-0.08} = 7.5$;
48. a. $C = 750 + 0.25x$,
b. $C = 750 + 0.25(500) = 750 + 125 = 875$,
c. $1000 = 750 + 0.25x$
$250 = 0.25x$
$x = \frac{250}{0.25} = 1000$, 1000 booklets can be produced.
d. $R = 0.75x$,
e. $R = C$
$0.75x = 750 + 0.25x$
$0.50x = 750$
$x = \frac{750}{0.50} = 1500$,
1500 booklets must be sold to break even. The cost and revenue would both be \$112.

f. $0.75x - (750 + 0.25x) = 500$
$0.75x - 750 - 0.25x = 500$
$0.50x - 750 = 500$
$0.50x = 1250$
$x = \frac{1250}{0.50} = 2500$
2500 booklets must be sold to make a $500 profit.
49. a. $1.50x$, **b.** $22 - x$,
c. $(22 - x)(0.20)(15) = 3(22 - x)$,
d. $C = 1.50x + 3(22 - x) = 1.50x + 66 - 3x$
$= 66 - 1.50x$,
e. $42 = 66 - 1.50x$
$-24 = -1.50x$
$x = \frac{-24}{-1.50} = 16$
Driving days = 22 − 16 = 6
I can drive for 6 days a month if I budget $42 per month.
f. $x = 11$, $C = 1.50(11) + 3(11)$, $C = 49.50$. I must budget $49.50 to be able to take the bus only half of the time.

Gateway Review

1. $-26 = 2x; x = -13$; **2.** $0.15x = 45.9$;
$x = \frac{45.9}{0.15} = 306$;
3. $1.6x = 39; x = 24.375$; **4.** $21 = 3x; x = 7$;
5. $4x + 20 - x = 80$
$3x + 20 = 80$
$3x = 60; x = 20$;
6. $38 = 57 - (x + 32)$
$38 = 57 - x - 32$
$38 = 25 - x; x = 25 - 38; x = -13$;
7. $5x + 3(2x - 8) = 2(x + 6)$
$5x + 6x - 24 = 2x + 12$
$11x - 24 = 2x + 12$
$9x = 36; \quad x = 4$;
8. $2.5x + 10 = 5.8x - 11.6$
$21.6 = 3.3x$
$x = \frac{21.6}{3.3} = 6.\overline{54}$;
9. x^9; **10.** $3x^{20}$; **11.** $-6x^4y^3$;
12. $x^3y^4x^8 = x^{11}y^4$;
13. $-3x^5 + 6x^4 - 3x^2$;
14. $10x^2 + 9x - 6x^2 = 4x^2 + 9x$;
15. a. $P = \frac{I}{rt}$ **b.** $\frac{P - b}{2} = a$ **c.** $\frac{E - S}{1.5n} = r$
d. $P = \frac{k}{V} - a$;
16. a. $2(-3)^2 - 3(-3) + 5 = 2(9) + 9 + 5 = 32$,
b. $2(17.3) + 11.8 = 34.6 + 11.8 = 46.4$,
c. $10 - (-6 + (-8)) = 10 - (-14) = 24$;
17. a. $S = 100 + 0.30x$, **b.** $S = 150 + 0.15x$,
c. $100 + 0.30x = 150 + 0.15x$,
d. $100 + 0.30x = 150 + 0.15x$
$0.15x = 50; x = \frac{50}{0.15} \approx 333.33$
For $333.33 in sales per week, the salaries for both options would be the same.
e. The salary is approximately $100 + 0.30(333.33) \approx$ \$200; **18. a.** $x + 10$, **b.** $C = x + 10 + 55 = x + 65$,
c. $x + x + 10 + x + 65 = 3x + 75$,
d. $120 = 3x + 75$
$45 = 3x$
$x = 15$ mi.,
e. Running: 15 + 10 = 25 mi.
Cycling: 25 + 55 = 80 mi.;
19. a.

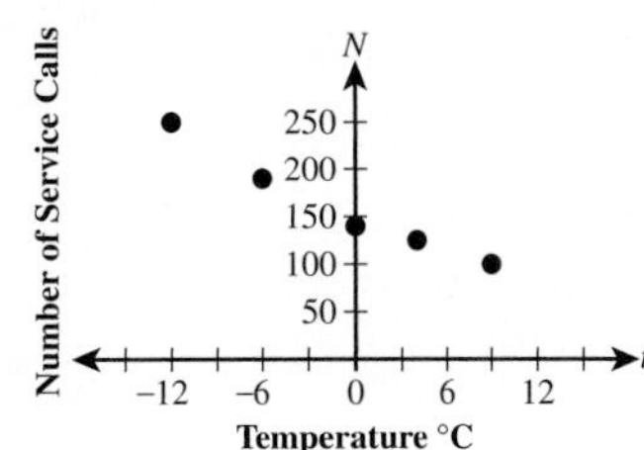

b. $C = -7.11(-20) + 153.9 = 296.1 \approx 296$ calls,
c. $50 = -7.11t + 153.9$
$7.11t = 153.9 - 50$
$t = \frac{103.9}{7.11} \approx 14.6$, about 15°C,
d. $\frac{C - 153.9}{-7.11} = t$, **e.** $t = \frac{50 - 153.9}{-7.11} \approx 14.6$;
20. a. ABC: $C = 50 + 0.20x$
Competition: $C = 60 + 0.15x$,
b. $50 + 0.20x = 60 + 0.15x$,
c. $50 + 0.20x = 60 + 0.15x$,
$0.05x = 10; x = 200$ mi.,
d. It would be cheaper to rent from the competition.
21. a. $C = 600 + 10x$, **b.** $R = 40x$,
c. $P = 40x - (600 + 10x) = 30x - 600$,
d. $30x - 600 = 0; 30x = 600; x = 20$ campers,
e. $30x - 600 = 600; 30x = 1200; x = 40$ campers,
f. $30(10) - 600 = 300 - 600 = -300$. The camp would lose $300.

GLOSSARY

absolute value of a number The size or magnitude of a number. It is represented by the distance of the number from zero on a number line. The absolute value is always nonnegative.

actual change The difference between a new value and the original value of a quantity.

acute angle An angle that is smaller than a right angle (measures less than 90°).

acute triangle All angles in the triangle measure less than 90°.

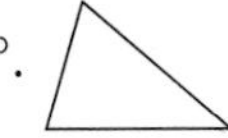

addends The numbers that are combined in addition. For example, in $1 + 2 + 7 = 10$, the numbers 1, 2, and 7 are the addends.

addition The arithmetic operation that determines the total of two or more numbers.

addition property of zero The sum of any number and zero is the same number.

algebraic expression A symbolic code of instructions for performing arithmetic operations with variables and numbers.

angle The figure formed by two rays meeting at a same point. The size of the angle is the amount of turn needed to move one ray to coincide with the other ray.

area A measure of the size of a region that is entirely enclosed by the boundary of a plane figure, usually expressed in square units.

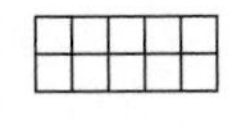

associative property of addition For all numbers a, b, and c, $(a + b) + c = a + (b + c)$. For example, $(2 + 3) + 4 = 5 + 4 = 9$ and $2 + (3 + 4) = 2 + 7 = 9$.

associative property of multiplication For all numbers a, b, and c, $(ab)c = a(bc)$. For example, $(2 \cdot 3) \cdot 4 = 6 \cdot 4 = 24$ and $2 \cdot (3 \cdot 4) = 2 \cdot 12 = 24$.

average (arithmetic mean) A number that represents the center of a collection of data values. The arithmetic mean is determined by adding up the data values and dividing the sum by the number of data values in the collection.

bar graph A diagram that shows quantitative information by the lengths of a set of parallel bars of equal width. The bars can be drawn horizontally or vertically.

base number The number that is raised to a power. In N^P, N is the base number and P is the exponent.

break-even point The point at which two mathematical models for determining cost or value will result in the same quantity.

Cartesian (rectangular) coordinate system in the plane A two-dimensional scaled grid of equally spaced horizontal lines and equally spaced vertical lines that is used to locate the positions of points on a plane.

circle A collection of points that are the same distance from a given point, called the center of the circle.

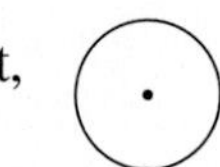

circumference of a circle The distance around a circle. If a circle has radius r, diameter d, and circumference C, then $C = 2\pi r$ and $C = \pi d$.

coefficient (numerical) A number written next to a variable that multiplies the variable. For example, $2x$ means that 2 multiplies the variable x.

commutative property of addition For all numbers a and b, $a + b = b + a$, which means that changing the order of the addends does not change the result.

commutative property of multiplication For all numbers a and b, $ab = ba$, which means that changing the order of the factors does not change the result.

composite number A whole number greater than 1 that is not prime.

cone A three-dimensional figure consisting of a curved surface that is joined to a circular base at the bottom and a fixed point at the top.

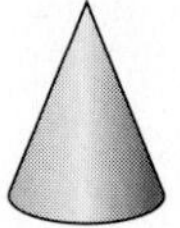

constant A term in an expression that consists of only a number.

cumulative effect of consecutive sequence of percent changes The product of the associated growth or decay factors.

coordinates of a point An (input, output) label to locate a point in a rectangular coordinate system.

counting numbers The set of positive whole numbers $\{1, 2, 3, 4, \ldots\}$.

cube root The cube root of a number N is a number M whose cube is N. The symbol for the cube root is $\sqrt[3]{\ }$. For example, the cube root of 8 is 2 $\left(\sqrt[3]{8} = 2\right)$ since $2^3 = 8$.

cylinder *See* **right circular cylinder.**

decay factor When a quantity decreases by a specified percent, the ratio of its new value to its original value is called the decay factor associated with the specified percent decrease.

$$\text{decay factor} = \frac{\text{new value}}{\text{original value}}$$

The decay factor is also formed by subtracting the specified percent decrease from 100% and then changing the percent into decimal form.

decimal number A number with a whole part, to the left of the decimal point, and a fractional part, to the right of the decimal point.

denominator The number written below the line of a fraction; the denominator represents the number of equal parts into which a whole unit is divided.

diameter of a circle A line segment through the center of a circle that connects two points on the circle; the measure of the diameter is twice that of the radius.

difference The result of subtracting the subtrahend from the minuend. For example, in $10 = 45 - 35$, the number 10 is the difference, 45 is the minuend, and 35 is the subtrahend.

digit Any whole number from 0 to 9.

dimensional analysis *See* **unit analysis.**

distributive properties For all numbers a, b, and c, $a(b + c) = ab + ac$ and $a(b - c) = ab - ac$.

dividend In a division problem, the number that is divided into parts. For example, in $27 \div 9 = 3$, the number 27 is the dividend, 9 is the divisor, and 3 is the quotient.

division by 0 Division by 0 is undefined.

division by 1 Any number divided by 1 is the number itself.

divisor In a division problem, the number that divides the dividend. For example, in $27 \div 9 = 3$, the number 27 is the dividend, 9 is the divisor, and 3 is the quotient.

equation A statement that says two expressions are equal, or represent the same number.

equiangular triangle All three angles in the triangle have equal measure, which is 60°.

equilateral triangle All three sides of the triangle have equal length.

equivalent expressions involving a single variable If for every possible value of the variable, the values of the expressions are equal, the expressions are said to be equivalent.

equivalent fractions Fractions that represent the same ratio. For example, $\frac{4}{6}$ and $\frac{10}{15}$ both represent the ratio $\frac{2}{3}$. Divide both the numerator and denominator of $\frac{4}{6}$ by 2 to obtain $\frac{2}{3}$. Similarly, divide both the numerator and denominator of $\frac{10}{15}$ by 5 to obtain $\frac{2}{3}$.

estimate of a numerical calculation An estimate is the result of rounding numbers in a calculation so that the numbers are easier to combine. The estimate is close to the actual result.

evaluate an algebraic expression Substitute a number for a variable and simplify the resulting numerical expression by applying the order of operations rules.

even number A whole number that is divisible by 2, leaving no remainder.

exponent If a number N is raised to a power P, meaning

$$N^P = \underbrace{N \cdot N \cdot \ldots \cdot N}_{(P \text{ factors})},$$

the power P is also called an *exponent.*

exponential form A number written in the form b^x, where b is called the base and x is the power or exponent. For example, 2^5.

factor A number a that divides another number b and leaves no remainder is called a factor of b. For example, 4 is a factor of 12 since $\frac{12}{4} = 3$.

factor, to To write an expression as a product of its factors.

factorization The process of writing a number as a product of factors.

formula An equation, or a symbolic rule, consisting of an output variable, usually on the left-hand side of the equal sign, and an algebraic expression of input variables, usually on the right-hand side of the equal sign.

fundamental principle of algebra Performing the same operation on both sides of the equal sign in a true equation results in a true equation.

graph A collection of points that are plotted on a grid, the coordinates of which are determined by an equation, formula, or table of values.

greatest common divisor (GCD) The greatest common divisor of two numbers is the largest number that divides into each of the numbers, leaving no remainder.

growth factor When a quantity increases by a specified percent, the ratio of its new value to its original value is called the growth factor associated with the specified percent increase.

$$\text{growth factor} = \frac{\text{new value}}{\text{original value}}$$

The growth factor is also formed by adding the specified percent increase to 100% and then changing the percent into decimal form.

histogram A bar graph with no spaces between the bars. *See* **bar graph.**

horizontal (x-) axis The horizontal number line in the Cartesian coordinate system that is used to represent input values.

hypotenuse of a right triangle The longest side, opposite the right angle, in a right triangle.

improper fraction A fraction whose numerator is greater than or equal to its denominator.

input values Replacement values for the variable(s) in the algebraic expression. For example, in the formula $P = 2l + 2w$, l and w may be replaced by the input values 5 and 10, respectively. Then, $P = 2(5) + 2(10) = 30$.

integers The collection of all of the positive counting numbers $\{1, 2, 3, 4, \ldots\}$, zero $\{0\}$, and the negatives of the counting numbers $\{-1, -2, -3, -4, \ldots\}$.

inverse operation An operation that "undoes" another operation. For example, subtraction is the inverse of addition, so if 4 is added to 7 to obtain 11, then 4 would be subtracted from 11 to get back to 7.

isosceles triangle A triangle that has two sides of equal length. The third side is called the base. The two base angles formed on the base are the same size.

least common denominator (LCD) The smallest number that is a multiple of each denominator in two or more fractions.

leg of a right triangle One of the sides that forms the right angle in a right triangle.

like terms Terms that contain identical variable factors, including exponents. For example, $3x^2$ and $4x^2$ are like terms but $3x^2$ and $3x^4$ are not like terms.

lowest terms Phrase describing a fraction whose numerator and denominator have no factors in common.

mathematical model Description of the important features of an object or situation that uses equations, formulas, tables, or graphs to solve problems, make predictions, and draw conclusions about the given object or situation.

mean The arithmetic average of a set of data. See *average*.

median The middle value in an ordered list containing an odd number of values. In a list containing an even number of values, the median is the arithmetic average of the two middle values.

minuend In subtraction, the number being subtracted from. For example, in $10 = 45 - 35$, the number 45 is the minuend, 35 is the subtrahend, and 10 is the difference.

mixed number The sum of an integer and a fraction, written in the form $a\frac{b}{c}$, where a is the integer, and $\frac{b}{c}$ is the fraction.

multiplication by 0 Any number multiplied by 0 is 0.

multiplication by 1 Any number multiplied by 1 is the number itself.

negative numbers Numbers that are less than zero.

number line A straight line that extends indefinitely in opposite directions on which points represent numbers by their distance from a fixed origin or starting point.

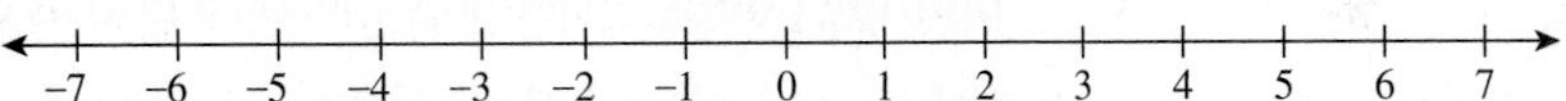

numeral A symbol or sequence of symbols called digits that represent a number.

numerator The number written above the line in a fraction; the numerator specifies the number of equal parts of a whole that is under consideration. For example, $\frac{3}{4}$ means there are three parts under consideration and each part is a quarter of the whole.

numerical expression A symbolic representation of two or more numbers combined by arithmetic operations.

obtuse angle An angle that is larger than a right angle (measures more than 90°).

obtuse triangle A triangle where one angle measures greater than 90°.

odd number Any whole number that is not even, that is, if the number is divided by 2, the remainder is 1.

operation, mathematical A procedure that generates a number from one or more other numbers. The basic mathematical operations are negation, addition, subtraction, multiplication, and division. For example, -3 is the result of negating the number 3. The number 8 is the result of adding 1, 2, and 5.

opposite numbers Two numbers with the same absolute value but different signs.

order of operations An agreement for evaluating an expression with multiple operations. Reading left to right, do operations within parentheses first; then exponents, followed by multiplications or divisions as they occur, left to right; and lastly, additions or subtractions, left to right.

ordered pair Pair of input/output values, separated by a comma, and enclosed in a set of parentheses. The input is given first and the corresponding output is listed second. An ordered pair serves to locate a point in the coordinate plane, with respect to the origin.

origin The point at which the horizontal and vertical axes of a rectangular coordinate system intersect.

output Values produced by evaluating the algebraic expression in a symbolic rule or formula.

parallel lines Lines in a plane that never intersect.

parallelogram A four-sided plane figure whose opposite sides are parallel.

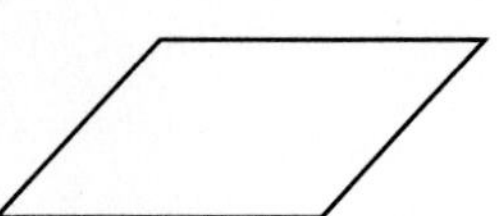

percent(age) A fraction expressed as a number of parts out of 100. For example, 25 percent, or 25%, means 25 parts out of a hundred.

percent change *See* **relative change.**

perimeter A measure of the distance around the edge of a plane geometric figure.

perpendicular lines Two intersecting lines that form four angles of equal measure (four right angles).

pi (π) The ratio of the circumference of a circle to its diameter. Pi has an approximate value of 3.14, or $\frac{22}{7}$.

plane A plane is a flat surface where a straight line joining any two points on the plane will also lie entirely in the plane.

plotting points Placing points on a grid as determined by the point's coordinates.

polygon A closed plane figure composed of three or more sides that are straight line segments.

For example,

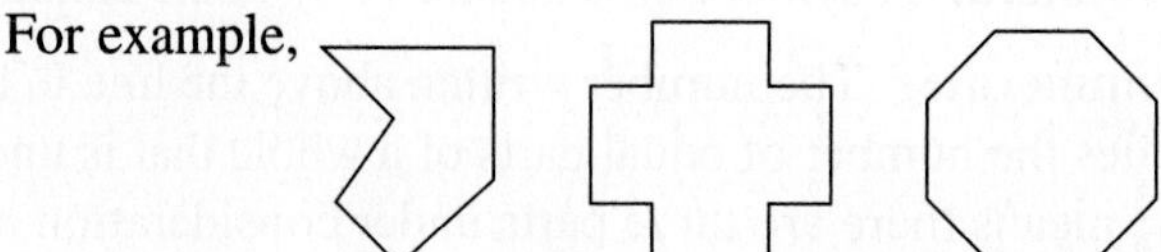

prime factorization A way of writing an integer as a product of its prime factors and their powers. For example, the prime factorization of 18 is $2 \cdot 3^2$. For each integer there is only one prime factorization, except for the order of the factors.

prime factor A factor of a number that is a prime number.

prime number A whole number greater than 1 whose only whole-number factors are itself and 1. For example, 2 and 5 are prime numbers.

product The resulting number when two or more numbers are multiplied.

proportion An equation stating that two ratios are equal.

proportional reasoning The thought process by which a known ratio is applied to one piece of information to determine a related, but yet unknown, second piece of information.

protractor A device for measuring the size of angles in degrees.

Pythagorean theorem The relationship between the sides of a right triangle: the sum of the squares of the lengths of the two perpendicular sides (legs) is equal to the square of the length of the side opposite the right angle (hypotenuse).

$$a^2 + b^2 = c^2$$

quadrant One of four regions that result when a plane is divided by the two coordinate axes in a rectangular coordinate system.

quotient The result of dividing one number by another.

radius of a circle The distance from the center point of a circle to the outer edge of the circle.

rate The comparison of an actual amount one quantity changes in relation to another quantity in different units; for example, 30 miles per hour.

ratio A quotient that represents the relative measure of similar quantities. Ratios can be expressed in several forms—verbal, fraction, decimal, or percent. For example, 4 out of 5 or 80% of all dentists recommend toothpaste X.

rational number A number that can be written in the form $\frac{a}{b}$, where a and b are integers and b is not zero.

ray Part of a line that has one end point but extends indefinitely in the other direction. Sometimes a ray is called a *half-line.*

reciprocal Two numbers are reciprocals of each other if their product is 1. Obtain the reciprocal of a fraction by switching the numerator and denominator. For example, the reciprocal of $\frac{3}{4}$ is $\frac{4}{3}$.

rectangle A four-sided plane figure with four right angles.

rectangular prism A three-dimensional figure with three pairs of identical parallel rectangular sides that meet at right angles; a box.

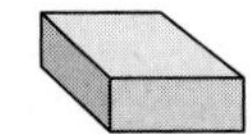

reduced fraction In lowest terms, a fraction whose numerator and denominator have no factor other than 1 in common.

relative change The comparison of the actual change to the original value. Relative change is measured by the ratio

$$\text{relative change} = \frac{\text{actual change}}{\text{original value}}.$$

Since relative change is frequently reported as a percent, it is often called *percent change.*

remainder In division of integers, the whole number that remains after the divisor has been subtracted from the dividend as many times as possible.

right angle One of four equal-size angles that is formed by two perpendicular lines, measuring 90°.

right circular cylinder A three-dimensional figure with two identical circular bases connected by sides perpendicular to the bases; a can.

right triangle A triangle in which one angle measures 90° (a right angle).

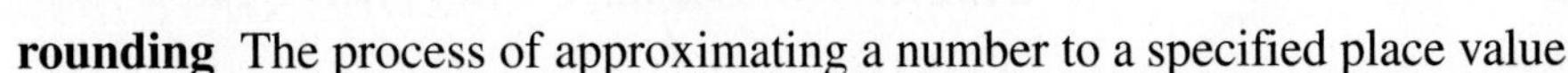

rounding The process of approximating a number to a specified place value.

scale The spacing of numbers on a coordinate axis, or number line.

scalene triangle A triangle whose sides all have different lengths.

sign of a fraction The sign of a fraction may be placed in one of three positions, preceding the numerator, preceding the denominator, or preceding the fraction itself. For example,

$$\frac{-a}{b} = \frac{a}{-b} = -\frac{a}{b}.$$

similar triangles Triangles whose corresponding angles are equal. The ratios of the lengths of the corresponding sides (sides opposite equal angles) are equal and are said to be proportional.

solution of an equation The value of a variable that makes the equation a true statement. For example, in the equation $x - 7 = 3$, the value 10 for the variable x is a solution, since $10 - 7 = 3$.

sphere A three-dimensional figure consisting of all points that are the same distance (measured by the radius) from a given point called its center; a ball.

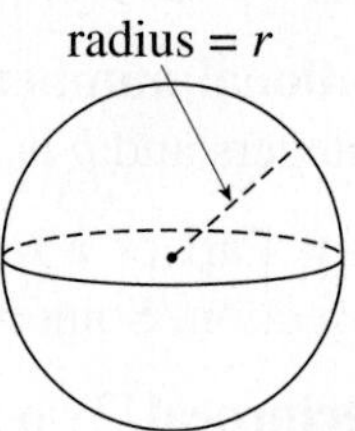

square A closed four-sided plane figure with sides of equal length and with four right angles.

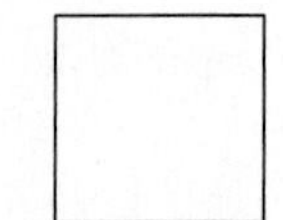

square number A number that can be written as the product of two equal whole-number factors. For example, 9 is a square number because $3 \cdot 3 = 9$.

square root The square root of a nonnegative number N is a number M whose square is N. The symbol for square root is $\sqrt{}$. For example, $\sqrt{9} = 3$ because $3^2 = 9$.

subtraction The operation that determines the difference between two numbers.

subtrahend In subtraction, the number that is subtracted. For example, in $10 = 45 - 35$, the number 45 is the minuend, 35 is the subtrahend, and 10 is the difference.

sum In addition, the total of the addends. For example, in $1 + 2 + 7 = 10$, the number 10 is the sum.

surface area The total area of all the surfaces of a three-dimensional figure.

table of input/output values A listing of input values with their corresponding output values.

terms of an expression Parts of an algebraic expression separated by addition or subtraction.

trapezoid A closed four-sided plane figure with two opposite sides parallel and the other two opposite sides not parallel.

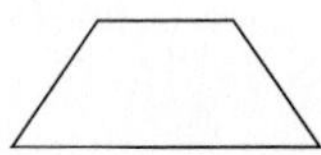

triangle A closed three-sided plane figure.

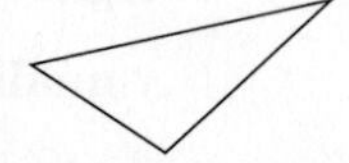

unit analysis Also called *dimensional analysis.* A process using measurement units in a rate problem as a guide to obtain the desired unit for the result. The process involves a sequence of multiplications and/or divisions that cancel units, leaving the desired unit.

variable A quantity, usually represented by a letter, that varies from one situation to another.

variable expression A symbolic code giving instructions for performing arithmetic operations with variables and numbers.

verbal rule A description of a rule using words.

vertex of an angle The point where the two rays of an angle intersect.

vertical (y-) axis The vertical number line in the Cartesian coordinate system that is used to represent the output values.

volume The measure, in cubic units, of the space enclosed by the surfaces of a three-dimensional figure.

whole numbers The collection of numbers that includes the counting numbers, 1, 2, 3, . . ., and the number 0.

zero power Any nonzero number raised to the 0 power is 1. For example, $10^0 = 1$.

INDEX

A

Absolute value on number line, 154–56
Actual change, 365
  relative change versus, 367–68
Actual measure, 350–52
Acute angles, 458
Acute triangles, 460
Addends, 20
  missing, 24
Addition
  associative property of, 22, 163
  commutative property of, 21, 163
  of decimals, 235–36
  distributive property of multiplication over, 32
  of fractions with same denominator, 203–4
  of integers, 159–61
  key phrases, examples. and arithmetic expressions, 26
  from left to right rule for, 61
  of mixed numbers, 204–5
  in order of operations, 60
  properties of integers, 163
  of whole numbers, 19–27
Algebraic equations, translating statements into, 134
Algebraic expressions
  evaluating, 277–78
  simplifying, 550
Algebra in calculating break-even point, 537–38
Angles, 418
  acute, 458
  corresponding, 471
  measuring, 458–60
  obtuse, 458
  right, 418, 457
  straight, 457
  vertex of, 418
Area
  of circle, 446–49
  of geometric figure, 94
  of parallelogram, 439
  of polygon, 442
  of rectangle, 93–95, 438
  of square, 437
  of trapezoid, 441
  of triangle, 440
Arithmetic expressions, 26
Assignment, due dates for, A-4–5
Associative property
  of addition, 22, 163
  of multiplication, 33
Astronomical distances, 49, 50
Attitude about mathematics, A-2–A-3
Axis
  horizontal, 12
  vertical, 12

B

Bar charts, 13
Bar graphs, 10–11
  horizontal direction, 10
  vertical direction, 10
Base
  of exponential expressions, 49, 63, 269
  of parallelogram, 446
Base-10 number system, 2
  reading numbers in, 2
Base 60 number system, 457
Bermuda Triangle, 426
Brahmagupta, 151
Break-even point, 537–38

C

Calculator, 59–64
Categories, 13
Celsius scale, 325
Circle
  area of, 446–49
  circumference of, 429–30
  diameter of, 429
  radius of, 429
Circumference of circle, 429–30
Classes, attending on time, A-3
Coefficients, 125
Combining like terms, 124, 126–27
Common denominator, 198
Commutative property
  of addition, 21, 163
  of multiplication, 33
Comparison
  of decimals, 223–25
  of fractions, 197–99
  of integers, 153–54
  of whole numbers, 3
Composite numbers, 5
Cones, 493
  volume of, 493
Consecutive decay factors, 384–85
Consecutive growth factors, 386–87
Consecutive percent decreases
  forming single decay factor for, 385–86
  forming single factors for, 387–88
Consecutive percent increases, forming single factors for, 387–88
Consecutive rates, using unit analysis to solve problems involving, 394–95
Constants, 92
Coordinates of point, 12, 103
Coordinate systems
  graphing and, 11–14
  rectangular, 175–77
Corresponding angles, 471
Corresponding sides of similar triangles, 471
Counting numbers, 2, 151, 152
Cross multiplication, solving proportions by, 401–2
Cube of number, 54
Cumulative grade point average, 315
Cylinders
  right circular, 483
  volume of right circular, 487

D

Data set, median value in, 5
Decay factors, 377–80
  consecutive, 384–85
  defined, 378
  determining from percent decrease, 378–79
  determining original value using, 380
  effective, 386
  forming single, for consecutive percent decreases, 385–86
  in percent decrease problems, 379–80
Decimal fraction, rounding, to specified, 227
Decimals, 223
  adding, 235–36
  comparing, 223–25
  converting, to a percent, 353
  converting, to fraction or mixed number, 224
  converting percent to, 354
  dividing, 319–20
  estimating quotients of, 321
  multiplying, 317
  reading and writing, 225–26
  rounding, 227
  solving equations involving, 241–42, 326
  subtraction of, 236–37
Degrees, 325
Denominators, 191
  adding fractions with same, 203–4
  common, 198
  least common, 198
  subtracting fractions with same, 203–4
Diameter of circle, 429
Difference, 25
Digits, 2
Distances, astronomical, 49, 50
Distributive property of multiplication over addition, 32, 327–29, 544
Dividend, 39
Division
  applying known rate directly, in problem solving, 392–93
  of decimals, 319–20
  estimating quotients in, 320–22
  of fractions, 291–93
  grade point average (GPA) and, 318–19
  of integers, 266–67
  involving zero, 270
  from left to right rule for, 61
  of mixed numbers, 299–300
  in order of operations, 60
  properties involving 0 and 1, 42
  of whole numbers, 39–42
Divisor, 39

E

Effective decay factor, 386
Effective growth factor, 387
Equality, fundamental principle of, 110
Equations
  as mathematical models, 557–58
  solving, 110, 168, 212
    of form $ax = b, a \neq 0$, that involves fractions, 293–94
    of form $ax = b$ that involve integers, 278–79
    of form $ax = b$ that involves decimals, 326
    of form $x - a = b$, 112–13
    involving decimal numbers, 241–42
    for unknown, *x*, 540
Equiangular triangle, 460
Equilateral triangle, 460
Equivalent fractions, 193–94
Estimation
  of products, 34–35, 320–22
  of quotients, 43, 320–22
  of sums of whole numbers, 23
Even numbers, 4–5
Exams
  creation of study cards in preparing for, A-10
  learning from, A-12
Exponential expressions
  base of, 49, 63, 269
  exponent of, 63
  order of operations involving, 63
  power as, 63
Exponential forms, 49
Exponentiation, 62–64
Exponents, 49, 269
  base of, 49
  of exponential expressions, 63
  first property of, 544–45
  negative integers and, 269
  second property of, 546–47
Expressions
  algebraic, 277–78, 550
  arithmetic, 26
  evaluating, that involves fractions, 302–3

F

Factorization of number, 51
Factors, 5
  multiplying series of, 545
Fahrenheit scale, 325
Formulas, 92
  graph of, 104
  as mathematical models, 558–59
  solving, for given variable, 170–71
Fractions, 191
  adding, with same denominators, 203–4
  comparing, 197–99
  converting, to a percent, 353
  converting decimal numbers to, 224
  denominator of, 191
  determining square of, 305
  dividing, 291–93
  equivalent, 193–94
  evaluating expressions that involve, 302–3
  improper, 195–97
  like, 203
  multiplying, 289–91
  numerator of, 191
  reducing, to lowest terms, 194–95
  sign of, 291
  solving equations of form $ax = b$, $a \neq 0$, that involves, 293–94
  subtracting, with same denominators, 203–4
From left to right rule
  for addition/subtraction, 61
  for multiplication/division, 61
Fundamental principle of equality, 110
Fundamental property of whole numbers, 51

G

Geometric figure, area of, 94
Geometry
  cubes in, 54
  squares in, 53
  of three-dimensional space figures, 482–99
  of two-dimensional plane figures, 417–81
Grade point average (GPA), 315
  determining, 318
  division and, 318–19
Graphing calculators, 42
Graphing grid, 14
Graphs
  bar, 10–11
  coordinate systems and, 11–14
  of formula, 104
  as mathematical models, 560–63
Greater than (>), 4, 154
Greatest common factor (GCF), 195
Growth factors, 370–73
  consecutive, 386–87
  defined, 371
  determining, from percent increase, 371–72
  determining original value using, 373
  effective, 387
  in percent increase problems, 372–73

H

Hangtime, 304
Help, need for extra, A-10
Horizontal (input) axis, 12, 176
Hypotenuse of right triangle, 464

I

Improper fractions, 195–97
  converting, to mixed number, 196
  converting mixed number to, 196
Inequalities, 3
  symbols for, 154
Input/output pairs, visual display of, 103–4
Input/output table, 102
Input value, 11, 12
Input variables, 92, 102
Integers, 151
  addition of, 159–61, 163
    more than two, 162
  comparing, 153–54
  division of, 266–67
  multiplication of, 266
  negative, 152
  number line and, 152–53
  positive, 152
  solving equations of form $ax = b$, 278–79
  subtraction of, 161–62, 163
    more than two, 162
  symbols for comparing, 154
Inverse operations
  addition and subtraction as, 25
  multiplication and division as, 40
Isosceles triangle, 460

J

Jordan, Michael, 304

K

Known rate, applying, directly by multiplication/division to solve problem, 392–93

L

Least common denominator (LCD), 198, 210
Legs of right triangle, 464
Length, 92
Less than (<), 4, 154
Like fractions
  addition of, 203
  subtraction of, 203
Like terms, 124
  combining, 124, 126–27
Lines
  parallel, 417
  perpendicular, 418

M

Mathematical models, 557
  equations as, 557–58
  formulas as, 558–59
  graphs as, 560–63
  in problem solving, 519–75
  tables as, 559–60
Mathematics
  attitude about, A-2–A-3
  using, A-9
Mathematics course
  appropriate for skilled level, A-1
  needing extra help, A-10
Measurement
  actual, 350–52
  metric system of, 231–33
  Old English system of, 231
  relative, 350–52
Measurement units, solving rate problems using, as guide, 393
Median, 5
Mental arithmetic, 59–60
Metric system of measurement, 231–33
  length, 231–33
  mass, 233
  volume, 233
Minuend, 25
Missing addend, 24
Mixed numbers, 195–97
  adding, 204–5
  converting, to improper fraction, 196
  converting decimal numbers to, 224
  converting improper fraction to, 196
  determining square of, 305
  division of, 299–300
  multiplication of, 299–300
  powers of, 300–302
  subtracting, 205
Multiplication
  applying known rate directly, in problem solving, 392–93
  associative property of, 33
  commutative property of, 33
  of decimals, 317
  distributive property of, over addition, 32
  estimation of products, 34–35, 320–22
  fractions, 289–91
  of integers, 266
  involving zero, 270
  from left to right rule for, 61
  of mixed numbers, 299–300
  in order of operations, 60
  quality points and, 315–17
  of series of factors, 545
  of whole numbers, 31–34

N

Negative integers, 152
  exponents and, 269
Negative numbers, 151
Number line
  absolute value on, 154–56
  integers and, 152–53
Numbers
  composite, 5
  counting, 2, 151, 152
  cube of, 54
  even, 4–5
  factorization of, 51
  negative, 151
  odd, 4–5
  positive, 151
  prime, 5
  rational, 151, 192
  square of, 53
  whole, 1–7
Numerals, 2
Numerator, 191

O

Obtuse angle, 458
Obtuse triangle, 460
Octagon, regular, 456
Odd numbers, 4–5
Old English system of measurement, 231
One, division properties involving, 42
Operations, calculations involving combination of, 267–69
Opposites, 155
Ordered pairs, 11
  of numbers, 104
Order of operations, 60–62
  involving exponential expressions, 63
Organization, skills of, A-3
Origin, 176
Original value, determining
  using decay factors, 380
  using growth factors, 373
Output value, 11, 12
Output variables, 92, 102

P

Parallel lines, 417
Parallelograms, 422, 438–39
  area of, 439
  base of, 446
  perimeter of, 422
Parentheses, order of operations and, 61
Percent change, 365
Percent decrease
  determining decay factor from, 378–79

using decay factors in problems, 379–80
Percent increase
determining growth factor from, 371–72
using growth factors in problems, 372–73
Percents, 352–55
converting, to a decimal, 354
converting fraction or decimal to, 353
Perfect square, 53
Perimeter
of parallelogram, 422
of polygon, 424
of rectangle, 91–92, 420
of square, 419
of trapezoid, 423
of triangles, 421
Perpendicular lines, 418
Place value, 2
rounding decimal fraction to, 227
rounding whole number to specified, 4
system of writing numbers, 3
Plane, 417
Plot, 13
Polygons, 423–24, 442–43
area of, 442
perimeter of, 424
regular, 423
Positive integers, 152
Positive numbers, 151
Positive whole numbers, 152
Power
of exponent, 49
as exponential expressions, 63
Prime factorization, 51
Prime numbers, 5, 51
Prism, rectangular, 482
Problem solving
applying known rate directly by multiplication/division, 392–93
basic steps for, 130–32
mathematical models in, 519–75
using unit analysis involving consecutive rates, 394–95
variables and, 91–150
with whole numbers, 5–7
Products, 31
estimating, 34–35, 320–22
of two integers, 266
Progress, keeping track of, A-6
Proportional reasoning, 349, 360–62
Proportions, 471
defined, 399
shortcut method for solving, 401–2
solving, by cross multiplication, 401–2
Protractor, measuring angles with, 458
Pythagoras, 465
Pythagorean theorem, 465–67
Pythagorean triples, 470

Q

Quadrants, 176
Quality points, multiplication and, 315–17
Quotients, 39
estimating, 43, 320–22
of two integers, 267

R

Radius of circle, 429
Ranking scale, 352
Rates
multiplication and division in problem solving involving, 392–93
solving problems using measurement units as guide, 393
Ratio, 351
Rational number, 151, 192
Ratio * total = part, setting up, 360
Ray, 417
Reciprocals, 291–92
Rectangles, 419–20, 438
area of, 93–95, 438
perimeter of, 91–92, 420
Rectangular coordinate system, 175–77
Rectangular prism, 482
volume of, 486
Regular octagon, 456
Regular polygon, 423
Relative change, 365
actual change versus, 367–68
Relative measure, 350–52
Right angles, 418, 457
Right circular cylinder, 483
total surface area of, 483
volume of, 487
Right triangles, 460
hypotenuse of, 464
legs of, 464
Pythagorean theorem and, 465–67
Rounding, 4
decimal numbers, 227
in estimating, 23
whole numbers, 4

S

Scalene triangle, 460
Scales, 103, 325
Scientific calculator, 42
Sectors, 446
Sides, corresponding, 471
Similar triangles, 471
corresponding sides of, 471
Single factors forming for consecutive percent increases and decreases, 387–88
Skill level, appropriate mathematics course for, A-1
Solution of equation, 110, 168
Spheres, 491–92
surface area of, 484
volume of, 491
Square roots, 304, 466
calculating, 466
determining of fraction or mixed number, 305
Squares, 418–19, 437
area of, 437
of number, 53
perfect, 53
perimeter of, 419
Square units, 437
Statements, translating, into algebraic equations, 134
Straight angle, 457
Study cards, creation of, in preparing for exams, A-10
Subtraction
of decimals, 236–37
of fractions with same denominator, 203–4
of integers, 161–62
key phrases, examples. and arithmetic expressions, 26
from left to right rule for, 61
of mixed numbers, 205
in order of operations, 60
properties of integers, 163
of whole numbers, 24
Subtrahend, 25
Sums, 20
estimating, of whole numbers, 23
Surface area of sphere, 484
Symbolic rule, 92
Symbols, inequality, 154

T

Tables
as mathematical models, 559–60
of values, 102
Terms, 124
reducing fraction to lowest, 194–95
Test-taking strategies, developing effective, A-11
Textbook, knowledge of, A-6–A-7
Three-dimensional space figures, geometry of, 482–99

Tick mark, 152
Time management, A-3, A-7–A-9
Total surface area of right circular cylinder, 483
Trapezoids, 423, 441
  area of, 441
  perimeter of, 423
Triangles, 421, 439–40
  acute, 460
  area of, 440
  classifying, 460–61
  equiangular, 460
  equilateral, 460
  isosceles, 460
  obtuse, 460
  perimeter of, 421
  right, 460
  scalene, 460
  similar, 471
Triples, 2
  Pythagorean, 470
Two-dimensional plane figures, geometry of, 417–81

U

Undefined, 42
Unit analysis, using, to solve problem involving consecutive rates, 394–95
Unit conversion, 394

V

Value
  input, 11, 12
  output, 11, 12
Variable expression, 92
  evaluating, 93
Variables
  defined, 92
  input, 92, 102
  output, 92, 102
  problem solving and, 91–150
  solving formulas for given, 170–71
Verbal rules, 92
  in making lists of paired input/output values, 522
Vertex of angle, 418
Vertical (output) axis, 12, 176
Visual display of input/output pairs, 103–4
Volume, 486
  of cone, 493
  of rectangular prism, 486
  of right circular cylinder, 487
  of sphere, 491

W

Whole numbers, 1–7
  addition of, 19–27
  classifying, 4–5
  comparing, 3
  division of, 39–42
  division properties of, 45
  estimating sums of, 23
  fundamental property of, 51
  multiplication of, 31–34
  positive, 152
  problem solving with, 5–7
  rounding, 4
  subtraction of, 24
Width, 92
Wind-chill, 529
Woods, Tiger, 275
Writing numbers, place value system of, 3

X

X-axis, 12
X-coordinate, 12

Y

Y-axis, 12
Y-coordinate, 12

Z

Zero
  division involving, 270
  division properties involving, 42
  multiplication involving, 270

Geometric Formulas

Perimeter and Area of a Triangle, and Sum of the Measures of the Angles

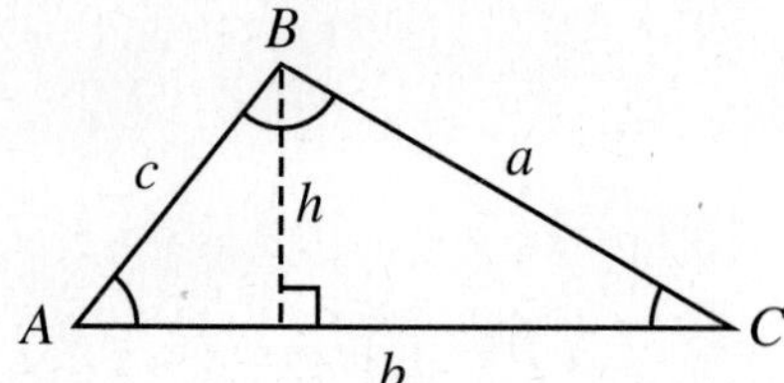

$P = a + b + c$
$A = \frac{1}{2}bh$
$A + B + C = 180°$

Pythagorean Theorem

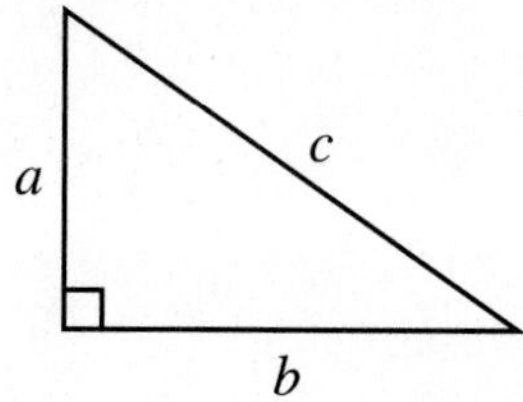

$a^2 + b^2 = c^2$

Perimeter and Area of a Rectangle

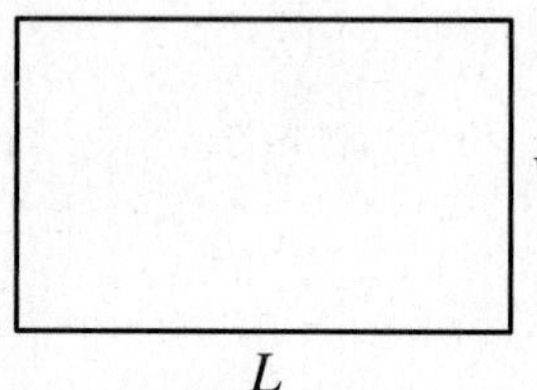

$P = 2L + 2W$
$A = LW$

Perimeter and Area of a Square

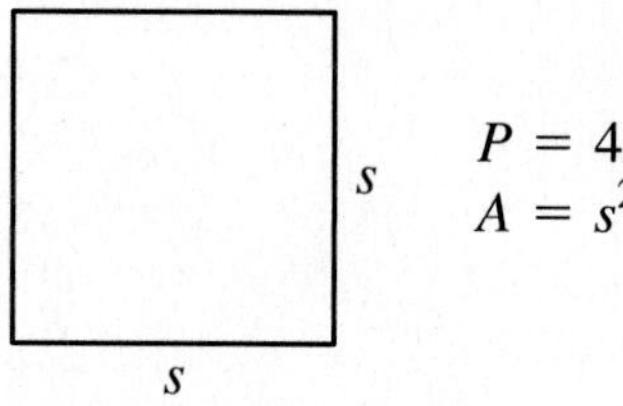

$P = 4s$
$A = s^2$

Area of a Trapezoid

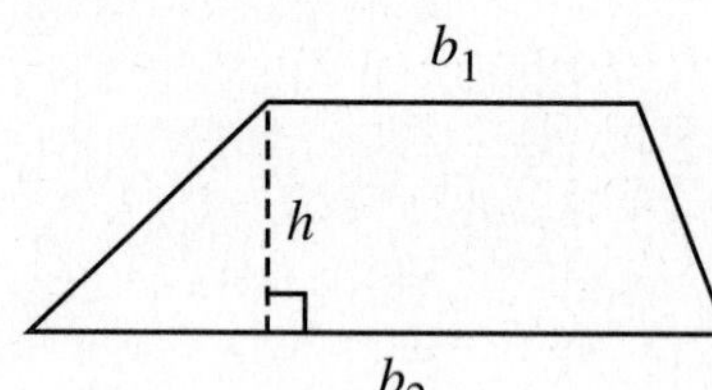

$A = \frac{1}{2}h(b_1 + b_2)$

Circumference and Area of a Circle

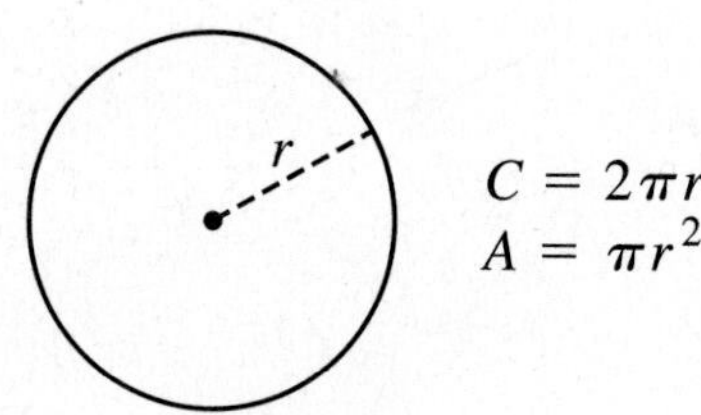

$C = 2\pi r$
$A = \pi r^2$

Volume and Surface Area of a Rectangular Solid

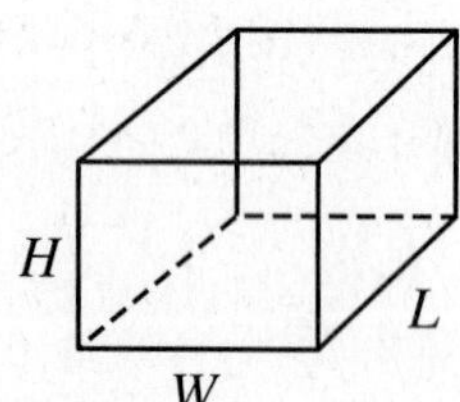

$V = LWH$
$SA = 2LW + 2LH + 2WH$

Volume and Surface Area of a Sphere

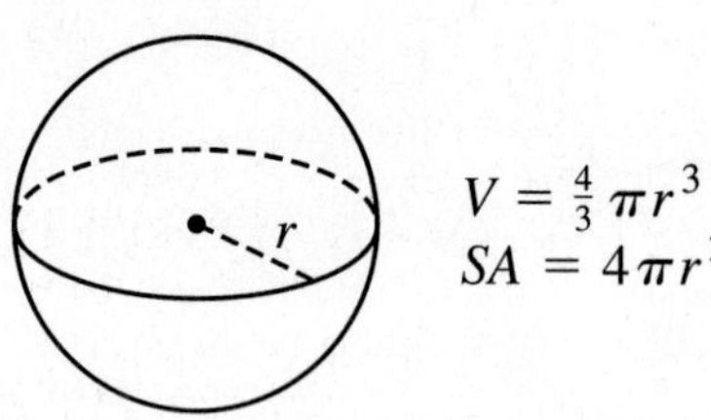

$V = \frac{4}{3}\pi r^3$
$SA = 4\pi r^2$

Volume and Surface Area of a Right Circular Cylinder

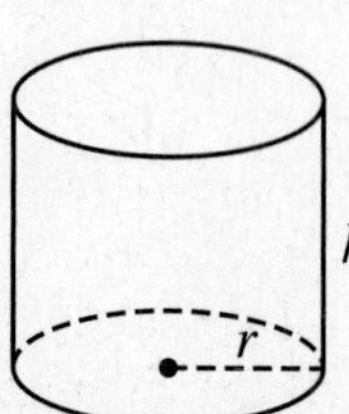

$V = \pi r^2 h$
$SA = 2\pi r^2 + 2\pi rh$

Volume and Surface Area of a Right Circular Cone

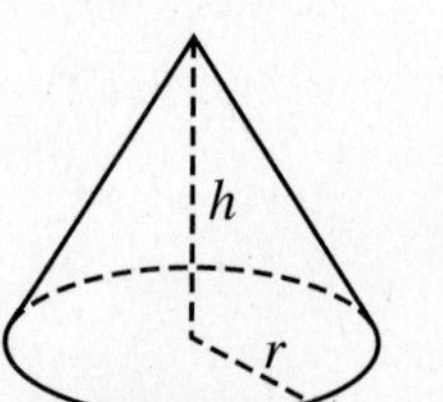

$V = \frac{1}{3}\pi r^2 h$
$SA = \pi r^2 + \pi rl$